1 MONTH OF
FREE
READING

at
www.ForgottenBooks.com

By purchasing this book you are eligible for one month membership to ForgottenBooks.com, giving you unlimited access to our entire collection of over 1,000,000 titles via our web site and mobile apps.

To claim your free month visit:
www.forgottenbooks.com/free902374

ISBN 978-0-265-87151-5
PIBN 10902374

U. S. DEPARTMENT OF AGRICULTURE,
BUREAU OF ENTOMOLOGY—BULLETIN No. 91.
L. O. HOWARD, Entomologist and Chief of Bureau.

THE IMPORTATION INTO THE UNITED STATES OF THE PARASITES OF THE GIPSY MOTH AND THE BROWN-TAIL MOTH:

A REPORT OF PROGRESS, WITH SOME CONSIDERATION OF PREVIOUS AND CONCURRENT EFFORTS OF THIS KIND.

BY

L. O. HOWARD,
Chief, Bureau of Entomology,

AND

W. F. FISKE,
*In Charge, Gipsy Moth Parasite Laboratory,
Melrose Highlands, Mass.*

· Iᴇsᴜᴇᴅ Jᴜʟʏ 29, 1911.

WASHINGTON:
GOVERNMENT PRINTING OFFICE.
1911.
C.

CALOSOMA SYCOPHANTA.

Adult eating gipsy-moth caterpillar, lower left; pupa, lower right; eggs, upper left; eaten chrysalides of gipsy moth, upper right; full-grown larvæ from above and from below. (Original)

AUG

9790

U. S. DEPARTMENT OF AGRICULTURE,

BUREAU OF ENTOMOLOGY—BULLETIN No. 91.

L. O. HOWARD, Entomologist and Chief of Bureau.

THE IMPORTATION INTO THE UNITED STATES OF THE PARASITES OF THE GIPSY MOTH AND THE BROWN-TAIL MOTH:

A REPORT OF PROGRESS, WITH SOME CONSIDERATION OF PREVIOUS AND CONCURRENT EFFORTS OF THIS KIND.

BY

L. O. HOWARD,

Chief, Bureau of Entomology,

AND

W. F. FISKE,

*In Charge, Gipsy Moth Parasite Laboratory,
Melrose Highlands, Mass.*

ISSUED JULY 29, 1911.

WASHINGTON:
GOVERNMENT PRINTING OFFICE.
1911.
C.

BUREAU OF ENTOMOLOGY.

L. O. HOWARD, *Entomologist and Chief of Bureau.*
C. L. MARLATT, *Entomologist and Acting Chief in Absence of Chief.*
R. S. CLIFTON, *Executive Assistant.*
W. F. TASTET, *Chief Clerk.*

F. H. CHITTENDEN, *in charge of truck crop and stored product insect investigations.*
A. D. HOPKINS, *in charge of forest insect investigations.*
W. D. HUNTER, *in charge of southern field crop insect investigations.*
F. M. WEBSTER, *in charge of cereal and forage insect investigations.*
A. L. QUAINTANCE, *in charge of deciduous fruit insect investigations.*
E. F. PHILLIPS, *in charge of bee culture.*
D. M. ROGERS, *in charge of preventing spread of moths, field work.*
ROLLA P. CURRIE, *in charge of editorial work.*
MABEL COLCORD, *in charge of library.*

PREVENTING SPREAD OF MOTHS.

PARASITE LABORATORY.

W. F. FISKE, *in charge;* A. F. BURGESS, C. W. COLLINS, R. WOOLDRIDGE, J D. TOTHILL, C. W. STOCKWELL, H. E. SMITH, W. N. DOVENER, F. H. MOSHER, *assistants.*

FIELD WORK.

D. M. ROGERS, *in charge;* H. B. DALTON, H. W. VINTON, D. G. MURPHY, I. L. BAILEY, H. L. McINTYRE, *assistants.*

2

LETTER OF TRANSMITTAL.

U. S. DEPARTMENT OF AGRICULTURE,
BUREAU OF ENTOMOLOGY,
Washington, D. C., April 12, 1911.

SIR: I have the honor to transmit herewith the manuscript of a report of progress on the importation into the United States of the parasites of the gipsy moth and the brown-tail moth. To this has been added some consideration of previous and concurrent efforts to handle the parasites of destructive insects in a practical way. The work with the foreign parasites of the gipsy moth and the brown-tail moth has been going on now for rather more than five years. It promises excellent results, and the present seems the proper time to present to the people interested a somewhat detailed account of what has been done and of the present condition of the work. I recommend that this manuscript be published as Bulletin No. 91 of this bureau.

Respectfully,

L. O. HOWARD,
Entomologist and Chief of Bureau.

Hon. JAMES WILSON,
Secretary of Agriculture.

3

CONTENTS.

ILLUSTRATIONS.

9

THE IMPORTATION INTO THE UNITED STATES OF THE PARASITES OF THE GIPSY MOTH AND THE BROWN-TAIL MOTH:

A REPORT OF PROGRESS,

WITH SOME CONSIDERATION OF PREVIOUS AND CONCURRENT EFFORTS OF THIS KIND.

INTRODUCTION.

By L. O. HOWARD,
Chief, Bureau of Entomology.

As will appear from the opening portion of this bulletin, which gives an account of previous work in the practical handling of natural enemies, carried on in various parts of the world, nothing comparable to the work which is to be described has ever before been undertaken. As will appear also, most of the successful work in this direction has been done with the fixed scale insects. The exceptions to this general statement among the measurably successful efforts have been the introduction of parasites of the sugar-cane leafhopper into Hawaii, some reported work in the introduction of South American natural enemies of fruit flies into Western Australia, and the introduction of one of the many European enemies of the codling moth from Spain into California; but it does not appear that practical results of any very great value have been achieved by the last two introductions, although information from Western Australia is scanty. At the time when the work began nothing practical had been accomplished with the natural enemies of any lepidopterous insects, and in the whole history of the practical handling of parasites no work of this character has ever been attempted upon anything like the large scale with which the present work has been carried on. Some studies had already been made both by the writer and by Mr. Fiske on the subject of the intensive parasitism of two native species of American moths, and for years the bureau had been keeping records of the rearings of parasites of lepidopterous insects as well as of others; moreover, the writer had made a careful study of the records of the rearings of hymenopterous parasites from host insects all over the world and had accumulated an enormous catalogue of such records. Nevertheless the initial work on such a scale was experimental in its character. It seemed to the writer that by attempting to reproduce in New

England as nearly as possible the entire natural environment of the gipsy moth and the brown-tail moth in their native homes, similar conditions of comparative scarcity could surely be reached, and this view he still holds with enthusiasm. Naturally, in the course of the work as it progressed year after year his ideas have been changed as to methods, and very great improvements have been made upon the earlier methods, largely through the intelligence and ingenuity of the junior author of this bulletin. Moreover, the careful, intensive studies which have been made at the gipsy-moth parasite laboratory by the junior author and a corps of trained assistants, aided by abundant material, funds, and supplies, have resulted not only in the ascertainment of very many facts new to science, but in the accumulation of such facts to such a degree as to enable generalizations of a novel character and of a sounder basis than could have been had under other conditions. Many points are brought out in this bulletin which will doubtless be entirely new to the trained scientific reader. Mistakes have been made and wrong conclusions have been drawn from time to time, but these have been corrected, and we are now in a fair way to see a favorable result from the long and expensive work.

The initial idea was that since a large percentage of gipsy-moth caterpillars or brown-tail moth caterpillars in Europe contains parasites each year, therefore if these caterpillars were brought to America in large numbers from every possible place we could not fail to rear from them an abundance of adult foreign parasites. This idea was sound, and in following it out we have constantly improved the methods—methods of collection, of packing, of shipment, and of subsequent rearing. Very large numbers of parasites have been reared.

It was first thought that when parasites had been reared in sufficient numbers they should be widely distributed in small colonies, on the theory that each colony would remain in substantially the same general locality and would increase and spread from that point. This idea was a natural one and was fully justified by previous work which had been done with parasites of other groups of insects, but in this case it proved to be erroneous, and valuable time and valuable specimens were lost. Eventually it was shown to be of prime importance, first to establish a given species of parasite in this country, and not until this has been accomplished to pay any attention to the matter of dispersion. It seems to be the first instinct of many species that have been imported to spread widely. Therefore, if the colony put out be a small one the individuals composing it spread rapidly beyond all means of meeting and of mating, and thus the colonies in many instances were lost. By rearing in the laboratory, however, until colonies of at least a thousand are to be had, such colonies while dispersing are much more likely to remain in touch, mate, and multiply.

By methods based upon the first idea, and by the subsequent modification of the second idea, some of the most important natural enemies of both species have been established in the United States to a certainty. It has been found with several species that they could not be recovered until after three years had elapsed from the time of the original colonization; hence it follows with a reasonable certainty that other species which have not been recovered will ultimately be recovered as a result of colonization one, two, and three, and even perhaps four years ago. It is deemed, however, at this time that nearly as much has been accomplished as can be accomplished by the earlier methods, and subsequent efforts will be devoted to a more specific attempt to import the species still lacking, several of which are known in their original homes to be of very great importance. As will be pointed out elsewhere, attempts will also be made to import the species which, while of lesser importance at home, may here fill in gaps and may possibly multiply to an unprecedented extent in the face of new conditions and a superabundance of host material.

The work has been going on since 1905. Nothing has been published concerning its progress except the short accounts in the annual reports of the writer submitted each year to the Secretary of Agriculture, and except a bulletin on the general subject prepared by the junior author and published by the State forester of Massachusetts. It is hoped that the present account will be deemed a satisfactory reply to all expressed desire for information as to progress.

The joint authorship of the bulletin is deemed desirable by both authors, but the writer takes it upon himself to sign this introduction for the explicit purpose of stating in his own way the conditions under which it has been prepared. The work from the beginning has been under the direct supervision of the writer, and he is therefore to be held responsible for any failures in the speedy accomplishment of results, but the greatest credit in bringing about the results which have been accomplished, he wishes frankly to state, belongs to Mr. Fiske. Following the breakdown in health of Mr. E. S. G. Titus in. the spring of 1907, as is shown in the bulletin, Mr. Fiske was stationed at the parasite laboratory and has since been given every freedom in the conduct of its affairs. Nearly every suggestion which he has made, while it has been fully discussed by the two of us, has been adopted. The ingenuity which he has displayed in matters of method and the broad grasp which he has shown of the whole phenomena of parasitism in insects, together with his competent and practical grouping of his ideas, deserve every praise. Such portions of the bulletin as were dictated by the writer have received the editorial criticism of the junior author, and the portions prepared by the latter

·have received a most careful consideration and editorial pruning of the writer. Mr. Fiske, by virtue of his practical residence at the field laboratory and of his intimate charge of all the field notes and laboratory notes, has prepared all of the matter in this bulletin relating to the laboratory and field end, subject, of course, to the writer's revision. The rest has been prepared by the senior author.

Acknowledgements of assistance should be made by the score. The State authorities of Massachusetts, the admirable corps of laboratory and field assistants, and above all the very numerous foreign officials, voluntary assistants, and paid observers have united to make the undertaking possible. Their individual names are all mentioned in the following pages in connection with the parts they played, but the Governments of Austria, France, Germany, Hungary, Italy, Japan, Portugal, Russia, and Spain should especially be thanked in an official publication like this for the assistance given by the officials of these Governments.

PREVIOUS WORK IN THE PRACTICAL HANDLING OF NATURAL ENEMIES OF INJURIOUS INSECTS.

Two very thorough and careful general papers on the subject of the practical handling of natural enemies of insects, treating the subject from the different points of view, including the historical side, have been published in the last few years. The first of these, entitled "The Utilization of Auxiliary Entomophagous Insects in the Struggle against Insects Injurious to Agriculture," by Prof. Paul Marchal, of the National Agronomical Institute of Paris, was published in 1907,[1] and was partly republished in English in the Popular Science Monthly in 1908.[2] The other, by Prof. F. Silvestri, of the Royal Agricultural School at Portici, Italy, entitled "Consideration of the Existing Condition of Agricultural Entomology in the United States of North America, and Suggestions which can be Gained from it for the Benefit of Italian Agriculture," was published in 1909.[3] This paper was in part translated into English and published in the Hawaiian Forester and Agriculturist for August, 1909. Both of these papers should be consulted by persons wishing to inform themselves thoroughly on this question. For the present purpose, treatment of the subject must be brief.

The study of parasitic and predatory insects is old. Silvestri has pointed out that Aldrovandi (1602) was the first to observe the exit of the larvæ of *Apanteles glomeratus* L. (which he supposed to be eggs) from the common cabbage caterpillar, and that Redi (1668) published the same observation and another on insects of different species born from the same pupa. A later writer, Vallisnieri (1661–

1 Annals of the National Agronomical Institute (Superior School of Agriculture), second series vol. 6, no. 2, pp. 281-354, Paris, 1907.
2 Popular Science Monthly, vol 72, pp 351-370, 407-419, April and May, 1908
3 Bulletin of the Society of Italian Agriculturists, vol 14, no. 8, pp. 305-367, Apr. 30, 1909.

1730) was apparently the first to discover the real nature of this phenomenon and to realize the existence of true parasitic insects. Réaumur (1683–1757) and De Geer (1720–1778) each studied the life histories of living insects with great care and among these worked out the biology of a number of parasites. Very many descriptive works on parasites were published in the closing years of the eighteenth century and the beginning of the nineteenth century, especially by Dalman (1778–1828), Nees ab Esenbeck (1776–1858), Gravenhorst (1777–1857), Walker (publishing from 1833 to 1861), Westwood (publishing from 1827 on through nearly the whole of the century), Förster (publishing from 1841 on), and Spinola (1780–1857).

Many later writers have contributed to the systematic study of these insects, among them Holmgren and Thomson, of Sweden; Mayr, of Austria; Motschulsky, of Russia; Ratzeburg, Hartig, and Schmiedeknecht, of Germany; Wesmael, of Belgium; Haliday, Marshall, and Cameron, of England; Rondani, of Italy; Brullé, Giraud, Decaux, and others in France; Provancher, of Canada; and, in America, Cresson, Riley, Howard, Ashmead, Crawford, Viereck, Brues, Girault, and others.

The best contribution appearing in Europe and devoted to the biology of hymenopterous parasites, and especially consideration of their relations to their hosts, was that by Ratzeburg, whose great work entitled "Die Ichneumonen der Forstinsekten," was a standard for many years. Ratzeburg understood the rôle played by parasites in the control of forest insects, but did not believe that this control could in any way be facilitated by man.

EARLY PRACTICAL WORK.

Froggatt has pointed out that probably the earliest suggestion made regarding the artificial handling of beneficial insects was printed in Kirby and Spence's entomology (1816), where the authors called attention to the value of the common English ladybird as destroying the hop aphis in the south of England. "If we could but discover a mode of increasing these insects at will, we might not only clear our hothouses of aphides by their means, but render our crops of hops much more certain than they are now." As a matter of fact, gardeners and florists in England for very many years have recognized the value of the ladybirds and have transferred them from one plat to another.

Prof. A. Trotter, of the Royal School of Viticulture at Avellino, Italy, has recently pointed out in an interesting paper entitled "Two Precursors in the Application of Carnivorous Insects," published in Redia, in 1908.[1] that probably the first person to make a practical application of the natural enemies of injurious species was Prof.

---

[1] Redia, vol. 5, pp. 126–132, Florence, 1908.

Boisgiraud, of Poitiers, France, in 1840. Prof. Trotter found this reference in a little-known paper by N. Joly, published in 1842, and entitled "Notice of the Ravages which *Liparis dispar* L. has made around Toulouse, followed by some Reflexions upon a Method of Destroying Certain Insects." It seems that Boisgiraud, about 1840, freed the poplars along a road near Poitiers of the gipsy moth by placing upon them the carabid beetle *Calosoma sycophanta* L., and destroyed earwigs in his own garden by placing with them a rove beetle (*Staphylinus olens* Müll). He also experimented against the same insect with the ground beetle *Carabus auratus* L. His experiment must have become rather well known at the time, since Prof. Trotter points out that in 1843 the technical commission of the Society for the Encouragement of Arts and Crafts of Milan offered a gold medal to be given in 1845 to the person who in the meantime should have undertaken with some success new experiments tending to promote the artificial development of some species of carnivorous insects which could be used efficaciously to destroy another species of insect recognized as injurious to agriculture. This offer drew forth a memoir from Antonio Villa, a well-known writer on entomology, who had previously confined himself to the Coleoptera, entitled "The Carnivorous Insects used to Destroy the Species Injurious to Agriculture." This memoir was presented December 26, 1844, and he advocated the employment of climbing carabid beetles for tree-inhabiting forms, rove beetles to destroy the insects found in flowers, and ground beetles for cutworms and other earth-inhabiting forms. The paper of Villa was praised in certain reviews and criticized in others. It seems to have been entirely lost sight of in later years.

A later Italian writer, Rondani, who devoted himself for the most part to systematic work, appreciated the practical importance of parasite work and published tables giving the host relations of different species. His work influenced many arguments in the dispute which sprang up in Italy about 1868 as to the usefulness of insectivorous birds to agriculture, and Silvestri calls attention to the fact that Dr. T. Bellenghi was referring to Rondani when, in 1872, he spoke what Silvestri calls "the prophetic words:" "Entomological parasitism has a future, and in it more than in anything else Italian agriculture must put its faith."

PERMITTING THE PARASITES TO ESCAPE

The earliest published suggestion as to the practical use of parasites of injurious insects, by permitting the parasites to escape while the host insect is killed, appears to have been made by C. V. Riley when State entomologist of Missouri. Writing of the rascal leaf-crumpler (*Mineola indiginella* Zell.) in his Fourth Report on the

Insects of Missouri,[1] he advocated the collecting of the winter cases of
the destructive insect and placing the cases in small vessels in the
center of a meadow or field, away from any fruit trees, with the idea
that the worms would be able to wander only a few yards and would
perish from exhaustion or starvation, while their parasites would
escape and fly back to the fruit trees. It is stated that this method
was put in practice later by D. B. Wier with success.

A French writer, F. Decaux, the following year made practically
the same suggestion with regard to apple buds attacked by Anthonomus. He advised that instead of burning these buds, as was generally done, they be preserved in boxes covered with gauze, raising the
latter from time to time during the period of issuing of parasites so as
to permit them to escape. In 1880 he put this method in practice,
and collected in Picardy buds reddened by the Anthonomus from
800 apple trees, and thus accomplished the destruction of more than
1,000,000 individuals of the Anthonomus, setting at liberty about
250,000 parasites which aided the following year in the destruction of
the weevils. The following year the same process was repeated, and,
the orchards being isolated in the middle of cultivated fields, all serious
damage from the Anthonomus was stated to have been stopped for
10 years.[2]

Practically the same suggestion was made later, in 1877, by J. H.
Comstock, in regard to the imported cabbage worm (*Pontia rapæ* L.).
Comstock deprecated the indiscriminate crushing of the chrysalids
collected under trap boards, on account of the large percentage which
contained parasites. He recommended instead the collecting of the
chrysalids and placing them in a box covered with a wire screen which
should permit the parasites to escape and at the same time confine the
butterflies so that they could be easily destroyed. The same author,
in his report upon cotton insects,[3] recommended a similar course with
the pupæ of the cotton caterpillar (*Alabama argillacea* Hübn.).

Riley later recommended the same plan for the bagworm (*Thyridop-
teryx ephemeræformis* Haw.); Berlese in Italy recommended it for the
grapevine Cochylis, and Silvestri for the olive fly (*Dacus oleæ* Rossi),
for *Prays oleellus* Fab., and for *Asphondylia lupini* Silv.

Writing on the Hessian fly, Marchal has pointed out that the
destruction of the stubble remaining in the field after harvest may
have unfortunate consequences, for if this is done a little late there is
a risk that all of the destructive flies will have emerged and abandoned the stubble, exposing to destruction only the parasites whose
part would have been to stop the invasion the following year. Marchal also points out that Kieffer has shown that one of the measures

[1] Riley, C. V. Fourth Report on the Insects of Missouri, p. 40, 1871
[2] An excellent article covering these general questions was published by Decaux in the Journal of the
National Horticultural Society of France, vol 22, pp. 158–184, 1899.
[3] Cotton Insects, pp. 230–231, Washington, 1879.

advised for the destruction of the wheat midge (*Contarinia tritici* Kirby), namely, burning the débris after thrashing, has only an injurious effect, for, while it is true that the pupæ of the midge are to be found in this débris, it should be remembered that the healthy nonparasitized larvæ of the midge transform in the ground, while those which remain in the heads are, on the contrary, parasitized.

Still another method of encouraging parasites is pointed out by Marchal and Silvestri. It is to cultivate in the olive groves various plants upon which allied insects live which are parasitized by the same species of parasites as the olive fly. This idea, independently developed in the United States, has been practically used by Hunter in the fight against the cotton-boll weevil. Allied insects feeding in certain weeds along the borders of the cotton fields have parasites capable of attacking the boll weevil. Careful study of the biology of these allied weevils and of their parasites resulted in the gaining of the information that if the weeds are cut at a certain time the parasites are forced to attack the cotton-boll weevil in order to maintain their existence; actual experimentation has resulted in the very considerable increasing of the percentage of parasitism of the cotton-boll weevil in this way.

THE TRANSPORTATION OF PARASITES FROM ONE PART OF A GIVEN COUNTRY TO ANOTHER PART.

In 1872 attempts were made by Dr. William Le Baron, at that time State entomologist of Illinois, to transport *Aphelinus mali* Le Baron, a parasite of the oyster-shell scale of the apple (*Lepidosaphes ulmi* L.) from one part of the State of Illinois to another portion of the same State where the parasite seemed to be lacking. Some slight success was reported, and at the end of the year it was stated that the parasite had become domiciled in the new locality, but, as this parasite subsequently proved to be one of general American distribution, the experiment can not be said to have been worth while except in a very small way.

In France, F. Decaux, above quoted, in 1872, made some experiments in the transportation of parasites from one locality to another.

Riley, in his third report as State entomologist of Missouri (1870), in considering two parasites of the plum curculio, stated that he intended the following year, if possible, to rear enough specimens of *Sigalphus curculionis* Fitch to send at least a dozen to every county seat in the State and have them liberated in someone's peach orchard. There seems, however, to be no record that this was ever done.

In 1880, in his report on the parasites of the Coccidæ in the collection of the Department of Agriculture,[1] the senior author called attention to the fact that with the parasites of scales the matter of trans-

---

[1] Annual Report U S. Department of Agriculture for 1880, p. 351.

portation from one part of the country to another becomes easy, since all that has to be done is simply to collect twigs bearing the scales, preferably during the winter months, and carry them to non-protected regions, the parasites being dormant and protected each by the scale of the coccid which it had destroyed; and it was specifically recommended that the important parasite of the black scale (*Saissetia oleæ* Bern.), described in the article as *Tomocera californica*, could be readily

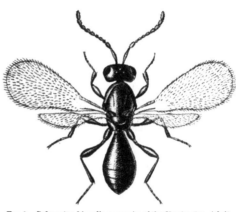

FIG. 1.—*Polygnotus hiemalis*, a parasite of the Hessian fly: Adult. Greatly enlarged. (From Webster.)

carried from California and utilized to destroy Lecanium scales in the Southeast.

FIG. 2.—*Polygnotus hiemalis*: Adults which have developed within the "flaxseed" of the Hessian fly and are ready to emerge. Much enlarged. (From Webster.)

Excellent work in this direction has been done of late years by the Bureau of Entomology. In the study of the Hessian fly (*Mayetiola destructor* Say), under Prof. F. M. Webster, early-sown plats of wheat at Lansing, Mich., and Marion, Pa., in 1906, were very seriously attacked by the Hessian fly, but when examined carefully at a later date fully 90 per cent of the flaxseeds (pupæ) were found to have been stung by a hymenopterous parasite, *Polygnotus hiemalis* Forbes (figs. 1, 2), and to contain its developing larvæ. A field of wheat near Sharpsburg, Md., was found to be infested by the fly, and examination indicated the absence of the parasite. On April 8, 1907, a large number of the parasitized flaxseeds from Marion, Pa., were brought to Sharpsburg and placed in the field. On July 8 an examination of the Sharpsburg field showed that the parasites had taken hold to such an extent that of the large number of flaxseeds taken and brought to the laboratory for investigation not one was found which had not been parasitized. Additional material secured from Sharpsburg in the spring of 1908 in the same locality showed all of the Hessian flies to be parasitized.

In the same way excellent results have been obtained in the investigation of the cotton-boll weevil, under Mr. W. D. Hunter. In the summer of 1906 a number of parasites were taken from Waco, Tex., and liberated in a cotton field near Dallas, Tex., and apparently by this means the mortality rate due to parasites was raised in a few ·weeks about 9 per cent. Later, parasites were introduced from Texas into Louisiana and increased the mortality of the weevil. Work of this character is still being carried on by Mr. Hunter, and elaborate, although as yet unsuccessful, experiments have been made by Webster in the transfer of the hymenopterous parasite *Lysiphlebus tritici* Ashm. (fig. 3) from southern points into Kansas wheat fields for the destruction of the spring grain aphis or so-called "green bug" (*Toxoptera graminum* Rond.), definite results being prevented by the occurrence of the parasite throughout the range of the destructive insect, parasitic, as it is, upon other species of plant lice.

Prof. S. J. Hunter, of the University of Kansas, however, in the Bulletin of the University (vol. 9, p. 2) states that he was able, in 1908, to hasten the destruction of the Toxoptera in Kansas by the importation of Lysiphlebus from some other point.

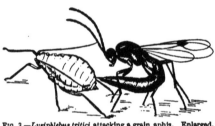

Fig. 3.—*Lysiphlebus tritici* attacking a grain aphis.   Enlarged.
(From Webster.)

In the last two years some very interesting work has been carried on by the State Horticultural Commission of California in the way of collecting Coccinellidæ on a large scale in their hibernating quarters, boxing them, and sending them to different parts of the State for use against plant lice upon truck crops. The biennial report of the commissioner of horticulture for 1907–8, published in Sacramento in 1909, for example, indicates that 50,000 specimens of the ladybird beetles *Hippodamia convergens* Guér. and *Coccinella californica* Mann. had been so collected. This, however, was very small compared to the scale upon which these insects were collected during the winter of 1909–10. Mr. E. K. Carnes, of the commission, writing to the Bureau of Entomology under date of March 14, 1910, makes the following statement:

We have quite a sight at the insectary now—over a ton of *Hippodamia convergens*, boxed in 60,000 lots each, screened cases, and in our own cold storage. We handle them in large cages, run them into a chute, and handle like grain. They are for the melon growers of the Imperial Valley.

This species collects in large numbers late in summer and early in the autumn at the bases of plants in the mountain valleys and can easily be collected by the sackful. The actual good accomplished by the distribution of these ladybirds among the melon growers has not yet been reported upon, but theoretically speaking the experiment should have excellent results.

THE TRANSFER OF BENEFICIAL INSECTS FROM ONE COUNTRY TO ANOTHER.

EARLY ATTEMPTS.

Dr. Asa Fitch, for many years State entomologist of New York, was probably the first entomologist in America, or elsewhere for that matter, to take into serious consideration the question of the transfer of beneficial insects from one country to another. In 1854, following a disastrous attack upon the wheat crop of the eastern United States by the wheat midge (*Contarinia tritici* Kirby), a species that had been accidentally introduced from Europe during the early part of that century, Dr. Fitch, who had made a careful study of the insect both in this country and from the European records, was struck with the fact that in Europe the insect in ordinary seasons did no damage, and that when occasionally it became so multiplied as to attract notice it was but a transitory evil which subsided soon and was not heard of again for a number of years. He was aware that in Europe certain parasites of this insect were found, and, comparing the insects taken from wheat in flower in France with those taken from wheat in flower in New York, he found that in France the wheat midge constituted but 7 per cent of the insects thus taken, while its parasites constituted 85 per cent; whereas in New York the wheat midge formed 59 per cent of the insects thus captured, and there were no certain parasites. He speculated as to the cause for this extraordinary difference and wrote:

There must be a cause for this remarkable difference. What can that cause be? I can impute it to only one thing; we here are destitute of nature's appointed means for repressing and subduing this insect Those other insects which have been created for the purpose of quelling this species and keeping it restrained within its appropriate sphere have never yet reached our shores. We have received the evil without the remedy And thus the midge is able to multiply and flourish, to revel and riot, year after year, without let or hindrance. This certainly would seem to be the principal if not the sole cause why the career of this insect here is so very different from what it is in the Old World.

Quite naturally after this train of reasoning had entered his brain, Dr. Fitch made an effort to introduce the European parasites of the wheat midge, and in May, 1855, addressed a letter to John Curtis, the famous English economic entomologist, and at that time president of the Entomological Society of London, informing him of the immense amount of damage done by the midge in America and suggesting the manner in which parasitized larvæ could be secured in England and transmitted alive to this country. Mr. Curtis was ill and on the point of starting for the Continent, but laid the letter before the Entomological Society of London, which resulted in the adoption of a resolution to the effect that if any member of the society should be able to find parasitized midges he should send them to Dr. Fitch.

Nothing ever came of this effort, but it is of interest on account of its apparent priority over other experimentation of this kind.

The next international attempt seems to have been made in 1873, when Planchon and Riley introduced into France an American predatory mite (*Tyroglyphus phylloxeræ* Riley) which feeds on the grapevine Phylloxera in the United States. The mite became established, but accomplished no appreciable results in the way of checking the famous grapevine pest.

In 1874 efforts were made to send certain parasites of plant lice from England to New Zealand, but without results of value, although *Coccinella undecimpunctata* L. is said to have become established.

In 1883 Riley imported the braconid *Apanteles glomeratus* into the United States from Europe, where it is an abundant enemy of the imported cabbage worm (*Pontia rapæ* L.). This species has since established itself in the United States and has proved a valuable addition to the North American fauna.

THE AUSTRALIAN LADYBIRD (NOVIUS CARDINALIS MULS (IN THE UNITED STATES).

But all previous experiments of this nature were completely overshadowed by the remarkable success of the importation of (*Vedalia*) *Novius cardinalis* Muls. (fig. 4), a coccinellid beetle, or ladybird, from Australia into California in 1889. The orange and lemon groves of California had for some years been threatened with extinction by the injurious work of the fluted or cottony cushion scale (*Icerya purchasi* Mask.) a large scale insect which the careful investigations of Prof. Riley and his force of entomologists at the United States Department of Agriculture had shown to have been originally imported, by accident, from Australia or from New Zealand, where it had originally been described by the New Zealand coccidologist, the late W. M. Maskell. The Division of Entomology had been for several years engaged in an active campaign against this insect, and had discovered washes which could be applied at a comparatively slight expense and which would destroy the scale insect. It had also in the course of its investigations discovered the applicability of hydrocyanic-acid gas under tents as a method of fumigating orchards and destroying the scale. The growers, however, had become so thoroughly disheartened by the ravages of the insect that they were no longer in a frame of mind to use even the cheap insecticide washes, and many of them were destroying their groves. In the meantime, through some correspondence in the search for the original home of the scale insect, Prof. Riley had discovered that while the species occurred in parts of Australia it was not injurious in those regions. In New Zealand it also occurred, but was abundant and injurious. He therefore argued that the insect was probably introduced from Australia into New Zealand, and that its abundance in the latter country and its relative scarcity in Australia were due to the fact that in its native home it was held in subjection by some parasite or natural enemy, and that in the introduction into New Zealand the scale insect had been brought in alone. The same thing, he argued, had occurred in the case of the introduction into the United States. He therefore, in his annual report for 1886, recommended that an effort be made to study the natural enemies of the scale in Australia and to introduce them into California; and the same year the leading fruit growers of California in convention assembled petitioned Congress to make appropriations for the Department of Agriculture to undertake this work. In February, 1887, the Department of Agriculture received specimens of an Australian parasite of Icerya from the late Frazier S. Crawford, of Adelaide, South Australia. It was a dipterous insect known as *Lestophonus iceryæ* Will., and for some time it was considered, both by Prof. Riley and his correspondents and agents, that the importation of this particular parasite offered the best chances for good results.

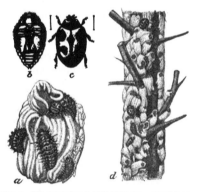

Fig. 4.—The Australian ladybird (*Novius cardinalis*), an imported enemy of the fluted scale: *a*, Ladybird larvæ feeding on adult female and egg sac; *b*, pupa; *c*, adult ladybird; *d*, orange twig, showing scales and ladybirds. *a–c*, Enlarged; *d*, natural size. (From Marlatt.)

Neither the recommendations of Prof. Riley nor of the then commissioner of agriculture, Hon. Norman J. Colman, nor the petitions of the California horticulturists gained the needed congressional appropriations, and, since there appeared at that time annually in the bills appropriating to the entomological service of the Department of Agriculture a clause preventing travel in foreign parts, it became necessary to gain the funds for the expense of the trip to Australia from some other source. A movement was started in California to raise these funds by private subscription, but it was never carried through. In an address given by Prof. Riley before the California State Board of Horticulture at Riverside, Cal., in 1887, he repeated his recommendations. During the summer of 1887 he was absent in Europe, and the senior author, who was at that time the first assistant entomologist of the department, by correspondence secured from Mr. Crawford numerous specimens of Icerya infested by the Lestophonus above mentioned. During the winter of 1887–88 preparations were being made for an exhibit of the United States at the Melbourne Exposition, to be held during 1888, and Prof. Riley, after interviewing the Secretary of State, who had charge of the funds appropriated for the exposition, was enabled to send an assistant, Mr. Albert Koebele, to Australia at the expense of this fund. This result was hastened, and Mr. Koebele's subsequent labors were aided by the fact that the commissioner general of the United States to the exposition was a California man, Mr. Frank McCoppin, and his recommendation, joined to that of Prof. Riley, decided the Secretary of State in favor of the movement. In order to partially compensate the exposition authorities for this expenditure, another assistant in the Division of Entomology, Prof. F. M. Webster, was sent out to make a special report to the commission on the agricultural features of the exposition. Mr. Koebele, who sailed from San Francisco August 25, 1888, was thoroughly familiar with all the phases of the investigation of the cottony cushion scale, and had for some time been stationed in California working for the Department of Agriculture. His salary was continued by the department and his expenses only were paid by the Melbourne Exposition fund. He made several sendings of the Lestophonus parasite to the station of the Division of Entomology of the Department of Agriculture at Los Angeles, where, under the charge of Mr. D. W. Coquillett, a tent had been erected over a tree abundantly infested with the scale insect; but it was soon found that the Lestophonus was not an effective parasite.

On October 15 Mr. Koebele found the famous ladybird (*Vedalia*) *Novius cardinalis* in North Adelaide, and at once came to the conclusion that this insect would prove effective if introduced into the United States. His first shipments were small, but others continued from that date until January, 1889, when he sailed for New Zealand and made further investigations. Carrying with him large supplies of *Vedalia cardinalis*, the effective ladybird enemy, he arrived in San Francisco on March 18, and on March 20 they were liberated under the tent at Los Angeles, where previous specimens which had survived the voyage by mail had also been placed.

The ladybird larvæ attacked the first scale insect they met upon being liberated from the packing cages. Twenty-eight specimens had been received on November 30 by Mr. Coquillett, 44 on December 29, 57 on January 24, and on April 12 the sending out of colonies was begun, so rapid had been the breeding of the specimens received alive from Australia. By June 12 nearly 11,000 specimens had been sent out to 208 different orchardists, and in nearly every case the colonizing of the insect proved successful. In the original orchard practically all of the scale insects were killed before August, 1889, and, in his annual report for that year, submitted December 31, Prof. Riley reported that the cottony cushion scale was practically no longer a factor to be considered in the cultivation of oranges and lemons in California. The following season this statement was fully justified, and since that time the cottony cushion scale, or white scale, or fluted scale, as it is called, has no longer been a factor in California horticulture. Rarely it begins to increase in numbers at some given point, but the Australian ladybirds are always kept breeding at the headquarters of the State Board of Horticulture at Sacramento, and such outbreaks are speedily reduced. In fact, it has been difficult for the State horticultural authorities to keep a sufficient supply of scale insect food alive for the continued breeding of the ladybirds.

The same insect was introduced direct from California into New Zealand at a later date, and the same good results were brought about. The Icerya is no longer a feature in horticulture in New Zealand.

NOVIUS IN PORTUGAL.

Still a third striking instance of the value of the Australian ladybird was seen later in the case of Portugal. *Icerya purchasi* was probably introduced into that country in the late eighties or early nineties from her colonies in the Azores, to which point it was probably introduced many years previously from Australia. The insect spread rapidly and threatened the complete destruction of the orange and lemon groves along the banks of the River Tagus. In September, 1896, persons in Portugal applied to the senior author for advice as to the most efficacious means of fighting the scale insect, and a reply was made urging them to make an effort to introduce (*Vedalia*) *Novius cardinalis* and sending information as to the success of the insect in California. In October, 1897, the chief of the bureau was able to secure, through the kindness of the State Board of Agriculture of California, about 60 specimens of the ladybird, which were sent by direct mail from Washington packed in moss. But five reached Portugal alive, but these were so successfully cared for that there was a numerous progeny. Another sending was made on the 22d of November following. These were received on the 19th of December and proved successful. Early in September, 1898, the statement was published in Lisbon newspapers that already colonies or stocks of the Vedalia had been established on 487 estates, whence naturally many others were formed by radiation; gardens and orchards that were completely infested and nearly ruined were already entirely clean or well on the way toward becoming so. Since that time the pest has almost entirely disappeared. The bureau would not have been able to assist the Portuguese Government to this admirable result had it not been for the enlightened policy of the State Board of Horticulture of California in continuing the breeding in confinement of these predaceous beetles long after the apparent great necessity for such work had disappeared in California, and had it not been for the courtesy of the board in promptly placing material at its disposal.

ICERYA IN FLORIDA.

The general effect of the California success on the horticultural world at large was striking, but not wholly beneficial. Many enthusiasts concluded that it was no longer worth while to use insecticidal mixtures, and that all that was necessary in order to eradicate any insect pest to horticulture or to agriculture was to send to Australia for its natural enemy. The fact that the Vedalia preys only upon Icerya and perhaps some very closely allied forms was disregarded, and it was supposed by many fruit growers that it would destroy any scale insect. Therefore the people in Florida whose orange groves were suffering from the long scale (*Lepidosaphes gloveri* Pack.) and the purple scale (*Lepidosaphes beckii* Newm.) sent to California for specimens of the Vedalia to rid their trees of these other scale pests. Their correspondents in California sent them specimens of the beetle in a box with a supply of Iceryas for food. When they arrived in Florida the entire contents of the box were placed in an orange grove. The result was that the beneficial insects died, and the Icerya gained a foothold in Florida, a State in which it had never before been seen. It bred rapidly and spread to a considerable extent for some years, and did an appreciable amount of damage before it was finally subdued.

NOVIUS IN CAPE COLONY.

Prior to the introduction of Novius into Portugal, *Icerya puschasi* having been established at the Cape of Good Hope, the beneficial ladybird was, after an unsuccessful attempt, carried from California to Cape Town by Mr. Thomas Low, member of the Legislative Assembly of Cape Colony, and on the 29th of January, 1892, living specimens were placed in perfect condition in the hands of the department of agriculture of Cape Colony. These specimens multiplied and were reenforced late in 1892 by a new sending from Australia made by Koebele. At the present time the Novius is perfectly naturalized at the Cape.

NOVIUS IN EGYPT AND THE HAWAIIAN ISLANDS.

At the same time, through the United States Department of Agriculture and the courtesy of the State Board of Horticulture of California, the Novius was sent to Egypt to prey upon an allied scale insect, *Icerya ægyptiaca* Dougl., which was doing great damage to citrus trees and to fig trees in the gardens of Alexandria, Egypt. Six adult insects and several larvæ arrived in living condition at

Alexandria. These multiplied so rapidly as to cause an almost complete disappearance of the scales. Later the latter began to increase, but the Novius had not died out and also increased. The Icerya is still held in check in a very perfect way. ,

In 1890 the Novius had been introduced into the Hawaiian Islands for work against *Icerya purchasi* with the same success.

ICERYA IN ITALY

In 1900 *Icerya purchasi* was found also in Italy, in a small garden at Portici, upon orange trees. By the autumn of 1900 it had multiplied so abundantly that the owner of the garden tried to stop the trouble by cutting down the trees most badly infested, without bothering himself with the others, so that the infestation continued. When Prof. Berlese's attention was called to it an attempt was first made to destroy it by insecticides without success, and then *Novius cardinalis* was imported from Portugal and from America. The following June the ladybird in both sexes was distributed in the garden, prospered wonderfully, and multiplied rapidly. In July the results were already evident; one could hardly find patches of Icerya which did not show the work of Novius, and at the end of the month it was difficult to find adult Iceryas with which to continue the rearing in the laboratory for food for the reserve supply of Novius. At the present time the multiplication of the scale insect has been reduced to the point of practically no damage, but the original infestation still persists and the area of distribution of the scale insect is slowly enlarging. It is found not only at Portici but in all the little towns around Vesuvius and in the gardens in Naples; but the presence of the ladybird allows the culture of oranges and lemons to go on without interruption.

ICERYA IN SYRIA.

The latest utilization of the beneficial Novius is recorded by Silvestri. It seems that about the year 1905 Icerya made its appearance in Syria, and in July, 1907, Selim Ali Slam wrote to Prof. Silvestri that it had spread so greatly about Beirut that it had almost destroyed the trees. Silvestri sent a shipment of Novius in July, 1907, and another one in August. The result was the same in Syria as it had been in other countries; the Novius multiplied greatly and produced the desired effect.[1]

THE REASONS FOR THE SUCCESS OF NOVIUS.

It thus appears that in the Novius we have an almost perfect remedy against Icerya. There have been no failures in its intro-

---

[1] Since the above was written (in the autumn of 1909) still another success with Novius has been by its carriage from California to Formosa by Dr. T. Shiraki, the entomologist of the Formosan Government, who writes, under date of Jan. 28, 1910: "To-day it has relieved the region from Icerya and has reduced their number to a practically negligible quantity."

duction to any one of the different countries to which it has been carried. Its success has been more perfect than that of any other beneficial insect that has so far been tried in this international work. There are good reasons for this—reasons that do not hold in the relations of many other beneficial insects to their hosts. In the first place, the Icerya is fixed to the plant; it does not fly, and crawls very slowly when first hatched, and later not at all. The Novius, however, is active, crawls rapidly about in the larval state, and flies readily in the adult. In the second place, the Novius is a rapid breeder, and has at least two generations during the time in which a single generation of the host is being developed. In the third place, the Novius feeds upon the eggs of the Icerya. And in the fourth place, it seems to have no enemies of its own. This is a very strange fact, since other ladybirds are destroyed by several species of parasites. For example, as will be shown later, native American ladybird parasites brought about a great mortality in the larvæ of the Chinese ladybird imported from China into America at a later date by Marlatt. The hymenopterous parasites of the widespread genus Homalotylus feed exclusively in ladybird larvæ, which are frequently also fairly packed with the minute hymenopterous parasites of the genus Syntomosphyrum, while the adults are often destroyed by Perilitus, Microctonus, and Euphorus.

The astonishing results of the practical handling of Novius drew attention more forcibly than ever before to the possibilities of this kind of warfare against injurious insects, and although its perfect success as an individual species has never been duplicated, very many efforts in this direction have been made, some of which have met with measurable success and some with very positive results of value.

INTRODUCTION OF ENTEDON EPIGONUS WALK. INTO THE UNTED STATES.

In 1891, with the assistance of Mr. Fred Enock, of London, Riley introduced puparia of the Hessian fly (*Mayetiola destructor* Say) infested with the chalcidid parasite *Entedon epigonus* Walk. into America. These were distributed among several entomologists during the spring of 1891. One American generation was carefully followed by Forbes in Illinois, and four years later (in May, 1895) the species was recovered by Ashmead at Cecilton, Md., where a colony had been placed in 1891. Thus the introduction was apparently successful, but if the species still exists in the United States it must be rare, since extensive rearings of Hessian-fly parasites have been made by agents of the Bureau of Entomology in many different parts of the country during the past few years and not a single specimen of the Entedon has been recognized. The Maryland locality, however, it should be stated, has not been visited by an entomologist since Ashmead's trip in May, 1895.

OTHER INTRODUCTIONS BY KOEBELE INTO CALIFORNIA.

Mr. Koebele took a second trip to Australia, New Zealand, and the Fiji Islands while still an agent of the Department of Agriculture, but at the expense of the California State Board of Horticulture, and in 1893 he resigned from the United States Department of Agriculture and was employed by the State Board of Horticulture of California for still another trip to Australia and other Pacific islands. He sent home a large number of beneficial insects, nearly all of them, however, coccinellids. Several of these species were established in California, and are still living in different parts of the State. The overwhelming success of the importation of *Novius cardinalis* was not repeated, but one of the insects brought over at that time, namely, the ladybird beetle *Rhizobius ventralis* Er. (fig. 5), an enemy of the so-called black scale (*Saissetia oleæ* Bern.), was colonized in various parts of California, and in districts where the climatic conditions proved favorable its work was very satisfactory, notably in the olive plantations of Mr. Ellwood Cooper, near Santa Barbara. Hundreds of thousands of the beetles were distributed in California and in some localities kept the black scale in check. Away from the moist coast regions, however, they proved to be less effective.

FIG 5.—*Rhizobius ventralis*, an imported enemy of the black scale: *a*, Adult ladybird; *b*, larva. Much enlarged. (From Marlatt.)

INTERNATIONAL WORK WITH ENEMIES OF THE BLACK SCALE.

It will here be convenient to drop the chronological sequence with which the subject in hand has been treated and to refer to the introduction of a very successful parasite of the black scale, whose work against this destructive enemy to olive and citrus culture in California for a time seemed second only to the success of the Novius against the Icerya. In 1859 Motchulsky described, under the name *Scutellista cyanea* (fig. 6), a very curious little hymenopterous parasite reared by Nietner from the coffee scale in Ceylon. Subsequently this parasite became accidentally introduced into Italy and was sent to the senior author for identification by Dr. Antonio Berlese as a parasite of the wax scale, *Ceroplastes rusci* L. As there are wax scales (*Ceroplastes floridensis* Comst. and *C. cirripediformis* Comst.) which are more or less injurious in Florida and the Gulf States, an attempt was made, with Berlese's assistance, to introduce this parasite at a convenient location at Baton Rouge, La., with the further assistance of Prof. H. A. Morgan at that place. Berlese's sending arrived in good condition, and the parasites issued at Baton Rouge and immediately began to attack the native species. The importation was successful for a time, but the introduced species was finally reduced to an insignificant number, presumably through the attacks of hyperparasites.

In the meantime Prof. C. P. Lounsbury, an American occupying the position of entomologist of the department of agriculture at the Cape of Good Hope, on his arrival at the Cape in 1895 and searching for the usual cosmopolitan scale insects on fruit trees, failed to find the black scale. He commented on this fact in one of his first-published papers, and alluded to the severity of the scale as a pest in California. Shortly afterwards he found the species, and sent the senior author specimens for identification in 1895, together with parasites which he had reared from it. Subsequent correspondence showed other species, and eventually *Scutellista cyanea* was forwarded. Writing to Mr. Lounsbury September 14, 1896, the chief of the bureau made the following suggestion: "I think parasitized black scales could be sent to California to advantage. Mr. Alexander Craw would be the proper person to to whom to send them." Mr. Lounsbury made further

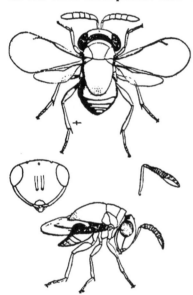

FIG. 6 —*Scutellista cyanea*, an imported parasite of the black scale Dorsal and lateral views of adult, with enlarged details. Greatly enlarged. (From Howard.)

studies, and commented in his 1898 report on the existence of parasites. When this report met the eye of Mr. E. M. Ehrhorn, of the State horticultural commission, Mr. Ehrhorn wrote Mr. Lounsbury, under date of December 22, 1899, asking him to send a colony of the parasite. Mr. Lounsbury had in the meantime, in a letter to the senior author, suggested that in order to gain authority to spend time over the matter and incur necessary expense it would be desirable for the Secretary of Agriculture of the United States to make a formal request for these parasites to the secretary of agriculture of Cape Colony. This was done, and in May, 1900, Lounsbury secured leave of absence and started for America, carrying with him a box of parasitized scales, and landed at New York on June 2. His box of parasites was at once forwarded to Washington, and the Bureau of Entomology notified Mr. Ehrhorn by telegram, repacked the box, and sent it to California. Mr. Ehrhorn succeeded in temporarily establishing the Scutellista indoors and out around his home at Mountain View, Cal. September 19, 1900, Mr. C. W. Mally, Lounsbury's assistant, sent two more boxes by post direct to California, addressing them to S. F. Leib, of San Jose, notifying the senior author to wire Mr. Ehrhorn to be on the lookout for them. A third lot was sent October 31 of the same year. These later sendings were small, and both failed to yield living parasites. More were requested, and on Lounsbury's return to South Africa a box was shipped in cool chamber to England and thence direct to California by express, Lounsbury's letter of February 28, 1901, to the bureau stating: "To avoid extra delay in transmission the box goes direct to California, but will you kindly have a message sent to Craw to advise him of its coming?" Unfortunately the box was detained by a customs officer at New York, but the bureau secured its release by the Government dispatch agent, Mr. I. P. Roosa. A few parasites emerged after arrival, but failed to propagate. October 1, 1901, Lounsbury started another sending by letter post to insure quick transit and noninterference by customs. These boxes were delivered to Mr. Craw on October 31. Only four females of the Scutellista were reared by Mr. Craw, and probably to these four females are due all of the Scutellistas subsequently occurring in California. This is the full story of the introduction of the species, taken from the letter files of the Bureau of Entomology and the letter files of Mr. Lounsbury in Cape Town.

Mr. Craw was remarkably successful in his rearings, and during the following three years constantly distributed colonies in different portions of California. By July, 1902, he had distributed 25 colonies. It was in the southern part of the State that the parasite did its best work, and there for a time it surpassed the most sanguine expectations of everyone. It was established in every county south of Point Conception and had become very plentiful in Los Angeles, Orange, and San Diego Counties. In the colonization districts by midsummer, 1903, it was estimated that over 90 per cent of the black scale had been destroyed. A year or so later there was great mortality among these parasites caused by a sudden increase in numbers of a predatory mite, *Pediculoides ventricosus* Newp. (fig. 7), which destroyed the larvæ in vast numbers. The Scutellista gradually recovered from this attack, and is at present to be found in very many localities in California, keeping the black scale partly in check.

Another enemy of the black scale was imported in 1901. It is a small moth, *Erastria scitula* Ramb. (fig. 8), the larva of which feeds in the bodies of mature scales, each larva destroying a number of scales. An effort had been made by Riley to import this insect from France in 1892, but without success. In 1901 Berlese sent the senior author living pupæ, which were at once forwarded to Craw and Ehrhorn in California. It was reported in 1902 that the insects had

been reared and liberated in Santa Clara, Los Angeles, and Niles, Cal., but if the species was established in the State it has not flourished and has not recently been found.

A similar lepidopterous insect, *Thalpochares cocciphaga* Meyrick, was brought over from Australia in the summer of 1892 by Koebele and left by him at Haywards, Cal., but the species evidently died out.

THE HAWAIIAN WORK.

FIG. 7.— *Pediculoides ventricosus.* Greatly enlarged. (From Marlatt.)

In 1893 Koebele resigned from the service of the State of California and entered the employment of the then newly established Hawaiian Republic for the purpose of traveling in different countries and collecting beneficial insects to be introduced into Hawaii for the purpose of destroying injurious insects. Before leaving California he had introduced a very capable ladybird, *Cryptolæmus montrouzieri* Muls., which feeds upon mealy bugs of the genus

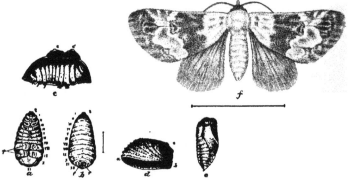

FIG. 8.—*Erastria scitula*, an imported enemy of the black scale *a*, Larva from below; *b*, same, from above; *c*, same, in case; *d*, case of full-grown larva; *e*, pupa, *f*, moth Enlarged (After Rouzaud.)

Pseudococcus. This insect flourished, especially in southern California, and on arrival in Hawaii he found that coffee plants and certain other trees were on the point of being totally destroyed by the allied scale insect known as *Pulvinaria psidii* Mask. He at once introduced this same Cryptolæmus, which is an Australian insect, with the result that the Pulvinaria was speedily reduced to a condition of harmlessness.

It may be incidentally stated that within the past year efforts have been made by the Bureau of Entomology to send the Cryptolæmus to Malaga, Spain, for the purpose of feeding upon a Dactylopius. The first attempt was unsuccessful, and the results of the last attempt have not yet been learned.

Another importation of Koebele's into Hawaii was the ladybird *Coccinella repanda* Thunb. from Ceylon, Australia, and China, which was successful in destroying plant lice upon sugar cane and other crops. Writing in 1896, Mr. R. C. L. Perkins stated that Koebele had already introduced eight other species which had become naturalized and were reported as doing good work against certain scale insects. Among other things he introduced *Chalcis obscurata* Walk. from China and Japan, which multiplied enormously at the expense of an injurious lepidopterous larva (*Omiodes blackburni* Butl.) which had severely attacked banana and palm trees.

Koebele's travels from 1894 to 1896 were through Australia, China, Ceylon, and Japan. In 1899 he left for Australia and the Fiji Islands, and sent many ladybirds and parasites to Hawaii, especially to attack the scale *Ceroplastes rubens* Mask. The Hawaiian Sugar Planters' Association, an organization which was responsible for Koebele's appointment, subsequently employed Mr. R. C. L. Perkins, Mr. G. W. Kirkaldy, Mr. F. W. Terry, Mr. O. H. Swezey, and Mr. F. Muir. By the close of 1902 sugar planters were especially anxious concerning the damage of an injurious leafhopper on the sugar cane, *Perkinsiella saccharicida* Kirk. This insect had been accidentally introduced from Australia about 1897, had increased rapidly, and by 1902 had become a serious pest. Koebele had made an effort to introduce parasites of leafhoppers from the United States into Hawaii, with unsatisfactory results, and consequently in the spring of 1904 Koebele and Perkins visited Australia and collected all possible parasites of different leafhoppers. Altogether they succeeded in finding more than 100 species. Of these the following hymenopterous parasites are said to have become acclimated in Hawaii: *Anagrus* (two species), *Paranagrus optabilis* Perk. and *P. perforator* Perk. and *Ootetrastichus beatus* Perk. These species are all parasitic upon the eggs of the leafhopper. By the end of 1906 observations upon a certain plantation indicated the destruction of 86.3 per cent of the eggs by these parasites. In addition to these egg parasites certain proctotrypid parasites of hatched leafhoppers have apparently become established, namely, *Haplogonatopus vitiensis* Perk., *Pseudogonatopus* (two species), and *Ecthrodelphax fairchildii* Perk. Three predatory beetles, namely, *Verania frenata* Erichs., *V. lineola* Fab., and *Callineda testudinaria* Muls.. were also distributed in large numbers.

The practical results of these importations seem to have been excellent. There seems to be no doubt that the parasites have been the controlling factor in the reduction of the leafhoppers.

The good work in Hawaii is still continuing. Koebele is now on a visit to Europe to import the possible parasites of the horn fly (*Hæmatobia serrata* Rob.–Desv.), Muir is trying to find an enemy to a sugar-cane borer (*Rhabdocnemis obscurus* Boisd.), and other similar work is under way.

AN IMPORTATION OF CLERUS FROM GERMANY.

An early attempt to import beneficial species into the United States was made in 1892 by Dr. A. D. Hopkins, then entomologist to the West Virginia Agricultural Experiment Station and now of the Bureau of Entomology. A destructive barkbeetle, *Dendroctonus frontalis* Zimm., was extremely injurious in that State in the years 1889 to 1892, and Hopkins made the effort to import from Europe another beetle, (*Clerus*) *Thanasimus formicarius* L., from Germany. In Germany he collected more than a thousand specimens of the Clerus, which he took with him to West Virginia and distributed in various localities infested by the barkbeetle. The following year, however, the barkbeetle disappeared almost completely from other causes, and the Clerus has not since been found.

MARLATT'S JOURNEY FOR ENEMIES OF THE SAN JOSE SCALE.

Another and later expedition was that undertaken by Mr. C. L. Marlatt, of the Bureau of Entomology, in search of the natural enemies of the San Jose scale. The question of the original home of the San Jose scale (*Aspidiotus perniciosus* Comst.) had been a mooted point. As is well known, it started in this country in the vicinity of San Jose, Cal., in the orchard of Mr. James Lick, who had imported trees and shrubs from many foreign countries. Mr. Lick died before the investigation started, and no records of his importations were to be found. The scale was not of European origin, since it does not occur on the continent. In the course of investigation it was found that it occurred in the Hawaiian Islands, in Japan, and in Australia, but in the case of Australia and the Hawaiian Islands it was shown that it had been carried on nursery stock from California. In 1897 plants entering the port of San Francisco from Japan were discovered by Mr. Craw to carry the San Jose scale. Correspondence, however, seemed to point to the conclusion that it had also been introduced into Japan from the United States. In 1901–2 Mr. Marlatt made a trip of exploration in Japan, China, and other eastern countries, lasting more than a year. Six months were spent in Japan, and after a thorough exploration the conclusion was reached that the scale is not a native of that country and that wherever it occurs there it has spread from a center of imported American fruit trees. Finally, as a result of this extended trip, the native home of the San Jose scale was found to be in northern China in a region between the Tientsin-Peking Road and the Great Wall, and its original host plant was found to be a little haw apple which grows wild over the hills. Into that region no foreign introductions of fruit or fruit trees had ever been made, and the fruits in the markets were all of the native sorts. Here in China was found everywhere present a little ladybird, *Chilocorus similis* Rossi (fig. 9), feeding in all stages upon the San Jose scale. One hundred and fifty

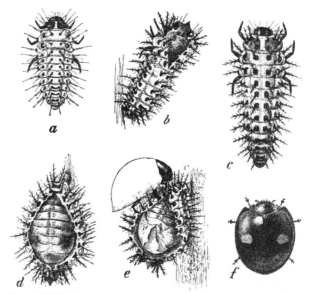

Fig. 9.—The Asiatic ladybird (*Chilocorus similis*), an imported enemy of the San Jose scale: *a*, Second larval stage; *b*, cast skin of same; *c*, full-grown larva; *d*, method of pupation, the pupa being retained in the split larval skin; *e*, newly emerged adult, not yet colored; *f*, fully colored and perfect adult. All enlarged to the same scale. (From Marlatt.)

or two hundred specimens of the beetle were shipped by Mr. Marlatt to Washington alive, but all but two perished during the winter. One at least of the two survivors was an impregnated female, and began laying eggs early in April. From this individual at least 200 eggs were obtained, the work being done in breeding jars. After some hundred larvæ had been hatched from these eggs the beetles were placed on a large plum tree in the experimental orchard and protected by a wire-screen cage covering the tree. The stock increased very rapidly, and during August shipments to various eastern experiment stations were begun, about 1,000 specimens being sent out. At the end of the first summer there remained of the stock at Washington about 1,000 beetles. Among the colonies sent out the best success was obtained in Georgia. An orchard at Marshallville in that State, containing some 17,000 peach trees and covering about 85 acres, adjoined a larger orchard containing about 150,000 trees, all scatteringly infested with the scale. The ladybirds were liberated in August, 1902, in the smaller orchard, and an examination made 11 months later indicated that they were rapidly spreading and would soon cover the orchard. The number of beetles in all stages present was estimated at nearly 40,000. Colonies established in the Northern States perished. In the South the almost universal adoption of the cheap and satisfactory lime-sulphur washes destroyed the possibility of rapid multiplication and destroyed the majority of the beneficial insects. This species has not been found recently, but probably exists in Georgia. The introduction and establishment of the species was successful, but it was practically killed out by the cheap and satisfactory washes in general use. Without the washes the probabilities are that the ladybird would be found at the present time occurring in great numbers in southern orchards.

THE PARASITES OF DIASPIS PENTAGONA TARG.

For a number of years the mulberry plantations of Italy had suffered severely from the attack of the insect known as the West Indian peach scale (*Diaspis pentagona* Targ.). This insect occurs in the United States and is widely distributed in other parts of the world. In the United States, however, it is not especially injurious. In 1905, at the request of Berlese, the writer sent parasitized Diaspis from Washington to Florence, Italy. One of the parasites which issued, *Prospaltella berlesei* How., was artificially reared in Florence by Berlese and his assistants, and at the time of present writing has been so thoroughly established in several localities that the ultimate reduction of the Diaspis to harmless numbers is confidently anticipated by Berlese. Similarly, Silvestri at Portici has introduced the same species from America, and also certain ladybirds, and is making the effort to import the parasites of this species from its entire range.

THE WORK OF MR. GEORGE COMPERE.

Mr. George Compere, employed jointly by Western Australia and California as a searcher for beneficial insects, for several years has been traveling in different parts of the world in search of beneficial insects which he has either sent or brought to California and Western Australia. One of the most interesting of his achievements was sending living specimens of *Calliephialtes messor* Grav., an ichneumon fly, from Spain to California. This species is a parasite of the codling moth.

In California this ichneumon fly has been reared with great success and has been sent out in large numbers from the headquarters of the State board of horticulture. In the field, however, it is apparently not succeeding, and there is no evidence that the numbers of the codling moth have been at all reduced by it. Nor is it, according to Froggatt, effective in Spain.

Mr. Compere has collected many beneficial species attacking many different injurious insects. He is an indefatigable worker, and his untiring qualities and his refusal to accept failure are well shown in his search for the natural enemies of the fruit fly of Western Australia, *Ceratitis capitata* Wied. He visited the Philippine Islands, China, Japan, California, Spain, returning to Australia, afterwards visiting Ceylon and India, and subsequently Brazil. In Brazil he succeeded in finding an ichneumon fly and a staphylinid beetle feeding upon fruit-fly larvæ. He collected some numbers and carried them to Australia in living condition, prematurely reporting success. The fruit fly is a pest in South Africa, and following the announcement of Compere's importations Claude Fuller and C. P. Lounsbury proceeded from Africa to Brazil to get the same parasites. The result of this journey was discouraging. They did not find the predatory staphylinid, but obtained a braconid parasite, *Opiellus trimaculatus* Spin.; they also concluded from information gained that the fruit fly had been introduced into South America more recently than into South Africa. The material carried home died. Compere left Australia again about the close of 1904; went to Spain for more codling-moth parasites, and then went on to Brazil, collecting more fruit-fly parasites and carrying them to Australia. The Brazilian natural enemies, however, did not succeed, and in 1906 he proceeded to India to collect parasites of a related fly of the genus Dacus, finding several and taking them to Western Australia. He arrived, however, in the middle of winter, and the insects perished. In May, 1907, once more this indefatigable man returned to India, and in a few months collected 70,000 to 100,000 parasitized pupæ, and brought them to Perth, Western Australia, in good condition on the 7th of December. It is reported that the parasites issued from this material in great numbers and in three distinct species. In April, 1908, it was reported that 120,000 parasites had been obtained and distributed, 20,000 of them having been sent to South Africa. The writer has not seen any definite reports of success in the control of the fruit fly by these parasites, but surely Compere deserves great credit for his efforts.

WORK WITH THE EGG PARASITE OF THE ELM LEAF-BEETLE.

In 1905 Dr. Paul Marchal, of Paris, published in the Bulletin of the Entomological Society of France for February 22 a paper entitled "Biological observations on a parasite of the elm leaf-beetle," to which he gave the name *Tetrastichus xanthomelænæ*.   In this very interesting article Dr. Marchal called attention to the fact that the elm leaf-beetle had multiplied for several years in a disastrous way about Paris, skeletonizing the leaves in the parks and along the avenues.   In 1904 the ravages apparently stopped, and Marchal's observations indicated that this was largely due to the work of this egg parasite.   He studied the life history of the parasite carefully during that year at Fontenay-aux-Roses and published his full account the following February.

Visiting Dr. Marchal in June, 1905, after the publication of this interesting article, the senior author asked him whether he had been able to make the further observations promised in the article, and he replied that the elm leaf-beetle had so entirely disappeared in the vicinity of Paris that he had not been able to do so.   The visitor urged him to make an effort through his correspondents to secure parasitized eggs of the beetle for sending to the United States in an effort to introduce and establish this important parasite on this side of the Atlantic. It was considered hopeless to attempt the introduction that summer, as the time was so late and it was not then known in what part of France the elm leaf-beetle could be found abundantly.   During 1906 practically the same conditions existed.   A locality was found, but the parasites did not seem to be present.   In 1907, reaching Paris about the 1st of May, the visitor again reminded Dr. Marchal of his desire to import the parasite into the United States, and meeting M. Charles Debreuil, of Melun, the subject was again brought up and M. Debreuil later in the season forwarded eggs of the beetle to the United States, which were promptly sent to the parasite laboratory at North Saugus, Mass., but the time was too late, and the parasites had emerged and died.

In April, 1908, the Entomological Society of France published in its bulletin (No. 7, p. 86) a request from the senior author that eggs of the elm leaf-beetle should be sent to the United States for the purpose of rearing parasites.   This notice brought a speedy and effective response. About the 20th of May Prof. Valery Mayet, of Montpellier, France, a personal friend, secured a number of leaves of the European elm carrying egg masses of the beetle, placed them in a tight tin box, and mailed them to Washington.   They were received May 28, and at once forwarded to the junior author at the parasite laboratory at Melrose Highlands.   On opening the box the junior author found a considerable number of active adults of the parasite.   Most of them were placed in a large jar containing leaves of elm upon which were newly deposited masses of the elm leaf-beetle eggs.   Probable oviposition was noticed within an hour after the receipt of the sending. There were probably somewhat more than 100 adults received in the shipment and very few emerged from the imported egg masses after the first day. The adults lived certainly for 35 days. Reproduction occurred in the experimental jars, and the adults secured by this laboratory reproduction were liberated in two localities near Boston and parasitized eggs were sent to Prof. J. B. Smith at New Brunswick, N. J., Prof. M. V. Slingerland at Ithaca, N. Y., and others to Washington. The first of the Massachusetts colonies consisted of about 600 parasites inclosed in an open tube tied to a tree in the Harvard yard, Cambridge, Mass., on June 22. Mr. Fiske thinks that more than 100 found their freedom on the same day, and almost certainly all of the rest within a week. A little more than a month later Mr. Fiske found parasitized eggs one-fourth of a mile away from this colony. At Melrose Highlands more than 1,200 were liberated on the 21st of June and the 8th of July; and on the 27th of July fresh native eggs in the neighborhood produced parasites, indicating the development of a generation on American soil. In the summer of 1909 none of the parasites was found, but this by no means indicates that the species has not become established. Both the eggs and the parasites are very small, and the writer expects that even from this first experiment good results will follow. Arrangements had been made for a repetition of the sending in May, 1909, from Montpellier, this southern locality allowing such an early sending as to insure the arrival of the parasitized eggs in the United States at the proper time of the year. Relying upon Prof. Mayet's promises and his great experience as an entomologist, no other arrangements were made. Most unfortunately, however, just before the time arrived Prof. Mayet died, and the introduction was not made. It should be stated that in the death of this admirable man France lost one of its most enlightened and able economic zoologists. It is hoped to repeat the introduction, through the kindness of Dr. Marchal in France and Prof. Silvestri in Italy. Silvestri has promised also to send other natural enemies of the elm leaf-beetle from Italy.

WORK WITH PARASITES OF TICKS.

In 1907 the senior author described the first species of a hymenopterous parasite ever recorded as having been reared from a tick. The name given to it was *Ixodiphagus texanus*, and it had been reared from the nymphs of *Hæmaphysalis leporis-palustris* Pack. collected on a cotton-tail rabbit in Jackson County, Tex., by Mr. J. D. Mitchell. In 1908 he described another, *Hunterellus hookeri*, reared by Mr. W. A. Hooker at Dallas, Tex., from *Rhipicephalus texanus* Banks taken from a Mexican dog at Corpus Christi, Tex., by Mr. H. P. Wood. Inasmuch as a closely allied if not identical tick, *Rhipicephalus sanguineus* Latr., is supposed to be a transmitter of a trypanosome disease in South Africa, sendings of the Hunterellus were made in the autumn of 1908 to Prof. Lounsbury at Cape of

Good Hope and to Mr. C. W. Howard, entomologist to the government of Lourenço Marques, Portuguese East Africa. In June, 1909, Mr. C. W. Howard reared parasites from engorged nymphs of *Rhipicephalus sanguineus* taken from dogs, with which transmission experiments with trypanosomiasis were being made. Examination showed them to be *Hunterellus hookeri*. Mr. C. W. Howard is of the opinion that these 1909 reared specimens could not have been the offspring of those sent over in the autumn of 1908, since, as he writes under date of September 3, 1909, the latter arrived while he was absent in the Zambesi country, and, as he was gone nearly three months, they remained on his desk unopened. When he returned they were all dead. He kept the ticks some time, however, in a sealed jar to see if any more parasites might emerge, but none did so. In his opinion there is absolutely no possibility that the 1909 specimens are the descendants of those sent from Texas. Of course Mr. C. W. Howard is probably correct in his surmise, but a most interesting question arises as to the original home of the parasite. Could it have been carried accidentally from Texas to Africa at an earlier date? As a matter of fact, during the Boer War thousands of horses and mules were shipped from southern Texas to Cape Town, and much of this stock came from the very region in which the Texas Rhipicephalus occurs. Banks, in his revision of the ticks,[1] records this species from horses as well as from dogs, the horse record coming from New Mexico. The suggestion regarding the importation of horses and mules from Texas to Cape Town during the Boer War was made to the writer by Mr. W. D. Hunter, who also suggests that as *Rhipicephalus sanguineus* occurs throughout Africa and Mediterranean Europe, and that as in 1853 several shipments of camels were brought to Texas from Tunis, being turned loose at Indianola and roaming wild throughout the territory around Corpus Christi for some years, it is possible that the Rhipicephalus was brought to Texas on these camels, and the parasite as well. This seems unlikely, however, since the parasite had never been found in Africa or Europe until the specimens referred to were reared by Mr. C. W. Howard in 1909.

MR. FROGGATT'S JOURNEY TO VARIOUS PARTS OF THE WORLD IN 1907-8.

As a result of a conference of Australian Government entomologists, held in Sydney, July 9, 1906, and of a conference of State premiers, held in Brisbane, June, 1907, it was agreed that Mr. W. W. Froggatt, entomologist to the Department of Agriculture of the State of New South Wales, should be dispatched to America, Europe, and India, to inquire into the best methods of dealing with fruit-flies and other pests, the expenses of the journey to be shared by Queensland, South

---

[1] U. S. Department of Agriculture, Bureau of Entomology, Technical Series No. 15, p. 35, 1908.

Australia, New South Wales, and Victoria. As a result of the trip following this authorization, Mr. Froggatt has published a report on parasitic and injurious insects, issued in 1909, in which he considers, (1) the commercial value of introduced parasites to deal with insects that are pests; (2) the range and spread of fruit-flies, and the methods adopted in other countries to check them; (3) the value of parasites in exterminating fruit-flies; (4) the habits of cosmopolitan insect pests. On his journey, which began the end of June, 1907, Mr. Froggatt visited Hawaii, the United States, Mexico, Cuba, Jamaica, Barbados, England, France, Spain, Italy, Austria, Hungary, Turkey, Cyprus (spending a day in Smyrna and two days at Beirut on the way), Egypt, India, Ceylon, and thence to Australia, stopping in Western Australia before his return to Sydney. In the course of this trip Mr. Froggatt not only studied the question of parasites and of economic entomology in general, but looked into a large number of matters of agricultural interest, and has given a report which can not fail to be interesting to every one occupied with any branch of agriculture.

With regard to the practical handling of parasites, and especially international work, he is inclined to be rigidly critical. His motive obviously was to look everywhere for accomplished results and where he could not find these to distinctly state the fact. He deprecates all claims that are not or have not been justified by practical results of value. Thus, while admitting the good work of the introduced parasites of the sugar-cane leafhopper in Hawaii, he states that the advocates of the parasite system do not take into account the alteration of methods of cultivation which occurred about the same time, namely, the burning of the refuse (probably containing many eggs and larvæ) instead of burying it as formerly, and the introduction of new varieties of cane more resistant to the leafhoppers. In California, he admits the value of the introduction of the Australian ladybird, but states that his observations show that no good has followed the introduction of the codling-moth parasite from Spain, although it had been claimed previously that this parasite would prove a perfect remedy for the apple pest, and pointing out that when he visited Spain he found that a very large percentage of the apple crop is always infested by the codling moth. He states that the promises of the advocates of the parasite method in California have not been fulfilled; that Western Australian claims that staphylinid beetles destroy the majority of the fruit-fly maggots in Brazil, and that nature's forces in that country control the destructive fruit flies are to be contrasted with the statement of the South African entomologists that only a few months after the visit of the West Australian entomologist to Brazil they found that "all along the Brazilian coast it was difficult to obtain a fruit that had not been punctured by a fly." The statement that nature controls the destructive fruit flies in India he opposes, as a result of his own observations in India. He does not contend that this work has not a great practical value, but insists that it should be done by trained entomologists, and that full information of the habits and life histories of both the pests and their parasites should be understood before liberation is attempted. As already stated, he especially deprecates premature claims, and points out that in New South Wales the passage of the very necessary vegetation diseases bill was delayed for some years by the outcry "Why should we be made to clean up our orchards and spend money, when the department can send out to other countries and get us parasites that will do all that is needed?" In conclusion he states:

Let the whole question be judged on its results. Allow that one or two experiments have shown perfect results; yet because mealy bugs or scale insects in a restricted locality have once or twice been destroyed by parasites, that can be no reason why the parasite cure alone should be forced upon anyone. Its admirers should be perfectly honest; and if a friendly introduced insect from which, rightly or wrongly, great things had been expected turns out upon further trial to be a failure, they should say so; and they should never proclaim results for a parasite till those results have actually been proved in its adopted country, for the wisest can never be sure of the results of any experiment. Economic entomology is a great commercial science, and those at work for its far-reaching interests could do it no greater harm than by misleading or unproved statements

OTHER WORK OF THIS KIND.

Reference has already been made to the importations of *Prospaltella berlesei* into Italy to attack the destructive mulberry scale, *Diaspis pentagona*, through cooperative arrangements between the senior author and Prof. Berlese, of Florence. Prof. Berlese has been successful in establishing the species, and believes that it is best to rely upon this species only, and not to attempt to introduce the predatory enemies of the scale, his idea being that coccinellids will feed indiscriminately upon parasitized and unparasitized scales and that thus the Prospaltella will not have a chance to multiply to its limit. The contrary view is taken by Prof. Silvestri, at Portici, in the south of Italy, and he has been making every effort to introduce from all parts of the world all of the enemies, whether parasitic or predatory, of the mulberry scale. He has brought over and has had breeding in his laboratory at Portici, as well as in an experimental olive orchard southeast of Naples, a number of species of Coccinellidæ brought from different parts of the world. At his request, in May, 1910, the senior author carried from Washington a box containing possibly 200 living specimens of *Microweisia misella* Lec. and a few specimens of *Chilocorus bivulnerus* Muls. These were carefully packed with plenty of food in a small paper-covered wooden box, approximating a 10-inch cube. He sailed from New York direct to Naples and, through the kindness of the officers of the Royal Italian Line steamship *Duca di Genova*, was enabled to suspend the box by a cord from a crossbeam in the ordinary cold room of the steamer. After an eleven days' passage, the box was opened in Prof. Silvestri's laboratory in Portici, and practically every coccinellid was found to be alive and in apparently good condition.

Efforts have been made by the Bureau of Entomology, in cooperation with the Pasteur Institute in Paris, to introduce a large bembecid wasp (*Monedula carolina* Fab.) from New Orleans into Algeria to prey upon the tabanid flies concerned in the carriage of a trypanosome disease of dromedaries. The wasps were sent in their cocoons in refrigerating baskets from New Orleans by direct steamer to Havre and from New York by direct steamer to Havre. There they were met by agents of the Pasteur Institute, carried to Marseilles by rail and thence by boat to Algeria, and were planted under conditions as closely as possible resembling those under which they were found in Louisiana, care being taken to simulate not only the character of the soil but the exposure to light, the prevailing wind directions, and the moisture conditions. Adults issued, but the species has not since been recovered, although it is quite possibly established.

In the same way an attempt was made to introduce the common bumblebee *Bombus pennsylvanicus* De Geer of the United States into the Philippine Islands for the purpose of fertilizing red clover. These were sent in refrigerating baskets, carried by hand by Filipino students returning from the United States to the Philippines, and for the most part in the pupal stage. These were properly planted upon arrival and reared, and a few specimens have been recovered.

In the summer of 1910 Dr. L. P. De Bussy, biologist of the Tobacco Planters' Association of Deli, Sumatra, visited the United States for the purpose of investigating damage to the tobacco crop by insects and disease and to make an effort to import into Sumatra the parasites of the destructive tobacco worm known as *Heliothis obsoleta* Fab. Already shipments of an egg parasite, *Trichogramma pretiosa* Riley, have been made to Sumatra via Amsterdam, but information as to the results of these preliminary shipments has not yet reached this country.

Prof. C. H. T. Townsend, an assistant in the Bureau of Entomology, receiving a temporary appointment as entomologist to the Department of Agriculture of Peru, especially to study the injurious work done by the scale insect *Hemichionaspis minor* Mask. on cotton, has during the past year, with the assistance of the bureau, imported a number of shipments of *Prospaltella berlesei* from Washington into Peru. It is too early to announce results.

In July, 1910, Mr. R. S. Woglum, an agent of the Bureau of Entomology, was sent abroad to find the original home of the white fly of the orange, *Aleyrodes citri* R. & H., and to attempt to find parasites or satisfactory predatory enemies. In November, 1910, he found the white fly at Saharampur, India, and discovered that it was killed by a fungous disease (lately determined as a species already occurring in the United States—*Ægerita webberi*—by Prof. H. S. Fawcett, of the Florida Agricultural Experiment Station). He also found that it was attacked by two species of Coccinellidæ (*Verania cardoni* Weise and *Cryptognatha flavescens* Motsch.). A preliminary shipment of the ladybirds by mail was apparently unsuccessful. Later shipments by direct steamer from Calcutta to Boston were also unsuccessful.

At Lahore, India, Mr. Woglum found his first evidence of parasitism by hymenopterous parasites. A certain proportion of *Aleyrodes citri* was found to contain the exit holes of a true parasite. The specimens on leaves sent in by Mr. Woglum were examined with great care. None of the full-grown larvæ or nymphs contained parasites, but five specimens of a very minute aphelinine of the genus Prospaltella were found dead and attached to the orange leaves in close vicinity to the perforated aleyrodids. The size of the specimens was such as to justify the conclusion that they had issued from the aleyrodids, and their juxtaposition and the known habits of the genus confirm this conclusion. The species was described by the senior author as *Prospaltella lahorensis* in the Journal of Economic Entomology for February, 1911, pages 130–132. Efforts will be made to import this parasite into Florida.

The occurrence of a European weevil, *Phytonomus murinus* Fab., in the alfalfa fields of Utah in alarming numbers and the difficulty of fighting the pest by mechanical or cultural means has started an investigation as to its parasites in its original home. Mr. W. F. Fiske, of the Bureau of Entomology, sent from Naples, Italy, on March 17, 1911, a large lot of stems of alfalfa containing eggs of an allied weevil parasitized by a minute mymarid, which at the time of this writing are on their way to Utah.

In the meantime the State board of horticulture of California has been continuing its efforts to import beneficial insects of different kinds. Mr. George Compere returned from a lengthy trip during the summer of 1910, bringing with him a number of interesting species, among them a new coccinellid enemy of mealy bugs in which he has great faith, and which promises to be a valuable addition to the insect fauna of the United States.

Entomologists and horticulturists all over the world have become greatly interested in this aspect of economic entomology and for the immediate future a great deal of experimental work has been planned by the officials of different countries.

EARLY IDEAS ON INTRODUCING THE NATURAL ENEMIES OF THE GIPSY MOTH.

Promptly with the discovery that the gipsy moth had become acclimatized in Massachusetts, in 1889 there was published by Prof. C. H. Fernald a special bulletin of the Massachusetts Agricultural College Hatch Experiment Station, in which he gave popular descriptions of the different stages of the insect and recommended spraying with Paris green. He stated that the insect is generally held in check by its natural enemies in Europe, but occasionally becomes very destructive, and stated that 11 species of hymenopterous parasites and several of dipterous parasites had been noticed in Europe. This bulletin was published in November, 1889. In January, 1890, an illustrated article on the gipsy moth, by Riley and Howard, was published in Insect Life,[1] and a list of 24 European hymenopterous parasites compiled by Howard was published.

Immediately following this publication, there was received at the Department of Agriculture, from Rev. H. Loomis, of Yokohama, Japan, a letter in which he stated that he had seen reports of the ravages of the gipsy moth in Massachusetts and had taken considerable interest in the matter. He also stated that he had seen the gipsy moth caterpillar on a wistaria vine near his house in Yokohama, and that it had been attacked and killed by a parasite. Several of the parasites were sent in an accompanying box, and proved to be Apanteles. Subsequent attempts were made by Mr. Loomis to send this parasite in living condition both to the Department of Agriculture and to the State of Massachusetts, but all arrived dead, for the most part having been killed by secondary parasites.

In March, 1891, a conference was held in the rooms of the committee on agriculture at Boston, at which were present Prof. N. S. Shaler, Gen. F. H. Appleton, and Mr. William R. Sessions, of the State board of agriculture; Prof. C. V. Riley, entomologist of the United States Department of Agriculture; Prof. C. H. Fernald, entomologist of the State Experiment Station; Mr. S. H. Scudder, a well-known entomologist; the mayors of Medford, Melrose, Arlington, and Malden, and others. In the course of the conference, which was held for the purpose of discussing the best measures to be taken against the gipsy moth by the State, Prof. Riley advocated an attempt at extermination by spraying. Mr. Scudder advocated the destruction of the eggs, and in the course of the discussion Prof. Riley made the following remark:

I would make one other suggestion, and that is, that as an auxiliary method it would be well to spend $500 or $600 in sending one or two persons abroad next summer with no other object than to go to some section of northern Europe to collect and transmit to authorized persons here a certain number of the primary parasites of this species,

[1] Insect Life, Division of Entomology, U. S. Department of Agriculture, vol. 2, pp. 208-211, 1890.

which are known to check its ravages over there. The insect was undoubtedly brought over by Trouvelot without any of its natural checks. In my judgment it would be well worth trying to import its parasites from abroad. The advantage would be this: If you failed to exterminate it by spraying, its parasites, seeking for this particular host, would be more apt to find the overlooked or escaped specimens than man would.

No action was taken upon this suggestion, and the State authorities, believing that such an attempt would be useless owing to the fact that their effort for some years was consistently devoted to the aim of absolute extermination of the gipsy moth, perhaps wisely saved the expense of a mission abroad for this purpose. Then, also, there was some hope that the native parasites, particularly the ichneumon flies and the native species of Apanteles, as well as tachina flies and some of the carabid beetles, might gradually accommodate themselves to the imported pest and prove prominent factors in the fight against it.

This last faint hope, however, was not justified. In the course of the careful work done by the State during the next seven or eight years, the better part of which is summarized in the admirable Report on the Gipsy Moth, by Forbush and Fernald, published in 1896, several native parasites and predatory insects were observed to attack the gipsy moth in its different stages, but at no time was the percentage of parasitism sufficiently great to have any value as a factor in the suppression of the pest. At no time was there a greater percentage of parasitism by native parasites than 10, whereas the condition in Europe is such that the percentage reaches frequently well above 80. It may be worth mentioning that parasitism by native species has never exceeded 5 per cent in any collections made since the present laboratory was established. It is nearer 2 per cent on the average.

In discussions among the Washington entomologists it was repeatedly pointed out by E. A. Schwarz and by B. E. Fernow (at that time Chief of the Division of Forestry of the United States Department of Agriculture) that one of the most important of European enemies of the gipsy moth, and the nun moth as well, is one of the tree-climbing ground beetles known as *Calosoma sycophanta* L. There exist a number of species of this same genus Calosoma in the United States, but none of them has the tree-climbing habit developed to the same extent as have *Calosoma sycophanta* of Europe and its relative *Calosoma inquisitor* L. Prof. Fernald, writing to the famous German authority on forest insects, Dr. Bernard Altum, early in 1895, asked his opinion as to the advisability of importing these tree-inhabiting ground beetles, but received the reply that such an importation would not give good results. Prof. Altum considered the services of the hymenopterous parasites of the old genus Microgaster as of much more importance.

In the report just cited Fernald disposed of the question of importing parasites in the following words:

No attempt has been made to import parasites thus far for the reason that the law requires the work to be conducted with direct reference to the extermination of the gipsy moth, and, therefore, the general destruction of the insect would also destroy the parasites. There is no reason why our native hymenopterous parasites may not prove to be quite as effective as those of any other country, since there is no parasite known which confines itself exclusively to the gipsy moth, and, as has been shown, we have several species which attack it as readily as any in its native country.

This position with regard to the nonimportation so long as extermination of the gipsy moth was the end, held until the State of Massachusetts ceased its appropriations, in the year 1900.

CIRCUMSTANCES WHICH BROUGHT ABOUT THE ACTUAL BEGINNING OF THE WORK.

During the five years that elapsed before the State again began to appropriate money for the suppression of the gipsy moth and the brown-tail moth, as is well known, the gipsy moth spread from a restricted territory of 359 square miles throughout an extended range of 2,224 square miles and even more. As soon as the effort to exterminate it was abandoned, owing to the lapse of the appropriations for the year 1900, the project of importing parasites was taken up by the Chief of the Bureau of Entomology, who began correspondence with a number of European entomologists with this end in view. Especial efforts were made to import the Calosomas, but failed, partly owing to a lack of interest in the matter on the part of the Europeans. In 1902 Mr. W. B. Alwood, entomologist of the Virginia Agricultural Experiment Station, went abroad for a series of months and was requested by the chief of the entomological service of the United States Department of Agriculture to endeavor to find, in some well-placed situation in Europe, one or more competent collectors of insects who would undertake systematically to send gipsy-moth parasites to America. This effort also failed, and Mr. Alwood was unable to find the proper persons. Finally, in December, 1904, Congress was asked to make a small appropriation for the distinct purpose of attempting the importation of these parasites, and the sum of $2,500 was appropriated for this purpose in the session of the winter of 1904–5. During the corresponding session of the Massachusetts State Legislature, State appropriations began once more. In 1904 it was apparent to everyone that the old areas had become reinfested and that the insect had spread widely. Private estates and woodlands in June and July of that year were almost completely defoliated. Kirkland wrote:

From Belmont to Saugus and Lynn a continuous chain of woodland colonies presented a sight at once disgusting and pitiful. The hungry caterpillars of both species of moths swarmed everywhere; they dropped on persons, carriages, cars, and automobiles, and were thus widely scattered. They invaded houses, swarmed into living and sleeping rooms, and even made homes uninhabitable * * *. Real estate in the worst-infested districts underwent a notable depreciation in value. Worst of all, pines and other conifers—altogether too scarce in eastern Massachusetts—were killed outright by the gipsy-moth caterpillars, while shade trees and orchards were swept bare of foliage.

There was a general demand upon the State legislature and an excellent bill was prepared and passed with the appropriation of $300,000, $75,000 to be expended during 1905, $150,000 and any unexpended balance during 1906, and $75,000 and any unexpended balance during 1907, up to May 1, 1907, inclusive. And to this appropriation there was added the clause "for the purpose of experimenting with natural enemies for destroying the moths, $10,000 is additionally appropriated for each of the years 1905, 1906, and 1907." There was then available in the spring of 1905 the appropriation of $2,500 by the General Government and that of $10,000 by the State of Massachusetts for work with the natural enemies. Mr. A. H. Kirkland was appointed superintendent for suppressing the gipsy and brown-tail moths, by Gov. Douglas, and immediately following his appointment, and with the approval of his excellency the governor, went to Washington, and by arrangement with the honorable the Secretary of Agriculture, Mr. James Wilson, arranged a cooperation between the State and the Department of Agriculture whereby the Chief of the Bureau of Entomology of the department was practically placed in charge of the details of the attempt to import parasites from abroad, in consultation with Mr. Kirkland.

The reasons which influenced Mr. Kirkland in entering into this cooperation between the State and the United States Department of Agriculture were expressed in his first annual report (p. 117).

At this time for more than 25 years the chief of the bureau had been devoting his especial efforts to the study of the parasitic Hymenoptera, and had especially interested himself in the subject of their biology and host relations. He had accumulated a card catalogue of more than 20,000 entries of records of the specific relations of
. parasites to specific insects, the great majority of these being European records and covering all of the published information regarding the parasites of the gipsy moth and the brown-tail moth. He also had the advantage of the personal acquaintance of most of the European entomologists interested in this kind of work. These facts were known to Mr. Kirkland and caused his action.

AN INVESTIGATION OF THE INTRODUCTION WORK.

From the beginning of the work, and even before, certain citizens of Boston, impressed by the claims of the State Board of Horticulture of California as to the results said to have been achieved by the agents of the board in the introduction of beneficial insects, urged the employment of these agents in the work of introducing the parasites of the gipsy moth and brown-tail moth. The arguments in favor of this proposal were duly considered by the superintendent of the Massachusetts work, who decided for many reasons to conduct the introduction experiments along the lines just described and not to call in the assistance of the California people. In his third report, submitted January 1, 1908, Mr. Kirkland expressed the situation as follows:

In spite of all the thought, energy, and skill that have been brought to bear on this most important problem of introducing the natural enemies of the moths—a problem entirely novel in the field of entomology—it was apparent during the winter of 1906-7 that several of our influential citizens had expected immediate results from the importation of the parasites, and were beginning to get restive because such results had not been obtained. Several expressed a doubt if everything possible was being done to secure the successful introduction of the parasites. Others became enthusiastic over the specious proposition put forward by a certain western horticulturist (not an entomologist), who offered to suppress the gipsy moth in Massachusetts by means of parasites for the sum of $25,000, "no cure, no pay." This state of affairs was no doubt a natural outcome of the desire to avoid a repetition of the great damage to property caused by the moth in past years. Again, men without any technical knowledge of entomology or of the life histories of the parasites, not realizing the difficulties in securing, shipping, breeding, and disseminating these beneficial insects, and equally ignorant of how long it takes an imported insect to become established even under the most favorable conditions, might well be pardoned for expecting almost immediate results from the introduction of the relatively small number of parasites—small indeed in comparison with the tremendous numbers of the moths.

Coming before the legislature during the session of 1906–7, this group of Boston citizens stated that it was their opinion that the work with parasites was not progressing with sufficient rapidity, and asked the legislature to appropriate funds and to instruct the superintendent to secure additional counsel and advice in the matter to determine whether the work was going on in the right way. The legislature agreed and appropriated the additional sum of $15,000 to enable the superintendent to secure such advice.

It was first suggested that he consult only with certain California men who had had experience in importing parasites of scale insects. He, however, considered that consultation with men whose experience had been confined to a single group of insects, not to the same group as the gipsy moth and the brown-tail moth, while possibly helpful, would not be broad enough to throw any great light on the Massachusetts problem. To use his own words—

It seemed much wiser and certainly more thoroughgoing, since this entire work might be called in question at any time, and in view of the large amount of money Massachusetts was expending in securing parasites, to consult not with the trained entomologists of a single State, but with as many entomologists of national or even world-wide reputation as possible. In other words, that a large number of entomologists of the highest possible scientific standing, and particularly those having practical experience in dealing with parasitic insects, should be invited to visit Massachusetts,

learn of our difficult problems on the spot, examine into the methods of importing,
rearing, and distributing parasites, and then give us the benefit of their criticism and
counsel, based on a full knowledge of the facts at hand He also suggested that,
since by some this movement might be taken as a criticism on his management and
on his judgment in placing the direction of the work in the hands of Dr Howard, it
would be well to have some outside board or commission take charge of the matter,
so that it should be entirely an ex parte affair free from any suggestion of influence
by the present administration of the work. The suggestion to authorize the superintendent to invite the entomologists was heartily indorsed by the legislative committee which had the matter under consideration, while the arrangement of the entire
affair was left in his hands

In his selection of experts, Mr. Kirkland was aided by Prof. C. H.
Fernald, of the Massachusetts Agricultural Experiment Station,
one of the oldest and best posted entomologists in the country; and
Mr. Kirkland himself, it must be remembered, had been engaged in
active entomological work for 15 years and had held official positions in the Association of Economic Entomologists, thus having a
very broad personal acquaintance with the best workers.   The list
selected, as quoted from Mr. Kirkland's report, was as follows:

Prof Edward M. Ehrhorn, deputy commissioner of horticulture, State of California,
a man of large practical experience in importing, breeding, and disseminating insect
parasites, particularly those of scale insects, and also a man well trained in applied
entomology.
Prof. Herbert Osborn, Ohio State University, one of the country's best known
teachers of entomology, and of large experience in investigation and laboratory work
Dr. John B. Smith, entomologist, New Jersey Agricultural Experiment Station,
an investigator of the highest order, a successful teacher, and the author of numerous
standard works on insects.
Prof. S. A. Forbes, State entomologist, Illinois, a most successful teacher and investigator, and one of the most prominent entomologists of the Middle West
Prof E. P. Felt, State entomologist of New York, a well-known writer on and investigator of insect pests, and particularly ingenious in devising laboratory methods.
Prof. H. A Morgan, director of the Tennessee Agricultural Experiment Station,
of large experience, and one of the best-known entomologists of the Southern States.
Prof. M. V Slingerland, Cornell University, New York, an investigator with hardly
an equal, and one who has had great success in studying life histories of beneficial
and injurious insects

In addition to these, the following well-known foreign entomologists, visiting Boston, were asked to investigate the situation carefully, to study the laboratory and field methods, and to report:

Prof. Charles P. Lounsbury, entomologist, Cape Town, South Africa, one who has
had great experience as well as great success in importing beneficial insects.
Prof. Walter W. Froggatt, government entomologist, New South Wales, and also
investigator for Victoria and Queensland. Prof. Froggatt's work has been practically along the same line as that of Prof. Lounsbury, and has met with a large measure
of success.
Dr. James Fletcher, dominion entomologist, Canada, well known for his success in
working out difficult points in the life histories of insects, and more particularly in
dealing with a wide range of injurious species
Prof. R. Blanchard, University of Paris, and member of the Academy of Medicine
Dr. G. Horvath, director of zoological section, National Hungarian Museum, mem-

ber of the Academy of Science of Hungary and formerly director of the entomological station of Hungary. The last two gentlemen are entirely familiar with the two moths and their parasites.

Dr. Richard Heymons, extraordinary honorary professor and custodian at the Zoological Museum of the Royal Institute of Berlin. Dr. Heymons has made a large study of the injurious insects of central Europe, and particularly of their natural enemies.

Prof. A. Severin, conservator at the Royal Museum of Natural History of Belgium, and member of the Superior Council of Forests. Prof Severin's position is naturally that of one of the best posted entomologists, particularly with reference to dangerous forest insects.

In addition to these foreign entomologists, Prof. Filippo Silvestri, of the Royal Agricultural School of Portici, Italy, visiting America on an official mission in the summer of 1908, visited Boston, and was asked to give his professional opinion of the work, his report being printed in the fourth annual report of the superintendent, issued January, 1909, by L. H. Worthley, acting superintendent.

It is worthy of note that Prof. Silvestri had been commissioned by the R. Accademia dei Lincei and by the royal minister of agriculture of Italy to investigate the work in economic entomology being done in the United States, and had visited all portions of the country, including California and Hawaii, studying with especial care all the work being done with parasites. It should be pointed out also that the California claims were perfectly well understood by all of the American experts, Mr. Ehrhorn himself being the second ranking officer in the California service, and the others having either visited California partly for the purpose of investigating this work, or being perfectly familiar with the situation by study of the publications and by correspondence. Moreover, of the foreign experts, Mr. Froggatt had just come from California on an investigating trip for the government of the Federated Colonies of Australia which subsequently carried him around the world, Mr. Lounsbury had visited California for the purpose of studying this work, and Dr. Fletcher had repeatedly visited that State.

The reports of all of these experts, with the exception of that of Prof. Silvestri, are published in the third annual report of the Massachusetts superintendent, Boston, 1908, Prof. Silvestri's report being published, as above stated, in the fourth annual report of the superintendent.

It will be entirely unnecessary to quote from these reports, since they may be found in full in the State documents mentioned. It will suffice to state that the work was commended, it is safe to say, with enthusiasm by every individual. Specific consideration was given to the California suggestion by Mr. Lounsbury, by Mr. Froggatt, and by Prof. Slingerland. Suggestions were made by several of them that the study of the fungous, bacterial, and protozoan diseases of the larvæ should be taken up. Dr. Felt and Dr. Smith recommended the importance of the introduction of the Japanese parasites, and Dr. Felt suggested the importance of careful biological studies of the parasites, not only in America but in Europe. All of these suggestions coincided with plans already made which were about to be entered upon, as indicated in following pages.

The subject of the study of the diseases of the caterpillars does not come under the range of the present bulletin, but since it has been mentioned, it should be stated that the State superintendent has for the past two years been having this subject investigated and that it is now going on under the expert supervision of Dr. Roland Thaxter, of Harvard University, and Dr. Theobald Smith, of the Harvard Medical School.

Mr. Kirkland's summary seems fully justified. It is as follows:

It will be seen from the foregoing that the work of importing parasites of the gipsy and brown-tail moths in Massachusetts has been thoroughly examined by practically a congress of the world's leading entomological experts And it is believed that their consensus of opinion, which is, in the main, that everything possible to secure the successful importation of these insects is being done, will be taken as authoritative and final. It would seem that the last word has been said on this matter, and that there should be no further occasion for that kind of adverse criticism, whose sole effect is to harass those who are giving their best thought and most sincere effort to the accomplishment of the desired result. Destructive criticism of scientific work, by the amateur or dilettante, is absolutely valueless. Constructive criticism, such as these reports make on certain minor details of this important work, is helpful and a public good.

NARRATIVE OF THE PROGRESS OF THE WORK.

Down to the time when this work was begun, all attempts at the international handling of beneficial insects had been done either by correspondence or by the sending of an individual collector to search for such insects and to forward them by mail or express or to bring them back himself in comparatively small numbers, the beneficial species being either at once liberated in the field or reared for a time in confinement and then liberated. In planning the present work the normal geographic ranges of both the gipsy moth and the brown-tail moth were well known and most of their parasites had been listed, so that the problem seemed to be a comparatively simple one. Owing to the fact that the most abundant of the Japanese gipsy moths (four of them are listed) presents rather marked differences from the European and New England form—so much so, in fact, as almost to justify the opinion that it is a distinct species—and as the ancestors of the New England gipsy moth came from Europe, it was decided to concentrate the effort, for a time at least and in the main, upon European parasites and natural enemies. From the outset the idea was to secure as many parasites belonging to as many different species as possible from all parts of Europe, in the hope of establishing in New England approximately the natural environment of the gipsy moth

and the brown-tail moth in so far as their natural checks are concerned. It was the aim to establish, not one or half a dozen of its
natural enemies, but all of them, aiming at the same time to avoid
the introduction of hyperparasites—that is, those species that prey
upon the true parasites of the injurious forms—thus, if possible, bringing about an even more favorable situation for the primary parasites
in New England than exists in Europe.

On account of the enormous numbers in which both gipsy and
brown-tail moths existed in Massachusetts, it was considered that
the simplest way to secure the true European parasites was to
collect caterpillars and chrysalids wherever they could be found
in Europe, box them, and ship them directly to Boston; this always
with the certainty that a certain percentage, high or low, would
contain living parasites which would probably issue in the adult
condition on the journey or after arrival in America, in which event
they could be cared for, reared until sufficiently multiplied, and then
liberated.

A temporary laboratory for the receipt and care of specimens
was immediately established by Mr. Kirkland at Malden, Mass.,
and a careful search was begun for a suitable location for a permanent laboratory for the care of parasites. It was considered desirable
that this laboratory should be placed in a region in which both the
gipsy moth and the brown-tail moth occurred in abundance, so
that there might be plenty of material for food for the parasites
at all times; and it was also considered of importance that a considerable area of land should be secured which could be controlled
for outdoor experiments. Mr. Kirkland finally found a small farm
with buildings in North Saugus, the location easily accessible by
electric cars and sufficiently isolated. (See Pl. II, fig. 1.) The
house was large enough to give ample room for laboratory use,
and at the same time furnished dwelling rooms for the state official
in charge. In the immediate vicinity there was a chain of large
woodland colonies of the gipsy moth and numerous orchards infested
by the brown-tail moth, as well as a large area of scrub-oak land
where the brown-tail moth occurred very abundantly. A portion
of the building occupied as a laboratory was fitted up by the State
with shelves, tables, rearing cages, and all necessary apparatus
and supplies, and the State employed Mr. F. H. Mosher, with Mr.
E. A. Back and Mr. O. L. Clark as assistants, to help care for the
parasites.

While, as just stated and for the reasons given, the main effort
was made with Europe, correspondence was begun with the Imperial
Agricultural Experiment Station at Nishigahara, Tokyo, Japan,
and the Imperial Agricultural College at Sapporo, in order to secure,
if possible, the services of expert Japanese entomologists in sending

Japanese parasites, and Prof. S. I. Kuwana immediately prepared an important sending, which, however, was not productive, through accidents in transportation. The method tried by Prof. Kuwana was interesting. A small tree carrying a number of infested gipsy-moth caterpillars was packed in a large wooden case with wire-gauze sides; another case of small elms was shipped with the insects, and they were thus supplied with fresh food from time to time as far as Hawaii. The case, however, shrunk in transit, making openings through which the parasites for the most part escaped.

In May, 1905, the Chief of the Bureau of Entomology visited Boston for conference with Mr. Kirkland, and on June 3 sailed from Boston to Naples. Landing in Naples on June 13, he at once proceeded to the Royal Agricultural School at Portici, some miles away, and held a conference with Prof. F. Silvestri, the entomologist of the college, and his principal assistant, Dr. G. Leonardi. By good fortune, Prof. Silvestri was able to point out a locality in Sardinia where, during 1904, there had been a severe outbreak of the gipsy moth and where, therefore, during 1905 parasites could with almost absolute certainty be predicted to occur in numbers. With true scientific enthusiasm, both Prof. Silvestri and Dr. Leonardi volunteered their assistance, and Dr. Leonardi was at once commissioned by his chief to proceed to Sardinia and to collect such caterpillars as he could find and forward them in tight wooden boxes, with a supply of food, to Boston. His expedition was a success, and there were received from him at Boston, on the 15th of July, 7 boxes, on the 26th of July 24 boxes, and on the 1st of August 7 boxes, all containing valuable material, the most important being a large series of living puparia of certain parasitic tachina flies.

This extremely cordial and profitable reception at Portici by Prof. Silvestri and Dr. Leonardi, both personally known to the chief of the bureau from former visits, was but a foretaste of the encouragement which was to be met at all points, and it may very properly be said in advance that throughout the whole of the work many European and Japanese entomologists, both officials and private individuals, have shown an extreme liberality in their offers of assistance in this great piece of experimental work, and the State of Massachusetts and the United States Government are under great obligations to them for their help and encouragement. For the work done by Dr. Leonardi, just described, and for similar work done in ensuing years, with Prof. Silvestri's permission, no compensation would be accepted, and the State of Massachusetts has paid simply for the expenses, such as packing, postage, small traveling expenses, and items of that general character.

PLATE II.

FIG. 1.—VIEW OF PARASITE LABORATORY AT NORTH SAUGUS, MASS. (ORIGINAL.)

FIG. 2.—VIEW OF PARASITE LABORATORY AT MELROSE HIGHLANDS, MASS. (ORIGINAL.)

After Portici, Florence was visited, where a conference was held with Prof. A. Berlese, of the Royal Station of Agricultural Entomology, and his assistants, Drs. Del Guercio and Ribaga. It seemed that no occurrences of either the gipsy moth or the brown-tail moth were known that season in Tuscany or adjoining portions of Italy. Prof. Berlese spoke of the destruction of an outbreak of the gipsy moth in southern Italy some years previously by a disease which he considered to be identical with the pébrine of the domestic silkworm. He promised to keep up a watch for occurrences of the pests and wherever possible to assist in the introduction of parasites. A few days were then spent in Lombardy, searching for the larvæ of either of the injurious species, but without success. Then, proceeding to Vienna, the celebrated Natural History Museum was visited and the well-known curator of Lepidoptera, Dr. Hans Rebel, was interviewed. Dr. Rebel stated that both the gipsy moth and the brown-tail moth were to be found rather commonly in parts of Austria, and it was decided to employ a professional collector to assist in the work of shipping larvæ to Boston. Upon Dr. Rebel's recommendation, Mr. Fritz Wagner was employed. Mr. Wagner was and is a resident of Vienna, is well versed in the subject of European butterflies and moths, and perfectly familiar with all the best collecting places for many miles about Vienna. Mr. Wagner accompanied the writer on several expeditions. The first trip was taken to the suburbs of Vienna, and there the first European specimen of the gipsy-moth larva was found. It was resting on the trunk of a locust tree by the side of the street, and further examination showed that there were a hundred or more caterpillars on the trunk and limbs of the same tree. There was some evidence of parasitism, and the white cocoons of a microgaster parasite (*Apanteles fulvipes* Hal.) were found here and there in the crevices of the bark. This particular tree and another one, to be mentioned later, indicate very well the condition of the gipsy moth in Europe. A hundred nearly full-grown larvæ were present, but there was hardly any evidence of defoliation. A trained entomologist walking by the tree would not have noticed that insects had been feeding upon it to any serious extent. On the other hand, a similar tree in any of the small towns about Boston would have carried not 100 larvæ, but probably some thousands, and at that time of the year would hardly have had a whole leaf. These specimens were collected and sent to Boston.

Later a trip was taken into the country to the battlefield of Wagram, and here on two roadside poplars was found another colony of the caterpillars ranging in size from the second stage to full-grown larvæ. There was here more extensive evidence of parasitism by microgaster parasites. Their white cocoons were found abundantly, and here again, although there must have been 250 or more larvæ on the trees, the evidences of defoliation were very slight—so much so that at a rather short distance the trees appeared in full leaf. During the remainder of June and July Mr. Wagner continued the search and sent considerable material to Mr. Kirkland, at Boston.

After Vienna, the city of Budapest was visited. At the Natural History Museum in that city Dr. G. Horvath, the well-known director, and Prof. Alexander Mocsary were consulted, Prof. Mocsary being one of the first authorities in Europe on the subject of parasitic Hymenoptera. Neither of these gentlemen, however, was able to give any new points in connection with the parasites of the gipsy moth and the brown-tail moth. The agricultural experiment station in the suburbs of Pesth was then visited, and Prof. Josef Jablonowski, the entomologist of the station, was consulted. By this time it was the 4th of July, and already the season in Hungary was far advanced, being about two weeks or more earlier there than at Vienna. Prof. Jablonowski stated that gipsy moths had been found in certain localities in Transylvania, but that the adults were already issuing and that the brown-tail moths had been flying for some time. He exhibited, however, a large box full of the previous winter's nests of brown-tail larvæ, and stated that in the early spring he had reared from these nests many hundreds of parasitic insects. This at once seemed to indicate a very easy way of importing such parasites, since these nests could be readily collected in the winter in large numbers and sent to Boston in great packages—a bushel or more in each package—in the late fall or winter season, and Prof. Jablonowski volunteered to make every effort the following winter to send over a large quantity. Taking into consideration the small size of the brown-tail moth caterpillars during hibernation, it seemed very strange that they should be so extensively parasitized as indicated by Jablonowski. The larger caterpillars in the late spring and early summer would seem to be much more likely to be extensively infested. These winter nests, remaining alone on the trees after the leaves have fallen, would seem to be an attractive place for small Hymenoptera of various kinds, in which they might seek shelter for hibernation, and, while of course there was a chance that some of the true parasites of later stages might thus be sheltered, it was with considerable doubts as to the ultimate result that the writer arranged for the importation of these nests in large quantity. Even if unsuccessful, however, it seemed that the experiment must be tried.

From Budapest, Dresden was reached, and, as in Vienna and Budapest, the principal museum (the Zoological Ethnological Museum) was at once visited. Dr. K. M. Heller, at that time acting director of the museum, was asked to recommend a good man who might be employed as a professional collector to undertake work in the same manner as that done by Fritz Wagner in Vienna. Dr. Heller recommended Mr. Edward Schopfer, who was at once engaged. Although at the date of the first visit to him the season was already considerably advanced (July 7), Mr. Schopfer had rearing cages in operation in his rooms, and in these cages were a number of nearly full-grown larvæ of the gipsy moth. He knew the localities about Dresden where these insects were to be found, and at once began sending specimens to Boston. The well-known Forest Academy at Tharandt, near Dresden, was visited, and Prof. Arnold Jacobi and his assistant, Mr. W. Baer, were interested and promised assistance, especially in the matter of sending specimens of *Calosoma sycophanta* (see Pl. I, frontispiece) and *C. inquisitor.* Other trips were made in the vicinity of Dresden, and then the journey was resumed to Zurich, where, through the kindness of Dr. Herbert Haviland Field, director of the Concilium Bibliographicum Zoologicum, the writer met Miss Marie Rühl, editor of the Societas Entomologica, a very well-posted entomologist, especially on matters relating to Lepidoptera, who had and has a large correspondence throughout northern Germany. She was engaged as the official agent of the investigation for that part of Germany and was able, through her own work and that of her correspondents, to send a large amount of material to Boston before the close of the season of 1905, and has since continued the work.

From Zurich the trip was resumed to Paris, where some time was spent in interviewing Dr. Paul Marchal, the entomologist of the agricultural school conducted under the ministry of agriculture, and other entomologists, and in visiting the scientific societies for the purpose of interesting naturalists in the work. Many trips were taken to towns around Paris in search of the pupæ of the gipsy moth and to visit local collectors in search of information, after which the return journey was made to America.

The result of this initial trip was to demonstrate that it is an easy matter and a comparatively inexpensive one to import certain of the parasites of both the gipsy moth and the brown-tail moth in living condition into the United States. The most important part of the European range of the two species was visited, and the entomologists were organized into an active body of assistants.

Mention has already been made of the number of boxes sent in by Dr. Leonardi from Sardinia. Ten boxes were shipped by Fritz Wagner from Vienna, 47 boxes from Schopfer in Dresden, and 36 from Miss Rühl in Zurich, all of these containing parasitized larvæ or pupæ of the gipsy moth or brown-tail moth.

Acting upon Prof. Jablonowski's observations concerning the existence of parasites in the wintering nests of the brown-tail moth,

arrangements were made with Miss Rühl, Mr. Schopfer, Prof. A. J.
Cook, who was then in Berlin, and several volunteer collectors to send
in numbers of the winter nests. During his visit to Paris in July, the
chief of the bureau had addressed a meeting of the Entomological
Society of France on the subject of his mission and asked the members
of the society to assist in the work. The most remarkable response to
this request came from Mr. Rene Oberthür, of Rennes, who, although
not present at the meeting, read the account in the bulletin of the
society, and placed himself and his services entirely at the disposal
of the United States authorities. During the autumn of 1905 and
the winter of 1905–6 he sent to Boston more than 10,000 winter
nests of the brown-tail moth. In all, 117,000 nests were received
and cared for during that winter.

In the autumn the laboratory house (Pl. II, fig. 1, p. 56) at North
Saugus was taken possession of by Mr. Kirkland, fitted up as previously described, and occupied by Mr. Mosher; the parasite material
from Malden was brought over and installed, and arrangements were
made for the receipt of the brown-tail winter nests. Very many
large boxes were constructed, somewhat on the plan of the California parasite-rearing cage, each one large enough to contain from 500
to 1,000 nests of the brown-tail moth, the front being pierced with
auger holes in which were inserted round-bottom glass tubes into
which the emerging parasites would come in search of light and
through which they might be examined to differentiate between the
primaries and the hyperparasites. Much carpenter work was done
during the autumn and winter months and on into the spring.
Double windows and double doors were provided, and every crack
in the laboratory rooms was sealed. Realizing that many different
kinds of insects might emerge from this large supply of silken nests,
including possibly species injurious to agriculture not previously
introduced into the United States, as well as dangerous parasites of
beneficial insects, every possible effort was made to prevent the escape
of any insect whatever from the laboratory rooms.

On account of the importance of a speedy detection of injurious
forms coming from these rearing cages, and on account of the
necessity for the most expert supervision of the laboratory end of the
experiment, Mr. E. S. G. Titus, an especially well trained expert
from the Bureau of Entomology, was assigned in the spring of 1906
to the charge of the laboratory end of the introduction.

In March, 1906, Mr. Titus, with the chief of the bureau and with
Mr. Kirkland and Mr. Mosher, visited the parasite laboratory, and
for the first time examined the contents of the imported nests.
There were in the different cages, well separated as to localities,
winter nests from almost the whole of the European range of the
brown-tail moth, from Transylvania on the southeast to Brittany

on the northwest, and from the Pyrenees on the southwest to the shores of the Baltic on the northeast. In spite of the voluntary assistance of such men as Rene Oberthür and Josef Jablonowski, the expense of getting these nests to Boston had been very considerable, and the moment when this examination was begun was considered to be rather a critical one. No published record of the rearing of parasites from these winter nests was recalled by the senior author or by any of his European correspondents, and the expensive experiment rested solely on the unpublished observation of Jablonowski, and he himself had simply seen parasites emerge from nests in the spring. Would they prove useless? Had the parasitic insects, even if useful, simply crawled into the nests for hibernation? Or were they, some of them, true parasites of the young larvæ? Representative nests were examined from a number of different localities, and the relief and joy were great when parasitic larvæ were found in considerable numbers in each of the nests examined, feeding within the nest pockets externally upon the brown-tail larvæ. This particular experiment was a success, and the expenditure of money and trouble was justified. About April 25 these parasites began to issue from the nests. The nests had been gathered in all from 33 different localities, and from some of them only a small number of parasites was reared. In all, about 70,000 issued, of which about 8 per cent were hyperparasites. In the rearing cages above mentioned it was a comparatively easy matter for Mr. Titus to separate the hyperparasites from the true parasites and to destroy the former. Of the species issuing in that spring—and they continued to issue until about June 15—there were two species which appeared to be important, namely *Pteromalus egregius* Först. and *Habrobracon brevicornis* Wesm. The latter species proved later to have entered the nests for hibernation only.

With the cooperation of Mr. Kirkland, several localities were found in which there was slight danger of forest fire and in which no work against the moths would apparently be undertaken for at least some months to come, and colonies of various sizes—the three principal ones including, respectively, 10,000, 15,000, and 25,000 parasites—were liberated in the open. Outdoor cages had been built over trees, and some smaller colonies of the parasites were placed in these cages. Both the outdoor experiments and the open experiments were seriously hampered, however, by the fact that the season proved to be one of extraordinary humidity, which caused the appearance of a fungous disease which destroyed a large proportion of the brown-tail moth larvæ in the vicinity of Boston.

Coincident with the issuing of these parasites from the nests, as the season grew warm the young larvæ swarmed from the nests and filled the glass tubes in the breeding cages and were constantly being destroyed by the assistants in the laboratory, and when the parasites ceased to issue the remaining nests and larvæ were burned. But later observations showed this destruction to have been a mistake. It was not considered likely that other parasites could be reared from these imported larvæ if they were fed and reared as far as possible, but such proved to be the case, as will be shown later.

During the winter of 1905–6 efforts were made to import in wintering conditions the two large European ground-beetles, *Calosoma sycophanta* (see Pl. I, frontispiece) and *C. inquisitor*. No success in importing living specimens was gained until March, 1906, but from that time on until July small consignments of living adult beetles were received, and in all 690 living specimens of *Calosoma sycophanta* and 172 of *C. inquisitor* arrived at Boston alive, some of them dying soon after arrival. Colonies were started in various localities about Boston. Consideration of the history of these two species will be given in Bulletin 101.

After visiting the parasite laboratory in March and determining the success of the importation of the brown-tail nests, the senior author sailed from New York on the 17th of the month for Europe, returning to America May 17.

Proceeding directly to Paris, Mr. Rene Oberthür was met by appointment, and the whole subject of the summer work was carefully considered. Mr. Oberthür is a man of affairs, proprietor of a large printing business, a learned amateur entomologist, and the possessor of one of the largest insect collections in the world. His advice and assistance throughout the whole work has been most important, and he assures the American representatives that he has highly appreciated the opportunity of being of assistance and of taking part in such an interesting piece of work. At his advice the writer proceeded to the south of France, after interviewing correspondents and agents in Paris, and visited Prof. Valery Mayet at the agricultural school at Montpellier, Dr. P. Siepi, of the Zoological Gardens in Marseilles, and Mr. Harold Powell, of Hyères. Both Prof. Mayet and Dr. Siepi stated that both of the injurious species of insects were rare in their vicinity, but both promised to assist in the importation of the Calosoma beetles. Mr. Powell proved to be a lepidopterist who had been employed professionally by Mr. Oberthür as a collector, and he was engaged to collect parasitized larvæ in Hyères and in the Enghadine district. He sent in much good material, and later, as will be shown in subsequent pages, organized a very efficient service in the summer of 1909. The visit to Prof. Mayet at Montpellier, moreover, was by no means devoid of results, since at a later date he was able to send a few specimens of carabid beetles, and in 1908, as a result of this personal interview, he was able to send to America the first living specimens of the European egg parasite of the imported elm leaf-beetle, *Tetrastichus xanthomelænæ* Marchal, which, as a result of this sending, is now possibly established in New England, although it was not recovered during the summers of 1909 and 1910.

While at Marseilles interviewing Dr. Siepi, April 10, the news was received of the eruption of Vesuvius and the partial destruction by lava flow of Boscatrecase and other villages on the slope of Vesuvius. Having to interview Prof. Silvestri and Dr. Leonardi at Portici, and fearing for their safety, the visitor proceeded at once to Naples, arriving there the day of the great market-house accident in which the roof fell in from the weight of volcanic ash and a number of persons were killed. Everything in Naples was in a state of confusion; the streets were filled with volcanic ash almost knee-deep, and it was with great difficulty that a conveyance could be secured to drive to Portici. Portici is almost on a direct line between Naples and Mount Vesuvius, and the agricultural college was found to be in bad condition; the gardens were utterly destroyed by ashes, and the roof of the old building was deeply covered. The accident happened the week before Easter, and the majority of the faculty and students had, on account of the catastrophe, anticipated their Easter vacations and had departed for their home, Silvestri and Leonardi among the rest. Letters were forwarded to them, however, giving detailed suggestions as to methods of packing and shipment of parasites.

As in 1905, Florence, Milan, Vienna, Budapest, Dresden, Tharandt, and Zurich were visited. Efforts were made to learn of localities where either the gipsy moth or the brown-tail moth might reasonably be expected to be abundant during the summer of 1906, and a number of such localities were learned and the information given to agents. All of the agents and correspondents were given full instructions regarding the work for the summer of 1906 and the winter of 1907. The experience of 1905 with regard to the best methods of packing and shipment and the best kinds of boxes used was related to all, and these points were fully discussed, with the result that the material received during the summer of 1906 was not only greater in quantity but better in condition than that received during the previous summer.

In Vienna the visitor had the good fortune to find Dr. Gustav Mayr, whom he had missed in the summer of 1905. Dr. Mayr (since deceased) was the European authority on several of the groups of parasites most intimately connected with the work in hand, and the writer had a long consultation with him concerning the systematic position of some of the forms already imported and concerning the practical possibilities of the whole series of Microhymenoptera. Through him was learned the probable importance of certain egg parasites of the brown-tail moth, which he himself had reared in Europe and had described. As a result of this information the agents visited later were instructed to send over egg masses of the brown-tail moth to Massachusetts in midsummer, and later to send over egg masses of the gipsy moth. From the brown-tail moth egg masses parasites were reared by Mr. Titus at North Saugus and were observed to oviposit in native eggs. Mr. Titus reared not only the species referred to by Dr. Mayr, namely, *Telenomus phalæ-narum* Nees, which came from eggs forwarded by Miss Rühl and collected in Croatia, but he also reared an interesting parasite of the genus Trichogramma from egg masses received from Würtemberg, Dalmatia, and Rhenish Prussia.

At Budapest the visitor was especially glad to be able to announce to Prof. Jablonowski the success of the rearings of parasites from the winter nests of the brown-tail moth, so many of which had been brought over from Europe the previous winter on the basis of Jablonowski's unpublished observations. At the time of this visit Prof. Jablonowski was too busy completing his important work upon the migratory grasshoppers invading Hungary to be able to promise much assistance beyond that of corresponding with foresters and other persons well located in Hungary in order to obtain information as to good places to secure material.

Returning to America about the end of May, the laboratory at North Saugus was again visited, with Mr. Kirkland and Mr. Titus, and the work of preparing indoor cages and field cages was pushed. In the course of the summer a number of outdoor houses were constructed, and in these houses it was hoped to study the breeding habits of the imported insects.

During the summer the number of shipments received from Europe was so large that Mr. Kirkland made no attempt to list them in his Second Annual Report published January 1, 1907. In June, in addition to egg masses previously mentioned, larvæ and pupæ of both the gipsy moth and the brown-tail moth were received in number from many different European localities, and from these a large number of parasites of several different species were reared, the most abundant having been tachina flies. In one lot received from Holland more tachinids were reared than there were gipsy moth caterpillars originally. Nearly 40,000 gipsy-moth larvæ and pupæ were received and more than 35,000 brown-tail moth larvæ and pupæ. The receipt of predatory beetles is recorded in a previous paragraph.

It will be noticed that in the work conducted so far the effort to import parasites was confined to the continent of Europe west of Russia, whereas the well-known occurrence at intervals in large numbers of the gipsy moth in parts of Russia, and especially in southern Russia (a very good account of which will be found in the Third Report on the Gipsy Moth, by Forbush and Fernald), seemed to render it desirable that search should be made in those regions for parasites. The fact, however, that during these two years the writer had been unable to secure answers to letters addressed to correspondents in Russia and the reported unsettled condition of affairs in that country deterred him during the 1905 and 1906 trips from visiting the Russian southern Provinces. In the late summer of 1906, however, advices were received from Prof. J. Porchinsky, of the ministry of agriculture at St. Petersburg, with the information that in the southern part of Russia both the gipsy moth and the brown-tail moth were at that time occurring in sufficiently great numbers to enable the collection of parasites and commending the writer to certain officials, trained entomologists, in Simferopol (Crimea), Kishenef (Bessarabia), and Kief. Prof. Porchinsky wrote that he had apprised these officials of the intended visit, and plans were therefore made to include southern Russia in the itinerary for the spring of 1907.

During the autumn of 1906 egg masses of the gipsy moth continued to be received from parts of Europe, and during the winter hibernating nests of the brown-tail moth were sent in. More than 111,000 nests were received from different portions of the European range of the species. These were placed in the especially constructed cages, and from many of them large numbers of parasites were reared, issuing mainly during the month of May, 1907. As it happened, the month of May in New England, as well as in other parts of the United States, was phenomenally cold and wet. As a result of this unlooked-for condition very many of the parasites refused to leave the nests until they were so weakened as to be unable to survive the close confinement and careful scrutiny to which they were necessarily subjected in order to eliminate the danger of introducing secondary parasites. As a result, a smaller number of *Pteromalus egregius* was colonized in the summer of 1906, but 40,000 specimens were put out in several localities, the principal colonies consisting, respectively, of 13,000, 11,000, and 7,000 individuals. At this time, as well as in the summer of 1906, although this fact has not as yet been stated, a number of important parasites of the genus Monodontomerus issued from the winter nests and were allowed to escape. As will be shown subsequently, this parasite has proved to be more important than the Pteromalus and has made a phenomenal spread.

In this important work with the introduced hibernation nests of the brown-tail moth it was early found most difficult to preserve the health of the laboratory assistants. The irritating and poisonous hairs of the brown-tail moth larvæ, of which the nests are full, soon penetrated the skin of the assistants handling them, entered their eyes and throats, and the atmosphere of the laboratory became almost filled with them. It was necessary that the rooms should be kept thoroughly closed; double windows and screens were used, and the doors of the rooms were doubled, in order that a possible secondary parasite, if accidentally liberated, should have no chance of escape. This made the rooms very warm and increased the irritating effect of the larval hairs. Some of the assistants employed could not stand the work and resigned. One of the best and most experienced helpers was induced to continue the second year only upon the promise that he would be relieved from this especial class of work. Spectacles, gloves, masks, and even headpieces were invented to avoid this difficulty, but these, while greatly increasing the suffering from the heat, were not entirely effective. The most serious result of this trouble was the breaking down in health of Mr. E. S. G. Titus of the bureau, in charge of the laboratory at Saugus, who was obliged to resign in May, 1907, on his physician's advice, in order to save his life. The difficulty in Mr. Titus's case was the intense irritation to his lungs from the entrance of the barbed hairs. Mr. Titus was soon after appointed entomologist of the Utah Agricultural Experiment Station, and the change of work and climate fortunately brought about a speedy recovery. His necessitated departure in the midst of important work, however, threw us into what appeared to be a serious dilemma, but fortunately it so happened that the services of the junior author, then occupying another position in the Bureau of Entomology in Washington, could be spared from the other work upon which he had been engaged, and, since he had made especial studies of the parasitic Hymenoptera and had done a large amount of rearing of parasites in the course of his other work, he was sent on from Washington to replace Mr. Titus in the parasite laboratory and has since had charge of the laboratory.

One of the early points to which the junior author devoted his attention was the invention of new methods of handling the brown-tail nests in order to avoid the serious effect upon the work of the breaking out of the rash on himself and his assistants. He soon devised an apparatus like the ordinary show cases that are seen in shops, the glass on one side being replaced by cloth with armholes, through which the gloved hands of the worker could be thrust and the brown-tail nests handled in full sight through the top glass. Most of the work with these nests, it has been found, can be done in these cases with a minimum escape of the barbed hairs. There still continued, however, considerable trouble from the rash, since much rearing of brown-tail larvæ must be carried on under conditions in which such cases can not be used, and this difficulty still exists. Miss Rühl, of Zurich, in handling and repacking the large number of nests sent to her by her European correspondents and forwarded by her to Boston, has been a great sufferer from the rash. She has made for herself a complete costume of an especially finely woven cloth, and has made a large light helmet covered with cloth and provided with a cape, the space opposite the eyes being fitted with a sheet of very transparent celluloid. Of course this costume would be very uncomfortable in the summer time on account of the heat, but since she handles her nests for the most part in the autumn and winter, she has been able to reduce the discomfort of the brown-tail rash to a minimum.

Sailing again for Europe on April 20, 1907, the senior author landed at Cherbourg and proceeded directly to Paris, and from Paris to Budapest by the Oriental Express. At Budapest, by prearrangement, he met Mr. Alexander Pichler, whom he had engaged as a guide and courier for the Russian trip. After a conference at Budapest with Dr. Horvath and Prof. Mocsary, of the Natural History Museum, and Prof. Jablonowski, of the agricultural station, he proceeded to Kief, via Lemburg. Prof. Porchinsky, of the ministry of agriculture, had arranged with Prof. Waldemar Pospielow, of the University of Kief, to consult with the Chief of the Bureau of Entomology about future arrangements, and a conference with Prof. Pospielow was held, in the course of which it was agreed that one of Pospielow's assistants, engaged especially for the purpose, at 34 rubles per month, should occupy himself throughout the summer, under Pospielow's directions, in collecting larvæ of the gipsy moth and brown-tail moth, forwarding material to Boston, rearing and studying the parasites, and conducting observations in an orchard in the suburbs of Kief, rented by the writer for the State of Massachusetts for the summer at the rate of 20 rubles per month. This procedure was novel in the work, but was later tried in another locality, as will be shown in subsequent pages.

From Kief, Pichler and the visitor proceeded to Odessa and from Odessa to Kishenef, at which point he had been recommended to Dr. Isaak Krassilstschik by Prof. Porchinsky. Through some misunderstanding as to dates, owing to the difference between the Russian calendar and the one in use in other parts of the world, Prof. Krassilstschik had mistaken the date of arrival announced in the letter sent in advance, and was absent from Kishenef on a brief visit to Germany. Full written instructions, however, were left for him at Kishenef, and the visitor returned to Odessa and thence by boat to Sebastopol, and by train to Simferopol. At Simferopol he was expected by Prof. Sigismond Mokshetsky, the director of the Museum of Natural History at that place and an enthusiastic economic entomologist, through whose efforts American methods in the warfare against insects had been introduced into southern Russia. Prof. Mokshetsky had done some rearing of the Russian parasites of both the gipsy moth and the brown-tail moth, and was able to furnish much valuable information. His hospitality and cordiality were of the most encouraging nature, and after consultation as to the best methods, he promised his hearty support to the work, refusing, however, to accept any compensation from the State of Massachusetts or from the United States Government.

The visitor then proceeded by boat from Sebastopol to Constantinople, but was unable to learn of any person in Turkey having any information on the subject of insect pests, nor was he able in the country about Constantinople to find any indication of the occurrence of either gipsy moth or brown-tail moth.

Leaving Constantinople, the expedition proceeded to Vienna, dropping Mr. Pichler at Budapest. At Vienna the Seventh International Congress of Agriculture was held, beginning May 22, 1907. The visitor met there a number of delegates from the different countries in Europe, with whom he discussed the question of parasite importation, receiving warm assurances of support, especially from Prof. Dr. Max Hollrung, of the Agricultural Department of the University of Halle, Prof. Dr. Karl Eckstein, of the Forest Academy at Eberswalde, and Prof. Dr. J. Ritzema Bos, director of the Phytopathological Station at Wageningen, Holland. While in Vienna arrangements were made with Mr. Fritz Wagner for continuance of the work, and a further consultation on the subject of parasites was held with Dr. Gustav Mayr.

After Vienna, Mr. Schopfer was visited in Dresden, Dr. Hollrung at Halle, Dr. R. Heymons in Berlin, Dr. Eckstein in Eberswalde, Miss Rühl at Zurich, and Prof. G. Severin at Brussels. Prof. Severin is connected with the Royal Natural History Museum at Brussels, is an admirably well-posted entomologist, and is connected with the Forest Conservation Commission of Belgium. He was able to give good advice in the parasite work and promised assistance.

Returning to France, an important conference was held with Mr. Rene Oberthür, and it was arranged to establish during the summer of 1908 a field station at Rennes, to be placed in charge of a special expert, Mr. A. Vuillet, chosen by Prof. Houlbert, of the University of Rennes. Through Mr. Oberthür's courtesy it was arranged to establish field rearing cages at a convenient point near the University of Rennes and to carry on the work in much the same way as it had been arranged for the present summer at Kief. The University of Rennes having a certain connection with the University of Paris, it was considered desirable that the cooperation of the scientific faculty of the University of Paris be gained by direct application. This was readily arranged, through the cordial and sympathetic cooperation of Prof. Alfred Giard, of the faculty of science of the University of Paris (since deceased).

In dealing with the European parasites reared at North Saugus, considerable difficulty was experienced in ascertaining their names. It was very desirable, of course, to have a definite name by which to designate each species, and by which to correlate it with published accounts of observations already made. With the assistance of Dr. O. Schmiedeknecht, of Cassel, Germany, a number of these forms had been named, but with others it seemed practically impossible to bring this about by correspondence. As a result, on the trip in question the writer made an effort, by studying the collections in some of the principal European museums, to determine a few of the unnamed forms reared in America from European material. The difficulty of this search was surprising. The Pteromalus, for example, which had been reared in Boston by scores of thousands and which, therefore, must be a very common European insect, was found to be absolutely unrepresented in the large natural history museums of Vienna, Dresden, Berlin, Brussels, and London; nor did it occur in the type collections of Ratzeburg carefully preserved by Dr. Eckstein at the Forest Academy at Eberswalde, where, on account of Ratzeburg's important work on the parasites of European forest insects, one would naturally expect to find it. At last, in a small special collection in the Museum of Natural History in the Jardin des Plantes at Paris, Mr. H. du Buysson of the museum found in the laboratory a box containing parasites reared many years ago by the French entomologist, Sichel, which had been named for him by the eminent authority on parasitic Hymenoptera, Arnold Förster, of Germany. In this box were specimens of the Pteromalus labeled "*Pt. egregius*" in the handwriting of Förster himself.

Especial efforts were made on the trip to arrange for the importation of large numbers of the egg parasites of both species and to introduce in living condition the important parasites of the genus Apanteles, which, according to the visitor's field observations, are among the most important of the European enemies of the gipsy moth. Previous importations of these parasites had failed, owing to the fact that they emerged and died on the journey. On this trip, however, specific directions were given to agents to send in young larvæ of the second stage, and by this means living specimens in considerable numbers were later reared in the laboratory at North Saugus. These on issuing laid their eggs in the gipsy-moth larvæ of the first stage, and from these caterpillars were secured the cocoons of adults of a second generation which was reared through all of its stages on American soil.

From Kief there were received two species hitherto unknown as parasites of the gipsy moth, and one of these, being a rapid breeder, promised to be of much assistance. This species, belonging to the genus Meteorus, seemed to produce cocoons in about 10 days after egg laying, and will be considered later in this bulletin.

We have previously referred to the destruction in 1906 of the great bulk of brown-tail caterpillars imported from Europe after the early appearance of adult parasites. Mr. Titus, in 1906, tried the experiment of rearing a very few of these imported larvæ, and found that in their later growth they gave out a second lot of parasites entirely different from those reared in May from the very young hibernating larvæ, indicating a delayed development of eggs which must have been laid by adult parasites the previous autumn. Among these were at least two species, one belonging to the genus Apanteles and the other a Meteorus. Before his resignation in 1907 he started an extensive series of rearing experiments with the end in view of securing these parasites in large numbers. Partly on account of his enforced absence from the laboratory during a critical period, and partly through the unsuitable character of the rearing cages which were employed, the project did not meet with entire success. Only about 1,000 of the parasites were reared, of which all but a small percentage were the Apanteles.

The importations of the summer following the trip above described were very large, and reasonably successful, and during June alone 872 boxes were received, many others following during July and into August, shipments of brown-tail eggs and gipsy-moth eggs following, and of brown-tail winter nests in the late autumn and during the winter. As in 1906, tachinids made up the great bulk of the parasites secured through the importation of pupæ and active caterpillars. Notwithstanding the improvement in methods of shipment over previous years, Apanteles invariably hatched en route, and only dead adults or secondary parasites were received.

Before the close of the summer it had become obvious that better quarters for the Massachusetts laboratory were necessary. The heating and lighting arrangements at North Saugus were insufficient; the building was not sufficiently commodious, and the location was not convenient. Therefore, after considerable search, Mr. Kirkland found and leased for a term of years a commodious house at Melrose Highlands (No. 17 East Highland Avenue) (see Pl. II, fig. 2, p. 56.) The building was remodeled so far as necessary to fit it for the work. The grounds back of the house were sufficiently ample to enable the building of several outdoor laboratories, properly screened and ventilated, which were planned and erected under the direction of the junior author. The building is well warmed, lighted with electricity, and, being close to fire protection, possesses many advantages over the old laboratory. Moreover, it is much nearer the central office in Boston, enabling an important saving of time in sending to the laboratory shipments of parasites received from abroad. The rental and the expense of construction were all borne by the State of Massachusetts. The new quarters are also within a stone's throw of a large area of waste land covered with scrub oak.

In planning the work for the season of 1908, several new features were introduced. The parasites constantly sent over by agents belong to three main groups, namely, those of the order Hymenoptera, including the ichneumon flies, the chalcis flies, and others; those of the Diptera, including the tachina flies, and those of the order Coleoptera, including the predaceous ground beetles. The amount of material received had been so great, and the character of the different life histories of the insects involved had been so diverse, that no one expert was able to do the fullest justice to the situation. Therefore, while the junior author was left in general charge of the whole mass of importations and retained his expert supervision of the work on the biology of the parasitic Hymenoptera, Mr. C. H. T. Townsend, of the Bureau of Entomology, was assigned to the work on the biology of the dipterous parasites, and Mr. A. F. Burgess, also of the Bureau of Entomology, was assigned to the expert charge of the ground beetles.

Owing to the fact that the condition of European sendings by mail and express during the summer of 1907 had been by no means uniformly good—those from eastern Europe, subjected to long railway journeys in addition to the sea voyage, frequently arriving in bad condition—the second innovation was made by establishing at Rennes, France, a general laboratory depot in addition to the field cages and rearing station mentioned in a previous paragraph. The expert assistant designated by Prof. Houlbert, of the University of Rennes, was Mr. A. Vuillet, who was placed in specific charge of the general laboratory depot under the general supervision of Mr. Rene Oberthür. Mr. Vuillet placed himself in relations with the steamship company agents at Cherbourg and Havre and was kept informed as to the dates of the sailings of steamers. Nearly all of the European sendings were shipped to Rennes, examined, repacked, and carried personally by Mr. Vuillet to Cherbourg or Havre on the known days of sailing of certain steamers and then placed in the hands of chief stewards of the vessels and carried in the cold rooms to New York, whence they were sent to Boston. Early in the course of the work the honorable the Secretary of the Treasury, upon request of the honorable the Secretary of Agriculture, had issued orders to the collector of the port of New York to admit all such packages without examination and to hasten their departure for Boston through the United States dispatch agent. The steamship officials showed themselves uniformly courteous, and as a result of this new arrangement the average condition of the material received proved to be much better.

With the installation of the new laboratory at Melrose Highlands, and with the added space afforded by the new structures in the garden, the junior author was able to carry out some new ideas with admirable results. The first of these was the carrying on of active winter work with parasites, especially those secured from the imported nests of the brown-tail moth, which began to come in from Europe in December. It was found quite possible to rear these parasites in artificially heated rooms, feeding them upon hibernating native brown-tail larvæ brought in in their nests from out of doors, feeding the latter upon lettuce and other hothouse foliage, and in the early spring securing more normal food for them by sending it up in boxes by mail from Washington and points south. In this way the rearing of the parasites of the genus Pteromalus was carried forward uninterruptedly throughout the winter, and, as during the rearing of successive generations they multiplied exceedingly, it was possible later in the year to liberate a vastly greater number of individuals than had the imported species been allowed to hibernate normally in the nests. In the course of this work the junior author invented a rearing tray which was of the utmost advantage and which has since greatly facilitated parasite rearing work. This tray will be described later.

With the importation of brown-tail moth eggs it often happened that they hatched too soon to be of use in America; or too late, arriving after the American eggs had all hatched. It was ascertained by the junior author during the summer and autumn that native eggs can be kept in cold storage until the arrival of the European egg parasites, which were found to lay their eggs and breed in these cold-storage eggs as freely as in those which they attack in the state of nature. It was found that this process can be carried on for a long time, and that successive generations of these egg parasites may be reared from eggs retarded in their development by cold storage. It was thus shown that it is easy to rear and liberate an almost infinitely greater number of these egg parasites, and under favorable conditions, than would have been possible from a simple importation of European parasitized eggs which would have to arrive in America at a specific time.

In the same way great advance was made in the rearing of the tachinid parasites in Mr. Townsend's charge. This expert devised methods and made observations that greatly added to our knowledge of the biology of these insects and resulted in the accumulation of a store of information of the greatest practical value, not only in the prosecution of the present undertaking but in any problem of parasite introduction or control that may arise later. Extraordinary and almost revolutionary discoveries were made in the life histories of certain of these flies, and without this knowledge the greatest success in handling them practically could not have been reached. Certain of these facts regarding the most important of these parasites are related in a later part of this bulletin, and many of them have been described in some detail in Technical Series No. 12, Part VI, Bureau of Entomology, United States Department of Agriculture (1908), by Mr. Townsend.

Similarly Mr. Burgess, in charge of the Coleoptera, succeeded in a very perfect way in rearing and liberating the important European predatory beetle, *Calosoma sycophanta*, as well as some other insects of the family Carabidæ.

While these extensive importations from Europe were going on, Japan had by no means been lost sight of. While it seemed probable that the European parasites in themselves would succeed in reestablishing the balance of nature in New England, and in spite of the somewhat dangerous nature of Japanese importations on the ground that the Japanese gipsy moth is probably a different species and might prove in New England even more voracious and destructive than the European moth, there was at no time any intention to neglect Japan in the search for effective parasites. Continuous correspondence had been carried on with Japanese entomologists, and some shipments had been made by correspondents which resulted unsuccessfully. For some time the Apanteles previously mentioned was the only gipsy-moth parasite known to occur in Japan. Later information was received from Prof. U. Nawa, of Gifu, Japan, to the effect that there exists in Japan an important egg parasite of the gipsy moth. During the previous annual trips of the Chief of the Bureau of Entomology to Europe the European service of collectors, agents, and advisers had been well organized and instructed, and the work during 1908 was reasonably sure to be well continued without further personal consultation; it was therefore decided to interrupt the European trip for 1908 and to send a skilled agent to Japan. In considering the appointment of such an agent, Prof. Trevor Kincaid, of the University of Washington at Seattle, was at once suggested to the mind of the writer, primarily on account of his extraordinary skill as a collector, as indicated in the remarkable results of his work on the Harriman expedition to Alaska in 1899, and also on account of his comparative proximity to Japan and the fact that he was personally acquainted with many persons in Japan. He was therefore recommended to the State officials of Massachusetts for appointment, and was commissioned by the State to undertake the expedition. At the same time he was formally appointed a collaborator of the Bureau of Entomology of the United States Department of Agriculture, and the Japanese Government was formally notified by the honorable the Secretary of Agriculture, through the Department of State, of the intended visit, the writer having also notified by personal correspondence some of the well-known Japanese entomologists. Prof. Kincaid sailed from Seattle on March 2, and the results of his expedition far more than justified the expense involved. A very large amount of parasite material was received from him in good condition at Boston, and very many parasites from Japan were colonized in the woodlands in New England. Prof. Kincaid was received with the most extreme courtesy and cordiality by the Japanese Government and by official and private entomologists everywhere. IIis work was facilitated in every possible way; assistants were placed at his disposal and in this way a large number of individuals occupied themselves in the collection of parasitized material. After consultation with the Japanese entomologists, whose great cleverness in manipulation and ingenuity in devising methods are well known, Prof. Kincaid was able to pack his shipments in such a way as to bring about a minimum of mortality on the journey. The steamship companies showed him every courtesy, and much of his material arrived at Melrose Highlands in better condition than corresponding sendings received from Europe. A single indication of the value of Prof. Kincaid's work may be mentioned: From one shipment of cocoons between 40,000 and 50,000 adults of the Japanese Apanteles were reared and were liberated directly in the open in Massachusetts, and this is the species which, although repeatedly sent by correspondents, had never arrived in New England in such condition that a single living adult could be reared.

The European importations in the meantime continued to arrive in numbers, and at the close of the summer it was found that the actual number of beneficial insects liberated had been far in excess of that for 1906 or 1907, and that the list included several species of apparently great importance and promise that had never before been received at the laboratory in living condition.

The successful European importations all came from western Europe, and unfortunately the few shipments sent from Russia arrived in very bad condition. This is considered to have been most unfortunate, since several of the Russian parasites were very promising, and the subject of improving the Russian service was taken into consideration.

With the great success of the summer's Japanese work, and the question of the great desirability of similar work in Russia in his mind, the senior author, visiting the Pacific coast in the autumn of that year (1908) on a tour of inspection of the field laboratories of the Bureau of Entomology, called on Prof. Kincaid at Seattle and discussed with him at length the plans for 1909. Although Kincaid expressed himself as charmed with Japan and anxious to repeat his visit to that most interesting country, his innate honesty compelled him to state that he considered the expense of the trip unnecessary; that he had found the Japanese entomologists, officials, and others so intelligent and so thoroughly competent, and at the same time so heartily interested in the experiment, that he considered them not only perfectly able, but perfectly willing to carry on the work by themselves. After this authoritative expression of opinion from one who knew the ground so well, the visitor asked Mr. Kincaid whether he would care to spend the early summer months of 1909 in Russia, and, upon his affirmative reply, later recommended his reappointment to the Massachusetts State authorities for that purpose.

During the autumn and winter shipments of eggs of the gipsy moth were received from Japan, principally from Prof. Kuwana. From these eggs were reared numerous specimens of *Anastatus bifasciatus* Fonsc., a previously known European parasite of these eggs, and of another parasite belonging to a genus and species new to science (since named by the senior author *Schedius kuvanæ*) which has turned out to be an important primary parasite and which is considered in later pages. During the winter, also, Prof. Jablonowski, of Budapest, sent over several thousand egg masses of the gipsy moth collected in various localities in Hungary. After they arrived in Massachusetts there were reared from them and liberated under the most favorable conditions more than 75,000 adult individuals of *Anastatus bifasciatus*. This was a surprising thing to the laboratory workers, since less than 1,000 parasites of this species had been received from all localities, the earlier ones having come from southern Russia and from Japan.

The winter of 1908–9 was spent at the laboratory, in additional rearing operations, some of them on a large scale, and in studying the parasites already reared, and planning for the coming summer.

As it happened, during the winter the brown-tail moth was introduced into the United States upon nursery stock from France in large numbers. Shipments of nursery stock bearing winter nests of this insect were sent to many States of the Union. Fortunately this was discovered early in the winter, and through prompt action and the cooperation of the customs officials and the railroads probably every sending was traced to its ultimate destination, and was there inspected and the nests destroyed either by State officials or by persons appointed for this purpose by the United States Department of Agriculture.

In the spring of 1909 it seemed necessary for the chief of the bureau to proceed to Europe for the purpose of making an investigation of the European methods of growing nursery stock, with a view to the prevention of similar introductions in the future either by general legislation by the United States Government or in some other way. On this trip he utilized the opportunity to consult further with European agents in the importation of the parasites and to arrange for the summer's work.

In the meantime Prof. Kincaid, whose appointment had been made by the State of Massachusetts, and who had again been made an official collaborator of the Bureau of Entomology of the United States Department of Agriculture, securing leave of absence from the University of Washington, proceeded to Russia, and stationed himself in Bessarabia for the purpose of collecting and sending parasitized material from that country to the United States. It had been noticed by Mr. Vuillet at Rennes during the preceding summer that all material coming from Russia had been opened on the journey and had deteriorated in consequence. Before Prof. Kincaid's departure from America, Russian officials had been communicated with through correspondence between the chief of the Bureau of Entomology and Prof. Porchinsky, of the ministry of agriculture, and also directly between the United States Department of State and the American ambassador at St. Petersburg through the instigation of the honorable the Secretary of Agriculture. The United States Government was assured that the Russian Government would welcome the expedition and would facilitate the sending of material in every way possible.

The chief of the bureau landed at Cherbourg May 12. He proceeded immediately to Paris, where a conference had been arranged in advance with M. Oberthür, M. Vuillet, and Mr. Henry Brown, the latter an English entomologist resident in Paris. At this conference it was decided to abandon the forwarding laboratory at Rennes and to station Mr. Vuillet, during the forwarding season, at Cherbourg. He was instructed to engage quarters at that seaport and to arrange for cold-storage facilities, with the intention that shipments from France, Switzerland, and Italy should be forwarded to him to be kept in cold storage until the date of sailing of vessels, and then should be transferred to the cold room of the next steamer, thus practically keeping all living specimens dormant from the time of arrival in Cherbourg until the time of arrival in New York, making the exposure to summer temperature practically only 24 hours or less in Europe and 24 hours or less in the United States. In the meantime Mr. Oberthür was authorized to arrange for an extensive service in the south of France, through Mr. H. Powell, of Hyères, one of the agents for the year 1906. The preparation of the requisite boxes was intrusted, as in previous years, to the superintendence of Mr. Oberthür, and Mr. Powell was authorized to engage as many collectors as the material would seem to need, with full instructions as to packing and shipping to Cherbourg.

The visitor then proceeded to Wageningen, Holland, where he arranged for further assistance from Prof. Dr. J. Ritzema Bos. From there he went to Hamburg, where he arranged with the American Express Co. to care for shipments coming from Germany, Russia, and Austria-Hungary, arrangements being made to keep the material on ice until the next steamer should sail, and in case of the breakage

PLATE III.

FIG. 1.—ROADSIDE OAK IN BRITTANY, WITH LEAVES RAGGED BY GIPSY-MOTH CATER-
PILLARS. (L. O. HOWARD, JUNE, 1909.) (ORIGINAL.)

FIG. 2.—M. RENE OBERTHÜR (IN CENTER), DR. PAUL MARCHAL (AT RIGHT), WITH ROAD-
SIDE OAKS (BEHIND) RAGGED BY GIPSY-MOTH CATERPILLARS. (L. O. HOWARD, JUNE,
1909.) (ORIGINAL.)

FIG. 1.—CATERPILLAR HUNTERS IN THE SOUTH OF FRANCE, UNDER M. DILLON, 1909.
(ORIGINAL.)

FIG. 2.—PACKING PARASITIZED CATERPILLARS AT HYÈRES, FRANCE, FOR SHIPMENT TO
THE UNITED STATES, 1909. (ORIGINAL.)

or other bad condition of packages arrangements were made with Dr. L. Reh, of the Hamburg Museum, to act as expert adviser of the express company.

From Hamburg he proceeded to Berlin for a short consultation with Dr. R. Heymons, and thence to St. Petersburg. At St. Petersburg he was assured by Mr. Montgomery Schuyler, the secretary of the embassy, that all arrangements had been made with the Russian Government, and the same assurance was given by Prof. Porchinsky. The Russian officials insisted that none of the 1908 packages going out of Russia had been opened by the Russian postal authorities, and stated that in their opinion the opening must have been done at the German frontier by German officials. A strong letter was then written to the Hon. David J. Hill, United States ambassador to Germany, reciting the facts, dwelling upon the importance to America of these importations, and urging him to secure from the German Government orders to postal officials to pass without opening boxes of these parasites addressed to the American Express Co. in Hamburg. Later, in Dresden, a reply was received from Ambassador Hill, stating that the German Government consented to issue the necessary instructions, but still later, in Paris, an additional communication from the ambassador requested detailed information as to the points on the German frontier where these sendings would enter the Empire. By telegraphic communication with Prof. Kincaid, in southern Russia, and the Austrian agents, this information was furnished, but there seems still to have been some opening of the Russian boxes with resulting damage to their contents.

After Russia, Dresden, Tetschen, Vienna, Budapest, Innsbruck, Zurich, and Paris were consecutively visited, and agents were instructed concerning the new arrangements for shipping material. At Innsbruck the visitor met for the first time Prof. K. W. von Dalla Torre, the author of the great catalogue of the Hymenoptera of the world, and got his views on the subject of the parasitic Hymenoptera and their practical handling.

From Paris he took a trip into Normandy and Brittany with Dr. Paul Marchal, of the ministry of agriculture of France, and Mr. René Oberthür, for the pupose of examining into the export nursery industry, and at the same time with a view of observing gipsy-moth and brown-tail moth conditions in that part of France. (See Pl. III, fig. 2.) It transpired that both of the injurious insects were unusually abundant in portions of this territory, and by good fortune a small oak forest covering some hundreds of acres was found not far from Nantes, in which there had been an outbreak of the gipsy-moth more serious than either Dr. Marchal or Mr. Oberthür had ever seen or had ever heard of in France. Practically every tree was defoliated (see Pl. III, fig. 1), and at the time of the visit, the last week in June, the larvæ were about full grown and making ready to spin. The natural enemies of the gipsy-moth were not abundant in this forest, although a few were seen on trees along the highway in this general region. Nevertheless the invariable experience in Europe is that following such an outbreak as this parasites congregate in the region the following year and multiply in enormous numbers. The finding of this area, therefore, seemed fortunate, since during the season of 1910 it seemed probable that parasites would be abundant at that point. This hope was not fulfilled, however, and in 1910 practically no gipsy-moth larvæ were to be found in that general region.

In the meantime the honorable minister of agriculture for Japan had at the request of the honorable the Secretary of Agriculture of the United States designated Prof. S. I. Kuwana, of the Imperial Agricultural Station at Tokyo, to be the official representative of the Japanese Government in the parasite work to be carried on during the spring and summer of 1909, and to conduct his operations in cooperation with and in correspondence with the chief of the Bureau of Entomology of the United States Department. Prof. Kuwana has shown himself in this, as in his previous work, a man of extraordinary intelligence and activity, and has sent in a number of interesting and valuable lots of parasitic material which were received at Melrose Highlands in uniformly good condition. This was due to the great care and intelligence shown by Prof. Kuwana in its collection and in his methods of packing and shipping.

The most nearly perfect European service during the summer of 1909 was secured in France, owing to the arrangement made at the May conference in Paris. In the south of France very many people were employed under Mr. Powell, and several thousand boxes of good material were received at the parasite laboratory from this region. (See Pl. IV, fig. 2.) In quantity it exceeded the total of all the importations of a similar character made since the inception of the work, and from it have been reared a greater number of important tachinid parasites than have been reared from all other importations of similar character taken together. The size of the French shipments is largely due to the intelligent energy of Mr. M. Dillon (see Pl. IV, fig. 1), with whom the bureau was placed in relations by Mr. Powell.

Quantities of miscellaneous material were also received, as formerly, from numerous collectors in Germany, Austria, Italy, Holland, Belgium, and Switzerland.

Prof. Kincaid's account of his Russian observations is as follows:

At the request of Dr. L. O. Howard, Chief of the Bureau of Entomology, United States Department of Agriculture, the writer visited the provinces of Russia bordering upon the Black Sea during the summer of 1909 with a view to the introduction into America of the parasites of the gipsy moth reported to exist in that part of Europe. Proceeding to St. Petersburg via New York and Paris, an interview was had with Prof. Porchinsky, of the Russian Bureau of Entomology, who supplied valuable information and suggestions for the furtherance of the investigation. Leaving the Russian capital on April 28, a journey of 48 hours brought the writer to the city of Kishenef and after making a survey it was decided to establish a base of operations in the forest of Gauchesty, an area of wooded hills adjacent to a village of that name about 30 versts [1] northwest from Kishenef. Since the accommodations in the village of Gauchesty were of an unsatisfactory character, Mr. Artemy Nazaroff, the manager of the estate of Prince Manook Bey, on the lands of which the more important infested areas existed, invited the writer and his interpreter to become his guests during the progress of the investigation. A suite of rooms in the guest house of Gauchesty castle was placed at our disposal, and Mr. Nazaroff did all in his power to forward our interests and to make agreeable our stay in that part of Russia. An outbuilding upon the farm of the estate was transformed into a laboratory in which was erected a set of rearing frames for the rearing of the parasites. During the first week of April systematic exploration of the adjacent wooded areas was begun. The forest cover was found to consist almost exclusively of young oaks, with a few scattering trees of other species. The ground beneath the trees was fairly free from underbrush and was carpeted with a rich profusion of shrubs and flowers. At a distance of 7 versts from Gauchesty was an area covered with trees of considerable age among which the underbrush was comparatively dense.

From the forester in charge of the timbered areas upon the estate it was learned that the gipsy moth had done great damage to the forest during the previous season, large areas having been completely defoliated. This statement was borne out by the immense number of egg masses attached to the trees. At the time we commenced our investigations the caterpillars had emerged from the eggs but were still resting upon the bark. Few signs of previous parasitic activity were observed beyond the discovery of a number of empty cocoons of *Apanteles solitarius* Ratz. attached to the bark of the trees. In the ancient forest mentioned above the egg masses were very numerous, but the number of larvæ upon the bark was remarkably small. From the abnormal appearance of most of these egg masses, and from the fact that several Microhymenoptera were discovered in them, it seemed probable that a considerable number of the eggs had been destroyed through this agency. In other parts of the forest no evidence was secured indicating the presence of egg parasites.

The brown-tail moth seemed to be practically absent from the forested areas, but in the open rolling country between Kishenef and Gauchesty many wild pear growing in cultivated fields were found to be completely defoliated. A large number of the larvæ were placed in rearing frames but yielded no parasites, not even Meteorus making its appearance.

By June 1 the caterpillars of the gipsy moth had passed into the second stage and the oak trees were showing obvious signs of damage, but up to this date there was no indication of the emergence of

---

[1] Verst: Russian measure of distance=3,500 English feet; 6 versts=approximately 4 English miles.

hymenopterous parasites either in the field or from the thousands of larvæ reared in rearing frames. It became apparent that the conditions were unfavorable for the purposes in mind of assembling parasites for export, and it was decided to shift our headquarters to a more promising locality.

On June 5 a new base of operations was established at the town of Bendery on the Dniester River. Quarters were selected in the principal hotel, the Petersburgia, and in a remote corner of the extensive grounds of the hostelry a temporary laboratory was constructed in which several tiers of rearing frames were erected. The forest conditions in this district were much more diversified than at Gauchesty. To the northeast of the town at a distance of 7 versts was the forest of Gerbofsky, occupying a dry elevated area of about 5,000 acres and consisting almost exclusively of mature oak trees. To the southward, on the banks of the river, was the forest of Kitzkany, composed largely of black poplar, maple, and willow. In both of these forests the caterpillars of the gipsy moth were found in immense numbers, and evidence of attack by both hymenopterous and dipterous parasites was readily obtained, although nowhere in the abundance hoped for. For two weeks the two forests, as well as the extensive orchards in the vicinity of Bendery and the neighboring town of Tiraspol, were scoured for parasites. A number of Russian boys were pressed into service and trained to assist in making collections, at which they became quite expert. Except for a few clusters of cocoons derived from *Apanteles fulvipes* Hal., the only hymenopterous parasite to appear in considerable abundance was *Apanteles solitarius*. Caterpillars of the gipsy moth attacked by this species crawl down to the trunk or lower branches of the tree and collect in colonies on the lower side of the branches, under bark, in cavities and other sheltered places. Here the larva of the parasite emerges and spins its cocoon beneath the body of its host. The task of collecting these scattered cocoons was a tedious one, since it was necessary to remove each one carefully from the bark without undue pressure and also to disentangle it from the hairy body of its host.

In the forest of Kitzkany, where the conditions were favorable for bacterial infection owing to excessive dampness, the caterpillars of the gipsy moth were swept away in vast numbers by a bacterial disease before any extensive defoliation took place. The search for hymenopterous parasites in this district soon become a vain one, since very few of the caterpillars appeared to have escaped the infection.

The forest of Gerbofsky, owing to its being elevated, open, and well drained, was not favorable for bacterial infection and no trace of disease was observed. This forest was therefore almost completely defoliated by the caterpillars, and multitudes of the insects, failing to find any further nourishment upon the oaks, descended to the ground, where they died in great numbers, apparently from starvation. Hymenopterous parasites seemed to play a relatively small part in the destruction of the caterpillars, since the attacks of *Apanteles solitarius* were of the most scattering character. In the shrubbery growths adjacent to the main forest, where new plantations had been recently established by the forester in charge, a considerable number of Calosoma were found at work destroying the caterpillars, but their operations did not appear to extend into the main forest, where the open grass-covered ground did not offer sufficient concealment for the beetles.

The principal check to the depredations of the caterpillars of the gipsy moth in this forest came with the advent of the tachinids, the latter appearing upon the scene after the trees had been almost or entirely defoliated. Chalcid flies also appeared at this time, but not in considerable numbers. The species of Limnerium, a few specimens of which had been previously received from Russia, and of which it had been hoped to secure a supply for transfer to America, proved to be exceedingly rare, only three specimens being found. The larva of this parasite on emerging from its host spins an elongated silken thread, at the end of which it spins a cocoon and transforms to the pupal state.

Considerable numbers of the cocoons of *Apanteles solitarius* were collected from the forest, from the extensive orchards of the neighborhood, and from clumps of willow bushes conmonly found at the edges of fields. For several weeks shipments were made almost daily to Hamburg, from which port the packages were shipped in cold storage to New York. Many difficulties arose in attempting to make rapid shipments. The postal connections were very unsatisfactory and caused annoying delays, while at the German frontier another cause for loss of time developed through the formalities of the customs authorities of the German Government.

The brown-tail moth seemed to be quite uncommon in the region about Bendery, and no parasites were observed upon the small number of larvæ collected at this point.

Since it seemed desirable to cover as extensive a territory as possible during the season, the writer, leaving an assistant in charge of the laboratory and collecting organization at Bendery, journeyed northward on June 17 and established a new center of exploration at the city of Kief, in the province of the same name. Through the courtesy of Prof. Waldemar Pospielow the writer was furnished with much valuable information in regard to the forests of this portion of Russia and concerning the areas in which the gipsy moth was known to exist. Several immense forested areas were traversed, but as they were for the most part purely coniferous in character the gipsy moth appeared to be quite a rare insect. Through information supplied by Prof. Pospielow it was ascertained that at Mechnigori, a monasterial institution on the banks of the Dnieper, several hours by steamer from Kief, an area of woodland existed which was infested to a moderate extent by caterpillars of the gipsy moth, among which the parasites were reported to be much in evidence. A visit to the locality showed an interesting condition. The monastery was surrounded by beautiful groves of elm and oak trees in which the gipsy moth had made considerable inroads, but the parasites had developed to a sufficient extent to practically clear the foliage of caterpillars. Almost the sole agency in bringing about this condition was *Apanteles rufipes*, which attacks the larvæ of the gipsy moth in a manner closely resembling *Apanteles japonicus*, as observed during the preceding season in Japan, but in the case of the latter the caterpillars usually die upon the leaves of the trees, whereas in the former the caterpillars descend to the trunk and lower branches to form colonies. On emerging from the caterpillars the parasites spin cocoons beneath the host, which are also attached ventrally to the bark of the tree, and as numerous caterpillars die in a restricted area a mass of Apanteles cocoons, often of considerable thickness, is formed. Such masses standing out as white patches against the dark tree trunks on which they rest may be seen for considerable distances. Cocoons of *Apantales solitarius* were also observed in the forest of Mechnigori, but were comparatively rare, so this species evidently did not represent a very important element in the control of the gipsy moth.

In the forested areas about Kief the caterpillars of the brown-tail moth were rarely met with, but in several of the parks on the outskirts of the city they were found in abundance. In the grounds of the military school a large number of magnificent oak trees were almost denuded of foliage, and some of the other deciduous trees and shrubs, such as poplars, rose bushes, and Cratægus, were severely damaged. The usual brown-tail parasites were found at work, the most effective being Meteorus. Almost every branch of the injured trees bore the suspended cocoons of this parasite. Tachinids were also active, so it was obvious that very few of the caterpillars would reach maturity.

On departing from Kief on July 9 the season was practically over, and gipsy moths were in flight.

Returning to Bendery, it was found that the season was over so far as *Apantales solitarius* was concerned, but large numbers of tachinid puparia were in evidence. As many as possible of these were assembled and shipped to America. The chrysalides of the gipsy moth were also forwarded in considerable numbers in the hope of securing pupal parasites.

These lines of work were continued till July 16, by which time the season was so advanced that the moths were beginning to deposit their eggs for the succeeding season. From the abundance of moths in flight it was obvious that unless the natural parasites multiplied sufficiently to control the situation the region would experience another visitation of the same character during the following year.

Leaving Bendery on July 16, the writer returned to Paris via Odessa, Constantinople, and Naples, arriving in New York August 28.

Owing to various unforeseen conditions, and principally owing to the deficient transportation facilities, the material received as the result of Prof. Kincaid's expedition proved to be unsatisfactory on the whole.

In May and June, 1910, the senior author went to Europe once more, visited agents and officials in Italy and France, and, through the courtesy of the Spanish and Portuguese Governments, was able to start new official services in each of these countries for the collection and sending of parasitized gipsy-moth larvæ to the United States. In Italy Prof. Silvestri at Portici and Dr. Berlese at Florence were visited and informed as to the latest ideas of the laboratory regarding methods of shipment. In Spain Prof. Leandro Navarro, of the Phytopathological Station at Madrid, volunteered his services with the approval of the minister of agriculture. In Portugal Senhor Alfredo Carlos Lecocq, director of agriculture, placed the visitor in relation with Prof. A. F. de Seabra, of the Phytopathological Station at Lisbon, and the latter gladly consented to act as the agent of the bureau in this work in Portugal. In France arrangements were made with Mr. Dillon as during the previous year in the south of France, and arrangements were renewed with Miss Rühl in Zurich and Mr. Schopfer in Dresden. The distributing agency in Hamburg was continued, and a new distributing agency was started at Havre, France, on account of its convenient proximity to the American line steamers starting from Southampton. In order to insure the best results, Mr. Dillon accompanied certain large shipments from Hyères to Havre, and personally saw that they were placed upon the channel steamer the night before the sailing of an American line steamer from Southampton.

Sendings from Japan were continued in the same manner as during the previous year. The minister of agriculture for Japan, at the request of the Secretary of Agriculture of the United States, again designated Prof. S. I. Kuwana, of the Imperial Agricultural Experiment Station at Tokyo, to be its official representative in this work, and he continued his extremely valuable sendings.

The amount received during the summer was larger than ever before, but the results obtained, owing partly to the condition of the material on receipt and owing to curious seasonal fluctuations and differences in the countries of origin and in the infested territory in America, the results by no means corresponded with the increased material. The work carried on in the laboratory during the season and the results obtained are mentioned later.

In the autumn the junior author visited France and Russia for the purpose of studying certain important points regarding the question of alternate hosts of the parasites and methods of hibernation. The results of his observations will be given in detail in the later section headed "The extent to which the gipsy moth is controlled through parasitism abroad."

At the close of the season of 1910, and in part owing to the preparation of the present bulletin, a general review of the whole work was undertaken, and a summing up of present conditions seemed to indicate that nearly as much had already been accomplished by present methods as could be expected. The great need at this time seemed to be a careful study in the countries of origin of the species of apparent importance which have been sent over but have not become established, in order to ascertain the reasons for the apparent failure; and, further, to see on the spot what can be done with regard to the importation of parasites of apparently lesser importance, but which, through the fact that they may fill in gaps in the parasitic chain and may at the same time increase beyond their native wont when confronted with American conditions, may be very desirable. Accordingly the junior author was commissioned to visit France, Italy, and Russia in the winter and early spring of 1911, and subsequently to spend the breeding season· if found desirable, in Japan. He was given authority to employ the necessary agents in each of these countries. He sailed January 5, 1911.

KNOWN AND RECORDED PARASITES OF THE GIPSY MOTH AND OF THE BROWN-TAIL MOTH.

When the work of introducing the parasites of the gipsy moth and of the brown-tail moth was begun in 1905, the available assets consisted of generous appropriations by the State of Massachusetts and the Federal Government, an abundant faith in the validity of the theory which was to be put to test, and a long bibliographical list of the parasites which were recorded as attacking these insects in Europe and Japan. Of these, the appropriations have withstood most effectively the ordeal of the years which have since passed. Our faith in the validity of the principle at stake has also stood out wonderfully well, when the numerous trials to which it has been subjected are taken into consideration. It is not too much to say that at the present time it is stronger than ever, notwithstanding that a good many facts have come to light in this period which are more or less flatly in contradiction to the theory of parasite control as generally accepted at the beginning. It has more than once been necessary to modify beliefs and ideas as previously held, in order to make them conform to the actual facts. To take a pertinent example, it was necessary to place an entirely different value upon the bibliographical list above mentioned than that which was placed upon it when the work was begun, and when the policies of the laboratory were first determined.

Nearly thirty years ago the present head of the Bureau of Entomology undertook the compilation of a card catalogue of references to the host relations of the parasitic Hymenoptera of the world. For more than twenty years the work was continued until some 30,000 such references were accumulated. From among them those in which the gipsy moth was mentioned as the host were collected and a list of gipsy-moth parasites was published in Insect Life.[1] With the exception of a comparatively few recent additions this list forms the basis of that which follows. That of the parasites which have been recorded as attacking the brown-tail moth is largely from the same source.

---

[1] U. S. Department of Agriculture, Division of Entomology, Insect Life, vol. 2, pp. 210–211, 1890.

HYMENOPTEROUS PARASITES OF THE GIPSY MOTH (*Porthetria dispar L*)

BRACONIDÆ.

Reared at laboratory.	Recorded as parasites.
Apanteles fulvipes (Hal.).	*Apanteles fulvipes* (Hal.). [1] [2]
Apanteles solitarius (Ratz.).	*Apanteles solitarius* (Ratz.). [1] [2]
	Microgaster calceata Hal. [1] [2]
	Apanteles tenebrosus (Wesm.). [1]
	Microgaster tibialis Nees. [1]
	(*Microgaster*) *Apanteles fulvipes liparidis* (Bouché). [1] [2]
	Apanteles glomeratus (L.). [1] [2]
	Apanteles solitarius var. *melanoscelus* (Ratz.). [1]
	Apanteles solitarius? *ocneriæ* Svanov.
Meteorus versicolor (Wesm.).	*Meteorus scutellator* (Nees). [1]
Meteorus pulchricornis (Wesm.).	
Meteorus japonicus Ashm. [3]	

ICHNEUMONIDÆ.

PRIMARY.

Pimpla (*Pimpla*) *instigator* (Fab.).	*Pimpla* (*Pimpla*) *instigator* (Fab.). [1] [2]
Pimpla (*Pimpla*) *porthetriæ* Vier. [3]	
Pimpla (*Pimpla*) *examinator* (Fab.).	*Pimpla examinator* (Fab.). [1]
Pimpla (*Pimpla*) *pluto* Ashm. [3]	
Pimpla (*Apechthis*) *brassicariæ* (Poda).	
Pimpla (*Pimpla*) *disparis* Vier. [3]	
Theronia atalantæ (Poda).	*Theronia atalantæ* (Poda). [1] [2]
Limnerium (*Hyposoter*) *disparis* Vier.	*Campoplex conicus* Ratz. [1]
Limnerium (*Anilastus*) *tricoloripes* Vier.	*Casinaria tenuiventris* (Grav.). [1]
Ichneumon disparis (Poda).	*Ichneumon disparis* (Poda). [1] [2]
	Ichneumon pictus (Gmel.). [1] [2]
	Amblyteles varipes Rdw. [2]
	Trogus flavitorius [sic.] *lutorius* (Fab.)? [1] [2]
	(*Cryptus*) *Aritranis amœnus* (Grav.). [1]
	Cryptus cyanator Grav. [1]

PROBABLY SECONDARY BUT RECORDED AS PRIMARY.

	Mesochorus pectoralis Ratz. [1] [2]
	Mesochorus gracilis Brischke. [1] [2]
	Mesochorus splendidulus Grav. [1] [2]
	Mesochorus confusus Holmgr. [1]
	Mesochorus semirufus Holmgr. [1]
	(*Hemiteles*) *Astomaspis fulvipes* (Grav.). [1] [2]
	= *A. nanus* (Grav.) according to Pfankuch.
	Hemiteles bicolorius Grav. [2]
	Pezomachus hortensis Grav. [2]
	Pezomachus fasciatus (Fab.) [1] = *Pezomachus melanocephalus* (Schrk.).

---

[2] Recorded by the senior author in a card catalogue of parasites kept in the Bureau of Entomology.
[2] Recorded by Dalla Torre in Catalogus Hymenopterorum.
[3] Japanese species.

CHALCIDIDÆ.

Reared at laboratory.

Recorded as parasites.
Pteromalus halidayanus Ratz.[1]
Pteromalus pini Hartig.[1]
Dibrachys boucheanus Ratz.[1] (Secondary.)
Eurytoma abrotani Panzer [1][2]=appendigaster Swed. (Secondary.)

Eupelmus bifasciatus Fonsc.   Eupelmus bifasciatus Fonsc.[1][2]
Monodontomerus æreus Walk.
Chalcis flavipes Panz.   Chalcis callipus Kby.[3]
Chalcis obscurata Walk.[5]
Schedius kuvanæ How.[4]

HYMENOPTEROUS PARASITES OF THE BROWN-TAIL MOTH (*Euproctis chrysorrhœa L.*).

BRACONIDÆ.

Reared at laboratory.   Recorded as parasites.
Meteorus versicolor (Wesm.).   Meteorus versicolor (Wesm.).[5]
        Meteorus ictericus (Nees).[5]
Apanteles lacteicolor Vier.   Apanteles inclusus (Ratz.)[2][5]
        Apanteles ultor Reinh.[1][2][3]
        Apanteles difficilis (Nees).[5]
        Apanteles liparidis (Bouché).[5]
        Apanteles vitripennis (Hal.).[5]
        Apanteles solitarius (Ratz.).[5]
        Microgaster consularis (Hal.)[5]= Microgaster connexa Nees.
        Microgaster calceata Hal.[1]
        Rogas geniculator Nees.[2][5]
        Rogas testaceus (Spin.).[1]
        Rogas pulchripes (Wesm.).[1]

ICHNEUMONIDÆ.

PRIMARY.

Pimpla (*Pimpla*) examinator (Fab.).   Pimpla (*Pimpla*) examinator (Fab.).[2][5]
Pimpla (*Pimpla*) instigator (Fab.).   Pimpla (*Pimpla*) instigator (Fab.).[1][2][3]
Pimpla (*Apechthis*) brassicariæ (Poda).
Theronia atalantæ (Poda).   Theronia atalantæ (Poda).[1][2][3]
        Campoplex conicus Ratz.[5]
        (*Campoplex*) Omorgus difformis (Gmel.).[5]
        Cryptus moschator (Fab.).[1]
        (*Cryptus*) Idiolispa atripes (Grav.).[1]
        Ichneumon disparis (Poda).[5]
        Ichneumon scutellator (Grav.).[2]

---

[1] Recorded by the senior author in a card catalogue of parasites kept in the Bureau of Entomology.
[2] Recorded by Dalla Torre in Catalogus Hymenopterorum.
[3] Reared by Dr. S. I. Kuwana.
[4] Japanese species.
[5] Recorded by Emelyanoff.

PROBABLY SECONDARY, BUT RECORDED AS PRIMARY.

Reared at laboratory. | Recorded as parasites.

Mesochorus pectoralis Ratz.[1][2][3]
Mesochorus dilutus Ratz.[2][3]
Hemiteles socialis Ratz.[3]

CHALCIDIDÆ.

Pteromalus sp. | Pteromalus rotundatus Ratz.[3]=Pt. chrysorrhœa D. T.[1][2]

Pteromalus nidulans Thoms.=Pt. egregius Först. | Pteromalus processioneæ Ratz.[1][2]

Diglochis omnivora Walk. | Pteromalus nidulans Thoms.[1][3]
Pteromalus puparum L.[3]
Dibrachys boucheanus (Ratz.).[3] (Secondary.)
· Chalcis scirropoda Först.[1]

Monodontomerus æreus Walk. | Torymus anephelus Ratz.[3]=Monodontomerus æreus Walk.[1][2]
Monodontomerus dentipes Boh.[1][2]
Anagrus ovivorus Rondani.[2]

Trichogramma sp. I.
Trichogramma sp. II.

PROCTOTRYPIDÆ.

Telenomus phalænarum Nees (?). | Telenomus phalænarum Nees.[1][2][3]

DIPTEROUS PARASITES OF THE GIPSY MOTH (Porthetria dispar L.).

The following are lists of the dipterous parasites reared and recorded from Porthetria dispar L. and Euproctis chrysorrhœa L. Each list is supplemented by a list of recorded hosts for each species enumerated.

These lists have been compiled from various sources, the principal being the "Katalog der Paläarktischen Dipteren," Brauer & Bergenstamm's "Die Zweiflügler des Kaiserlichen Museums zu Wien," Fernald and Forbush's "The Gipsy Moth," and the senior author's "List of parasites bred from imported material during the year 1907" (3d annual report of the superintendent for suppressing the gipsy and brown-tail moths).

In the choice of names of the foreign tachinids the Katalog der Paläarktischen Dipteren has been followed with the exception of a few cases in which other names have been in use at the Gipsy Moth Parasite Laboratory; in these few cases, to avoid confusion, no change has been made.

---

[1] Recorded by Emelyanoff.
[2] Recorded by Dalla Torre in Catalogus Hymenopterorum.
[3] Recorded by the senior author in a card catalogue of parasites kept in the Bureau of Entomology.

FOREIGN TACHINID PARASITES ON PORTHETRIA DISPAR.

Reared.	Recorded.
Blepharipa scutellata R. D.	Argyrophylax atropivora R. D.
Carcelia gnava Meig.	Carcelia excisa Fall.
Compsilura concinnata Meig.	Compsilura concinnata Meig.
Crossocosmia sericariæ Corn.	Echinomyia fera L.
Dexodes nigripes Fall.	Epicampocera crassiseta Rond.
Parasetigena segregata Rond.	Ernestia consobrina Meig.
Tachina larvarum L.	Eudoromyia magnicornis Zett.
Tachina japonica Towns.	Exorista affinis Fall.
Tricholyga grandis Zett.	Histochæta marmorata Fab.
Zygobothria gilva Hartig.	Lydella pinivoræ Ratz.
	Meigenia bisignata Schin.
	Parasetigena segregata Rond.
	Phryxe erythrostoma Hartig.
	Ptilotachina larvincola Ratz.
	Ptilotachina monacha Ratz.
	Tachina larvarum L.
	Tachina noctuarum Rond.
	Zenillia libatrix Panz.
	Zygobothria gilva Hartig.
	Zygobothria bimaculata Hartig.

N. B.—It is interesting to note that only four species are common to both lists.

RECORDED HOSTS OF FOREIGN TACHINID PARASITES OF PORTHETRIA DISPAR REARED
AT THE GIPSY MOTH PARASITE LABORATORY.

BLEPHARIPA SCUTELLATA R. D.:·
   Acherontia atropos L.; Vanessa antiopa L.

CARCELIA GNAVA Meig.:
   Malacosoma neustria L.; Orgyia antiqua L.; Stilpnotia salicis L.

COMPSILURA CONCINNATA Meig.:
   See list of recorded parasites of P. dispar

CROSSOCOSMIA SERICARIÆ Corn.:
   Antheræa yamamai Guér.; A. mylitta Moore; Sericaria mori L.

DEXODES NIGRIPES Fall.:
   Ascometia caliginosa Hb.; Agrotis candelarum Stgr.; Bupalus piniarius L.;
   Cucullia asteris Schiff.; Deilephila euphorbiæ L.; Eurrhypara urticæ L.; Heliothis
   scutosa Schiff.; Hybernia sp.; Mamestra pisi L.; Miana literosa Hw ; Ortholitha
   cervinata Schiff.; Phragmatobia fuliginosa L ; Plusia gamma L.; Porthesia
   similis Fussl.; Tapinostola elymi Tr.; Tephroclystia virgauriata Dbld.; Thaumetopœa pinivora Tr.; Vanessa io L.; V. polychlorus L.; V. urticæ L.; Lophyrus
   sp.; Nematus ribesii Scop.

PARASETIGENA SEGREGATA Rond.:
   (See list of recorded parasites of P. dispar)

TACHINA LARVARUM L. :
   (See list of recorded parasites of P. dispar.)

TACHINA JAPONICA Towns.:
   Porthetria dispar L.

TRICHOLYGA GRANDIS Zett.
   Arctia caja L.; Mamestra oleracea L.; M. pisi L.; Saturnia pavonia L.; S. pyri
   Schiff.; Sphinx ligustri L ; Thaumetopœa pityocampa Schiff.; Vanessa io L.

RECORDED HOSTS OF FOREIGN TACHINID PARASITES RECORDED ON PORTHETRIA
DISPAR.

ARGYROPHYLAX ATROPIVORA R. D.:

P. dispar L.; *Acherontia atropos* L.; *Notodonta trepida* Esp.; *Vanessa io* L.

CARCELIA EXCISA Fall.:

Abrostola tripartita Hufn.; *A. triplasia* L.; *Arctia caja* L.; *A. hebe* L.; *A. villica*
L.; *Bupalus piniarius* L.; *Callimorpha dominula* L.; *Cucullia scrophulariæ*
Cap.; *Dasychira pudibunda* L.; *Endromis versicolora* L.; *Hyloicus pinastri* L.;
P. dispar L.; *P. monacha* L.; *Malacosoma castrensis* L.; *M. neustria* L.; *Orgyia
antiqua* L.; *Phragmatobia fuliginosa* L.; *Pterostoma palpina* L.; *Pygæra
curtula* L.; *Saturnia pyri* Schiff.; *Sphinx ligustri* L.; *Stilpnotia salicis* L.;
Thalpochæres pannonica Frr.; *Thaumetopœa processionea* I..

COMPSILURA CONCINNATA Meig.:

Abraxas grossulariata L.; *Acronycta aceris* L.; *A. alni* L.; *A. cuspis* Hb.; *A.
megacephala* F.; *A. rumicis* L.; *A. tridens* Schiff.; *Araschinia levana* L.; *A.
prorsa* L.; *Arctia caja* L.; *Attacus cynthia* L.; *Catocala promissa* Esp.; *Cranio-
phora ligustri* Fab.; *Cucullia lactucæ* Esp.; *C. verbasci* L.; *Dasychira pudibunda*
L.; *Dilina tiliæ* L.; *Dilobia cæruleocephala* L.; *Dipterygia scabriuscula* L.;
Drymonia chaonia Hb.; *Euproctis chrysorrhœa* L.; *Hyloicus pinastri* L.; *Liby-
tha celtis* Laich.; *Porthetria dispar* L.; *P. moncha* L.; *Macrothylacia rubi* L.;
Mamestra brassicæ L.; *M. oleracea* L.; *M. persicariæ* L.; *Malacosoma neustria* L.;
Oeonistis quadra L.; *Phalera bucephala* L.; *Pieris brassicæ* L.; *P. rapæ* L.; *Plusia
festucæ* L.; *P. gamma* L.; *Pœcelocampa populi* L.; *Porthesia similis* Füssl.;
Pygæra anachoreta Fab.; *Pyrameis atalanta* L.; *Smerinthus populi* L.; *Spilo-
soma lubricipeda* L.; *S. menthastri* Esp.; *Stauropus fagi* L.; *Stilpnotia salicis* L.;
Tæniocampa stabilis View.; *Thaumetopœa processionea* L.; *T. pityocampa*
Schiff.; *Timandra amata* L.; *Trachea atriplicis* L.; *Vanessa antiopa* L.; *V. io* L.;
V. urticæ L.; *V. xanthomelas* Esp.; *Yponomeuta padeila* L.; *Cimbex humeralis*
Fourcr.; *Trichiocampus viminalis* Fall.

ECHINOMYIA FERA L.:

Agrotis glareosa Esp.; *Arctia aulica* L.; *Leucania obsoleta* Sb.; *Porthetria dispar* L.;
P. monacha L.; *Mamestra pisi* L.; *Oeonistis quadra* L.; *Panolis griseovariegata*
Goeze.

EPICAMPOCERA CRASSISETA Rond.:

Porthetria dispar L.; *Thaumetopœa processionea* L.

ERNESTIA CONSOBRINA Meig.:

Cucullia artemisiæ Hufn.; *Porthetria dispar* L.

EUDOROMYIA MAGNICORNIS Zett.:

Agrotis sp. ind.; *Hadena adusta* Esp.; *Porthetria dispar* L.

EXORISTA AFFINIS Fall.:

Acronycta alni L.; *Arctia caja* L.; *Porthetria dispar* L.; *Pachytelia villosella* O.;
Saturnia pavonia L.; *S. pyri* Schiff.

HISTOCHÆTA MARMORATA Fab.:

Arctia caja L.; *A. quenselii* Payk.; *A. villica* L.; *Cucullia verbani* L.; *Malacosoma
neustria* L.; *Porthetria dispar* L.; *Goniarctena rufipes* Payk.

LYDELLA PINIVORÆ Ratz.:

Porthetria dispar L.:

MEIGENIA BISIGNATA Schin.:

Porthetria dispar L.:

PARASETIGENA SEGREGATA Rond.:

Porthetria dispar L.; *P. monacha* L.; *Lophyrus pini* L.

PHRYXE ERYTHROSTOMA Hartig:

Dendrolimus pini L.; *Haloicus pinastri* L.:

PTILOTACHINA LARVINCOLA Ratz.:
　Porthetria dispar L.:
PTILOTACHINA MONACHA Ratz.:
　Porthetria dispar L.:
TACHINA LARVARUM L.:
　　Acronycta rumicis L.; Agrotis præcox L.; Arctia caja L.; A. villica L.; Catocola
　　fraxini L.; Cosmotriche potatoria L.; Cucullia prenanthis B.; Dasychira fascellina
　　L.; Deilephila gallii Rott.; D. euphorbiæ L.; Dendrolimus pini L.; Gastropacha
　　quercifolia L.; Lasiocampa quercus L.; Porthetria dispar L.; P. monacha L.;
　　Macroglossa stellatarum L.; Macrothylacia rubi L.; Malacosoma castrensis L.;
　　M. neustria L.; Mamestra brassicæ L.; Melitæa didyma O.; Melopsilus porcel-
　　lus L.; Ocneria detrita Esp.; Olethreutes hercyniana Tr.; Orgyia ericæ Germ.;
　　O. gonostigma F.; Orthosia humilis F.; Panolis griseovariegata Goeze; Papilio
　　machaon L.; Plusia iota L.; Saturnia pyri Schiff.; Stilpnotia salicis L.; Vanessa
　　antiopa L.; Vanessa io L.; V. polychloros L.; V. urticæ L.; Yponomeuta
　　evonymella L.; Lophyrus pini L.; Pamphilius stellatus Christ.
TACHINA NOCTUARUM Rond.:
　Cosmotriche potatoria L.; Porthetria dispar L.:
ZENILLIA LIBATRIX Panz.:
　　Abrostola asclepiadis Schiff.; Brephos nothum Hb.; Dasychira pudibunda L.; Laren-
　　tia autumnalis Strom.; Porthetria dispar L.; Malacosoma neustria L.; Pygæra
　　pigra Hufn.; Thaumetopœa processionea L.; Yponomeuta evonymella L.; Y.
　　padella L.
ZYGOBOTHRIA GILVA Hartig:
　　Porthetria dispar L.; Stauropus fagi L.; Lophyrus laricis Jur.; L. pallidus Klug.;
　　L. pini L.; L. rufus Latr.; L. variegatus Hartig.
ZYGOBOTHRIA BIMACULATA Hartig:
　　Lymantria monacha L.; Lophyrus pallidus Klug.; L. pini L.; L. rufus Latr.; L.
　　socius Klug ; L. variegatus Hart.; L. virens Klug.

NATIVE DIPTERA REARED FROM PORTHETRIA DISPAR.

Tachinidæ: [1]
　　Exorista blanda O. S.　　　　　Exorista fernaldi Will.
　　Exorista pyste Walk.　　　　　 Tachina mella Walk.

Other than Tachinidæ: [2]
　　Aphiochæta setacea Aldr.　　　 Phora incisuralis Loew
　　Aphiochæta scalaris Loew.　　　Sarcophaga sp.
　　Gaurax anchora Loew.

---

[1] These have only been reared very occasionally at the Gipsy Moth Parasite Laboratory.
[2] At the Gipsy Moth Parasite Laboratory these have been recorded only as scavengers and not as parasites.

DIPTEROUS PARASITES OF THE BROWN-TAIL MOTH (*Euproctis chrysor-rhœa* L.).

FOREIGN TACHINID PARASITES OF EUPROCTIS CHRYSORRHŒA.

Reared.	Recorded.
Blepharidea vulgaris Fall.	*Compsilura concinnata* Meig.
Compsilura concinnata Meig.	*Echinomyia præceps* Meig.
Cyclotophrys anser Towns.	*Erycia ferruginea* Meig.
Dexodes nigripes Fall.	*Pales pavida* Meig.
Digonichæta setipennis Fall.	*Tachina latifrons* Rond.
Digonichæta spinipennis Meig.	*Zenillia fauna* Meig.
Eudoromyia magnicornis Zett.	*Zenillia libatrix* Panz.
Masicera sylvatica Fall.	
Nemorilla sp.	
Nemorilla notabilis Meig.	
Pales pavida Meig.	
Parexorista cheloniæ Rond.	
Tachina larvarum L.	
Tricholyga grandis Zett.	
Zenillia libatrix Panz.	
Zygobothria nidicola Towns.	

N. B.—It is interesting to note that only three species are common to both lists.

RECORDED HOSTS OF FOREIGN TACHINIDS REARED FROM EUPROCTIS CHRYSORRHŒA AT THE GIPSY MOTH PARASITE LABORATORY.

BLEPHARIDEA (PHYRXE) VULGARIS Fall.:

Abraxas grossulariata L.; *Adopæa lineola* O.; *Aporia cratægi* L.; *Araschinia levana* L.; *A. prorsa* L.; *Argynnis lathonia* L.; *Arctia hebe* L.; *Boarmia lariciaria* Dbld.; *Brotolomia meticulosa* L.; *Calymnia trapezina* L.; *Cosmotriche potatoria* L.; *Cucullia anthemidis* Gn.; *C. asteris* Schiff.; *C. verbasci* L.; *Dendrolimus pini* L.; *Ephyra linearia* Hb.; *Epineuronia cespitis* F.; *Euchloë cardamines* L.; *Euplexia lucipara* Hb.; *Hybernia defoliaria* Cl.; *Hyloicus pinastri* L.; *Hylophila prasi-nana* L.; *Leucania albipuncta* F.; *L. lythargyria* Esp.; *Mamestra advena* F.; *M. persicariæ* L.; *M. reticulata* Vill.; *Melitæa athalia* Rott.; *Metopsilus porcellus* L.; *Nænia typica* L.; *Parasemia plantaginis* L.; *Pieris brassicæ* L.; *P. dapli-dice* L.; *P. rapæ* L.; *Plusia gamma* L.; *Thamnonona wavaria* L.; *Thaumetopœa pityocampa* Schiff.; *T. processionea* L.; *Toxocampa pastinum* Tr.; *Vanessa antiopa* L.; *V. io* L.; *V. urticæ* L.; *V. xanthomelas* Esp.; *Zygæna achilleæ* Esp., ab. *janthina*; *Z. filipendulæ* L. (?); *Procrustes coriaceus* L.

COMPSILURA CONCINNATA Meig.:

See host list of tachinid parasites of *P. dispar*.

CYCLOTOPHRYS ANSER Towns.:

No records other than at the Gipsy Moth Parasite Laboratory

DEXODES (LYDELLA) NIGRIPES Fall.:

Ascometia caliginosa Hb.; *Agrotis candelarum* Stgr.; *Bupalus piniarius* L.; *Cucullia asteris* Schiff.; *Deilephila euphorbiæ* L.; *Eurrhypara urticata* L.; *Helio-this scutosa* Schiff.; *Hybernia* sp.; *Mamestra pisi* L.; *Miana literosa* Hw.; *Ortholitha cervinata* Schiff.; *Phragmatobia fuliginosa* L.; *Plusia gamma* L.; *Porthesia similis* Füssl.; *Tapinostola elymi* Tr.; *Tephroclystia virgaureata* Dbld.; *Thaumetopœa pinivora* Schiff.; *Vanessa io* L ; *V. polychloros* L.; *V. urticæ* L.; *Lophyrus* sp.; *Nematus ribesii* Scop.

DIGONICHÆTA SETIPENNIS Fall.:

Grapholitha strobilella L.; *Notodonta trepida* Esp.; *Pheosia tremula* Cl. (?); *Forfi-cula auricularia* L.

DIGONICHÆTA SPINIPENNIS Meig.:
  Lasiocampa quercus L.; *Panolis griseovariegata* Goeze.
EUDOROMYIA MAGNICORNIS Zett.:
  Agrotis sp.; *Hadena adusta* Esp.; *Porthetria dispar* L.
MASICERA SYLVATICA Fall.:
  . *Apopestes spectrum* Esp.; *Cucullia verbasci* L.; *Deilephila euphorbiæ* L.; *D. gallii*
  . Rott.; *D. vespertilio* Esp.; *Dilina tiliæ* L.; *Gastropacha quercifolia* L.; *Lasio-
  campa quercus* L.; *Nonagria typhliæ* Thbg.; *Pieris brassicæ* L.; *Saturnia pav-
  onia* L.; *S. pyri* Schiff.; *S. spini* Schiff.; *Sphinx ligustri* L.
NEMORILLA NOTABILIS Meig.:
  Notocælia uddmanniana L.; *Plusia festucæ* L.; *Sylepta ruralis* Scop.; *Tachyptylia
  populella* Cl.
PALES PAVIDA Meig.:
  Acronycta tridens Schiff.; *Agrotis stigmatica* Hb.; *A. xanthographa* F.; *Attacus
  cynthia* L.; *A. lunula* Fab.; *Eriogaster catex* L.; *Emphytus cingillum* Klug;
  Euproctis chrysorrhœa L.; *Orgyia ericæ* Germ.; *Panolis griseovariegata* Goeze;
  Plusia gamma L.; *Thaumetopœa processionea* L.
PAREXORISTA (EXORISTA) CHELONIÆ Rond.:
  Ammoconia cæcimacula Fab.; *Arctia caja* L.; *A. hebe* L.; *A. villica* L.; *Hadena
  secalis* L.; *Macrothylacia rubi* L.; *Orthosia pistacina* Fab.; *Phragmatobia fuli-
  ginosa* L.; *Rhyparia purpurata* L.; *Spilosoma lubricipeda* L.; *Stilpnotia salicis* L.;
  Cimbex femorata L.; *Pamphilius stellatus* Christ.
TACHINA LARVARUM L.:
  See host list of foreign tachinids recorded from *P. dispar.*
TRICHOLYGA GRANDIS Zett.:
  Arctia caja L.; *Mamestra oleracea* L.; *M. pisi* L.; *Saturnia pavonia* L.; *S. pyri*
  Schiff.; *Sphinx ligustri* L.; *Thaumetopœa pityocampa* Schiff.; *Vanessa io* L.
ZENILLIA LIBATRIX Panz.:
  See host list of foreign tachinids recorded as parasites of *P. dispar.*
ZYGOBOTHRIA NIDICOLA Tn.:
  No record other than at the Gipsy Moth Parasite Laboratory.

RECORDED HOSTS OF FOREIGN TACHINIDS RECORDED AS PARASITIC ON EUPROCTIS
CHRYSORRHŒA.

COMPSILURA CONCINNATA Meig ·
  See list of recorded hosts of foreign tachinids recorded as parasitic on *P. dispar.*
ECHINOMYIA PRÆCEPS Meig.:
  Hemaris fuciformis L.; *Euproctis chrysorrhœa* L
ERYCIA FERRUGINEA Meig.:
  Euproctis chrysorrhœa L.; *Melitæa athalia* Rott.; *M. aurinia* Rott.; *Porthesia
  similis* Füssl.; *Vanessa io* L.
PALES PAVIDA Meig.:
  See list of recorded hosts of foreign tachinids reared from *E. chrysorrhœa* L.
TACHINA LATIFRONS Rond.:
  Euproctis chrysorrhœa L.; *Zygæna filipendulæ* L.
ZENILLIA FAUNA Meig. ·
  Acronycta rumicis L.; *Cossus cossus* L.; *Euproctis chrysorrhœa* L.: *Smerinthus
  ocellatus* L.
ZENILLIA LIBATRIX Panz. ·
  See list of recorded hosts of foreign tachinids reared from *E. chrysorrhœa* L.

NATIVE (AMERICAN) TACHINIDS REARED FROM EUPROCTIS CHRYSORRHŒA L., AT THE GIPSY MOTH PARASITE LABORATORY.

Blepharipeza leucophrys Wied.	? *Phorocera leucaniæ* Coq.
Euphorocera claripennis Macq.	*Sturmia discalis* Coq.
Exorista griseomicans V. de Wulp.	*Tachina mella* Walk.

N. B.—The above species have only been reared very occasionally. The species, however, doubtfully referred to *Phorocera leucaniæ* Coq. has been reared through to the pupal stage in considerable numbers These pupæ have always been imperfect and "larviform" and at the time of writing none has been reared through to the adult.

The compilation of the catalogue of parasites was originally undertaken in the expectation that it would prove of great service upon exactly such occasions as the present, when the application of the theory of control by parasites should be put to the test. Its value naturally depended upon the accuracy of the original records, and it was only right to suppose that in the majority of instances these could be depended upon. It was equally natural to suppose that the parasitic fauna of such common, conspicuous, and widely distributed insects as the gipsy moth and the brown-tail moth would be well represented in these lists, which were based upon a thorough overhauling of European literature; and it was not expected that any parasites of particular importance would be found which were not thus recorded, unless, indeed, they were confined to Continental Asia or to Japan.

In the fall of 1907, as soon as the turmoil of his first summer's work permitted, the junior author attempted to make use of the numerous bibliographical references for the purpose of learning as much as possible of the insects with which he was to deal. One after another, various species were taken up, until he was in possession of practically all of the published information concerning perhaps half of the Hymenoptera listed. Then he stopped, because the information thus gained was obviously not worth the labor. It was not so much that recorded information was scanty, or lacking in interest, but it was because in a great many instances it was contradictory to the results of the actual rearing work which had been carried on in the laboratory throughout the summer. It was obviously impossible to accept everything at its face value, and apparently next to impossible to choose between the true and the false. But one thing remained to be done, and that was to determine at first hand everything which it was necessary to know concerning the numerous species of parasites which it was desired to introduce into America.

If the list of parasites which have been reared at the laboratory from imported eggs, caterpillars, and pupæ of the gipsy moth and the brown-tail moth be compared with the lists which have already been given, the numerous and obvious differences which are immediately apparent will serve better than words to illustrate the situation which confronted us at the close of the season of 1907.

ESTABLISHMENT AND DISPERSION OF THE NEWLY INTRODUCED PARASITES.

In the beginning we were very far from accrediting to that phase of the project which has to do with the establishment and dispersion of the newly introduced parasites the importance which it deserved. Many widely diverse species of insects were known to have been introduced from the Old World and firmly established in America. Presumably they were accidentally imported, as was the case with the gipsy moth and the brown-tail moth; presumably, also, they had spread and increased from a small beginning, at first very gradually and later more rapidly, until they had become component parts of the American fauna over a wide territory. The circumstances under which the gipsy moth was imported were well known, and a good guess had been made as to those which resulted in the introduction of the brown-tail moth. But these were and are rare exceptions in this respect, and for the most part the preliminary chapters in the story of each of the insect immigrants never have been and probably never will be written.

Because the two very conspicuous instances of the gipsy moth and the brown-tail moth were constantly and automatically recurring whenever the probable future of the intentionally introduced parasites was considered, it was, perhaps, taken a little too much for granted, that they were to be considered as typical and significant of what to expect. In each instance the invasion started from a small beginning, and while the subsequent histories were different, the more rapid spread of the brown-tail moth was directly due to the fact that the females were capable of flight, and the relatively slow advance of the gipsy moth into new territory to the reverse. Even the brown-tail moth was for some years confined to a comparatively limited area, and it was rather expected that the parasites, if they established themselves at all, would remain for a similar period in the immediate vicinity of the localities where they were first given their freedom.

Accordingly, in accepting this theory without submitting it to a test, attempts were made to encompass the rapid dissemination of the parasites coincidently with their introduction. In 1906 and 1907 the parasites which were reared from the imported material were mostly liberated in small and scattered colonies. In a few instances this procedure was the best which could have been adopted; in others the worst. Small colonies of Calosoma, for example, remained for several years in the immediate vicinity of the point where the parent beetles were first liberated before any material dispersion was apparent (see Pl. XXIV), and the small colony was thus justified. The gipsy-moth egg parasite Anastatus, as was later determined, spreads at a rate of but a few hundred feet per year, and if it is to become generally distributed throughout the gipsy-moth-infested area within a reasonable time, natural dispersion must be assisted by artificial.

These, however, are both exceptions. In the case of Monodontomerus, and perhaps of other parasites, gregarious in their habit, it is not only conceivable but probable that a single fertilized female would be sufficient to establish the species in a new country, because the union between the sexes is effected within the body of the host in which they were reared. No matter how far a female may range and no matter how widely separated the victims of her maternal instincts, her progeny will rarely die without each finding its mate. Species having such habits are eminently well fitted to establish themselves wherever they secure foothold, even in the smallest numbers, and the small colony is again justified.

Many of the hymenopterous parasites, and very likely all of them, are capable of parthenogenetic reproduction, and here again is a factor which becomes of considerable importance in this connection. Some few of these are thelyotokous (bearing females only) and as such are eminently well fitted to establishment in a new country under otherwise unsatisfactory conditions. Most are arrhenotokous (bearing males only), and such are probably better fitted to establishment than would be the case if the species were wholly incapable of parthenogenetic reproduction. It has been proved, for example, that a single female of a strictly arrhenotokous species, may, through fertilization by her own parthenogenetically produced offspring, become the progenetrix of a race the vigor of which appears not to be immediately affected by the fact that their continued multiplication must be considered as the closest form of inbreeding.

Whenever opportunity has offered the ability of the various species to reproduce pathenogenetically has been studied, and many interesting and some peculiar facts have been discovered which, it is hoped, will serve as the subject for a technical paper later on. This power appears to be confined to the Hymenoptera, however, and the tachinid parasites, like their hosts, are rarely or perhaps never parthenogenetic.

When continued existence of an insect in a new country is dependent upon the mating of isolated females it is at once evident that it is also dependent upon the rapidity of dispersion and upon the number of individuals which are comprised in the original colony. One of the most constant sources of surprise is in the rapidity with which the parasites disperse. One, Monodontomerus, has undoubtedly extended its range for more than 200 miles in the course of the five years which have elapsed since its liberation, and there is no reason to believe that others among the introduced species will not disperse at an equal rate, once they are sufficiently well established.

But Monodontomerus is eminently well fitted for dispersion, and its case is altogether different from that of a tachinid which is dependent upon sexual reproduction for the continuation of the species. A few hundred individuals, spreading rapidly toward all points of the compass, soon become widely scattered, and it is, and will remain for a long time, a question just how rare an insect may be and each individual still be able to find its mate. That the individuals of the first colonies of many of the tachinid parasites scattered so widely as to make the mating of the next generation purely a matter of chance and of rare occurrence is now accepted as well within the bounds of probability.

The first serious doubts as to the wisdom of the policy of the small colony were felt in 1907, and beginning with June of that year larger colonies were planted in the instance of every species than had been the practice up to that time. In the fall of 1908 the recovery of Monodontomerus over a wide territory lent strength to these half-formed convictions, and when, during 1909 and 1910, one after another of the various parasites were recovered under circumstances which were in most cases essentially similar, all doubts vanished as to the wisdom of the course finally adopted. At the present time there is no more inexorable rule governing the conduct of the laboratory than that establishment of a newly introduced parasite is first to be secured, while dispersion, if later developments prove that it can be artificially aided, comes as a wholly secondary consideration. For the most part, however, dispersion may be left to take care of itself.

An even larger appreciation of the necessity for strong colonies has been reached during the present winter (1910–11), coincidently with the results of the scouting work for Monodontomerus and Pteromalus in the brown-tail moth hibernating nests. (See maps, Pls. XXII, XXV.) The details will not be given in this immediate connection, but they will be found later on in connection with the discussion of these species. It is sufficient at this time to say that the circumstances under which the Pteromalus was recovered after the lapse of two years following its colonization were such as to cast doubts upon the conclusions which had been tentatively reached concerning the inability of certain other species to exist in America, and their possible significance had something to do with the decision to continue the work of parasite importation along wholly different lines in 1911. It may be, after all, that 40,000 individuals of *Apanteles fulvipes* are not enough to make one good colony.

DISEASE AS A FACTOR IN THE NATURAL CONTROL OF THE GIPSY MOTH AND THE BROWN-TAIL MOTH.

In continuing this work consideration must be given to the probable effect which the prevalence of disease would possibly have in the reduction of the gipsy moth and the brown-tail moth to the ranks of ordinary rather than of extraordinary pests. In America, as is generally well known, the brown-tail moth is annually destroyed to an extraordinary extent as the result of an epidemic and specific fungous disease, while the gipsy moth is frequently subjected to very material diminution of numbers through a much less well known affection popularly known as "the wilt," apparently similar to the silkworm disease "flacherie."

In more respects than one the prevalence of these diseases has been inimical to the prosecution of the parasite work. In the beginning, when it was expected that the parasites would remain in the immediate vicinity of the localities where they were first given their freedom, great pains were taken to provide colony sites in situations where the caterpillars were not only common but where there was reason to believe that they would remain healthy for at least one or two years. This was an exceedingly difficult matter, and one which was the cause of more troubles, doubts, and fears during 1907 and 1908 than almost any other phase of the parasite work.

With the final recognition of the great superiority of the large colony, which came about through a better knowledge of the powers of rapid dispersion possessed by the parasites, this seeming obstacle to success wholly disappeared, except in the case of such parasites as Anastatus, which actually did remain in the spot where they were placed, and which could not travel beyond a certain limited radius, no matter how great the necessity.

At the present time the association with the parasite problem of the otherwise wholly separate question of disease as a factor in the control of the gipsy moth and the brown-tail moth is entirely confined to speculations as to the probable future of these pests, provided their control is left to disease alone. If, as is conceivable, effective control is exerted through disease, further importation of parasites is rendered not only needless but wholly undesirable. If, on the contrary, such control is likely to be inefficient, from an economic standpoint, every effort should be exerted to make the parasite work a success. In other words, the decision as to the adoption of a policy for the future conduct of the activities of the laboratory depended very largely upon whether or not disease seemed likely to become effective in the case of the more important of the two pests. The fact that the present plans provide for the continuation of the work along even more energetic lines than in the past indicates sufficiently well the character of the decision finally reached.

This is not the place for, nor are the writers prepared to enter into, a discussion of the caterpillar diseases of the gipsy moth and the brown-tail moth, but it is perhaps not out of place to recount some of the incidents which have been taken into consideration in the present instance. In this, as in many others similar in character, the brown-tail moth has largely been ignored, owing to its being generally considered as the lesser pest of the two. As frequently before, the work upon the brown-tail moth parasites, although pursued quite as actively as that upon the parasites of the gipsy moth; was relegated to a secondary position.

Very little has been published concerning the gipsy-moth caterpillar disease previously to 1907, when the junior author first had opportunity to familiarize himself with the situation at first hand. It was to him a novelty when, early in the summer of that year, wholesale destruction of the half-grown caterpillars was first noticed in numerous localities before they had succeeded in effecting the complete defoliation of the trees and shrubs upon which they were feeding. In all its essential characters the disease was similar to that which had swept over the army of tent caterpillars which were defoliating the apple and cherry trees in southern New Hampshire in 1898, as recounted in the bulletin upon the parasites of that insect, published as No. 5 in the Technical Series of the New Hampshire Agricultural Experiment Station. It was believed of this disease, at the time when these investigations were being conducted, that it was infectious, since the inhabitants of whole nests would all perish simultaneously. At the same time, its infectious or contagious nature was not established.

On the supposition that the disease of the tent caterpillar was infectious, and that that of the gipsy-moth caterpillars was similar in character, it looked for a time as though the parasite work was destined to an untimely end through the destruction of the gipsy-moth caterpillars before the parasites had opportunity to establish themselves and increase to the point of efficacy. Neither was there anything observed during the summer of 1907 to render this supposition untenable, except (and from an economic standpoint the exception was one of grave importance) the fact that, taking the infested area as a whole, there was a tremendous increase in the number of egg masses of the gipsy moth in the fall of 1907 over the number which had been present the previous spring.

There did not seem to be any particular reason why the disease should not increase in effectiveness as time passed on, however, and when in the spring of 1908 myriads of caterpillars in the first stage were found "wilting" in the forests in Melrose, and when just a little

later practically every caterpillar was destroyed in one particular
locality which had been selected as a good place for the very first
colony of *Apanteles fulvipes*, there seemed to be reason to hope for
speedy relief through disease. About this time these hopes were
rudely shattered by the failure of several attempts to demonstrate
the infectious or contagious nature of the disease through experiments carried on at the laboratory. Its noncontagious nature was
further indicated by the fact that it did not spread across a narrow
roadway near the laboratory, one side of which was swarming with
dying caterpillars, while the other was peopled with an alarming
but not destructive abundance of healthy ones. It appeared, after
all, as though the views often expressed by Mr. A. H. Kirkland (at
that time superintendent of the moth work in Massachusetts) to the
effect that the disease was nothing more than the natural concomi-
tant of overpopulation, and that an insufficient or unsuitable food
supply was the true explanation of its prevalence, were right. That
it was not to be depended upon for immediate results was certain
when, at the close of 1908, a further alarming and apparently an
unaffected increase in the distribution and abundance of the gipsy
moth in Massachusetts and New Hampshire was found to have taken
place wherever conditions were not such as to render destruction
through disease the only thing which saved the gipsy moth from
extinction through starvation, or where active hand suppression work
had not been undertaken.

In 1909, and again in 1910, observations upon the progress of the
disease were made almost daily throughout the caterpillar season.
It was no longer looked upon as a serious obstacle to the success of
the parasite work, except as it interfered (as it frequently did most
seriously) with the work of colonizing Anastatus, and to a lesser
extent Calosoma. It was also, as ever, the cause of serious trouble
whenever attempts were made to feed caterpillars in the laboratory
in confinement.

The disease acquired new interest, however, through the gradual
accumulation of evidence tending to support the theory that it was
either transmissible from one generation to another through the egg
or that a tendency to contract it was thus transmitted.

Recognition of this characteristic through cumulative evidence
resulting from more occasional or specific observations than it would
be possible to review at this time, was accompanied by the almost
equally apparent fact that the disease was becoming slightly more
effective at a somewhat earlier stage in the progress of a colony of
the gipsy moth following its establishment in a new locality. It
was found, for example, in New Hampshire in colonies which had
barely reached the stripping stage. A few years before the caterpillars composing such colonies would naturally have migrated from

the stripped trees to others in the vicinity, and it is an unmistakable fact that such migrations, which have several times been mentioned in the earlier reports of the State superintendent of moth work, are now decidedly less frequent, even without taking into consideration the greater territory throughout which the moth is now present in destructive abundance. Although the junior author has personally visited large numbers of outlying colonies of the moth in the course of 1908, 1909, and 1910, he has yet to see one in which the disease had not appeared coincidently with the development of the colony to the stripping stage, if not slightly in advance of that time.

It is probably safe to say that such conditions as are described in the first annual report of the superintendent of moth work as prevailing over a large territory in the old infested section during 1904 and 1905, will probably not immediately recur in the history of the gipsy moth in eastern Massachusetts. That something approaching this may result in parts of New Hampshire is well within the bounds of probability, and that the conditions will be very bad in that State during the course of the next few years as well as in some of the towns in Massachusetts may be accepted as most probable. Whatever may be the condition presented by the older infested sections in eastern Massachusetts five or ten years from now, the only hope of preventing an ever-increasing wave of destruction from spreading over western Massachusetts, across New Hampshire and Vermont, and over the border into the State of New York, seems to lie, as always, in an increasing expenditure for hand suppression or in the success of the experiment in parasite introduction. Through the methods now in operation it is probable that the pest will very largely be prevented from making long "jumps," which would otherwise have been of frequent occurrence, but the slower and more steady natural spread, through the agency of wind, and probably, when the headwaters of the Connecticut, Hudson, and Ohio are reached, by water, must be considered in every attempt to discount the future. It was taken into consideration when the future of the parasite work was decided upon.

In the course of the studies of the parasites and parasitism of native insects which have been undertaken in connection with those of the parasites of the gipsy moth and the brown-tail moth, no less than three species have been encountered which are controlled to some extent by a disease which bears a very close superficial resemblance to the "wilt" of the gipsy moth. These are the white-marked tussock moth, the tent caterpillar, and the "pine tussock moth."

The white-marked tussock moth (*Hemerocampa leucostigma* S. & A.) is well known as a defoliating pest in cities, and has been so abundant as at times to become a rival of the gipsy moth in its destructive capacities in certain of the larger cities in southeastern New England.

It is very subject to a "wilt" disease, and no colony has been observed which has reached such proportions as to threaten the defoliation of street trees in which the disease has not appeared. In one instance the disease was so prevalent as to destroy practically all of the caterpillars, and, as in the case of the gipsy moth, the scattering caterpillars which hatched from the eggs deposited by the few survivors were seriously affected the following year, notwithstanding the presence of an abundance of food. Furthermore, caterpillars hatching from eggs collected in this and similar colonies removed to the country where the few native caterpillars to be found have always been remarkably healthy. perished through the "wilt" exactly as though they had hatched in the city.

Nevertheless, the white-marked tussock moth has been for long, is now, and probably will remain the worst defoliating insect enemy of such trees as the horse chestnut, maple, sycamore, etc., in strictly urban communities in the Eastern States generally. It does not appear unreasonable to suppose that the gipsy moth may similarly continue to be a pest in spite of the disease. As a matter of fact, every observation which has been made upon either the fungous disease of the brown-tail moth caterpillars, the wilt disease of the gipsy-moth caterpillars, or diseases of other defoliating caterpillars, such as that of the white-marked tussock moth, the tent caterpillar, and the " pine tussock moth," has tended to confirm the conclusion that such insect epidemics rarely play more than the one rôle in the economy of nature. They do not *prevent* an insect from increasing to an extent which renders it a pest, but they may, and frequently do, render very efficient service in effecting a wholesale reduction in the abundance of such insects when other agencies fail. When the insect in question is ordinarily controlled by parasites, as appears to be the case with the white-marked tussock moth, the "pine tussock moth," etc., it is probable that a long time will elapse before it will again encounter the combination of favorable circumstances which make possible abnormal increase.

When, as with the white-marked tussock moth in cities, the tent caterpillar in southeastern New England, or the brown-tail moth and gipsy moth in America, adequate control by parasites is lacking, reduction in numbers through disease is not likely to result in more than temporary relief. The more complete the destruction wrought by the insect the longer the period which must necessarily elapse before it again reaches the state of destructive abundance and, looking at it from this standpoint, it is not unlikely that the gipsy moth is much more abundant at the present time than it would have been had it not been for the prevalence of disease. There are, each year, an abundance of localities where the destruction of a great majority of the caterpillars by disease has been the only thing which has saved the whole race from complete extinction in that locality through consumption of the entire supply of available food before growth was completed. Under such circumstances the disease has been of positive benefit to the gipsy moth, rather than the reverse.

STUDIES IN THE PARASITISM OF NATIVE INSECTS.

Among a considerable number and variety of native insects studied at the laboratory which resemble the gipsy moth in habit, or which are more or less closely allied to it in their natural affinities, no two have been found in the economy of which parasitism has played an exactly similar rôle. There is this to be said, however, that only one amongst them, and this the tent caterpillar, appears to be ineffectually controlled by parasitism, except under unusual circumstances.

Several very beautiful examples of control by parasites have been encountered in the course of these investigations, and, comparatively speaking, the exceptional instances in which parasites lose control through one reason or another are exceedingly rare. Such instances are usually, if not inevitably, accompanied by a conspicuous outbreak of the insect in question.

The destructiveness of the white-marked tussock moth in cities is apparently due to the fact that it is peculiarly adapted to life under an urban environment. It is an arboreal insect, and one which is prevented through the winglessness of its females from dispersing over the country as the brown-tail moth, for example, would do under similar circumstances. Its parasites, on the other hand, are not always fitted for a peculiarly arboreal existence. Many of them are partially terrestrial, and in addition they are strong upon the wing.

Most of the introduced parasites of the gipsy moth and brown-tail moth which are known to have established themselves in America are known to be dispersing at a rapid rate. Several of them have been reared as parasites of the white-marked tussock moth from caterpillars or pupæ collected under urban surroundings, and since we have positive proof of their wandering habits there is every reason to believe that the native parasites of the tussock moth possess similar characteristics. That is to say, instead of staying within the limited area in which their host abounds, they are likely to scatter throughout the country immediately following the completion of their transformations. They are neither fitted for continued existence in the city to the degree which is characteristic of their host, nor are they compelled, like it, to accept it when they find themselves city-born through chance ancestral wanderings.

Every season's observations (and for four consecutive years the tussock moth has received more than a modicum of attention) has added arguments to support the contention that the white-marked tussock moth is controlled in the country through parasitism and not by birds or other predators. In any event it is controlled to such an extent as to have made a study in parasitism under strictly rural conditions very difficult, except when eggs or caterpillars have been artificially colonized for the purpose.

The outbreak of the Heterocampa in New Hampshire and Maine is another exceptional instance. In many respects the results of the relatively limited study given to this insect were the most remarkable of any, since there was offered what, to the writers, was the unique spectacle of unrestricted increase being checked through starvation without the intervention of disease. Notwithstanding the fact that the abundance of this insect was so great as to bring about complete defoliation of its favored trees over a very wide area, not a sign of disease was observed in the fall of 1909 in forests where millions of caterpillars were literally starving to death. The final, thoroughly effective, and miraculously complete subjugation of the outbreak, which resulted in the insect dropping from the abundance above mentioned to what is perhaps less than its normal numbers in the course of a single year, has already been described in a paper which appeared in a recent number of the Journal of Economic Entomology. There is every reason to believe that it was entirely the result of insect enemies, including both parasites and a predaceous beetle, which latter, through its ability to increase abnormally at the direct expense of that particular insect, played a rôle exactly comparable to that of the true facultative parasites. Such another outbreak of Heterocampa has never been known, and it is probable that it will be very many years before a combination of conditions makes its repetition possible. It is altogether probable that during this period the parasites will remain in full control.

A third exceptional instance is the present outbreak of the "pine tussock moth" in Wisconsin. This interesting and, as it has proved itself, potentially destructive insect is decidedly rare in Massachusetts, but notwithstanding its scarcity a sufficient number was collected in 1908 and 1909 to make possible a study of its parasites. Parasitism to an extent rarely exceeded amongst leaf-feeding Lepidoptera was found to be existent, and it is safe to say that had it not been for its parasites the host would have increased at least fivefold or sixfold in 1909 over the numbers which were present in 1908. Such a rate of increase, if continued, would have placed it among the ranks of destructive insects in a very few years, and it appears that something of this sort actually occurred in northern Wisconsin. There, some years ago, it reached a stage of abundance which resulted in partial or complete defoliation of pine throughout a considerable territory and, as was expected, a relatively small percentage of the caterpillars and pupæ were found to be parasitized. Existing parasitism in 1910 was not sufficiently effective to prevent its increase to a point which would have made complete defoliation of its food plant, and consequently its death through starvation, an accomplished fact had its abundance not been reduced through the prevalence of a disease superficially similar to the "wilt" of the gipsy moth.

The fall webworm is generally a common and abundant insect in New England, but rarely as common or abundant as it frequently becomes in the South. An elaborate study of its parasites and the effect which parasitism apparently played in effecting its control was made in the fall of 1910, with interesting results. It was found that the prevailing percentage of parasitism was sufficient to offset an increase of no less than fourfold annually, and even at that there is reason to believe that our results err on the side of conservatism. The elimination of these parasites for a very short period of years would undoubtedly be followed by an increase of the host comparable to that of the gipsy moth.

The one insect studied at the laboratory which appears habitually and under its normal environment to become so unduly abundant as to invite destruction through disease at regular intervals is the tent caterpillar.

In the report upon its parasites,[1] it was contended that they played a part subservient to that taken by the disease, and this statement drew forth some criticism at the time of its publication. It is a satisfaction to note that the original contention appears to be upheld by the results of studies conducted at the gipsy-moth parasite laboratory. These results seem to justify the further contention that the present status of the tent caterpillar is, in a way, prophetic of that which would result were the gipsy moth to be left to the control of its disease.

At frequent but irregular intervals the tent caterpillar increases to such an extent as to become a pest, and unless artificially checked it defoliates fruit trees in southern New England. That it never reaches the destructiveness characteristic of the gipsy-moth invasion is seemingly due to difference in habit. As is well known, the gipsy-moth caterpillar is almost an omnivorous feeder and the female moth is incapable of flight. Its eggs are deposited indiscriminately in every conceivable place to which a caterpillar or moth can gain access. The adult of the tent caterpillar is in no way restricted to the immediate vicinity of the locality where it chose to pupate as a caterpillar, but, instead, uses what really amounts to an unwise amount of discretion in its selection of a place for oviposition. Cherry first and then apple is selected in preference to all other food plants, and with the exception of a limited number of other rosaceous trees and shrubs, its eggs are almost never found elsewhere. · As a result, when it is at all abundant its caterpillars, which have not the

---

[1] Technical Bulletin 5, New Hampshire Agricultural Experiment Station.

wandering characteristics of those of the gipsy moth, but rather the opposite, find themselves crowded in excessive numbers upon a limited variety of shrubs and trees; complete defoliation of these comparatively few host plants quickly follows, and weather conditions being favorable to the development of disease, wholesale destruction is all that intervenes between an unnatural migration or starvation. Such reduction is followed by a period of years during which the parasites check but do not overcome the tendency to increase, and it is only a little while before the process is repeated.

There were, in certain localities in eastern Massachusetts in the summer of 1910, continuous strips of roadside grown up to a variety of trees and shrubs, the most of which were defoliated by tent caterpillars, all of which had hatched from eggs deposited upon the occasional wild-cherry tree which was present. Several such strips were visited at about the time when the caterpillars elsewhere were beginning to pupate, and not a single living caterpillar or pupa could be found amongst the thousands of dead and decomposing remains of the victims of overpopulation. These were but a repetition of conditions as observed a few miles north in New Hampshire 12 years ago. How frequently similar conditions occurred during the intervening period is not known.

In addition to those species mentioned in the preceding pages, quite a number of other leaf-feeding Lepidoptera have been more or less casually studied in a less comprehensive but at the same time a careful manner.

PARASITISM AS A FACTOR IN INSECT CONTROL.

In reviewing the results of these studies, the fact is strikingly evident that parasitism plays a very different part in the economy of different hosts. Some habitually support a parasitic fauna both abundant and varied, while others are subjected to attack by only a limited number of parasites, the most abundant of which is relatively uncommon. No two of the lepidopterous hosts studied, unless they chanced to be congeneric and practically identical in habit and life history, were found to be victimized by exactly the same species of parasites. Neither are the same species apt to occur in connection with the same host in the same relative abundance, one to another, year after year in the same locality, nor in two different localities the same year.

At the same time there are certain features in the parasitism of each species which are common to each of the others, whether these be arctiid, liparid, lasiocampid, tortricid, saturniid, or tineid, one of the most common of which is that each host supports a variety of parasites, oftentimes differing among themselves to a remarkable degree in habit, natural affinities, and methods of attack. Departures from this rule have not been encountered among the defoliating Lepidoptera as yet, and while exceptions will probably be found to exist, they will doubtless remain exceptions in proof of the rule. From this the rather obvious conclusion has been drawn, that to be effective in the case of an insect like the gipsy moth or the brown-tail moth, parasitic control must come about through a variety of parasites, working together harmoniously, rather than through one specific parasite, as is known to be the case with certain less specialized insects, having a less well-defined seasonal history. To speak still more plainly, it is believed that the successful conclusion of the experiment in parasite introduction now under consideration depends upon whether or not we shall be able to import and establish in America each of the component parts of an effective "sequence" of parasites. This belief is further supported by the undoubted fact, that in every locality from which parasite material has been received abroad, both the gipsy moth and the brown-tail moth are subjected to attack by such a group or sequence of parasites, of which the component species differ more or less radically in habit and in their manner of attack.

In the case of the gipsy moth and the brown-tail moth abroad, as well as in that of nearly every species of leaf-feeding Lepidoptera studied in America, there are included among the parasites species which attack the eggs, the caterpillars, large and small, and the prepupæ and pupæ, respectively. Frequently, but not always, there are predatory enemies, which, through their ability to increase at the immediate expense of the insect upon which they prey, whenever this insect becomes sufficiently abundant to invite such increase, are to be considered as ranking with the true facultative parasites when economically considered.

It is, therefore, our aim to secure the firm establishment in America of a sequence of the egg, the caterpillar, and the pupal parasites of the gipsy moth and brown-tail moth as they are found to exist abroad, and until this is either done or proved to be impossible of accomplishment through causes over which we have no control, we can neither give up the fight nor expect to bring it to a successful conclusion.

It was stated a page or two back that some species of insects support a parasitic fauna both numerous and varied, while others are subjected to attack by only a limited number of parasites, none of which can be considered as common. Notwithstanding the fact that somewhat similar differences are discernible between the parasitic fauna of the same insect at different times or under different environment, it is perfectly safe to elaborate the original statement still further and to say that some species are habitually subjected to a much heavier parasitism than others. Unquestionably the average percentage of parasitism of the fall webworm in eastern Massachusetts, taken over a sufficiently long series of years to make a fair average possible, is the same as the average would be over another similar series of years in the same general region. This could be said of the larvæ of any other insect as well as of that of the fall webworm, but the average percentage of parasitism in another would most likely not be the same, but might be very much larger or very much smaller. To put it dogmatically, each species of insect in a country where the conditions are settled is subjected to a certain fixed average percentage of parasitism, which, in the vast majority of instances and in connection with numerous other controlling agencies, results in the maintenance of a perfect balance. The insect neither increases to such abundance as to be affected by disease or checked from further multiplication through lack of food, nor does it become extinct, but throughout maintains a degree of abundance in relation to other species existing in the same vicinity, which, when averaged for a long series of years, is constant.

In order that this balance may exist it is necessary that among the factors which work together in restricting the multiplication of the species there shall be at least one, if not more, which is what is here termed facultative (for want of a better name), and which, by exerting a restraining influence which is relatively more effective when other conditions favor undue increase, serves to prevent it. There are a very large number and a great variety of factors of more or less importance in effecting the control of defoliating caterpillars, and to attempt to catalogue them would be futile, but however closely they may be scrutinized very few will be found to fall into the class with parasitism, which in the majority of instances, though not in all, is truly "facultative."

A very large proportion of the controlling agencies, such as the destruction wrought by storm, low or high temperature, or other climatic conditions, is to be classed as catastrophic, since they are wholly independent in their activities upon whether the insect which incidentally suffers is rare or abundant. The storm which destroys 10 caterpillars out of 50 which chance to be upon a tree would doubtless have destroyed 20 had there been 100 present, or 100 had there been 500 present. The average percentage of destruction remains the same, no matter how abundant or how near to extinction the insect may have become.

Destruction through certain other agencies, notably by birds and other predators, works in a radically different manner. These predators are not directly affected by the abundance or scarcity of any single item in their varied menu. Like all other creatures they are forced to maintain a relatively constant abundance among the

other forms of animal and plant life, and since their abundance from
year to year is not influenced by the abundance or scarcity of any
particular species of insect among the many upon which 'they prey
they can not be ranked as elements in the facultative control of such
species.   On the contrary, it may be considered that they average
to destroy a certain gross number of individuals each year, and
since this destruction is either constant, or, if variable, is not correlated in its variations to the fluctuations in abundance of the insect
preyed upon, it would most probably represent a heavier percentage
when that insect was scarce than when it was common.   In other
words, they work in a manner which is the opposite of "facultative"
as here understood.

In making the above statement the fact is not for a moment lost
to sight that birds which feed with equal freedom upon a variety of
insects will destroy a greater gross number of that species which
chances to be the most abundant, but with the very few apparent
exceptions of those birds which kill for the mere sake of killing they
will only destroy a certain maximum number all told.   A little
reflection will make it plain that the percentage destroyed will
never become greater, much if any, as the insect becomes more common, and, moreover, that after a certain limit in abundance is passed
this percentage will grow rapidly less.   A natural balance can only
be maintained through the operation of facultative agencies which
effect the destruction of a greater proportionate number of individuals as the insect in question increases in abundance.

Of these facultative agencies parasitism appears to be the most
subtle in its action.   Disease, whether brought about by some
specific organism, as with the brown-tail moth, or through insufficient or unsuitable food supply without the intervention of any
specific organism, as appears at the present time to be the case with
the gipsy moth, does not as a rule become effective until the insect
has increased to far beyond its average abundance.   There are
exceptions to this rule, or appear to be, but comparatively only a
very few have come to our immediate attention.   Finally, famine
and starvation must be considered as the most radical means at
nature's disposal, whereby insects, like the defoliating Lepidoptera,
are finally brought into renewed subjugation.

With insects like the gipsy moth and the brown-tail moth disease
does not appear to become a factor until a degree of abundance has
been reached which makes the insect in question, *ipso facto*, a pest.
Whether in the future methods will be devised for artificially rendering such diseases more quickly effective, remains to be determined
through actual experimental work continued over a considerable
number of years.

RATE OF INCREASE OF GIPSY MOTH.

In effect, the proposition is here submitted as a basis for further discussion that only through parasites and predators, the numerical increase of which is directly affected by the numerical increase of the insect upon which they prey, is that insect to be brought under complete natural control, except in the relatively rare instances in which destruction through disease is not dependent upon super-abundance.

The present experiment in parasite introduction was undertaken and has been conducted on the assumption that there existed in America all of the various elements necessary to bring about the complete control of the gipsy moth and the brown-tail moth, except their respective parasites. Believing that this stand was correctly taken, much time has been devoted to a consideration of the extent to which these pests are already controlled through natural agencies already in operation. The fact that both insects have increased steadily and rapidly in every locality in which they have become established and where adequate suppressive measures have not been undertaken, until they have reached a stage of abundance far in excess of that which prevails in most countries abroad, renders superfluous further comment upon the present ineffectiveness of these agencies. The difference between the rate at which they have averaged to increase in localities where they have become established and their potential rate of increase as indicated by the number of eggs deposited by the average female should indicate very accurately the efficiency of such agencies, and the difference between the actual rate of increase and no increase similarly indicates the amount of additional control which must be exerted by the parasites if their numbers are to be kept at an innocuous minimum.

THE RATE OF INCREASE OF THE GIPSY MOTH IN NEW ENGLAND.

The potential rate of increase as determined by the number of eggs deposited by the average female of the gipsy moth varies considerably under different circumstances, and affords an interesting example of a phase of facultative control not touched upon in the last chapter. When the exhaustive studies into its life and habits were conducted under the general supervision of the Massachusetts State Board of Agriculture during the final decade of the last century, it was determined that the number was between 450 and 600.

In the opinion of some, the fecundity of the gipsy moth has distinctly decreased during the 14 years which have elapsed since the publication of the report in which these figures were given, and in order to determine the point a considerable number of egg masses was collected during the winter of 1908–9 and the eggs carefully counted. It was found that in those from the older infested territory or from outlying colonies where the moth was particularly abundant, the number of eggs to a mass averaged considerably less than 300. In egg masses from outlying districts where the infestation was new, and where the moth had never reached its maximum abundance, the average in a few masses counted was slightly in excess of 500. The number is, however, very variable, and the character of food and the meteorological conditions during the feeding period of the caterpillars are doubtless important features. Hot weather during June forces the development of the caterpillars and they do not become large. Small moths deposit fewer eggs rather than smaller eggs. It is possible that there is actually a decrease in the fecundity of moths brought about by our short and ardent summers, but for the present it is not proved, and it is believed that whenever abundance of the insect is sufficiently reduced the original rate of multiplication will prevail. The point is one well worthy of further investigation, but for the present the potential rate of increase, provided no controlling factors whatever are operative, will be considered as 250-fold annually.

The best information available as to the rate of increase of the gipsy moth actually prevailing in Massachusetts is contained in the report entitled "The Gypsy Moth," by Mr. Edward H. Forbush and Dr. C. H. Fernald, which was published under the direction of the board of agriculture. These authorities, in their discussion of the matter, say as follows:

The study of the increase and dissemination of the gipsy moth in Massachusetts is most interesting. Perhaps there never has been a case where the origin and advance of an insect could be more readily traced. As the moth appears to be confined as yet to a comparatively small area, and as the region has been examined more or less thoroughly for five successive years, the opportunities offered for the study of the multiplication and distribution of the insect have been unequaled.

When it is considered that the number of eggs deposited by the female averages from 450 to 600, that 1,000 caterpillars have been seen to hatch from a single egg cluster, and that at least one egg cluster has been found containing over 1,400 eggs, there can be no doubt that the reproductive powers of the moth are enormous. Mr. A. H. Kirkland has made calculations which show that in eight years the unrestricted increase of a single pair of gipsy moths would be sufficient to devour all vegetation in the United States. This, of course, could never occur in nature, and is mentioned here merely to give an idea of the reproductive capacity of the insect.

It seems remarkable at first sight that an insect of such reproductive powers, which had been in existence in the State for 20 years, unrestrained by any organized effort on the part of man, did not spread over a greater territory than 30 townships, or about 220 square miles. Some of the causes which at first checked its increase and limited its diffusion in Medford have already been set forth. Most of the checks which at first served to prevent the excessive multiplication of the gipsy moth in Medford operate effectively to-day wherever the species is isolated. True, it has now become acclimated But any small isolated moth colony still suffers greatly from the attacks of its natural enemies and from the struggle with other adverse influences which encompass it. The normal rate of increase in such isolated colonies as are found to-day in the outer towns of the infested district seems to be small. The annual increase can be readily ascertained by noting the relative number of egg clusters laid in successive years, the unhatched or latest clusters being easily distinguished from the hatched or "old" clusters, and the age of these latter, whether one, two, three, or more years, being indicated by their state of preservation. The ratio of the average annual increase of 10 such colonies was found to be 6.42; that is, six or seven egg clusters on an average may be found in the second season to one of the first season.

If the number of eggs deposited by the average female moth be set at 500, and if the sexes of her progeny are equally divided, a potential increase of 250-fold for each annual generation is provided for. Under complete control only one pair of moths would average to be produced from each mass of eggs deposited, and since each egg represents an individual embryo, all but 2 of each 500 must fail to reach full maturity. Reduced to percentage this is equivalent to the survival of 0.4 per cent and the destruction of 99.6 per cent of the gipsy moths in one stage or another every year. Since the total number of gipsy moths in any locality can not possibly be computed, the only method by which mortality through any cause may be expressed is on this basis.

It will surprise many who have not given the matter consideration to learn what an extraordinary *apparent* mortality it requires to offset a potential increase of 250-fold. The gipsy-moth caterpillars molt five or six times after they hatch and before they change to pupæ, making the number of caterpillar stages six or seven. If through natural controlling agencies 50 per cent of the young caterpillars were destroyed in the first stage before they had molted, and this was followed by similar destruction of another 50 per cent in the second stage, and so on through the third, fourth, fifth, sixth, and seventh stages, respectively, and in addition 25 per cent of the pupæ and 25 per cent of the adults before depositing their eggs were similarly destroyed, it would still permit of a slight annual increase.

The following table (if the incongruity of fractions as applied to insects may be overlooked) indicates the number of survivors of each stage resulting from the hatching of a mass of 500 eggs:

	Stage.	Number.	Loss.	Number remaining.	Potential increase.
			Per cent.		
Eggs		500	0	500	250 fold.
Caterpillars	First	500	50	250	125 fold.
Do	Second	250	50	125	62 fold.
Do	Third	125	50	62	31 fold.
Do	Fourth	62	50	31	15.5 fold.
Do	Fifth	31	50	15.5	7.75 fold.
Do	Sixth	15.5	50	7.75	3.875 fold.
Do	Seventh	7.75	50	3 875	2 906 fold.
Pupæ		3.875	25	2 906	2.179 fold
Adults		2.906	25	2 179	1.634 fold.

To give another illustration: The life of the gipsy-moth caterpillar is approximately seven weeks. If beginning on the first day after hatching and on every day thereafter during this period a decrease in numbers of 10 per cent should be brought about through natural causes, there would still be enough survivors to permit of a substantial increase in the abundance of the insect.

Twelve gipsy moths (6 pairs) from each egg mass would be sufficient to provide for a sixfold annual increase. If reference be made to the preceding table, it will be seen that if all destruction ceased after the caterpillars had reached the fourth stage, the survivors would permit of slightly in excess of sixfold increase; that is, the mortality during the first to the fourth stage, inclusive, with a part of that which resulted during the fifth stage, would be sufficient to account for all of the control at present exerted by natural agencies in New England, and this gives, at the same time, an idea as to the amount of additional control which the parasites must accomplish if they are to become effective.

The conditions under which the gipsy moth was studied at the time when the material for the report just quoted was accumulated were, for the most part, abnormal. In only relatively few localities was it allowed to increase undeterred, and there were relatively very few examples of unrestricted increase to the point when defoliation resulted. This, in part, explains what seems to be an element of indecision concerning the character of conditions which favored more rapid increase of the moth, as quoted below, from the same source.

CONDITIONS FAVORING RAPID INCREASE.

When any colony under average normal conditions has grown to a considerable size and then received an added impetus from exceptionally favorable conditions, its power of multiplication and its expansive energy are greatly augmented, and its annual increase arises above all calculations [1] Under such influences hundreds of egg clusters will appear in the fall where few were to be seen in the spring, and thousands are found where scores only were known before. It is probable that the season of 1889 was particularly favorable for the moth's increase. The season of 1894 and that of 1895 appear also to have furnished conditions especially favorable for an abnormal multiplication of the insect.

The operation of the causes of these sudden outbreaks is not understood. It is evident, however, that the warm, pleasant spring weather of the past two years (1894 and 1895) hastened the development of the caterpillars, thereby shortening their term of life. The length of life of the caterpillars varies from six to twelve weeks. During cold, rainy weather the caterpillars eat little and grow slowly. During warm, dry weather they consume much more food and grow with great rapidity. In the unusually warm spring and early summer of 1895 many of the caterpillars molted a less number of times than usual, and their length of life did not exceed six or seven weeks. Under these conditions they proved more quickly injurious to foliage than in a more normal season, and were more completely destructive within any given area in which their numbers were great. And they were not so long exposed to the attacks of their

---

[1] The increase of these large colonies seems to be limited only by the supply of food. Whenever food becomes scarce many of the moths are less prolific. The larvæ which do not find sufficient food either die or develop early, and the female moths lay fewer eggs than those which transform from well-nourished caterpillars.

enemies. While it may be true that the parasitic enemies of the moth will also develop rapidly under conditions that hasten the growth of their host, birds and other vertebrate enemies will secure fewer of the moths in 6 or 7 weeks than in 10 or 12. It is believed that dry weather is unfavorable for vegetable parasites of insects, but to what extent the caterpillars are affected by them in a humid season it is impossible to say.

The past two years have been "cankerworm years" in the infested region. Many of the birds which habitually feed on the caterpillars of the gipsy moth have been largely occupied during May and the early part of June in catching cankerworms, which they seem to prefer, turning their attention to the gipsy-moth caterpillars in the latter part of June and July, when the cankerworms have disappeared. The birds, therefore, have not been as useful in checking the increase of the gipsy moth as in years when the cankerworms were less numerous.

A few of the restraining influences which have been less active than usual during the past two years have been mentioned, and possibly many others have escaped observation, but those given serve in a measure to explain the unusual increase of the moth. It is during such seasons that its destructiveness is most apparent. It is then that the groves and forests are stripped of their leaves, and whole rows of trees in orchards and along highways appear to have been stripped in a single night.

The conditions as described seem to be comparable to those prevailing at the present time, and at the same time to be inadequately explained. Repeatedly personal observations have been made which indicate beyond the shadow of a doubt that under certain circumstances the gipsy moth has increased at a rate very far in excess of sixfold annually at the present time. Counts of old egg masses as compared with those newly laid, in several localities, in the spring of 1908 and each spring subsequently, have shown positively that an increase of at least twentyfold was not uncommon. In fact, unless an unduly large number of old egg masses was concealed, it could be said with equal certainty that increase sometimes amounted to fiftyfold in the course of a single generation. The arguments presented by Forbush and Fernald, who evidently observed something very similar, and who were inclined to credit it to seasonal or climatic conditions (in part at least), do not stand, in view of the fact that the rate of increase differs extraordinarily in localities nearly adjacent to where the conditions are practically identical, saving only the varying abundance of the moth; this latter, it may be noted, has in each instance corresponded roughly and in direct ratio to the rate of increase. The fact was not considered to be of more than coincidental interest at first, but later, when an attempt was made to classify according to their manner of operation, the various factors which were already responsible for the partial control of the gipsy moth in New England, the correlation between relative abundance and rate of increase recurred and seemed to afford excellent support to the contention which has been made as to the part which birds and most other predators play in bringing this about.

Without attempting to go into the details of rather elaborate calculations, which were made for the purpose of bringing out this point more graphically, attention is merely called to the three divisions into which the elements operative in the natural control of any insect naturally fall as they were outlined in the preceding section. These are, first, the catastrophic (storms, etc.), which result in the destruction of a certain fixed percentage, irrespective of the abundance of the insect; second, that represented by the birds, most other predators, and a part of the parasites which encompass the destruction of a certain gross number, rather than of any given percentage each year or generation; third, the facultative agencies, of which certain parasites are considered to be typical, which increase in efficiency as the insect increases in abundance.

The elements composing this last group are absent in New England, or, rather, those elements which are present (disease and starvation) and which do not properly belong to it, are inoperative until a state of extreme abundance is attained.

Such control as is effected by existing agencies would therefore fall into one or the other of the first two groups mentioned, and since both groups together are obviously inefficient, even when the moth is scarce, that due to the operation of the elements falling in the second group would become relatively less efficient as the time went on and the moth increased. This, it is believed, is actually what has happened and what is happening each year in each of the very numerous outlying colonies of the gipsy moth throughout the more recently infested territory, and thus the larger rate of increase is explained.

As a matter of fact, there is reason to believe that the average rate of increase during the first few years *immediately* following the introduction of the moth in a new locality is actually less than sixfold annually, and that it may even be as low as threefold, or perhaps less. In any case, there is a stage in the progress of the moth in which the average is no greater than that recorded by Forbush and Fernald, and there is no longer room for doubt that the lowest rate of increase is in localities where the moth is relatively a rare or uncommon insect for the time being, while the highest occurs in localities where the moth is rapidly approaching its maximum abundance.

AMOUNT OF ADDITIONAL CONTROL NECESSARY TO CHECK THE INCREASE OF THE GIPSY MOTH IN AMERICA.

It was evident in 1907, as it is now, that the problem of the introduction of parasites was far from being as simple as it might appear to be upon its surface and as it evidently did appear to be to some who were at that time agitating for a radical change in the methods adopted for its solution. It was plain that the expense incident to

the actual work of importation was going to be considerably more
than had been expected two years before and that practical results
could not possibly be achieved until long after the time originally
predicted. Additional information upon the biology and habits of
each of the several parasites, if not necessary in every instance, was
necessary in some and desirable in all, and here again additional
expenditures became imperative. Furthermore, the situation was
such as to make of very doubtful advisability the indiscriminate
importation of very large quantities of parasite material before a
better knowledge of the parasites themselves had been secured. The
repetition of the very large shipment of brown-tail hibernating nests
winter after winter, as will be described in another chapter, is an
instance in point. Had we been in possession of a complete knowledge of the parasites hibernating in those webs at the beginning, perhaps one winter's importation would have been sufficient.

There was no certainty that the results of the technical studies as
conducted at the American laboratory would be sufficiently full and
complete to answer our purposes and make possible the intelligent
continuation of the work. Should we fail in this respect, the only
alternative to a discontinuation of the introduction work in advance
of its logical conclusion was the establishment of a laboratory abroad,
at a considerable expenditure.

With these several reflections, it was inevitable that the advisability
of continuing the work beyond the time limit originally set should
come into question. Accordingly, in anticipation of the necessity for
making a decision when the time for it should arrive, the whole
proposition was subjected anew to the closest sort of scrutiny from
every point of view.

The successful consummation of the work involved, first of all, the
establishment in America of a group of parasites or other natural
enemies sufficiently powerful to meet and offset the prevailing rate of
increase of the gipsy moth. This, as determined by Forbush and
Fernald, was at least sixfold annually; as determined by actual
observation in the field, it was often far in excess of sixfold.
Before the continuation of the work could be recommended it was
absolutely necessary to arrive, first, at some conclusion as to the
amount of parasitism (gauged on the percentage basis) which would
be required in order to offset this increase and maintain the gipsy
moth at an innocuous minimum; and, second, whether parasitism
to such an extent actually prevailed abroad or whether natural control in those localities where it was obviously effected was due to the
increased efficiency of other agencies.

In so far as the first proposition was concerned, it was obvious from
the beginning that if enough egg masses could be destroyed each fall
so that the number remaining would be no greater than that which had

been present in the spring the insect could never, by any possibility,
increase beyond the abundance then prevailing. If the increase each
year was 6 egg masses for each 1 of the year before, it was merely
necessary to destroy 5 out of every 6 in order to maintain the status
quo. If the increase was tenfold, the destruction of 9 out of every 10
egg masses would be required, etc. The same would obtain if 5 out
of every 6 or 9 out of every 10 eggs in each and every mass were
similarly destroyed.

Reduced to percentages, this would be equivalent to the destruction
of 83.33 per cent or of 90 per cent, respectively, a rate of parasitism
which was physically an impossible accomplishment for the egg
parasites alone. Additional parasitism of the caterpillars or pupæ
would be a requisite to success, and such parasitism would of necessity be similarly limited in many instances through circumstances as
completely beyond our control as the physical inability of Schedius or
Anastatus to parasitize more than the uppermost layer of eggs in each
mass attacked. Without attempting to go into any of the details
of the processes by which conclusions were reached, it was finally
determined, beyond any doubts arising through arguments which
have been presented up to the present time, that an aggregate parasitism of 83.33 per cent would be absolutely necessary if a sixfold
increase was to be met, but that it made no difference whether this
was brought about by one species or two or a dozen, or whether they
attacked the egg, the caterpillar, or the pupæ. It was also determined that the aggregate percentage necessary could not be secured
by simply adding together the figures representing the parasitism
resulting through attack by each of two or more species. It was
going to be necessary to combine these several aggregates in a different manner. To illustrate: A 50 per cent parasitism of the eggs,
if it could possibly be secured, followed by another 50 per cent parasitism of the caterpillars, could not by any possibility be considered as
resulting in 100 per cent parasitism or complete extinction, but only in
50 per cent parasitism added to 50 per cent of what remained, which
amounted, in effect, to 25 per cent of the whole. In this manner an
aggregate of 75 per cent only is secured.

As is illustrated by the table on page III, it requires the combination of an imposing array of figures representing relatively small percentages of parasitism in each instance to acquire a sufficiently large
aggregate.

It was further determined that any specific amount of parasitism,
as 20 per cent of the eggs, was neither more nor less, but exactly as
effective as 20 per cent parasitism of the caterpillars or pupæ, in so far
as its value in constructing the final aggregate was concerned.

It can not be denied that when the validity of these conclusions
became established and when in addition the possibility that a much

greater rate of increase than sixfold would have to be met and offset before the much-to-be-desired consummation could reasonably be expected the prospects looked rather discouraging. Recognition of the correlation which existed between increased abundance and rate of increase served more than anything else to allay the doubts which these reflections created. Field work in 1908, 1909, and 1910 showed pretty conclusively that a rate of increase of not in excess of sixfold and possibly considerably less prevailed whenever the moth was in that state of innocuousness incident to the scarcity which it was hoped to bring about and maintain through the introduction of the parasites. An aggregate parasitism of 85 per cent will almost certainly be sufficient, and it may well be that 80 per cent or even 75 per cent will answer equally well. Much less than 75 per cent will probably not be effective.

THE EXTENT TO WHICH THE GIPSY MOTH IS CONTROLLED THROUGH PARASITISM ABROAD.

While it is true that the work which has been done for the purpose of determining the prevailing rate of increase of the gipsy moth in America leaves considerable to be desired in the way of exactness, in the main the statements made by Forbush and Fernald, as confirmed and modified by later observation, may be accepted as essentially accurate. It is fortunate that the situation is no worse than it appears to be, for if it were necessary to undertake the work of parasite introduction with the idea that the maximum rate of increase exhibited by the gipsy moth must be met by the parasites, such an unreasonable percentage of parasitism would be demanded as to make the proposition of introducing them a decidedly difficult task. As it is, there seems to be good reason to believe that a parasitism amounting to 75 per cent will be sufficient, provided that it can be maintained during the periods when the moth is relatively rare. It may be that less than that will answer equally well, but it would require actual test or else a much more careful study of the actual rate of increase of the moth under favorable conditions to justify such prophecy. In any event, 85 per cent will probably be amply efficient, if it can be established and maintained during all stages in the abundance of the moth. Such a rate would undoubtedly prevent the moth from increasing to destructive abundance in new territory or from regaining ground lost through the activities of disease in older infested regions.

Granted that it is sufficient, the question naturally arises as to whether such a degree of parasitism is to be found abroad in countries where the gipsy moth is present without being considered as a serious pest.

There is very little published information at hand bearing upon this subject, and that which is available is general rather than definite in its tenor. Anyone who for the first time encounters a tree covered with caterpillars of the gipsy moth dead and dying through the effects of the "wilt" disease is very apt to think that at last the gipsy moth has met its Waterloo, and disillusionment has only come in the present work as the result of several years' consecutive observations in the same or similar localities. In like manner the observations of foreign entomologists, or of American entomologists traveling abroad, as to the actual effectiveness of the parasites in accomplishing the control of the moth have to be taken with a grain of conservatism. Parasitism by a species as conspicuous as *Apanteles fulvipes* to the extent of 50 per cent would undoubtedly create a most favorable impression and the more conspicuous parasitized caterpillars would easily appear to outnumber the healthy. This amount of parasitism would certainly be inefficient in America unless it were supplemented by a much larger amount of parasitism by other species.

On this account it has been necessary to depend very largely upon the study of the parasite material imported from abroad as a source for information of this sort. From the beginning accurate notes have been kept of the many thousands of boxes of eggs, caterpillars, and pupæ of the gipsy moth and brown-tail moth, in which are recorded the locality from which each lot came, its condition on receipt, and the number and variety of parasites reared in each instance. The records are necessarily based in most instances upon such information as may be gained through a study of the condition of the material on receipt and the parasites reared, but in a few careful dissection work has been carried on to determine the true conditions, and thus to check up the results of the rearing work. It has been found that the amount of dependence which can be placed upon the rearing records is relatively small, and that nothing more than a general idea of conditions actually prevailing can be gleaned from them. · Nearly always some of the caterpillars or pupæ are dead or dying upon receipt as a result of the ordeal through which they have passed. On the average, taking the gipsy-moth material from all localities, not more than 25 per cent has arrived in good condition (when the shipment of eggs is excepted). The brown-tail moth material has averaged very much better, and probably 75 per cent has been in good condition on receipt.

It has been found that sometimes a larger percentage of parasites than of caterpillars or pupæ died en route, while at other times these conditions are entirely reversed, and since dissections can not be made in every instance, it has been necessary to consider the parasitism indicated by the notes upon two different bases, i. e., that of the number of hosts originally involved and that of the number successfully completing their transformations. If from a lot of 1,000 brown-tail caterpillars 250 individuals of *Parexorista chelonix* and 250 moths are reared, it is perfectly safe to assume that parasitism by Parexorista amounts to more than 25 per cent and less than 75 per cent. Further than this nothing absolutely definite may be said. Exactly the same is true of the determination of prevailing rates of parasitism of native insects through rearing work.

On account of the inadequacy of these methods when it comes to the point of securing absolutely authentic information, not nearly so much is known of the parasitism of the gipsy moth or of the brown-tail moth abroad as is needed to carry on the work to its best advantage. This much, however, can be said definitely, that in some instances existing parasitism is sufficient to answer the requirements of the situation in America; in others it is obviously insufficient; in most the results of the study of imported material are not sufficiently reliable to support either contention.

Here, again, was food for serious consideration when it came to the point of making definite recommendations concerning the continuation of the work. Would the foreign parasites certainly meet the demands which would be made upon them in America?

This has been answered in the affirmative through its consideration from quite a variety of different viewpoints. For one thing, the lack of accurate information as to the conditions under which the parasite material was originally collected, has rendered the results of its study in America of difficult analysis. No one, for example, would seriously question the statement that the white-marked tussock moth is under well-nigh perfect control in America except in cities. Nevertheless, if it was desired to transport caterpillars or pupæ of this insect to Europe in order that its parasites might be reared, the agent intrusted with the collection of the material for exportation would certainly go to the city for it, and the person who received and studied it upon the other side would find so few parasites present as to justify exactly the same doubts concerning the parasitism of the tussock moth in America as have actually arisen concerning the parasitism of the gipsy moth and the brown-tail moth in Europe. The tussock moth is not often subjected to the full extent of parasitism necessary to effect its control in any locality from which caterpillars can be secured in quantity. It is reasonable to suppose that something of the same sort is true of the gipsy moth or of the brown-tail moth.

Furthermore, the study of the tussock moth has resulted in demonstrating another fact which is of peculiar interest in this connection, which is, that the parasites which assist in effecting its control in country districts where this control is perfect are sometimes entirely

absent in the city. Something of the same sort may be true of the
parasites which assist in effecting the control of the gipsy moth in
many localities in Europe where it is so uncommon as to make collection of material for exportation in any quantity impossible. Some
of the most interesting lots of caterpillars or pupæ which have been
received were from such localities, and it may well be that there are
parasites abroad which have not been received at the laboratory in
Massachusetts in sufficient quantity for colonization, and which can
never be received there until new methods for collecting and importing them are devised, but which at the same time are actually among
the important species. This fact can only be determined definitely
by careful study of the gipsy moth in localities where it was not
sufficiently abundant to permit of its collection in large quantities.
These studies, it is hoped, will be instituted in 1911, and so long as
the gipsy moth continues to be a serious pest in America the investigation of its parasites abroad ought to be continued.

The ramifications of the parasite work have been so many and so
diverse and have led so far afield, both literally and metaphorically
speaking, as to make it practically impossible to report upon it as a
whole as fully as would be desirable and practicable were it less
extensive and varied. A chapter might be written upon the parasitism of the gipsy moth and another upon that of the brown-tail
moth in each of the several countries in Europe from which the
parasite material has been imported, but it is wholly impracticable
to do so. At the same time, now that a new phase of the work is
being entered upon, it will not be out of place to review in some
slight detail the results of the work which has been carried on in a
few of the localities which, for one reason or another, may be selected
as of more than general interest in this immediate connection, but
which are at the same time and in another sense typical.

PARASITISM OF THE GIPSY MOTH IN JAPAN.

From the viewpoint of gipsy-moth parasitism Japan possesses a
peculiar interest, because, if we are to judge from the reports of
those who have been there and incidentally or critically studied the
situation, the Japanese gipsy moth is pretty thoroughly controlled
through natural agencies, and among these its parasites appear to
rank very high. This is the more interesting and encouraging
because the Japanese race is notably larger and at the same time
more fecund than the European, judging from counts as made at the
laboratory of the number of eggs in a mass.

In 1908, after several unsuccessful attempts which had been made
to import its parasites had served to demonstrate the futility of any
less radical course, Prof. Trevor Kincaid, of the University of Washington, Seattle, was delegated to spend the summer there in the

interests of the work. As a result numerous large shipments of parasite cocoons and puparia, as well as of caterpillars in various stages and of pupæ were received at the laboratory. The condition of the material on receipt compared more than favorably with the average of similar shipments from Europe, and for the first time opportunity was afforded for the actual first-hand investigation of the parasitic fauna of the gipsy moth in Japan.

Similar shipments were made in 1909 and 1910, with even better results in so far as the condition of the material on receipt was concerned, and several of the more important parasites have now been liberated in the field in America under conditions which are apparently ideal and which ought to encompass their introduction and establishment, if such a thing is possible. · ·

TABLE 1.—*Sequence of gipsy-moth parasites in Japan.*

A total of 14 species of parasites has been reared from the imported material, of which 7 were present in sufficient abundance to indicate that they were of real importance in effecting the control of the moth. Two species are of such doubtful host relationship as to have been omitted from Table I.

Specimens of one species, *Meteorus japonicus*, the importance of which is not indicated by the examination of the imported material, have been sent to us by Mr. Kuwana with the statement that it is sometimes, locally at least, a common parasite, but none for colonization has been received. Still another is of possible importance, judging from the very limited opportunity which we have had for its investigation, but none of the others is of proved worth. Since nothing is actually known of the conditions under which particular lots of parasite material were collected, it can not be stated as confidently as the circumstances render desirable that some among the others are not incidentally of value in keeping reduced the numbers of the moth in localities where it is too rare to permit of collection of material for shipment to America.

Prof. Kincaid's reports upon the effectiveness of the parasites, even when taken with more than the prescribed grain of conservatism, have been so consistently optimistic as to leave no room to doubt that the parasitism to which the moth is subjected in Japan, even in localities where it is more than normally prevalent, is sufficient to meet and overcome the rate of increase of the gipsy moth in America.

How these parasites work together in bringing about the control of the moth in Japan is indicated in Table I, which, with its explanation, was published in a somewhat abbreviated form in the popular bulletin by the junior author which was issued through the office of the State forester of Massachusetts a year ago.

The addition of the names of the species marked with an asterisk makes the list complete, so far as it may be completed through the information now available. The species so designated are those which have never been received in sufficient abundance to make their colonization possible, and among them are some which are doubtless of wholly insignificant importance from an economic standpoint, while others may, upon investigation, prove to be of more than sufficient importance to justify an attempt to secure their introduction into America.

Opposite the name of each parasite, extending across a certain number of the vertical columns, is a dotted line. The vertical columns indicate different stages in the development and transformations of the gipsy moth, as the egg, the caterpillar, and the pupa, and these are still further divided into caterpillars of different sizes and eggs and pupæ of different ages and conditions. At the head of each column is stated the approximate number of days during which the individual gipsy moth remains in that particular stage.

The dotted line following the name of the parasite indicates those stages in the life of the gipsy moth during which the latter is likely to be attacked by the parasite in question, and it will be seen that in a number of instances, as, for example, Chalcis and Theronia, this period is exceedingly short. The solid line indicates the stages in the life of the gipsy moth during which it is likely to contain the parasite in its body. This, it may also be noted, varies considerably. Crossocosmia, for example, gains lodgment in the active caterpillar while it is only about half grown, and the extension of the solid line across all of the columns which stand for the later caterpillar stages, as well as for all of the pupal stages, indicates that the larvæ of this parasite do not leave the host caterpillar until after it has transformed to a pupa, and until the moth would naturally have emerged had the pupa remained healthy and unparasitized:

It will be noted that the parasites not designated by the asterisks, and which are therefore to be considered as of some importance in effecting the control of the moth, form, when taken together, a perfect sequence, and that every stage of the moth from the newly deposited egg to the pupa is subjected to attack. It is furthermore of interest to note in this connection that, so far as may be determined from the scanty information available, all of these parasites are present in more or less efficient abundance within a limited area in the vicinity of Tokyo, from which a part, and presumably the greater part, of the material was collected for exportation.

PARASITISM OF THE GIPSY MOTH IN RUSSIA.

The earliest first-hand knowledge of the gipsy moth and its parasites in Russia was secured as the result of the visit paid to that country by the senior author in the spring of 1907. Through his instrumentality several of the Russian entomologists were interested in the parasite-introduction work to such a practical extent as to collect or cause to be collected and forwarded to America several small and a few large shipments of the eggs, caterpillars, and pupæ of both the gipsy moth and the brown-tail moth. The difficulties attending the importation of material from Russia proved to be considerably more real and less easily surmountable than those which were so successfully overcome in the instance of the Japanese shipments, and for the most part the Russian material was of more interest from a technical than from an economic standpoint by the time it arrived at the laboratory.

From a technical standpoint it was exceedingly interesting and valuable, since there were found to be present in the boxes of young gipsy-moth caterpillars the cocoons of several species of hymenopterous parasites which had either not been received from other sources or which were not known to be sufficiently abundant in any other part of Europe to make possible their collection in large quantities. Prof. Kincaid's successful prosecution of the Japanese work encouraged his selection as the best and most experienced agent available for the decidedly more difficult proposition of visiting Russia and attempting to secure an adequate supply of the several species of parasites which could only be secured in that country to advantage, so far as could be determined from the information then at hand.

The manner in which he was impressed by the gipsy-moth situation which he encountered there is best described in the following extracts from his letters, in the course of which occasional comparisons are made between Russian and Japanese conditions.

BENDERY, BESSARABIA, RUSSIA, *June 11, 1909.*

The season here is in full swing, but the situation causes me considerable anxiety, as the whole business is so utterly different from my experience in Japan. The damage wrought by *dispar* in the forests and orchards of Bessarabia this season is enormous and parasite control seems to be most inefficient in checking the depredations of the caterpillars. When I think of the masterly and well-ordered attack of the Japanese parasites and the splendid fashion in which they wiped out the caterpillars in large areas before depredation took place I am surprised by what I see here. When I left Gauchesty on May 31 the forests of that district were thoroughly riddled over thousands of acres, and yet I had seen no sign of insects, fungous, or bacterial attack except a couple of clusters of Apanteles cocoons. * * * *Dispar* seemed to be having its own way as fully as in America, so far as could be seen on the surface. * * * Here at Bendery the season is slightly more advanced, owing to the lower altitude, and the prospects of securing parasites of *dispar* seem better. Even here, however, the situation seems to me quite remarkable. I have available three extensive and very different collecting grounds. The great forest of Gerbofsky about 6 versts from Bendery is composed almost exclusively of oaks. An immense area is covered by trees of this species, forming magnificent groves of fine trees 80 to 100 years old. Thousands of these great trees are completely defoliated, so that no sign of foliage remains. In the same forest are groves of young trees of the same kind, also greatly damaged. Examining the myriads of caterpillars in the field, I have found no sign of parasite attack or of fungous disease except the work of Apanteles. * * * The percentage of attack by Apanteles is so small that rearing in trays is of little practical importance. One would have to have billions of caterpillars to do any good. Collecting in the field is very difficult, as the caterpillars creep into crevices or suspend themselves to branches at some height from the ground, where they are hard to reach. No sign of bacterial disease has appeared in this area, nor have I seen any evidence of tachinid attack in the field or in my rearing trays. Another great forest of about 500 acres is at Kitzkany, about 7 versts from Bendery on the banks of the Dniester, a low damp situation. This forest has no oaks but is much mixed. The principal tree is *Populus nigra*, but there are many other trees, as Ulmus, Acer, Salix, etc. Here again the damage is tremendous, with almost no sign of parasite attack. Prolonged search yielded a few cocoons of Apanteles. On the other hand, thousands of the older caterpillars were found in the pendulous condition so characteristic of bacterial attack. The third condition I found in the numerous orchards adjacent to both of these forested areas. These orchards have been almost overwhelmed by *dispar*. The more progressive peasants have protected their trees by rings of axle grease or by strips of cotton wool, but others have done nothing and the trees are quite stripped. * * * The sight of a tree covered with hundreds of dead caterpillars bearing clusters of Apanteles cocoons such as I saw in Japan seems not to be hoped for.

KIEF, *June 26, 1909.*

From what I can see in the field and from what I can gather from Prof. Pospielow, *dispar* was almost exterminated in this district last year through the activity of the parasites. Only a few isolated colonies seem to have survived, the most important of these being at Mishighari, a small place on the river about two hours by steamer from Kief. In this place, which is perhaps 100 acres in extent, the trees are plastered with cocoons of *Apanteles fulvipes*. The attack of the parasite was so thorough that the first generation seems to have been sufficient to wipe out the caterpillars, as I can find no large caterpillars about the place, and a few days will doubtless witness the complete wiping out of *dispar*. * * * Tachinids also appear to be very active, as I find many eggs, but as these are laid upon caterpillars suffering from the attacks of Apanteles it would seem as if the emaciated caterpillars could not supply sufficient nourishment to bring the tachinids to maturity. * * * From the standpoint of parasite control the situation at Kief is most inspiring, but as a field in which to gather a quantity of material it is evidently not very hopeful. The whole situation is in violent contrast to what I found at Bendery and Gauchesty, where *dispar* is vastly more abundant this year than last, with little sign of the multiplication of parasites.

BENDERY, RUSSIA, *July 10, 1909.*

In the forest of Kitzkany where *dispar* caterpillars prevailed to an incredible extent three weeks ago, not a single caterpillar or pupa is to be found. An epidemic of a bacterial nature swept them away in millions. In the forest of Gerbofsky, among the great oak trees, the number of caterpillars that have formed pupæ is surprisingly small. Vast numbers of caterpillars swarmed over the trees, completely stripping them of leaves. Deserting the trees, the caterpillars swarmed over the ground in search of other food and vast numbers died of starvation and disease. These trees are now putting forth new leaves which promise to sustain the life of the forest.

After the close of the "caterpillar season" in 1910 the junior author took a vacation trip to Europe and, thanks to an extension of leave for the purpose and still more to the kindness of Mr. N. Kourdumoff, entomologist of the experiment station in Poltava, was enabled to spend about 10 days in the field in Kief and Kharkof Provinces. In Kief the forest at Mishighari, which is mentioned by Prof. Kincaid as the one locality where he found the parasites in control, was visited, as well as several other localities in that province. This portion of Kief Province, topographically, meteorologically, and otherwise, is radically different from Massachusetts, and much more like portions of Minnesota than any other part of the United States with which the visitor is at all familiar. The forests, which are limited in extent as compared with those of Massachusetts, are less diversified. For the most part they are of pine, mingled with a small quantity of oak, wild pear, birch, and occasionally other trees. Everywhere the gipsy moth was rare or at least uncommon, and everywhere the cocoon masses of *Apanteles fulvipes* were at least as abundant as the egg masses of their host.

At Mishighari the conditions remained much as described by Prof. Kincaid, except that the cocoon masses of Apanteles were even more abundant than his letters would indicate. Upon some trees they were litterally matted together by the thousands in such semiprotected situations as are selected by the caterpillars at the time of molting. The forest in this particular locality was varied to an extent not noticed elsewhere. In addition to the generally distributed oak, birch, and poplar were quantities of beech, alder, Carpinus, maple, elm, and other species, while the shrubs were equally varied and abundant. The forest was situated upon the steep bluffs overhanging the Dnieper, running down on one side to its banks, where great willows bore evidence of the high water which sometimes covered their trunks, and on the other extending back some little distance until it was met by a wide stretch of treeless prairie.   Here, at least, parasite control of the gipsy moth appeared to be pretty thoroughly effective, since there were in evidence a vastly larger number of old cocoon masses of the Apanteles than there were of old egg masses of its host, and the new egg masses were enormously outnumbered by the old. According to Prof. Pospielow, who was a member of the party, the forest in this locality was almost completely defoliated in 1908, but there was no indication of damage to any of the trees composing it, and every indication that the parasites alone were responsible for the disappearance of the moth.

In several other localities along the banks of the Dnieper the conditions encountered were essentially the same, differing principally in the lesser abundance of both egg masses and parasite cocoons.   In one locality quite near to Kief, fresh egg masses were more common than at Mishighari, and cocoon masses of the season of 1910 were also more common.   It seemed to offer opportunities for the collection of a sufficiently large quantity of this parasite to make an experiment in importation and colonization possible and practicable in 1911, which it is hoped may be carried out.

In Kharkof conditions, both as regards the gipsy moth, its parasites, and the country at large, were essentially different from those in Kief.   Numerous localities from 5 to 20 miles out of the city in different directions were visited, and everywhere indications of the recent presence of the gipsy moth were found in abundance.   Old egg masses were massed around the base of the trees in a manner exceedingly suggestive of uncared for woodland in Massachusetts, and mingled with them were a very few fresh masses; so few, relatively, as to indicate most conclusively that the moth had encountered very adverse conditions during the season of 1910, with the result that its abundance had been most materially reduced.

In every locality the conditions were the same, although the character of the forest varied to a material extent.   For the most part the province of Kharkof is devoid of forest, and quite suggestive of parts of North Dakota in appearance.   Such forest as does occur is mostly confined to the valleys in the neighborhood of streams, and though it may be fairly extensive, it is rarely very diverse.   No pine was seen.   Oak predominates very largely and, with the exception of some birch, forms practically pure forests away from the lowlands, except in the best watered localities.

Everywhere, irrespective of the character of the forest, the gipsy moth was found under the circumstances recounted above.   Everywhere there had been an abundance of eggs in the spring, everywhere there had been an abundance of caterpillars, a considerable proportion of which had gone through to pupation, and everywhere the number of fresh egg masses was very much smaller than was that of the old. Nowhere was there evidence of parasitism by *Apanteles fulvipes* to anything like the extent which prevailed in the vicinity of Kief. Cocoon masses were occasionally found, nearly always old, sometimes very old and so discolored as to be with difficulty distinguished from the bark to which they were attached. In Kief the number of cocoon masses was everywhere in considerable excess over the number of egg masses. Here the number of egg masses was enormously in excess of the number of cocoon masses.

Examination of the pupal shells for evidences of parasitism was unavailing. It could be said with assurance that pupal parasites were certainly not common and that the death of the pupæ (for proportionately very few of them hatched) was not due to any of the pupal parasites which were known from western European localities. The earth beneath the cocoon masses was examined for evidences of tachinid puparia. For a time none was found, but search was finally rewarded by the discovery of *Blepharipa scutellata* in most extraordinary abundance in a single one among the numerous localities visited. This particular forest, which was very near to the village of Rhijhof and about 8 miles from Kharkof, was unique among the others visited in the variety of its trees. The soil was rich, the trees were larger, and the undergrowth was more abundant and varied, but at the same time there was less diversity than was encountered in the forest at Mishighari. Unfortunately, the presence of the puparia could not be considered as of much significance, because they were practically all hatched and obviously dated back more than one year. The parasite had surely not been responsible for the reduction in numbers of the gipsy moth which had taken place in the season of 1910, and neither had it prevented the moth from increasing to such numbers as to bring about partial defoliation of the forest in 1910 before disaster in one form or another had overtaken it.

Of other tachinids there were practically none, and it is certain that they would have been found had they been present. *Compsilura concinnata* is even now so abundant as a parasite of the gipsy moth in Massachusetts as to bring about an appreciable percentage of destruction in 1910, and its puparia are recovered from the field with ease. Had it been one-tenth as common in Russia it could not have failed of detection. The same is true of Tachina, which, although it effects a parasitism of less than 1 per cent in Massachusetts, is not difficult of detection, and it is safe to say that not much if any more than this amount of parasitism prevailed in Kharkof. All told, not enough parasites were found to indicate that they had played any important part in the reduction of the moth from a serious menace to the well-being of the forest to such small numbers as to require several years at least before it would be possible for such conditions to recur.

The investigations having been conducted in September, some time after the death of all of the caterpillars and pupæ, it was no longer possible to determine with assurance the cause for the peculiar conditions, but everything conspired to indicate that nothing less than an epidemic of disease had been responsible. The condition of the pupal shells which hung upon the trees in countless thousands was in every respect identical with the condition of the pupal shells which are to be found in Massachusetts in every locality where the disease has prevailed to a destructive extent the season before. Among the old egg masses which plastered the extreme base of nearly every tree in most of the localities visited were found a variable, and sometimes a very large, proportion which had hatched only in part or not at all. The appearance of these unhatched masses was identical in every respect with the appearance of similarly large numbers which are frequently found in Massachusetts. The reason for the nonhatching of the eggs is not yet plain, but it is the consensus of opinion that this is probably associated with the "wilt" disease. It is known that affected caterpillars may pupate before death, and it seems not illogical to suppose that slightly affected caterpillars may pupate and produce moths which are able to deposit their eggs, but that these eggs fail to hatch as the direct result of the taint in the blood of their parent.

These Russian experiences seem, on the whole, to indicate that in that country the gipsy moth is not controlled by its parasites to an extent which serves to remove it from the ranks of a destructive pest. But as one day after another in the field at Kharkof served more and more indelibly to deepen this conviction, it served equally, first to create, and finally in retrospect to confirm, the observer in another, which was, in effect, that if this was the best that could be expected of disease as a factor in the control of the gipsy moth in its native home then something better than disease must be found to control it in America. Just so long as conditions similar to those seen in Kharkof or pictured in the letters of Prof. Kincaid are allowed to prevail in Massachusetts just so long will the incentive remain to see the parasite-introduction experiment carried on until success is either achieved or proved impossible. Conditions similar to those prevailing in Russia emphatically do not prevail in western Europe, nor, according to all accounts, in Japan. Natural conditions in western Europe and in Japan are in many respects more like those of our own Eastern States than are those of Kharkof Province. Conditions in Kief Province, even, are much more like those of Massachusetts than are those of Kharkof, and in Kief parasite control seemed to be an accomplished fact, although of course there is no assurance that it is continuous and perfect.

The final outcome of the Russian experience was, therefore, the opposite of what might have been expected, and it resulted in a firmer determination than ever to carry the work through to its end.

PARASITISM OF THE GIPSY MOTH IN SOUTHERN FRANCE.

Following the 10 days in Russia a shorter period was spent by the junior author in somewhat similar field work in southern France, where, with the aid of M. Dillon, he was enabled to visit the localities from which the largest, and in that respect the most satisfactory, shipments of parasite material ever received at the laboratory were collected. As the direct result of the senior author's visit to Europe in 1909 some thousands of boxes containing hundreds of thousands of gipsy-moth caterpillars had been collected in the vicinity of Hyères, about 50 miles to the eastward of Marseilles. These caterpillars were largely living upon receipt, and in the winter of 1909–10 Mr. W. B. Thompson dissected several hundred preserved specimens and the actual percentage of parasitism was thus determined. Some few pupæ which had also been received from the same locality made possible a fair understanding of the extent to which the pupæ were parasitized.

The results of these investigations, taken in connection with the actual rearing work, were disappointing. It was evident that the moth was fairly common in the region from which the material was collected—as common, perhaps, as it would need to be in Massachusetts to provide for an increase of sixfold annually. Nevertheless, the amount of parasitism which was indicated by this, the most thorough study of parasitism of the gipsy moth abroad which was ever undertaken in the laboratory, was less than enough to offset a twofold, much less a sixfold, increase.

For this reason much curiosity was felt as to the conditions which prevailed in a country where parasitism of such comparatively insignificant proportions was sufficient.

No sooner was the character of the country districts in this portion of France seen than the wonder which had been felt at the small percentage of parasitism which was sufficient to hold it in check was replaced by a much greater astonishment that the gipsy moth should exist under such conditions at all. It was a country of olive orchards and vineyards, with a strip along the littoral which was so nearly frostless as to permit the culture of citrus fruits, and even of date palms. The hills were semiarid, with the soil exceedingly scanty and often covered by loose stones. The principal forests consisted largely of cork oak and pine, except in the low and well-watered valleys and bottom lands where other trees in considerable variety occurred.

The slopes of the higher mountains were fairly well forested and a larger variety of trees and shrubs thrived than on the lower elevations. Over much of the country the soil was either too dry or too scanty or both to permit of cultivation, even in a land so densely populated by so thrifty a race, and here were found occasional thickets of scrub oak, apparently of a deciduous species, which sometimes reached the dignity of a small tree.   For the most part such country was covered with a scanty growth which would be called chapparal in some of our States, composed of a variety of uninviting looking shrubs, judging them from the probable viewpoint of a gipsy-moth caterpillar in search of food.   Taken altogether, the country may more aptly be compared with southern California than with any other part of the United States.

It seemed to the visitor that if the gipsy moth were to be found in any portion of this region it would most likely be within the rich and well-watered bottom lands, where occasional hedges, or rows and groups of large trees in considerable variety, seemed to offer fairly acceptable conditions for its existence.   But to his surprise and amazement he was assured by M. Dillon that it was from the chapparal covered, arid, and uncultivated elevations that most of the enormous quantities of caterpillars had been collected.   In support of this assertion, after the visitor had searched in vain in what would be the most likely situations in Massachusetts for the concealment of egg masses, pupal shells, or molted skins, M. Dillon proceeded to turn over a few loose stones among those which fairly covered the ground, and thereby disclosed sufficient indication of the presence of the moth in fair abundance to convince the most skeptical.   In this particular locality in the vicinity of the little provençal town of Meoun, in a thicket of deciduous oak surrounding and concealing the ruins of an ancient chapel, there were sufficient egg masses of the moth to represent a fair degree of infestation, but eggs, pupal shells, and molted larval skins were all so completely hidden as to evade completely the eyes of one who had been trained to look for first evidences in sheltered places on the bark or in the knot holes and hollow trunks of trees.

As a matter of fact, as was abundantly evidenced by that day's experiences, as well as of the several days which followed, the gipsy moth departed most materially from its characteristic habits in the cooler, better watered and forested localities in which it is present as a pest in America.   Instead of being a typically arboreal insect, it is rather terrestrial, and thereby becomes subjected to a variety of natural enemies to which it is practically immune so long as it remains arboreal.   In the course of the several years past a variety of species of the larger European Carabidæ has been studied at the laboratory for the purpose of determining their availability and probable worth

as enemies of the gipsy moth. Of them all, not one refused to attack
and devour the caterpillars and pupæ of the gipsy moth with businesslike dispatch, once given an opportunity, but with one or two exceptions none has shown a disposition to climb trees in search of its prey.
Being essentially terrestrial in habit, they were essentially unfitted
to prey upon an essentially arboreal insect.

We know little of the predatory beetles which are to be found in
that part of France which was visited upon this occasion, nor does
this lack of knowledge vitiate the strength of the argument to any
great extent. The fact was that if present (and undoubtedly some
species are to be found) any of the numerous forms which have been
studied at the laboratory and discarded as unfit for the purposes desired in Massachusetts would immediately assume high rank as enemies of the gipsy moth. In other words, the conditions under which
the gipsy moth exists in southern France are wholly incomparable with
those under which it exists in New England, and the agencies which
are effective in accomplishing its control are likewise incomparable.
The unimportant rôle obviously played by the parasites immediately
loses its significance. . Those species of true parasites which assist in
this control are practically the same as those which assist in other
localities, but the demand upon them and their opportunities for multiplication are insignificant compared to those existing in Massachusetts, if they are ever established there. True to their character as
agencies in facultative control, they do not increase in efficiency to an
extent which would practically mean the extinction of their host.

The results of the rearing and dissection work carried on at the
laboratory indicated that a parasitism varying from 25 per cent to
something in excess of 40 per cent prevailed in this locality. After
seeing the conditions under which the gipsy moth struggled for existence, real wonder was felt that it should be able to survive, and the
trip resulted in a firmer conviction than ever in the efficacy of parasitism, and the validity of the theory upon which the parasite-introduction work was conceived.

SEQUENCE OF PARASITES OF THE GIPSY MOTH IN EUROPE.

The parasitic fauna of the gipsy moth varies considerably in various
faunal divisions of Europe, and no attempt has been made to prepare
separate lists of the parasites peculiar to those regions which have
been represented in the material imported. In Table II, which is constructed in accordance with that representing the sequence of parasites in Japan, as explained on page 122, all of the various species reared
from the European material are listed. As in the table of Japanese
parasites those species which are of no consequence in the control of
the moth (so far as known) are marked with an asterisk.

THE BROWN-TAIL MOTH AND ITS PARASITES IN EUROPE.

Reference has already been made to the fact that in those sections of Massachusetts in which both the gipsy moth and the brown-tail moth occur, the latter is considered as the lesser pest of the two. This opinion, as held by those who are thoroughly familiar with the comparative noxiousness of the two, speaks quite plainly of the character of the gipsy moth as a pest, in view of the very considerable agitation which has come about on account of the brown-tail moth in

TABLE II —*Sequence of gipsy-moth parasites in Europe.*

localities into which it has preceded the gipsy moth or where the latter has not as yet reached a state of destructive abundance.

On account of the lesser interest aroused in the brown-tail moth in Massachusetts, its parasites have not been given quite the consideration, in some respects, that has been given to those of the gipsy moth, but this lack of consideration has had entirely to do with the question of the future policy of the laboratory, and has not extended to the actual handling of the parasites themselves. In every respect other than as a basis for calculations as to future policies of the laboratory, they have received as much and as careful consideration as have the parasites of the more dangerous pests.

So far as known the brown-tail moth does not occur in Japan, and in consequence no determined efforts have been made to secure, from Japanese sources, parasites likely to attack it. It has an ally and congener there in *Euproctis conspersa* Butl., which is attacked by a variety of parasites, some of which may be expected to attack the brown-tail moth if given an opportunity. A few of them have been collected and forwarded to the laboratory through the great kindness of Mr. Kuwana, but unfortunately have arrived in such condition, or at such time of the year, as to make their colonization impossible. It is intended in the near if not in the immediate future to devote some time to the investigation of the Japanese parasites likely to be of service in this respect, and, if any can be found of promise, to attempt their importation into America.

In Europe the brown-tail moth appears to be the more common of the two insects under consideration and, taken all in all, it is probably the more injurious as well. Neither in Europe nor in America does it bring about the wholesale defoliation characteristic of an invasion of the gipsy moth, but its injury is of a more insidious character and more evenly distributed throughout the years. In Russia, in the fall of 1910, the junior author was astounded at the tremendous abundance of its nests in many localities, notably on the irregular hedgerows planted as a windbreak alongside the railroad in the midst of an otherwise open prairie. Occasionally small scrubs of Cratægus, or wild pear, completely isolated by what seemed to be miles of open prairie, would be fairly covered with the nests.

In gardens in the vicinity of Kief pear and apple trees were frequently injured to a considerable extent by its caterpillars, and sometimes to a greater extent by the caterpillars of *Aporia cratægi* L., which are similar in their habit, and were constructing their own hibernating nests side by side with those of the brown-tail moth. In the forests round about it was common, but except occasionally not quite so common as in southeastern New England. On one occasion in excess of 50 nests were noted upon a small hawthorn which stood at the edge of an oak forest. This was just a little worse than anything which has been seen in America.

In southern France the circumstances under which it occurred were as surprising as those under which the gipsy moth was encountered, in respect to their departure from that which past experience led the visitor to consider as the normal. M. Dillon, who had collected and forwarded to the laboratory a considerable quantity of the winter nests, undertook to guide the visitor to the locality where they were collected. The way led through a rich and fertile valley, with many sorts of trees, including apple and pear, as well as hawthorn and oak, every one of which is a favored host plant in other regions, but not a single brown-tail nest was seen and, according to M. Dillon, it was never found upon these trees.   Farther on an elevated plain was passed, with occasional ridges of uncultivated land upon which were growths of a deciduous scrub oak.   Gipsy-moth eggs, pupal shells, etc., could usually be found by a little search under the stones on these ridges, but the brown-tail moth was conspicuous by its total absence.

The next day the route selected passed through an extensive forest of cork oak, mingled with pine, and finally up the sides of the mountain, until great plantations of aged chestnut trees were indicative of a change in climatic conditions brought about by the considerable altitude.   Various shrubs and a few trees unknown or rare in the lower elevations became a feature of the forest, and among them the arbusier (*Arbutus* sp.), closely resembling in its growth, in the appearance of its evergreen foliage, and in its habitat the mountain laurel of our own southern mountains.   It is very beautiful and unusual in its appearance, partly on account of its flowers (which are suggestive of Oxydendron) but more particularly because of its fruit.   This was globular, about the size of a marble, and hung pendant on long stems in more or less profusion, and in all stages of ripening.   In the course of this process it passed from green through a sequence of vivid yellows to orange, and finally intense scarlet.   It was at once recognized as the host plant of the hundreds of nests which had been collected and shipped by M. Dillon.   Although it was occasionally met with sufficiently far down the mountain side to mingle with orchards and hawthorn hedges, according to M. Dillon the brown-tail moth invariably seeks it out, even there.   The selection of a food plant representing a totally different order from any selected in other parts of Europe or in America, and this in spite of the fact that what are ordinarily its most favored hosts were frequently much the more abundant, was considered to be quite as remarkable as the assumption of terrestrial habits by the gipsy moth.[1]

In central and western Europe generally the brown-tail moth finds a stronghold in the dense Cratægus hedges which are commonly planted in many localities, and upon them as well as upon oak and fruit trees it is frequently abundant.   In these regions, also, not only the food plants, but the seasonal and feeding habits are quite like those in New England.   Occasionally an apple tree or an oak will be found carrying an abundance of nests and, as noted by the senior author in northwestern France in 1909, the moths are sometimes so

---

[1] It has since been learned that in the warmer parts of the region visited, the brown-tail moth caterpillars not only remain active but feed to some extent during the winter   In the middle of January, 1911, the nests were found commonly, always upon Arbutus, in parts of the coast regions near Hyères, and in nearly every instance the caterpillars were active and in most they were feeding   In this particular locality the nests were very different from those typical of the caterpillars in northern localities, being loosely woven, and not at all designed for hibernation in its stricter sense.

numerous as to lay their eggs in quantities on growing nursery seedlings and low-growing plants.

Among the very many lots of caterpillars and cocoons which have been received at the laboratory there is occasionally one in which a fungous disease is present. Usually, when it is present at all, the majority of the caterpillars received from that particular locality will be found dead and "shooting" the ascidiospores upon receipt. According to Dr. Roland Thaxter, to whom specimens have been several times been submitted, it is specifically identical with the fungus which is so effective in America as to have largely assisted in reducing the moth from the preeminent place which it would otherwise have occupied as a pest. Its presence under these conditions, as it was, for example, in 1909, in practically every box out of a large number which were forwarded to the laboratory from lower Austria, is strongly indicative of the importance of this disease.

Looked at from one standpoint, the brown-tail moth situation in America is less satisfactory than is the gipsy-moth situation. In numerous localities throughout western Europe as well as in eastern Europe it frequently increases to such an extent as to become a pest. It hardly seems as though more could be expected of the European parasites in America than is accomplished by them in Europe, but if even this much can be secured it will aid materially in reducing the frequency of the outbreaks. At the same time, it must be admitted that from nearly every point of view the prospects of unqualified success with the gipsy-moth parasites are better than with the parasites of the brown-tail moth.

SEQUENCE OF PARASITES OF THE BROWN-TAIL MOTH IN EUROPE.

The accompanying table (Table III), in which are listed all of the parasites of the brown-tail moth which have been definitely associated with that host in the course of the studies of imported European material, is constructed in the same manner as the tables of parasites of the gipsy moth in Japan and in Europe (see pp. 121, 132). It will be noted that the number and variety are slightly larger than of European gipsy-moth parasites, and that the species which are or which appear to be promising as subjects for attempted importation are also slightly more numerous. Very rarely, however, does any one among them become as relatively important as any one of several among the gipsy-moth parasites which might be mentioned. Neither has any lot of brown-tail material produced so many parasites of all species (as high a percentage of parasitism) as have several lots of gipsy-moth material.

PARASITISM OF THE GIPSY MOTH IN AMERICA.

Although the gipsy moth is attacked by a considerable variety of American parasites the aggregate effectiveness of all the species together is wholly insignificant, so far as has been determined by the rearing work which has been conducted on an extensive scale at the laboratory. Actual effectiveness may be greater than indicated, however, because it is possible that the caterpillars or pupæ may be attacked by parasites, the larvæ of which are unable to com-

TABLE III.—*Sequence of brown-tail moth parasites in Europe.*

PARASITES.	EGG.	FALL STAGES. FIRST	SEC-OND	THIRD	WIN-TER STAGE	SPRING STAGES. FIRST	SEC-OND	THIRD	FOURTH	PRE-PUPA	FRESH	OLD	ADULT
*TRICHOGRAMMA SP	▪▪▪												
*TRICHOGRAMMA PRETIOSA-LIKE	▪▪▪												
*TELENOMUS PHALAENARUM	▪▪▪												
APANTELES LACTEICOLOR		▪▪▪											
METEORUS VERSICOLOR		FALL GENERATION ▪▪▪				SPRING GENERATION ▪▪▪▪▪							
ZYGOBOTHRIA NIDICOLA		▪▪▪											
*PTEROMALUS EGREGIUS			▪▪▪										
*LIMNERIUM DISPARIS				▪▪▪▪▪▪▪									
PAREXORISTA CHELONIAE				▪▪▪▪▪▪									
DEXODES NIGRIPES				▪▪▪▪▪▪									
COMPSILURA CONCINNATA 1		..1...		▪▪▪▪▪▪									
*BLEPHARIDEA VULGARIS					▪▪▪▪▪								
*CYCLOTOPHRYS ANSER					▪▪▪▪▪								
*MASICERA SYLVATICA					▪▪▪▪▪								
EUDORONYMA MAGNICORNIS.					▪▪▪▪▪								
ZENILLIA LIBATRIX					▪▪▪▪▪								
PALES PAVIDA					▪▪▪▪▪								
TACHINA LARVARUM					▪▪▪▪▪▪▪▪								
*TRICHOLYGA GRANDIS					▪▪▪▪▪▪▪▪								
*PIMPLA BRASSICARIAE								▪▪▪▪▪▪					
*PIMPLA INSTIGATRIX								▪▪▪▪▪▪					
*PIMPLA EXAMINATRIX								▪▪▪▪▪▪					
*THERONIA ATALANTAE								▪▪▪▪▪					
MONODONTOMERUS AEREUS 2				2					▪▪▪▪				
*CHELONIS OMNIVORA									▪▪▪▪				
*PTEROMALUS SP									▪▪▪▪				

1, ATTACKS YOUNG CATERPILLARS BEFORE HIBERNATION, BUT LARVAE APPARENTLY FAIL TO MATURE.
2, ADULT FEMALES HIBERNATE IN WINTER NESTS.
*, SPECIES NOT CONSIDERED TO BE OF MUCH IMPORTANCE ECONOMICALLY.

plete their transformations under the conditions in which they find themselves. This is known to be true in the instance of what would otherwise be a very important parasite, *Tachina mella.*

In such instances the host usually remains unaffected and the parasite perishes. At other times, as proved through a series of experiments carried on by Mr. P. H. Timberlake, of the Gipsy Moth Parasite Laboratory, in the spring of 1910, the host may perish without exhibiting any external symptoms of 'its condition. No serious attempt to determine whether this actually happens in the field has been made, but undoubtedly it does occasionally result when the parasite larva finds itself under unnatural surroundings. It is thus well within the bounds of possibility that effective parasitism should pass unnoticed in the course of investigations in which reliance is placed entirely upon the results of rearing work.

As will be shown in another place, death of the host through superparasitism by a species fitted to attack it may similarly occur without the true cause becoming apparent.

A sufficiently large quantity of the native caterpillars of the gipsy moth has been dissected at the laboratory to indicate that such concealed parasitism, if it is ever a factor in the control of this insect, is of rare occurrence, or else of insignificant proportions. This can not be said of the pupæ of the moth in America, which have not been studied sufficiently well as yet.

The following native parasites have been reared from the gipsy moth in Massachusetts:

THERONIA FULVESCENS CRESS.

This, the most common American parasite completing its transformations upon the gipsy moth, was mentioned by Forbush and Fernald in their comprehensive report upon "The Gypsy Moth" under the name of *Theronia melanocephala* Brullé. The true *T. melanocephala* appears not to have been reared from this host. The importance of *T. fulvescens* as a gipsy-moth parasite is indicated by the summarized results of the rearing work conducted in 1910.

In his account of the parasites of the forest tent caterpillar (*Malacosoma disstria* Hübn.) in New Hampshire by the junior author it was credited as being a secondary parasite of *Pimpla conquisitor* Say, and was not recognized as a primary parasite. Investigations at the laboratory have served to throw considerable light upon its life and habits, and it is now known to be a true primary parasite, but one which, like *Pimpla conquisitor* itself, is able to complete its transformations under a variety of circumstances. The supposed secondary parasitism, in this instance, is to be classified rather as "superparasitism" and is believed to result through the circumstance that the primary host chances to contain the larva of Pimpla, rather than through the deliberate searching out by the parent Theronia of pupæ thus parasitized. In its relations to the gipsy moth, which is not successfully attacked by Pimpla at all frequently, Theronia has always been a primary parasite so far as known.

PIMPLA PEDALIS CRESS.

One or two specimens have been reared from the pupæ of the gipsy moth collected in the field, but it is of extremely rare occurrence as a parasite of this host, so far as recent rearing work indicates. It was mentioned as one of the more common parasites by Forbush and Fernald, but it is possible that the next following species is intended.

PIMPLA CONQUISITOR SAY.

Judging from observations made from time to time in the field the pupæ of the gipsy moth are frequently attacked by this species, but, unfortunately, the young larvæ of the Pimpla appear not to thrive upon this host and rarely complete their transformation. It is safe to say that more female Pimplas will be found attacking the gipsy-moth pupæ in the course of a day's observations in the field at the proper season of the year than would be reared if that day were to be spent in collecting pupæ instead. It is believed that the affected host usually dies, but the subject has not received the attention which it deserves. If it is true, *Pimpla conquisitor* may prove to be of some assistance in the control of the moth.

PIMPLA TENUICORNIS CRESS.

Recorded as a parasite by Forbush and Fernald, but never reared at the laboratory. Possibly *P. conquisitor* was actually the species reared.

DIGLOCHIS OMNIVORA WALK.

Mentioned by Forbush and Fernald as of some consequence as having been reared from this host, but during late years it has been so rare that only a single pupa has been found in which it has completed its transformations.

ANISOCYRTA SP.

Mentioned by Forbush and Fernald, but the record has not been confirmed by later rearing work.

LIMNERIUM SP.

A single cocoon, which was directly associated with the remains of the host caterpillar, was collected by Mr. R. L. Webster in 1906 during his association with the laboratory. It was very likely that of *L. fugitiva* Say, but the fact will never be known, because a specimen of *Hemiteles utilis* Norton, a hyperparasite, actually issued.

APANTELES SP.

In 1910 a colony of the caterpillars of the white-marked tussock moth was established upon some shrubbery in a locality where the gipsy moth was fairly common. The young caterpillars were sparingly attacked by a species of Apanteles, possibly *A. delicatus* How., although the fact was not determined. At the same time and place a young gipsy-moth caterpillar was found from which an Apanteles larva had issued, and spun a cocoon identical in appearance with that of the species from the tussock moth. This is the only known instance of the parasitism of the gipsy moth by an American Apanteles, and it is believed that it resulted through the fact that the parasites were first attracted, and subsequently excited into oviposition, by the tussock caterpillars.

A considerable number of a minute black and yellow elachertine secondary parasite was reared from this cocoon, so that the specific identity of the Apanteles originally constructing it will forever remain in doubt.

Syntomosphyrum esurus Riley.

In July, 1906, Mr. R. L. Webster, who was at that time associated with the parasite laboratory, found a pupa of the brown-tail moth from which he reared a number of Syntomosphyrum, probably *S. esurus* Riley. On the same date, July 18, the pupa of a gipsy moth was found to contain the early stages of a chalcidid parasite, presumed to be the same as that reared in connection with the brown-tail moth.

At about the same time several chalcidids, apparently of Syntomosphyrum, were found ovipositing in pupæ of the gipsy moth, but in no instance was the oviposition successful, so far as the notes indicate.

Tachina mella Walk.

In their report on the gipsy moth Forbush and Fernald speak of having collected no less than 300 caterpillars of the gipsy moth bearing Tachina eggs which were reared through in the laboratory. The most of these produced moths and the remainder died. No parasites were reared.

In 1907 and subsequently large numbers of caterpillars have been found in the open, bearing tachinid eggs, and many hundreds have been kept under observation in confinement with results substantially the same as those above mentioned. In one or two instances, however, the tachinids have completed their transformations and in each instance the species was *Tachina mella*. It is believed, therefore, that this is the species which deposits its eggs so freely and injudiciously.

The fact that effective parasitism failed to result was attributed by Forbush and Fernald to the molting off of the eggs before they had hatched, and this doubtless does occasionally happen. Mr. C. H. T. Townsend reinvestigated the subject and came to the conclusion that the explanation was to be found in the inability of the newly hatched larvæ to penetrate the tough integument of the caterpillars, since he actually observed such failure in one instance, and found many caterpillars upon which the eggs had actually hatched, but from which no parasites were reared.

That this explanation may serve in part to elucidate the mystery is also true, but still later observations have shown conclusively that the parasite larvæ may gain entrance into their host and yet fail to mature. Two explanations have grown where one was deemed sufficient, as the result of certain technical studies which have been made at the laboratory during the past year. Mr. P. H. Timberlake and, later, Mr. W. R. Thompson have thoroughly demonstrated the fact that a parasite larva gaining lodgment in an unsuitable host may die, and its body may be in great part absorbed through action of the phagocytes without causing the host obvious inconvenience. This very likely takes place with *Tachina mella* in its relations with the gipsy moth and is probably a better explanation of its failure to become an eTective parasite than any other which has yet been put forward.

Mr. Thompson also discovered another most remarkable and peculiar phenomenon in connection with parasitism by those tachinids the larvæ of which inhabit integumental funnels similar in character to those formed by Tachina. These funnels appear to be formed as a direct result of the tendency of the skin to grow over and heal the wound caused by the entrance of the tachinid maggot into the body of the host. This wound is kept open by the larva itself, and as a result the growing integument takes the form of an inverted funnel, more or less completely surrounding the parasitic maggot, which continues to breathe through the minute orifice in its apex. When the caterpillar molts the old skin is usually torn away from around this opening, leaving the maggot in situ and unaffected, but occasionally its attachment to the funnel may remain so strong as to result disastrously for the maggot, and the whole funnel, maggot included, may be withdrawn. Thus, not merely the eggs may be molted off, but the internal feeding maggots which have hatched from the eggs may be molted out and perish.

ACHÆTONEURA FRENCHII WILL.

A very few specimens of this species have been reared from time to time in the course of the work at the laboratory. It is probable that the species is synonymous with that mentioned by Forbush and Fernald under the name of *Achætoneura fernaldi.*

EXORISTA BLANDA O. S.

Occasionally reared as a parasite of the gipsy moth.

UNDETERMINED TACHINIDS.

Dr. S. W. Williston, in reporting upon a collection of Diptera reared from the gipsy moth in Massachusetts and sent to him in 1891 by Dr. Fernald, stated that there were present two species of Exorista and four ·of Phorocera. Unfortunately these specimens appear to have been lost before being definitely determined. No such variety of tachinids has been reared from this host in Massachusetts during recent years, but several species as yet undetermined, or represented only by unfamiliar puparia, are in our collection.

SUMMARY OF REARING WORK CARRIED ON AT THE LABORATORY IN 1910.

In Tables IV and V are the condensed results of a part of an extensive series of rearing experiments primarily instituted for the purpose of determining the present status of the introduced parasites of the gipsy moth in America in 1910. They also serve excellently as an indication of the effectiveness of parasitism by native species. In this respect they are typical of the results secured from similar work in previous years.

TABLE IV.—*Results of rearing work in 1910 to determine progress of imported parasites and prevalence of parasitism by native parasites of gipsy-moth pupæ.*

Laboratory No.	Localities.	Number of dispar. pupæ.	Moths reared.		Parasites reared.				Scarcophagid puparia.
			Male.	Female.	Blepharipa scutellata.	Theronia fulvescens.	Compsilura concinnata.	Miscellaneous tachinids.	
2185	North Andover...	2,700	1,222	259	2				
2186	Melrose..........	500	185	159	1				
2187	North Andover...	1,800	683	299	2	3			2
2188	Stoneham........	1,400	482	461		5	1		
2189	Saugus..........	1,000	412	250		2	2		3
2190	Wakefield.......	800	126	217					1
2190A	....do..........	[1]106	4	38	1		[1]12		
2191	Swampscott......	300	159	68					
2192A	North Andover...	300	106	92					
2192B	Woburn..........	282	69	115		1			
2192C	North Andover...	100	67	22					
2193	Saugus..........	1,050	274	172			2		1
2194	Melrose [2]......	1,700	312	211	1				1
2196	Stoneham........	1,047	379	127					
2197	Beverly.........	1,000	81	144					13
2199	....do..........	1,000	147	78		2			5
2280	Saugus..........	1,250	253	209		2	1	[5]1	10
2281	North Andover...	580	232	102		1			
2282	....do..........	1,700	442	218	8	17		[4]1	6
2283	Beverly.........	500	261	125		1			
2284	....do..........	400	152	133					
2285	Gloucester......	1,700	354	346		3		[5]2	5
2286	Beverly.........	600	82	87					1
2287	....do..........	300	85	49					
2289	North Andover...	1,000	143	461		1			
2290	....do..........	1,100	118	323		3			1
2291	....do..........	900	198	215		4			2
2293	Beverly.........	500	229	102					
2294	....do..........	500	81	104		2			2
2295	Middleton.......	1,100	193	156					
	Total......	27,215	7,531	5.342	15	47	18	4	53

[1] Prepupæ.
[2] Partly from another locality.
[3] Tachina-like species, puparium.
[4] *Exorista blanda.*
[5] One tachina-like and one unknown native species, puparia.

Table IV includes the results of rearing work in which caterpillars collected in the open were used. A large proportion of the collections, probably half or more, was from localities in which *Compsilura concinnata* is not as yet established, although it will probably have extended its range to include these by the end of another year. It will be observed, however, that the number of the Compsilura reared is very far in excess of that of any native tachinid or other parasites attacking this caterpillar as a host. This is especially encouraging in view of the fact that one year ago less than one one-hundredth of the proportionate number of Compsilura were reared under similar circumstances, and only from material collected over a much more restricted area.

TABLE V.—*Results of rearing work in 1910 to determine progress of imported parasites and prevalence of parasitism by native parasites of gipsy-moth caterpillars.*

Laboratory No.	Localities.	Number of caterpillars.	Number of pupæ reared.	Number of moths reared.		Number of parasites reared.					
				Male.	Female.	Compsilura concinnata.	"Tachina-like."[1]	Exorista blanda.	Apanteles fulvipes.	Theronia fulvescens.	Blepharipa scutellata.
4317	Wellesley	1,029	158	15	13						
4330	Stoneham	535	298	43	81	9					
4332	Wellesley	2,110	716	419	69		8				
4333	...do	3,000	962	371	115		4				
4334	Stoneham	2,000	599	59	177	108					
4337	...do	800	300	20	88	73					
4340	Wellesley	1,250	846	234	245		3				
4341	Swampscott	600	308	124	144	1					
4349	Wellesley	2,000	902	363	317		6			[2]1	
4351A	Medford	300	234	18	202	6					
4351B	Winchester	100	54	5	34						
4352	Danvers	1,500	566	30	199	6					
4353	Wilmington	125	69	7	25	2					
4354	Winchester	110	14	1	6	12		1			
4356	Wellesley	1,000	818	245	344		2		[2]3	[2]8	
4359	...do	1,125	934	150	301		1			[2]4	
4360	Stoneham (?)	500	207	11	125	59	1				1
4362	Beverly	230	56	4	30			1			
4366	Lexington	400	20	1	5	2					
4367	Salem	1,000	338	29	139	15					
4369	Marblehead	250	61	4	22	5					
4370	Salem	800	286	21	106	14				[2]1	
4372	Stoneham-Woburn	150	103	11	43	21					
4373	Wellesley	150	65	9	11				[2]1		
4377	Nahant	425	181	29	114	1					
4378	Winthrop	300	191	18	128	1					
4386	Burlington	100	37	2	22	1					
4387	Lexington	53	6		4	1					
4392	Manchester	341	281	21	42	1					
	Total	22,283	9,610			337	25	2	4	14	1

[1] The identity of this tachinid, the puparia of which resemble those of tachina, but which hibernates as a pupa, is wholly unknown. The adult has never been reared.
[2] From near site of colony of 1910. Recovery of no significance. Masses of cocoons counted.
[3] A few pupæ included in this collection.

In Table V are included results of rearing work in which collections of gipsy-moth pupæ were used. These were largely made in localities near the center of colonization of *Blepharipa scutellata* of the same year. The number of Blepharipa secured is higher in proportion to the number of pupæ collected than would be the case had these collections been made irrespective of the localities where the species was so recently colonized.

Parasitism by Theronia is somewhat less on the average than in some other years when similar studies have been made. At times it has amounted to as much as 2 per cent.

Sarcophagids are not considered as parasites, but rather as scavengers. Their true status is yet to be determined, however.

Compsilura concinnata is not commonly secured from the pupæ, and in one instance in which more than an insignificant number of this species was recovered the collection consisted of caterpillars, which had prepared for, but not undergone, pupation.

No Monodontomerus were reared from any of the collections of pupæ included in Table IV, nor has the species ever been recovered from counted lots of pupæ collected in the open. It was found in 1910, as in 1909, by the examination of unhatched pupæ after the most of those remaining healthy had produced the moths, and issued in unsatisfactory numbers from collections of pupæ made at the same time. These were not counted at the time of collection, and on that account were not included in the table.

PARASITISM OF THE BROWN-TAIL MOTH IN AMERICA.

The brown-tail moth in America is subjected to a considerably higher percentage of native parasitism than is the gipsy moth, but at the same time, as will appear in the summarized results of the rearing work in 1910, the aggregate is scarcely sufficient to be considered as consequential.

The following species have been reared, and doubtless the list will receive additions in the near future.

TRICHOGRAMMA PRETIOSA RILEY.

A very considerable percentage of the egg masses collected in the open is parasitized by this species, but because of the inability of the parasite to attack any but the more exposed eggs in a mass, the actual percentage of parasitism is insignificant.

LIMNERIUM CLISIOCAMPÆ WEED.

In 1907 a single specimen of this common parasite of the tent caterpillar was reared from a brown-tail caterpillar collected in Exeter, N. H. One or two other rather doubtful records have been made since. It is unquestionably not an important parasite of the brown-tail moth.

ANOMALON EXILE PROV.

Quite commonly reared as a parasite of the tent caterpillar and not infrequently as a parasite of the brown-tail moth, apparently attacking the caterpillars before pupation and probably while they are still very young. Its frequency as a parasite of the brown-tail is well indicated in Table VI (p. 147), which records the results of the summer's rearing work of 1910.

THERONIA FULVESCENS CRESS.

This is probably, as in the case of the gipsy moth, the most common native hymenopterous parasite. No attempt has been made to determine whether it is commonly primary or secondary in this connection, but it is presumably primary in the majority of instances.

PIMPLA CONQUISITOR SAY.

The pupæ of the brown-tail moth seem to afford much more suitable conditions for the development of the Pimpla larvæ than do the pupæ of the gipsy moth. In consequence this Pimpla is frequently reared and is probably about as important as a parasite of the brown-tail moth in America as are the European species, *Pimpla examinator* and *Pimpla instigator*, abroad.

PIMPLA PEDALIS CRESS.

This species is never so common as *Pimpla conquisitor* in its association with the brown-tail moth. It is apparently identical in habit with the more common species, but if results in studies in parasitism of other hosts are to be excepted, it is more apt to occur in forests and woodland than in open country.

DIGLOCHIS OMNIVORA WALK.

At times Diglochis is a common parasite of the brown-tail moth pupæ, but in 1910 it was unexpectedly scarce in Massachusetts, although it seemed to have been much more common in Maine, judging from the small amount of material which has been received from that State.

SYNTOMOSPHYRUM ESURUS RILEY.

Of irregular occurrence as a parasite of the brown-tail moth, but among the more effective of the native species in 1910. It was first reared in 1906 by Mr. R. L. Webster, while associated with the laboratory, and not again encountered until 1910, when large numbers issued from material collected in certain localities, as will be seen in Table VI which shortly follows.

CHALCIS sp.

Upon several occasions specimens of Chalcis have issued from cocoons of the brown-tail moth collected in the open. The species has not been definitely determined nor compared with *Chalcis ovata*, because it is thought likely at the present time that two species may be confused under that name. One of them is believed to be a primary parasite of lepidopterous pupæ and the other to be essentially a secondary parasite attacking tachinid puparia.

EUPHOROCERA CLARIPENNIS MACQ.

Several times reared from brown-tail caterpillars collected in the field, but always, apparently, rare in this connection.

TACHINA MELLA WALK.

Never a common parasite in connection with this host. Usually about on a par with Euphorocera and without economic significance.

PHOROCERA SAUNDERSII WILL.

A single specimen thus determined by Mr. Thompson was reared in 1910 under circumstances which quite conclusively indicate this host relationship.

EXORISTA BOARMLÆ COQ.

Like the above, only a single individual of this species has been reared from brown-tail caterpillar collections, but under circumstances which were not so decisive as in the last-mentioned instance. Mr. Thompson is authority for this determination also.

UNDETERMINED TACHINID: "NATIVE PARASITE OF CHRYSORRHŒA."

One of the most remarkable instances of attack on an unsuitable host by a tachinid parasite is that of the species which in the laboratory notes is always referred to as above, upon the caterpillars of the brown-tail moth. The history of this species, the identity of which is very much in doubt, is in certain respects comparable to that of *Tachina mella* in its relations to the gipsy moth. As may be seen by the summarized results of the rearing work in 1910 (Table VI), it is not infrequently a parasite of some little consequence, and in all many hundreds of its larvæ have been secured from field collections of brown-tail caterpillars.

Invariably, however, these larvæ died without forming perfect puparia. For a long time it was thought that this was due to unfavorable surroundings at the time when pupation was attempted and that if the larvæ were allowed to enter damp earth as soon as they issued from the host, better results might be obtained. In order to provide for this, the style of rearing cage which is shortly to be described and figured, was devised. With its aid we were enabled to secure a large proportion of the parasite maggots within a few moments after they had finally separated themselves from the cocoon which the host invariably spins before dying, and these were given every advantage which could be afforded to assist in the successful completion of their transformations. The results, however, were always the same and not one perfect puparium has been secured.

The reasons for this may not be far to seek, but the chances are that it will be a long time before an adequate explanation is afforded. When it was found that the larvæ failed to pupate under the most favorable conditions, they were carefully examined on the supposition that death might possibly accrue through the action of the poisonous spines into which they must necessarily come in direct contact in leaving the host cocoon. It was at once discovered that there was invariably a number of minute reddish spots scattered irregularly, and more or less abundantly, over the whole or a part of the body. It looked, at first, as though these spots might be the result of contact with the poisonous spines, but upon further examination it was found that they were of a character which could hardly be attributed to this cause. They are somewhat variable in size and seem to consist of a thickening of the epidermis which becomes slightly raised, shining, and brick-red in color. No attempt has been made as yet to determine whether they are present in the maggot before it leaves the body of its host, but little doubt is felt that they will be found when such examination is made.

That these spots are directly or indirectly responsible for the failure of the maggot to pupate is well indicated by the study of the numerous half-formed puparia which result from the attempt on the part of the larva of the parasite to do so. These are all more or less larviform, but occasionally one is found one end of which is smooth and rounded exactly as though pupation had successfully resulted, while the other is shrunken and withered, resembling a dead larva. Careful examination revealed that in such specimens the reddish spots were absent from the perfect portion and present in the withered.

SUMMARY OF REARING WORK IN 1910.

The accompanying tabulated results (Table VI) of an extensive series of rearing experiments for the purpose of determining the progress of the imported parasites of the brown-tail moth also indicate the extent to which that host is attacked by native parasites.

TABLE VI.—*Results of rearing work in 1910, to determine progress of imported parasites and prevalence of parasitism by native parasites of the brown-tail moth.*

Laboratory No.	Localities.	Estimated number of chrysorrhœas.	Moths reared.		Imported parasites reared.				Native parasites reared.										
			Male.	Female.	Compsilura concinnata.	Zygobothria nidicola.	Meteorus versicolor.	Monodontomerus æreus.	Undetermined Tachinid "N.P.C."	Pimpla conquisitor.	Pimpla pedalis.	Theronia fulvescens.	Anomalon exile.	Syntomosphyrum esurus.	Diglochis omnivora.	Chalcis.	Phorocera saundersii.	Exorista boarmiæ.	Tachina mella.
2100	Saugus	500	80	9	7	1											1		
2101	Melrose	500	79	127	6														
2102	Andover	500	2	9															
2104	Saugus	500	29	60	2				1									1	
2105	Lynnfield	550	10	20	2				2										
2109	Stoneham	300			21			5											
2110	Andover	200	10	3			3		1										
2111	Tewksbury	1,000	83	49					10										
2112	Reading	350	14	15				1	4	1									
2113	Bedford	500	67	65															
2114	Billerica	500	50	35					3	1									
2115	Woburn	500	10	17					7										
2116	Lexington	500	18	14						1									
2122	Saugus	400	12	17	13				1										
2123	Waltham	300	47	63					6										
2125	Arlington	500	9	13					2	1									
2126	Chelmsford	560	10	19				1	1										
2127	Drayout	500	4	8	1				12	1			1						
2131	Melrose	450	198	177					6										
2132	Manchester	130	8	24	2	3			4										
2133	Saugus	500	221	70	3				1										
2134	Melrose	100	16	26												2			
2135	Reading	300	88	59			4		19				1						
2136	Concord	500	26	48					2	1									
2137	Lincoln	90	12	18	9		1					2	4						
2138	Stoneham	250	68	45				20	3	3				136					
2139	Gloucester	150	21	24						3									
2140	Saugus	200	2	2															
2141	Beverly	200	116	85						1									
2142	Merrimac	200	82	71					6	2									
2143	Middleton	200	76	71				9		3									
2144	Groveland	500	104	93															
2145	Littleton	1,000	17	15					7	1		7							
2146	Westford	200	50	59					54	2	2		50	59					1

TABLE VI.—Results of rearing work in 1910, to determine progress of imported parasites and prevalence of parasitism by native parasites of the brown-tail moth—Continued.

Laboratory No.	Localities	Estimated number of chrysorrhoea	Moths reared: Male	Female	Imported parasites reared: Compsilura condinata	Zygobothria nidicola	Meteorus versicolor	Monodontomerus aereus	Undetermined Tachinid "N.P.C."	Native parasites reared: Pimpla conquisitor	Pimpla pedalis	Theronia fulvescens	Anomalon exile	Syntomosphyrum esurus	Diglochis omnivora	Chalcis	Phorocera saundersii	Exorista boarmiae	Tachina mella
2147	Beverly	250	22	16															
2149	Methuen	250	63	67															
2150	Andover	250	32	12															
2151	do	250	18	22															
2152	do	150	19	21															
2152A	do	100	35	38					10										
2152B	do	100	14	10					11										
2152C	do	150	15	23					16										
2152D	do	150	11	23					1										
2152E	do	200	30	64					5										
2156	Met Hun	150	31	3				3				2							
2157	do	50	31	42					9				1						
2158	do	150	46	52					15			1							
2159	do	200	47	34				7	9			1	2	7					
2160	Melrose	90	74	70			5	14	10	2		1	2						
2161A	do	100	49	50	1		1		12	1		2	12						
2161B	do	150	96	62			3	6				6							
2161C	do	250	9	15											53				
2161D	Andover	500	87	91				3	6	1		1	1	38					
2162	Boxford	500	121	119									7	60					
2164	Salisbury	500	46	42	1			2	2	3		2	7						
2166	Lynnfield	150	42	63		1						2	1						
2167	Peabody	175	70	74								1							
2168	Reading	220	22	33				12	2					40					
2169	Saugus	180	89	141		1	5			1		10	1	53					
2170	Reading	400	27	27	4			4	2	7		1							
2171	Lynnfield	200	64	29				57	17	1		1							2
2172	Salem	200	37	39															
2173	Melrose	1,000	304	314				24											
2174	do	150	45	57				19											
2175	do	200	38	63															
2176	do	150	13	17															
2177	do	175	24	54															

2179	Reading	200	50	48															
2180	Lynnfield	175	68	63						2	2		1						
2181	Melrose	325	217	240			19	1	2	1	3								
2182	Swampscott	325	79	87			19		1	4				54					
2183	do	225	96	96			2		4	5									
2184	Lynn	240	64	59			86		1										
	Total	24,350	3,955	3,851	72	7	19	313	276	43	4	58	32	432	55	2	1	1	4

As may be noted by reference to the table, the imported parasites are beginning to become sufficiently abundant so that parasitism by them will compare favorably with that by American species, but are not as yet so abundant as to exceed the American species in relative effectiveness. The table as presented does not indicate at all accurately the actual status of the several species of parasites mentioned, on account of the difference in the condition of the material at the time of collection.

Compsilura, for example, is much more apt to leave the caterpillars before they spin for pupation, and the same is true of Meteorus. Monodontomerus, Pimpla, Theronia, etc., never attack caterpillars before spinning, and Monodontomerus and Theronia frequently reserve attack until some little time after the host has pupated. As it stands, parasitism by Monodontomerus is about equal to that of Theronia and in excess of that by Pimpla or Anomalon. Parasitism by Compsilura is distinctly more effective than that by all of the other native tachinid parasites of the caterpillars. Meteorus is much more common than indicated in the limited territory over which it is now known to exist, and the specimens reared represented the second generation of adults to develop upon the brown-tail moth in 1910.

Apanteles lacteicolor Vier. is not represented in these collections, since it does not attack caterpillars so large as those involved.

In carrying on this work several styles of rearing cages were used, of which one was devised for the special purpose of securing the tachinid parasites with the minimum of exposure to the effects of the irritating hairs of the brown-tail caterpillar. This worked very satisfactorily, and since it may possibly be found of service in conducting similar work elsewhere, the following description is presented:

The basis of this cage (see fig. 10) consisted of a box of stiff pasteboard 8 inches square and 12 inches high. About 4 inches from the top a stiff paper funnel (a) was fitted and held in position by the cleats (b), which, in turn, were fastened to the sides of the box by broadheaded upholsterer's tacks driven in from the outside. These cleats served to support the tray (c), which just fitted into the cage. The bottom of this tray was covered, in some instances with coarse mosquito netting, and in others with a wire screen of ½-inch mesh. Two holes in the side of the tray corresponded with two 1-inch holes in the side of the box, and these in turn with similar holes in a wooden strip (d), which was fastened on the outside. When the tray was in position, paper cones (h) and large glass tubes (g) were inserted in these holes.

The stiff paper funnel (a) had its apex inserted into another hole bored diagonally in a similar wooden strip which was fitted in the bottom of the cage. Inside of this hole a stiff paper cone (formed like h by rolling up a section of a strip of paper cut to a circular shape) was held in position by a tack which passed through it into the wooden strip. The end of this cone, passing through the bottom of the cage, permitted a third glass tube (*f*) similar to the two above mentioned, to be held in position. No further support to this tube was needed than that afforded by the cone itself.

In using this cage a mass of cocoons of the brown-tail moth was placed in the tray, and the cover was put on with the several tubes in position. Tachinid maggots issuing from the prepupal caterpillars, or pupæ contained in the cocoon mass, in attempting to seek the earth would pass through the bottom of the tray and be conducted by the stiff paper funnel into the lower tube, where they were quickly noticed and easily re-moved. All other parasites, as well as the brown-tail moths themselves, when they emerged, were at-tracted by the light into the two upper tubes, and could be similarly removed with little difficulty. (See Pl. V, fig. 1.)

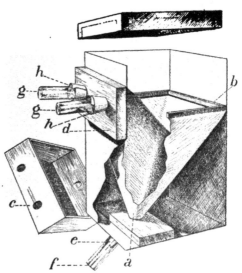

By the aid of this contrivance we were enabled to secure a quantity of the larvæ of the unknown tach-inid, already men-tioned, within a few minutes after they had issued from the host, and thereby de-termined that the fail-ure of this species to

Fig. 10.—Rearing cage for tachinid parasites of the brown-tail moth; *a*, Paper funnel; *b*, cleats holding paper funnel in position; *c*, tray; *d*, wooden strip on outside of cage, *e*, paper cone connecting paper funnel *a* and glass tube *f*; *h*, *h*, paper funnels supporting glass tubes *g*, *g*. (Original.)

pupate was in no way due to the unnatural surroundings. Some-times the tubes were partly filled with damp earth, in order that these larvæ might immediately come in contact with it, and at other times the larvæ were removed as soon as they dropped and placed upon earth similar to that which they would naturally have encoun-tered had they issued from cocoons in the field under wholly natural conditions.

The use of these cages also saved a large amount of exceedingly painful work which would otherwise have been necessary in determin-ing whether or not *Parexorista cheloniæ* was present in any of the field collections.

IMPORTATION AND HANDLING OF PARASITE MATERIAL.

Since insects like the gipsy moth and the brown-tail moth are subjected to the attack of different species of parasites at different stages in their development, it has been necessary, in order to secure all of these, to import the host insects in as many different stages as possible and practicable. If the present experiment in parasite introduction is brought to a successful conclusion, it will undoubtedly encourage the undertaking of other experiments in which similarly imported pests are involved. Even should it fail, from a severely practical standpoint, and the complete automatic control of neither the gipsy moth nor the brown-tail moth should be effected, it seems to us that the technical results already achieved are sufficient to give encouragement rather than the opposite to similar undertakings in the future. It is therefore desirable to describe in some detail the various methods employed for the importation and subsequent handling of the parasite material.

With very few exceptions the methods first employed proved more or less unsuitable. Sometimes they were entirely discarded; usually they were modified to suit the exigencies of the occasion. Sometimes these modifications were in comparatively unimportant particulars which would scarcely be pertinent to any other insect than the gipsy moth or the brown-tail moth, and realizing this there will be no attempt in such cases to enter into lengthy descriptions. At other times radical modifications have been found necessary on account of unforeseen difficulties which would be likely to occur in pretty nearly any other undertaking along anything like similar lines.

EGG MASSES OF THE GIPSY MOTH.

The importation of egg masses of the gipsy moth (see Pl. VI) from European sources has been attended with no difficulty whatever, beyond that of securing the collection of these eggs in sufficiently large quantities. Any style of package, provided that it were sufficiently tight to prevent loose eggs from sifting out, was as good as another, and any one of the established means of transportation served the purpose.

In the case of shipments from Japan serious difficulties were encountered. One of the parasites peculiar to that country and unknown in Europe invariably issued en route and died without reproducing. Various attempts to overcome this difficulty without having recourse to cold storage failed and it was only after cold-storage facilities were perfected and used that living parasites of this species were secured in numbers.

As in the instance of similar shipments from Europe, no special form of package was required, but at the same time a word of appre-

FIG. 1.—VIEW OF INTERIOR OF ONE OF THE LABORATORY STRUCTURES, SHOWING
REARING CAGES FOR BROWN-TAIL PARASITES. (ORIGINAL.)

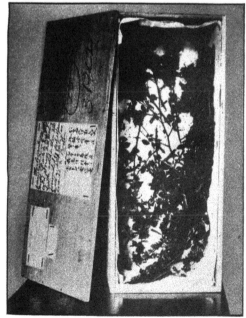

FIG. 2.—BOX USED IN SHIPPING IMMATURE CATERPILLARS OF
THE GIPSY MOTH FROM JAPAN. (ORIGINAL.)

ciation must be said for the wonderful care with which the Japanese entomologists packed the egg masses for shipment. Good-sized and wonderfully well-constructed wooden boxes were used and each mass was wrapped separately in a small square of soft rice paper.

Considering the ease with which egg masses of the gipsy moth ought, theoretically, to be obtained and shipped to the laboratory, the number received in response to the requests which were made for their collection and shipment was astonishingly small during the first two winters. Up to that time only a very few dead parasites of an undescribed genus and species had been received from Japan, and none at all had issued from any of the few European importations.

In 1908 the several lots of eggs were placed in small tube cages of the ordinary type and the caterpillars killed as they issued. Some time after the eggs had hatched a few parasites began to appear simultaneously from the European and Japanese material, which proved upon examination to be *Anastatus bifasciatus* in each instance. Later a few *Tyndarichus navæ* were reared from the Japanese eggs and supposed, rightly enough, to be secondary, although probably not, as at first supposed, upon Anastatus. This hyperparasitism was by no means certain, and it was resolved to determine the fact definitely the following fall and winter, provided additional importations could be secured. The desired material was imported and an exhaustive study of the parasites which were present was made, with the result that the five species were reared and their host relations as well as their relations one to another definitely determined. The execution of this project proved to be much more tedious than was expected, and was, in fact, the feature of that winter's work. Further mention of the investigation will appear in the discussion of *Schedius kuvanæ*.

GIPSY-MOTH CATERPILLARS, FIRST STAGE.

In the spring of 1907 an attempt was made to import the caterpillars of the gipsy moth in their first instar, and a considerable number was received from several different localities. The experiment was not a success and was not repeated. The mortality was heavy en route and only a small proportion of the caterpillars would feed after receipt. Some few were carried through to maturity, but no parasites were reared.

It is very probable that if recourse were had to cold storage, caterpillars could very successfully be transported in this stage, but the importation of slightly larger caterpillars indicates that the percentage of parasitism would average to be very small at the best, and it is probable that the best would rarely be achieved.

GIPSY-MOTH CATERPILLARS, SECOND TO FIFTH STAGES.

EUROPEAN IMPORTATIONS.

The first importations of gipsy-moth caterpillars in the second to fifth stages were made in 1907. Small wooden boxes, each with a capacity of about 40 cubic inches, were used for the purpose, and all shipments were by mail. The caterpillars, usually to the number of 100, were inclosed in these boxes, together with several twigs bearing fresh foliage.

The method was of doubtful utility, and at the same time no improvement upon it could be devised short of cold storage en route. On receipt the twigs would usually be stripped bare of foliage. Some of the caterpillars were invariably dead—whether from starvation or from injuries received at the time of collection or subsequently could not be determined. The remaining caterpillars were in all stages of emaciation and many of them, though still living, were too weak to recuperate.

Parasites in considerable variety but always in very small numbers issued, for the most part en route, but occasionally from the caterpillars after receipt. Nothing could be decided as a result of these importations and their repetition was resolved upon.

It was planned to import much larger numbers in 1908 without modifying the methods employed the year before. In this respect success was not achieved, principally, it would appear, on account of the difficulty of collecting these small caterpillars in numbers, especially in localities where the gipsy moth was not very abundant. Furthermore, it became increasingly evident that the percentage of parasitism (so far as it could be determined by the actual number of parasites secured) was so insignificant as to make the task of importing sufficiently large numbers of any one parasite for the purpose of colonization wholly impracticable. Many of the lots of caterpillars which were received in the best condition produced no parasites at all. It was therefore evident that if extensive operations in any locality should be determined upon, complete failure might result through the absence of the parasites in that particular locality during that particular season. Nothing less than an improvement in the service of several thousand per cent over that of 1907 or 1908 would answer, and this was altogether out of the question, except at an expenditure which even the generous funds appropriated by the State and Federal Governments could not cover. Further importations from Europe were regretfully decided to be impracticable.

·It has already been told how Prof. Kincaid spent the summer of 1908 in Japan in the interests of the parasite work. While there, in cooperation with the Japanese entomologists, he evolved a wholly new method for the transportation of the immature caterpillars of the gipsy moth, which would have been applicable in the case of European importations if it had seemed to be worth while to continue these importations in 1909. Large oblong wooden boxes having a capacity of about 1½ cubic feet were used. Like all the boxes received from Japan, they were most excellently constructed of a sort of wood which was less affected by dampness than most. The success of the work was very largely dependent upon both the character of the wood and the excellence of construction. It is certain that ordinary packing boxes would have warped to such an extent as to permit the escape of the small caterpillars.

These boxes (see Pl. V, fig. 2) were first lined with several thicknesses of absorbent paper, which was then thoroughly dampened. Small branches of a species of Alnus were attached to the sides, so that the interior was a mass of green foliage; the caterpillars to the number of several hundred were introduced, the cover tightly attached, and the whole sent in cold storage from Yokohama to Boston with scarcely an interruption en route. Sometimes the ends of the branches were thrust into a piece of succulent root (radish or potato), but this proved unnecessary, and rather a detriment than otherwise.

The condition of these boxes on receipt was usually good, and in some instances surprising. In some of the best of them scarcely a leaf was withered or even discolored, and in one in particular it seemed almost as though the branches had been freshly collected, with the early morning dew still clinging to the leaves. This illusion was almost instantly destroyed, for within an hour practically every leaf had dropped from the stem and was already beginning to blacken, as though struck by a sudden blight.

There was a good deal of difference in the condition of the caterpillars. Those which had been shipped in the second and third stages almost invariably arrived in the best condition. There was scarcely any mortality en route, and physically they were all in perfect health and ready to feed voraciously. Larger caterpillars did not survive their journey so well, and among those that had reached the fifth stage there was always a heavy mortality, and the survivors were never very healthy and would mostly die without feeding. It would appear that they were so heavy as to be thrown to the bottom of the box while dormant through cold, and thus become injured.

While technically a success, these attempts were practically failures. No parasites were secured in anything more than the most insignificant numbers, which could not be secured much more easily in other ways, and no further importations were attempted in 1909.

Relatively such small quantities of this class of material have.been received as to make unnecessary any specially devised methods for their economical handling.   With very few exceptions the boxes were opened immediately upon receipt and most carefully sorted for parasites and living caterpillars.   A few of the large Japanese boxes were not opened immediately, but holes were bored in the end, cones and tubes inserted, and living insects of all sorts thus attracted to the light and removed.   The living caterpillars were placed in cages or trays and fed, and occasionally a few parasites were thus secured in addition to those present in the boxes upon receipt.

It is very much to be regretted that the dead and dying caterpillars were not preserved for subsequent examination and dissection, but it was only in 1909, after the shipments of this sort of material had been discontinued, that the wholesale dissection of caterpillars was attempted for the purpose of ascertaining the proportion of parasitized individuals.   At the best, even after long experience, it is a tedious process, especially in the case of material which has been killed and preserved.

A few caterpillars, accidentally imported in their early stages with Apanteles cocoons in 1910, were saved and dissected with good results from a technical standpoint.

GIPSY-MOTH CATERPILLARS, FULL-FED AND PUPATING.

Importations of large caterpillars (Pl. VI) ready or nearly ready to pupate were first made in 1905, and it was demonstrated during that year that they could be brought to America with a fair degree of success, and that at least a proportion of the parasites with which they were infested could be reared.

Ever since 1905 we have been attempting to improve upon the methods first used during that year and have experimented with scores of modifications of the most successful, some of which were intentional while others were incidental to the fact that there have been many different collectors, each of whom has displayed some individuality in his methods of collecting and packing.   It would be tedious and is probably unnecessary to go into detailed descriptions of even a part of these various intentional or accidental experiments.

The most successful method yet devised involves the use of rather shallow wooden boxes having a capacity of from 40 to 70 cubic inches.   (See Pl. VIII, fig. 3.)   Quite a large number of shipments has been made in much larger boxes, but their condition on receipt has almost invariably been very bad.   The boxes *must* be tight to prevent the escape of tachinid larvæ, which can apparently pass through any

PLATE VL

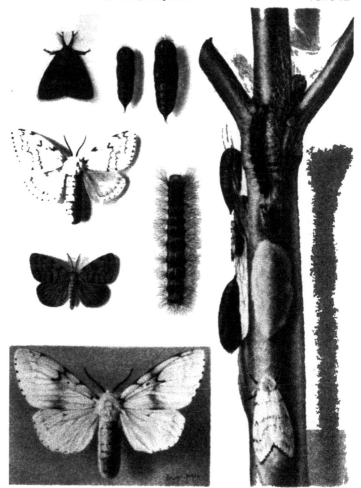

DIFFERENT STAGES OF THE GIPSY MOTH (PORTHETRIA DISPAR).

Egg mass on center of twig; female moth ovipositing just below, female moth below, at left, enlarged; male moth, somewhat reduced, immediately above; female moth immediately above, somewhat reduced, male moth with wings folded in upper left, male chrysalis at right of this; female chrysalis again at right; larva at center. (Original.)

opening large enough to accommodate the head. No special provision for ventilation is necessary, but it is necessary to construct the boxes of soft and absorbent wood in order to secure best results. This will not only prevent too rapid evaporation, but superfluous moisture will first be absorbed and subsequently will evaporate. Tin boxes are wholly unsuitable and paper or pasteboard have never been at all satisfactory.

Twigs with foliage should be included in each box, and these must be long enough to remain firmly braced in case the caterpillars eat the foliage. Some very bad results have followed the use of loose foliage, a practice which certain collectors have been persistent in following.

The fewer the caterpillars included in each box the better the results. The number has gradually been reduced from 100 at first to 20 during the past few years. Undoubtedly 10 would be better yet, but not enough better to make the added expense an economy.

The more nearly the caterpillars are ready to pupate when packed the better. If collected just a few days before pupation, they usually arrive in good shape, provided conditions otherwise are as they should be.

Shipments by mail have generally been successful when the boxes were not smashed, as has sometimes happened, or when something else was not wrong. Shipments by express without cold storage have been equally successful when the boxes have been properly packed. As has been said, there is no need to provide for the ventilation of the interior of the box, but the exterior must be exposed to the air on at least one side to permit the evaporation of the moisture absorbed by the wood. Otherwise, as nearly always happens, when a part of the caterpillars or pupæ die, they decompose, and as a result of their presence a similar fate usually overtakes the remainder. Some very large shipments were a complete loss in 1907, merely because a European agent, prevented by newly enforced postal regulations from making shipments by mail, packed the boxes tightly in large packing cases and forwarded them by express. When the lids were removed from these cases the sides of the boxes were found to be thoroughly damp, and the whole exhaled an ammoniacal odor so strong that it would seem of itself alone sufficient to destroy any ordinary form of insect life.

Bundles of boxes wrapped in thick, glazed paper have almost invariably been received in bad condition. If the paper is soft and absorbent it is generally satisfactory. One collector wrapped several packages in a thick fabric composed of tarred paper strengthened by muslin, and the contents rotted.

Cold storage in the case of shipments of this character would never have been either necessary or even desirable had it not been for the

difficulties experienced in the importation of *Blepharipa scutellata.*
All other species of parasites could be secured equally as well or better
from shipments under normal conditions, but because Blepharipa
differed from all the others in this apparently minor characteristic,
it was found necessary to make use of cold storage for practically
all of the very large shipments made during 1909 and 1910.

No large shipments of full-fed caterpillars have been made from
Japan. They were rendered unnecessary in the first place on account
of the very excellent and intelligent service rendered by the Japanese
entomologists. There are but three important parasites to be secured
from these large caterpillars in Japan, and the cocoons and puparia
of these have been reared and forwarded to us in specially devised
packages, with almost uniformly good results.

There have been a few very valuable lots of material of this character shipped otherwise than as above described. It is by no means
certain that if sufficient time and experimentation were to be devoted
to the subject some of these occasional and successful modifications
·might not be developed into something better than has yet been
tested. Any deviation is apt to prove disastrous, however, as witnessed in 1910, when failure resulted because the quality of the paper
used for wrapping the bundles of boxes was changed in several instances
from that employed at any time previously. It is very difficult, and in
practice impossible, to foresee such minor contingencies and provide
against them. The really serious phase of the situation lies in the
fact that such a slight modification may result not only in the complete loss of the shipment itself, but in a year's delay before it can be
remedied. By the time the first shipments are received and the
trouble recognized, it is apt to be too late to apply a remedy that
year, even by the use of the cable.

The laboratory methods in use for the handling of the parasite
material of this sort have been modified in various ways, more especially for the express purpose of overcoming the difficulty in hibernating the puparia of *Blepharipa scutellata.* Such of these modifications as have been primarily made for this special purpose will be
discussed in the account of Blepharipa which will be found elsewhere.

In general it has been the practice to open the boxes immediately
upon their receipt, and to sort the contents in accordance with their
character. The tachinid puparia were always carefully counted,
and of late years they have been sorted to a certain extent into
species.

In 1907 all the puparia were placed in jars, without sorting, with a
little very slightly dampened earth which was kept from drying by
the use of a wet sponge. In 1908 they were sorted to species, so far
as this was practicable, and all were kept dry. In 1909 the Blepharipa
puparia were sorted out and placed in earth as soon thereafter as

possible. The remainder were placed in small tube cages, and taken to the field where they were to be liberated. An attendant counted the number of flies issuing, and watched for secondary parasites. Since no secondary parasites issued in the summer from puparia secured in this manner, in 1910 the puparia were merely placed in cages which were taken to the colony site and left unattended until the flies had ceased to issue.

After the adults of the summer-issuing forms have all ceased to emerge, the sound puparia are more or less carefully sorted. Those supposed to be *Parasetigena segregata*, indistinguishable externally from dead Tricholyga or Tachina, are buried in damp earth for the winter. Mr. J. D. Tothill, one of the assistants at the laboratory, devised an ingenious method for separating the puparia containing the healthy pupæ of Parasetigena from those containing dead Tachina or Tricholyga, by holding them so that they were viewed against a narrow beam of very strong light. The method was not infallible, but served its purpose fairly well, and was the first of many which had been experimented with which was at all successful.

The living caterpillars removed from the boxes have been placed in cages or trays and fed, but only an insignificant proportion of them has ever lived long enough to be killed by the parasites which many of them have contained. Large numbers of them have been dissected for the purpose of determining the proportions parasitized.

The dead caterpillars not infrequently contain the puparia of Tricholyga, when this parasite happens to be common in the locality from which the shipment originated. Under such circumstances they are placed in tube cages for the emergence of the flies. If Tricholyga is not present in the boxes in the form of free puparia the dead caterpillars may as well be discarded, since it is only very rarely that any other tachinid pupates in this manner.

Pupæ, both living and dead, nearly always contain a considerable number of the larvæ of *Blepharipa scutellata*. They are, therefore, placed over damp earth in order that the larvæ may pupate under natural conditions. Other tachinids may occur in the pupa, but never in anything but insignificant numbers.

GIPSY-MOTH PUPÆ.

It would seem as though it ought to be an easy matter to import the pupæ (Pl. VI) of the gipsy moth in good condition, but for reasons which are not altogether clear in every instance the vast majority of the importations of pupæ have been worthless, or worse than worthless, since the handling of worthless material involves an additional waste of labor. Too often the cause of failure is directly and obviously the result of careless packing, and the number of lots of pupæ which have been received packed with the care which is essential to success is very small. The most successful shipments ever received were carefully packed in slightly dampened sphagnum moss, so arranged that the individual pupæ rarely touched each other. One or two successful importations thus received in 1908 were used as the basis for future instructions to collectors, and in every instance in which the directions were carefully followed in 1909 the results were equally good. In 1910 additional material apparently packed with the same care was received from the same source and via the same route. For no apparent reason whatever it was worthless when received.

In consequence of this another method which has occasionally been followed, will be recommended for shipments in 1911. This has been employed successfully upon various occasions, although without anything like uniform success, and is in effect the same as that used for the shipment of full-fed and pupating caterpillars and the same precautions must be used. It will probably be better to place a somewhat larger quantity of foliage in the box to prevent the pupæ from being thrown about too much in transit.

So very few shipments of gipsy-moth pupæ have been received at the laboratory as to have rendered unnecessary any special devices for their handling after receipt. Each year a few of the lots of caterpillars have contained a few individuals which were collected as prepupæ or as pupæ, and from such, an occasional parasite of one or another of the species peculiar to the pupæ has been reared. These have been so few, however, as to be of entirely inconsequential value, except from a technical standpoint.

The actual shipments of pupæ collected as such have been handled exactly as though they consisted of active caterpillars which pupate en route, with the one difference that the pupæ received have usually been inclosed in darkened cages with tubes attached, in order more easily to remove the parasites as they issued.

In 1908 pupæ were received in satisfactory condition for the first time since the preliminary shipments were made in 1905, and it was not until then that the host relations of several of the parasites, notably *Chalcis* spp. and *Monodontomerus æreus*, were finally determined. These lots were studied with the greatest care, each individual pupa being opened on receipt and for the most part isolated in a small vial, in order that it might be dissected after the contained parasite had issued.

BROWN-TAIL MOTH EGG MASSES.

No difficulty has ever been experienced in the importation of the egg-masses of the brown-tail moth (Pl. VII), except that when cold storage is not used a portion of the parasites are apt to hatch, and either escape or die en route. Wooden boxes of various sizes and

DIFFERENT STAGES OF THE BROWN-TAIL MOTH (EUPROCTIS CHRYSORRHŒA)

Winter nest at upper left; male and female adults, lower right; another winter nest, upper right: male and female chrysalides above, male at left; full-grown larva in center, somewhat reduced, young larvæ at its left; egg mass, the eggs hatching, at lower left; female ovipositing on leaf, egg mass also on same leaf. (Original.)

construction have been employed with uniformly good results, and nearly all shipments have been made by mail. All of the parasites have been found amenable to methods of laboratory control, and their reproduction has been undertaken as an economic venture in each instance. Under such circumstances there is no need to import large quantities except for the purpose of discovering other forms of parasites, should they exist.

HIBERNATING NESTS OF THE BROWN-TAIL MOTH.

The importation of hibernating nests (Pl. VII) of the brown-tail moth has been attended with very good success, as a rule, but by no means invariably. If they are sent too early in the winter and subjected to long continued high temperature before shipment, or while in transit, the caterpillars will die instead of resuming activity in the spring.

If sent too late in the spring, exposure to abnormally high temperatures en route results in premature activity of the caterpillars and they will arrive in bad and sometimes in worthless condition.

If sent in the middle of the winter they will be very nearly ready to resume activity on receipt, but if again exposed to cold they will become dormant and remain so until about the time when they would normally have become active. This seems not too prejudicial to them, if one is to judge by their activities during the first few weeks after the resumption of activity, but in some subtle manner a change has been wrought, and they do not commonly go through to successful pupation. This phenomenon, previously observed with other insects, is discussed at some length in the account of the tachinid *Zygobothria nidicola*, which hibernates within the caterpillars, but which does not destroy its host until after it has spun for pupation.

The failure to rear these caterpillars beyond a certain stage in the spring was at first attributed to some fault in the methods employed, and when it was finally apparent that the fault lay elsewhere there was no longer need to seek to remedy it. The one parasite desired was found to be already introduced and apparently well established as the result of a colonization some three years before.

The methods for handling the imported nests have varied from season to season in accordance with the habits of the parasites which it was desired to rear from them. These methods will be more fully described in the discussions of the several parasites involved: *Pteromalus egregius, Apanteles lacteicolor,* and *Zygobothria nidicola.*

IMMATURE CATERPILLARS OF THE BROWN-TAIL MOTH.

Taken all together, importations of active brown-tail caterpillars (Pl. VII) in the second to fourth spring stages aggregate a considerable number. These importations were undertaken on the supposition that there were parasites which attacked them as soon as they resumed activity in the spring, and left them before they reached their last stage. There are, indeed, several species which have such habits, but there is none amongst them which may not be secured equally well from either the hibernating caterpillars or from the full-fed and pupating caterpillars, or from both. In consequence the importation of partly grown caterpillars in the spring has never been attempted except experimentally.

Of the considerable number which has been received, nearly all have been packed in the same manner as are the full-fed caterpillars. They are nearly always alive on receipt, and usually feed voraciously when given an opportunity. Undoubtedly a considerable percentage contains parasites of those species which only emerge after the cocoon is spun, but every attempt to rear these parasites by feeding the caterpillars has resulted in failure. They will feed once, at least, but usually not more than a few times, and then die sooner than would have been the case had they not fed at all. Some among them will live for a long time, feeding a little but scarcely growing at all, and sometimes a very small percentage will complete growth and pupate. The percentage is so very small, however, and the labor and pain of handling the caterpillars is so great, as to render the work of feeding them of much more than doubtful economy, in every instance in which it has been attempted.

FULL-FED AND PUPATING CATERPILLARS OF THE BROWN-TAIL MOTH.

The temptation is strong to use the present opportunity for the purpose of giving vent to certain poorly suppressed and heartfelt expressions of opinion concerning the infliction known euphemistically and very inadequately as the brown-tail "rash." It is a very living subject of discussion during most of the year at the laboratory, but never more so than while the boxes of full-fed and pupating brown-tail caterpillars are being received from Europe.

Aside from the fact that the handling of this sort of parasite material has been productive of most acute physical anguish, it has been altogether the most uniformly satisfactory of any received. No modifications in the methods of packing have been suggested during the past five years, other than a slight modification in the form of the box in order to alleviate the trouble to which reference is made above.

Rather shallow boxes, having a capacity of about 50 cubic inches, are used for the purpose. (See Pl. VIII, figs. 1, 2.) The caterpillars are collected, preferably just before they spin their cocoons, and 100 are placed in each box, together with a few twigs with foliage attached, which serve less as food than as a support for the cocoons. When collected at the proper time practically all will spin and pupate en route, and the cocoons are so strong as to prevent the pupæ from becoming injured.

There has never been any trouble experienced through these boxes sweating en route, as has so frequently happened when boxes containing gipsy-moth caterpillars have been too closely confined in box or bundle, but at the same time those which have been exposed to free circulation exteriorly are noticeably in better condition than others.

Much the larger proportion of the material of this sort has been received in perfect condition at the laboratory, and large quantities of parasites have been reared from nearly every lot thus received. Occasionally boxes have been used which were not sufficiently tight to prevent the escape of the tachinid larvæ, and some loss has accrued in consequence. In a number of instances the boxes have been infected with fungous disease, and all or nearly all of the caterpillars or freshly formed pupæ have died in consequence. In rather an unnecessary number of instances, or so it would seem, caterpillars have been collected too young, and have failed to pupate en route. Such shipments properly fall in the class last mentioned, and are worthless for the purposes desired.

Through a misunderstanding nearly all of this class of material was sent in cold storage in 1909, with the result that the caterpillars failed to pupate en route, as would have been the result otherwise, and a good many of them failed to pupate after receipt. Considerable loss resulted on this account before the collectors could be notified to return to the original method of shipping by mail.

In handling the boxes of caterpillars and pupæ, a variety of methods has been employed, of which the most satisfactory appears to be simply boring a hole in the one end, and introducing a paper cone and tube. Even the removal of the covers from a dozen or more boxes without protection is accompanied by painful results, and owing to the difficulty of boring the holes without splitting the wood, after the box has been received at the laboratory, collectors are now instructed to prepare the boxes for the reception of the tube at the time of their manufacture.

In 1906 and 1907, when the first shipments of this character were received, and when little was known of the character of the parasites which were likely to be reared from them, it was thought necessary not only to open each box, but to sort it over, and remove the tachinid puparia which were always present in a larger or smaller number. It was known that there were always present certain species of tachinids which would only complete their final transformations successfully when their puparia were kept more or less moist, and it was expected that among the parasites of the brown-tail moth would be some possessing this characteristic. Opening the boxes without some sort of protection was utterly impossible.   Automobile goggles were used to protect the eyes, various forms of respirators to prevent the inhalation of the spines, the hands were protected by rubber gloves, and the neck and face were swathed in accordance with the fancy of the operator.

Two ingenious types of headdress (Pl. IX, fig. 1) were devised by Mr. E. S. G. Titus in the hope that they would solve the difficulty, but it was found that they were not only unbearably hot, but that the glass fronts would quickly become covered with moisture which could not be removed.

In 1907 a much larger quantity of this sort of material was received than during the previous summer, and it was practically a necessity that some method be devised which would do away with at least part of the trouble.   After some little experimentation the arrangement shown in the illustration was the result.   (Pl. IX, fig. 2.)   It consisted of an ordinary show case, with sides and top of glass and with a wooden slide in the back.   The two ends were removed and replaced with boards in which armholes had been cut.   Thick canvas sleeves were attached to these, through which the gloved hands of the operator were thrust, and it was found that the work could be done with what was, comparatively speaking, a minimum of discomfort and danger.

In 1908, for several reasons which need not be entered into here, it was thought desirable to discontinue, temporarily, the importation of large quantities of the pupating caterpillars, and it was also demonstrated that all of the parasites which were secured from them would complete their transformations without being kept moist.   The work of sorting over the boxes of parasite material was thus demonstrated to be unnecessary, and, consequently, in 1909, when large importations were resumed, the covers were simply removed from the boxes, which were then stacked up in the large wooden tube cages (Pl. X, fig. 1), which had originally been constructed for the rearing of parasites from the imported hibernating nests.

BROWN-TAIL MOTH PUPÆ.

Several attempts have been made to ship the pupæ (Pl. VII) of the brown-tail moth, packed in moss, as was at one time recommended for the shipment of the pupæ of the gipsy moth.   Such attempts have usually been more or less satisfactory, but never as satisfactory as when the cocoons were collected in the field and placed loose in the boxes together with the active caterpillars.   If only a small portion of the pupæ is collected in the field, the only sure method of detecting their presence is by the occurrence of the pupal parasites

FIG. 1.—BOXES USED IN 1910 FOR IMPORTATION OF BROWN-TAIL MOTH CATERPILLARS, WITH TUBES ATTACHED DIRECTLY TO BOXES. (ORIGINAL.)

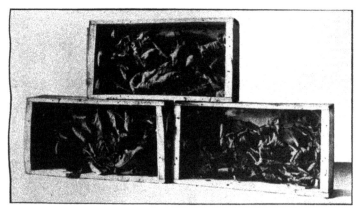

FIG. 2.—INTERIOR OF BOXES IN WHICH BROWN-TAIL MOTH CATERPILLARS WERE IMPORTED, SHOWING CONDITION ON RECEIPT. (ORIGINAL.)

FIG. 3.—BOXES USED IN SHIPPING CATERPILLARS OF THE GIPSY AND BROWN-TAIL MOTHS BY MAIL. (ORIGINAL.)

FIG. 1.—HEADGEAR DEVISED BY MR. E. S. G. TITUS FOR PROTECTION AGAINST BROWN-TAIL RASH. (ORIGINAL.)

FIG. 2.—SHOW CASE USED WHEN OPENING BOXES OF BROWN-TAIL MOTH CATER-PILLARS RECEIVED FROM ABROAD. (ORIGINAL.)

FIG. 1.—LARGE TUBE-CAGE FIRST USED FOR REARING PARASITES FROM IMPORTED BROWN-TAIL MOTH NESTS AND LATTERLY FOR VARIOUS PURPOSES. (ORIGINAL.)

FIG. 2.—METHOD OF PACKING CALOSOMA BEETLES FOR SHIPMEMT. (ORIGINAL.)

among those reared. A large number of the shipments has produced small or large percentages of these parasites.

Shipments of pupæ collected as such would preferably be made in cold storage. The most of the parasites, including those which are or which appear to be of the most importance, emerge coincidently or nearly so with the moths themselves, and if sent by ordinary mail they are apt to issue and die en route.

COCOONS OF HYMENOPTEROUS PARASITES.

There is only one hymenopterous parasite of demonstrated importance which attacks the gipsy moth, and which spins a cocoon outside of the host. This is *Apanteles fulvipes*, of Europe and Japan, and it is probable that the numbers of its cocoons imported as such have amounted to at least 1,000,000.

Little care is necessary in packing these for shipment, other than that they must not be crushed, nor yet too damp. A considerable degree of dampness has been sustained without injury, but upon one occasion in which they were packed between sheets of damp blotting paper, there was sufficient moisture present to thoroughly soak the cotton and some loss resulted.

The Japanese have displayed no little ingenuity in devising new methods for sending these, and with one exception, just noted, all have been good so far as packing was concerned. One method, which possessed a certain advantage over the others in permitting the adults which chanced to emerge en route a certain amount of very advantageous freedom, was used in a single shipment, which, partly on that account but principally on others, ranks as immeasurably the best ever received. The cocoons, to the number of about 1,000, were inclosed in a little wicker cage, which in turn was inclosed in an envelope of mosquito netting which prevented the cocoons from scattering out, but did not hinder the escape of the adults. This cage was supported in the very center of a large, otherwise empty wooden box by means of strings which were passed through screw eyes in the middle of each side and drawn taut. There was nothing loose in the box to crush the delicate parasites, no matter how roughly it was handled, and they were not only given ample space to expand and stretch their wings, but they were kept inactive by the perfect darkness (or at least were presumably so). It would be a simple matter to spray a portion of one side of the box with a very fine dew of honey, and if this were done the life of the adults would probably be considerably prolonged.

Cold storage is an absolute necessity if the cocoons of this parasite are to be as much as a week en route. The transformations are apt to be concluded in considerably less than one week after the spinning of the cocoon if the weather is hot, and ever under the best of conditions which can be devised for keeping them alive the mortality is heavy. Even in the ordinary temperature of a steamship's cold room development continues.

TACHINID PUPARIA.

The importation of tachinid puparia is by no means so simple as the importation of Apanteles cocoons, but at the same time it is easy as compared with the difficulties attending the importation of live gipsy-moth caterpillars from which to rear these puparia in America.

In all, quite large numbers have been received from both Europe and Japan. A variety of methods has been tested in the hope of hitting upon one that would be applicable for the purpose. Shipping in damp earth was early attempted, and seems to be the very first method which suggests itself to anyone wishing to ship a quantity of them, but of all ways it is very nearly the worst. It would probably be the best, if the larvæ could be allowed to enter the earth naturally and if they were left there wholly undisturbed throughout the time they were in transit, but mingled with damp earth and placed in a box to be sent by mail or express, disaster is pretty sure to result. Cotton has also been used several times, and it is usually as bad and sometimes worse. With the exception of excelsior, cotton is about the worst packing for living insects that has come under observation at the laboratory, although gritty moss, of a sort which dries brittle, is also bad. Presumably there are other worse substances, but they have not been discovered at first hand.

Probably the best packing material is slightly damp and preferably living sphagnum moss. The live moss retains its moisture in a manner wholly different from the moss which has been killed, dried, and subsequently dampened. Test shipments, which were sent to France and back without being opened, returned to the laboratory in good and almost unaltered condition, in the case of those which were inclosed in tight boxes. Even when fully exposed to the air the living moss seems to dry much more slowly and to hold its moisture more naturally. Sphagnum possesses the great additional advantage of being much softer when dry than most other kinds of moss.

One disadvantage attending the shipment of puparia, no matter how they are packed, is that of secondary parasites. A single colony of Dibrachys, issuing en route from one of a lot of puparia, will result in the parasitism of a large proportion of the remainder. This might possibly be prevented by packing in sand or earth, but this appears to be about the only advantage possessed by that method.

The puparia of certain tachinids must be kept damp, but this is not at all necessary in the case of all. Methods of packing and shipment will depend upon the characteristics of the particular species under consideration. So far as known, no tachinid which forms a free puparium outside of its host is injured by exposure to moisture.

The pupal period of the majority of the tachinid parasites of the gipsy moth and the brown-tail moth is quite short, usually lasting less than two weeks. It is therefore necessary to make use of cold storage en route, in order to make certain that the adults will not hatch before arrival. By far the larger part of the puparia which have been received at the laboratory have thus hatched, except when they were of species which naturally hibernated unless they were shipped in cold storage.

CALOSOMA AND OTHER PREDACEOUS BEETLES.

Quite a variety of the large carabid beetles has been imported from abroad for experimentation as to their serviceability as enemies of the gipsy moth, or for liberation in the field after this point had been demonstrated satisfactorily. At first some difficulty was experienced in accomplishing their importation successfully, but later it was found to be a simple matter if proper care was used in packing. The great majority of them have come in ordinary safety-match boxes (Pl. X, fig. 2), each box containing one beetle and a wisp of sphagnum moss. Usually one or two caterpillars or other sort of *succulent insect* have been included for the purpose of lunch en route, but the practice is of rather doubtful value in the case of those species which have been handled in the largest numbers in the laboratory. Should the beetle not fancy the quality of the sustenance provided, or refuse to eat for any other reason, death and decomposition of the victim may result disastrously and be prejudicial to the health of the beetle.

These small match boxes have been packed in larger wooden boxes and sent through the ordinary mails with little loss of life. Other small wooden or paper boxes have similarly been used with equal success.

Cold storage has occasionally been employed in a few minor shipments from Europe with very good results. In 1910 a large shipment of living beetles was received in cold storage from Japan in most excellent condition.

QUANTITY OF PARASITE MATERIAL IMPORTED.

In Mr. Kirkland's first report as superintendent for suppressing the gipsy moth and the brown-tail moth in Massachusetts, several pages were devoted to a detailed account of each shipment of parasite material received from abroad. After the first year no attempt to continue this practice was made, and if it were now attempted to treat each separate shipment with the same attention to detail, several hundred additional pages would be required.

Accordingly, in order that at least a rough idea of the quantity of material handled at the laboratory may be had, Table VII, which, without being absolutely accurate, is very approximately so, has been prepared by Mr. R. Wooldridge, an assistant at the laboratory.

TABLE VII.—*Table showing number of boxes received at the laboratory since beginning of work.*

	1905	1906	1907	1908	1909	1910
Porthetria dispar egg masses............ .....boxes.		1	1	18	32	1
Porthetria dispar larvæ and pupæ...... ... .do....	131	923	1,539	307	8,391	5,956
Euproctis chrysorrhœa egg masses............do....		46	87	17	1	0
Euproctis chrysorrhœa webs. . . . .webs..		117,259	55,082	32,830	29,295	29,696
Euproctis chrysorrhœa larvæ and pupæ.. . .boxes..		313	1,159	160	1,167	381
Apanteles fulvipes and Apanteles lacteicolor..do....				13	21	63

LOCALITIES FROM WHICH THE PARASITE MATERIAL HAS BEEN RECEIVED.

Mr. Wooldridge has also prepared the accompanying map (fig. 11) showing the various localities from which parasite material has been received each year from 1905 to 1910, inclusive. It will indicate the thoroughness with which the more accessible parts of the world have been searched for parasites of these pests.

THE EGG PARASITES OF THE GIPSY MOTH.

ANASTATUS BIFASCIATUS FONSC.

The first individuals of this species (fig. 12, female) were reared at the laboratory in the spring of 1908 from eggs imported the previous winter from Europe and Japan. The dissimilar sexes were not immediately recognized as of the same species, and for a few days there was some doubt as to whether one, two, or four were represented among the few scattering specimens emerging. The number was soon reduced to two through the obvious attraction between the sexes, and soon after to the one, when the senior author had an opportunity to examine and compare series from European and Japanese sources.

Their issuance had been anticipated a long time before, and a quantity of gipsy-moth eggs had been collected in the summer, before embryonic development had progressed beyond its initial stages, and placed in cold storage. It was thought possible that some species of parasite might be reared from imported eggs during the fall or winter which habitually and necessarily oviposited in undeveloped eggs, and it was hoped that those collected in the summer might be kept fresh enough to serve as host material for laboratory reproduction.

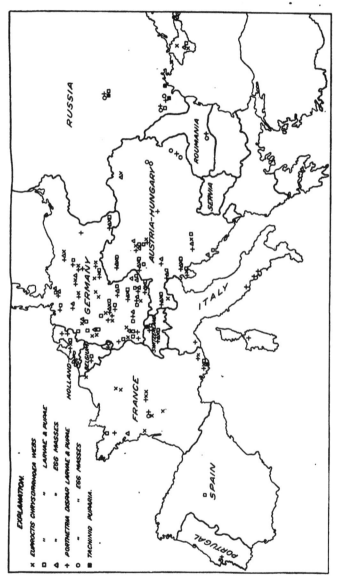

Fig. 11.—Map showing various localities in Europe from which parasite material has been received. (Original.)

As soon as females of Anastatus were secured, some of these eggs were removed from storage and found to be dead, with the contents partially decomposed. Nevertheless an attempt was made to use them, and the parasites were given their choice between them and others which contained embryonic caterpillars.

A few days after their emergence the females began to betray an interest in both sorts of eggs, and were several times observed in the act of oviposition or attempted oviposition. Apparently this was successfully accomplished, but without further results, for no second generation resulted. The experiment served one purpose, however, in indicating beyond reasonable doubt that the insect actually was a parasite upon the eggs of the gipsy moth, and not upon any chance form of insect life accidentally included.[1]

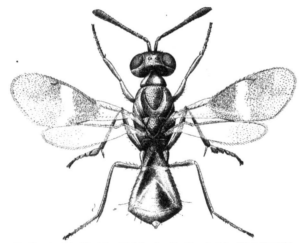

Fig. 12.—*Anastatus bifasciatus:* Adult female. Greatly enlarged. (From Howard.)

The exposure of the imported eggs to warmth for the purpose of hastening the emergence of any parasites which they chanced to con-

---

[1] How great is the likelihood of error when parasites are reared from unbroken egg masses has several times been demonstrated in the course of the investigations and rearing work at the laboratory. Upon several occasions small Lepidoptera have been reared from egg masses, and more than once their parasitized pupæ have been found. Eggs of other species of insects, and occasionally parasitized scale insects, have also been found attached to bits of bark to which egg masses were attached. Very frequently cocoons of *Apanteles fulvipes* are found, wholly or partially covered by the egg mass, and from them several species of hibernating secondaries have been reared. There is a record of a minute eulophid, allied to Entedon, having issued from a small lot of eggs which had been separated from nearly every trace of foreign matter. It was thought then and is still believed that these came from the eggs themselves, and that they were actually parasitic upon either Anastatus or Schedius, but when the material from which they issued was examined two or three cocoons of *Apanteles fulvipes* were found mingled with it, and what might otherwise have been a clear record was spoiled.

There is in Japan a limacodid moth—*Parasa sinica* Moore (*hilarula* Staud), as determined by Dr. H. G. Dyar—which appears habitually to seek out the gipsy-moth egg masses as a site for pupation. The larva buries itself in the mass before spinning its cocoon, and from outward appearances its presence is hardly noticeable. More than 25 of these moths have been reared under these circumstances from imported egg masses, or their cocoons have been found and destroyed.

tain in the hope that laboratory reproduction could be secured was soon recognized to be a mistake, and as the Anastatus continued to emerge considerably ahead of the time when they would obviously have issued under more natural conditions, it was resolved to remedy the evil, if possible, by placing the parasitized material in cold storage. This experiment was successful. The further transformations of the parasites were retarded without any apparent prejudicial effects upon their vitality, and in July some 500 were reared and colonized in the field.

Fig. 13 *Anastatus bifasciatus* Uterine egg. Greatly enlarged. (Original.)

Coincidently with the height of their emergence and subsequent to its close, a considerable number of a small black encyrtid, later described by the senior author as *Tyndarichus navæ*, issued, and all were destroyed on the supposition that they might be secondary. This was not by any means certain, and it was resolved to investigate their habits thoroughly so soon as opportunity should offer.

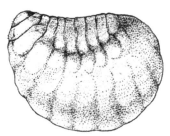

Fig. 14.—*Anastatus bifasciatus:* Hibernating larva. Greatly enlarged. (Original.)

Accordingly, in the fall of 1908, following the receipt of several considerable shipments of egg masses from Japan, an exhaustive investigation of the gipsy-moth egg parasites was inaugurated. These investigations were more intimately associated with the work upon Schedius, and more will be said of them in the discussion of that species. So far as Anastatus was concerned, its life and probable habits stood revealed from the start. Almost in the beginning its larvæ were found (fig. 14) and identified correctly, as was later proved. They were almost invariably found in eggs which had been destroyed before embryonic development had taken place, which showed conclusively that these eggs were attacked within a very short time after their deposition. It was known that the adults did not issue until after the caterpillars had hatched from healthy eggs in the spring, and the fact that the species was single brooded, with a life cycle that was correlated perfectly with that of the gipsy moth, was as

Fig. 15.—*Anastatus bifasciatus:* Pupa from gipsy-moth egg. Greatly enlarged. (Original.)

certainly evident then as now, after two years' observation of its progress in the field has given ample confirmation.

The egg of Anastatus has not been seen after deposition, but its appearance before is indicated by figure 13. The full-fed larva removed from host egg is well represented by figure 14, and the pupa by figure 15. This latter is very beautifully colored, the creamy ground color being set off by darker abdominal bands and wing covers, and by the delicately tinted reddish eyes.

It was soon demonstrated by a careful study of the European eggs that no other parasite and no secondaries were present. These eggs were therefore kept in confinement until after the caterpillars had all hatched in the spring. Then those which remained were examined, and the number which contained parasites carefully estimated, and found to be about 80,000, nearly all of which were contained in a very large shipment received during the winter through Prof. Jablonowski, and collected from various Hungarian localities.

The Japanese eggs, which contained numerous secondary parasites as well as Anastatus, were all carefully rubbed clear of their hairy covering, and those which contained Anastatus larvæ (Pl. XI, fig. 2) carefully and painstakingly picked out by hand, one by one. In this manner enough to make a total of nearly 90,000 of the parasites were secured.

One exceedingly important characteristic of the parasite was not considered with sufficient attention at a time when this might have been done. Several observations upon the activities of the females in the summer of 1908 had led the observer to question their ability to fly; but when several of them were placed upon a large sheet of paper and stirred into action, they disappeared with sufficient celerity to banish any doubts which may have been entertained. In considering these crude experiments in retrospect and in the light of subsequent developments, it would appear that their jumping abilities were rather underestimated, because it is now certain that they are either unwilling or else, like the female of their host, are unable to fly.

A most careful examination of egg masses in the vicinity of the locality where the colony of about 500 had been liberated the summer before had failed to discover the presence of parasitized eggs. It is now known that this was due to the accident of placing this colony in a locality where the gipsy-moth "wilt" disease proved later to be so destructive as to kill all the pupæ which were present when the first of the parasites were liberated, and which it was then thought would produce moths enough to deposit a sufficiency of eggs for attack. As a result of this extreme percentage of pupal mortality, there were practically no eggs within a radius of several hundred feet.

The cause of failure not being apparent, it was guessed that it might be due to the extremely rapid rate of dispersion rather than to the reverse, and to provide against loss through too rapid dispersion at first, very large colonies were decided upon as most advisable.

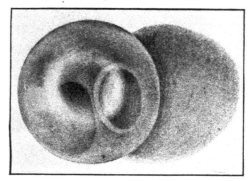

FIG. 1.—EGG OF GIPSY MOTH, CONTAINING DEVELOPING CATERPILLAR OF THE GIPSY MOTH. GREATLY ENLARGED. (ORIGINAL.)

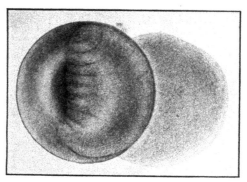

FIG. 2.—EGG OF GIPSY MOTH, CONTAINING LARVA OF THE PARASITE ANASTATUS BIFASCIATUS. GREATLY ENLARGED. (ORIGINAL.)

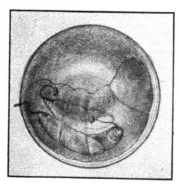

FIG. 3.—EGG OF GIPSY MOTH, CONTAINING HIBERNATING LARVA OF ANASTATUS BIFASCIATUS, WHICH IN TURN IS PARASITIZED BY THREE SECOND-STAGE LARVÆ OF SCHEDIUS KUVANÆ. GREATLY ENLARGED. (ORIGINAL.)

Fig. 2.—Views of Cage Prepared for Use in Colonization of Anastatus bifasciatus in 1911. A, Front View of Cage; B, Bottom of Cage. (Original.)

Fig. 1.—View of Cage Used for Colonization of Anastatus bifasciatus in 1910. (Original.)

The 90,000 parasitized eggs were divided into five lots and placed in the field at the proper time in localities where an abundance of eggs was certain.

The parasites hatched in due course and were found attacking the egg masses in a businesslike manner that was quite encouraging, but only within a very short distance of the center of the colony. The species thus spreads slowly.

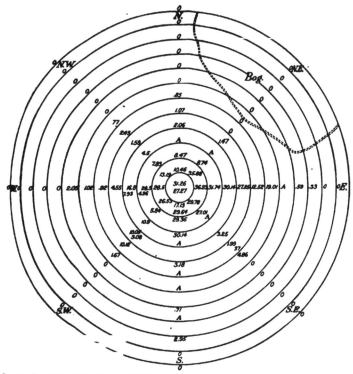

Fig. 16.—Diagram showing two years' dispersion of *Anastatus bifasciatus* from colony center. Each concentric circle represents a distance of 50 feet from the smaller or larger circle next it. *A* indicates parasitism of gipsy-moth egg-masses by *Anastatus bifasciatus* and *O* indicates absence of parasitism by Anastatus. The figures give percentages of parasitism. (Original.)

The accompanying diagram (fig. 16), which has been prepared by Mr. Wooldridge largely from the results of his own work, together with Table VIII, will serve as well as words to tell the story of the dispersion of this parasite in one of the 1909 colonies, and results of similar studies in various other colonies are substantially the same.

TABLE VIII.—*Average percentage of parasitism of gipsy-moth egg masses at different distances from center of colony.*

Distance from center.	Number of egg masses collected.	Percentage of parasitism.	Distance from center.	Number of egg masses collected	Percentage of parasitism
At center	20	29.26	350 feet........................	50	0.41
50 feet..........................	66	24.68	400 feet........................	70	.18
100 feet........................	78	21.75	450 feet........................	60	.055
150 feet........................	80	14.43	500 feet........................	70	.42
200 feet........................	60	8.61	550 feet........................	70	.00
250 feet.....	100	3.59	600 feet........................	70	.00
300 feet........................	85	3.44			

When in the fall of 1909 it had become rather certain that the rate of dispersion of Anastatus was only going to be about 200 feet per year, plans for colonization along very different lines in 1910 were immediately put into execution. In four of the five colonies all of the egg masses which could be easily secured were collected and brought to the laboratory, where the eggs were separated from their hairy covering. This is best effected by gently rubbing them over a piece of cheesecloth stretched on a frame. (Pl. XX, fig. 2.) The hairs pass through and the eggs are left.

The number of parasitized eggs present was then estimated, and found to be very close to 90,000. In the spring, after all of the healthy eggs had hatched, those remaining, including all which were parasitized, were divided into 100 lots, each of which was supposed to contain approximately 900 parasite larvæ. An equal number of small, wire-screen cages was prepared (Pl. XII, fig. 1), and about the middle of June, when the male parasites began to issue, and when it was becoming possible to determine with some degree of assurance just where there were likely to be large numbers of gipsy-moth eggs a little later, the work of placing these cages in the open was begun. (See also Pl. XII, fig. 2, showing front and bottom of cage prepared for use in Anastatus colonization in 1911.) They were finally placed, each in a separate locality, and each, so far as has been determined by subsequent investigation, in localities where the parasites had an excellent opportunity to work to the best advantage as soon as they issued. Not all of these colonies have since been visited, and probably some of them never will be seen again, but all that have been examined have been found in the best of condition.

Early in the fall of 1910 the dispersion studies of 1909 were repeated, with results which have already been indicated in Mr. Wooldridge's diagram, and the egg collections were also repeated for the purpose of securing material for additional colonization work in 1911. With little difficulty some 270,000 parasitized eggs have been secured, and were it not for the fact that the proper care in placing the number of colonies thus provided for will probably tax all available resources at the time when they must be placed, if placed to advantage, more could easily be collected.

The rate of increase in the field, as indicated by the work which has been done, is not excessive, but probably amounts to something like sixfold per year. The extreme limit of dispersion discernible in 1909 was not quite half that of the extreme for two years, as indicated in the diagram. It is possible that it may become more rapid as time goes on, and it is rather expected that a high wind, at an opportune time, will assist materially in the dispersion of the species. Should it not, it will require a very long time for it to become generally established everywhere through the infested area. Even though there were a colony planted to each square mile, something like 16 years would elapse before all of them met and fused, unless the present rate of dispersion were accelerated.

It has been pretty definitely proved of Schedius that it can only attack the uppermost layer of eggs in each mass, and the same is equally well proved in the case of Anastatus. Since there are two layers of eggs, and usually three in all but the very smallest masses, it is evident that the usefulness of Anastatus is still further reduced through its physical limitations. The figures of percentages given in the diagrams probably represent about the maximum which can ever be expected. None the less, this means a distinct benefit, and with all its faults, Anastatus stands high in favor at the present time.

In its distribution abroad, Anastatus is, as might be expected, of quite local occurrence. It has been received from about half of the localities represented by the European importations, and in very variable abundance. The numbers found in five lots of what was estimated as 1,000 egg masses each, received from five different localities in Hungary through Prof. Jablonowski in the winter of 1908–9, is rather typical in this respect. As estimated through careful examination and counts, these numbers were as follows:

Laboratory Number.	Locality.	Number of Anastatus.
3017.....	Lippa (Temes).....	34,000
3018.....	Bustyhaza (Maramoros)...............	0
3019.....	Huszt (Maramoros)...................	208
3020.....	Dorgos (Temes)	6,099
3021.....	Sistarobecz (Temes).................	39,000
	Total....................	79,307

In Japan it is also unevenly distributed. The most which were received from that country were in a lot of eggs from Fukuoka Ken, received during the same winter as those above mentioned from Hungary. It is not at all common from the vicinity of Tokyo, and while it is present in nearly every lot of Japanese eggs which has been received, in every instance but one (the shipment above mentioned) the number

present has not been sufficiently large to make the rearing of the
parasite economically worth while.   It is interesting and possibly
significant that there was no Schedius in the one locality where Anastatus was sufficiently common to be considered as a parasite of consequence, while in the other localities, where Anastatus was rare,
Schedius abounded.   More than one instance has been observed in
which parasites having similar habits alternate but rarely or never
occur simultaneously in anything like equal abundance in one locality.
Two fairly consistent examples of this sort will receive further mention later on, in which the tachinids *Dexodes nigripes* and *Compsilura
concinnata,* and *Tachina larvarum* and *Tricholyga grandis* are respectively involved.

SCHEDIUS KUVANÆ HOW.

Only one species of gipsy-moth egg parasite has been received at
the laboratory from Europe, but in Japan there are two, and, so far

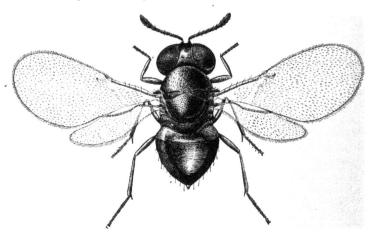

FIG. 17.—*Schedius kuvanæ:* Adult female.   Greatly enlarged.   (From Howard.)

as may be determined from their comparative abundance in the
material from that country which has been studied, *Schedius kuvanæ*
(fig. 17) is the more common and important as a factor in the control
of its host.   It resembles Anastatus in its choice of host, and in the
fact that it is similarly limited through physical inability from attacking more than a limited percentage of the eggs in each mass.   In every
other respect the two species are widely different.

Anastatus is a true egg parasite, and rarely attacks successfully
the eggs in which the young caterpillars have begun to form.   She-
dius, on the contrary, is strictly speaking an internal parasite of the
unhatched caterpillar.   Anastatus passes through but one genera-

tion annually, and its seasonal history is closely correlated with that of its host. Schedius, on the contrary, will pass through a generation per month, so long as the temperature is sufficiently high, and its seasonal history is in no way correlated to that of the gipsy moth. It appears not to hibernate in the gipsy-moth eggs, and it is quite probable that an alternate host is necessary to carry it through the summer months after the gipsy-moth eggs have hatched in the spring, and before the moths begin depositing eggs for a new generation.

At the time when the popular account of the parasite-introduction work was prepared for publication through the office of the Massachusetts State Forester it was considered to be much the more promising of the egg parasites, and its history in America was spoken of as one "of the most satisfactory episodes in the work of parasite introduction." The account of the first successful importation of living specimens as given at that time is included in the two following paragraphs, which are quoted verbatim.

As long ago as the spring of 1907 a few dead adults were secured in an importation of gipsy-moth egg masses received during the winter from Japan, but none was living on receipt. During the winter next following, large importations were made, and many thousands of eggs, from which some parasite had emerged, were found, but not a single living specimen was obtained. It was evident that it completed its transformations and issued in the fall, and that, if it hibernated in the eggs, it was warmed to activity while the packages were in transit to America, and the adult parasites either died or escaped en route.

In the fall, winter, and spring of 1908-9 a large quantity of eggs of the gipsy moth were received from Japan, the shipments beginning early in the fall and continuing until nearly time for the caterpillars to hatch in the spring. The first, received in September, contained hundreds, possibly thousands, of the parasites, which had issued from the eggs en route, and all of which, as usual, had died; not a single living individual was received. Specimens were referred to Dr. Howard, who found that they represented an entirely new and hitherto undescribed species, which he named after Prof. Kuwana, who collected and sent the eggs from which they had issued. A single pair of living specimens rewarded the careful attention which was lavished upon the importations received later in the fall and during the winter, and it was not until April, 1909, that a mated pair could be secured. During that month a total of 11 individuals issued from cages containing Japanese eggs recently received.

These 11 individuals served as the progenitors of a numerous and prolific race, but the story of the investigations which were made upon the various shipments of egg masses received at the laboratory from September, 1908, to April, 1909, which was not touched upon in an earlier account, is perhaps worthy of a place here.

LIFE OF SCHEDIUS AND ITS RELATIONS TO OTHER EGG PARASITES, PRIMARY AND SECONDARY.

Mention has already been made of the rearing of a small encyrtid parasite from Japanese eggs in company with Anastatus in the summer of 1908, of the doubts which were felt as to its true character, and of the resolve to investigate the matter thoroughly when the opportunity should arise. In accordance with this resolve an intensive study of the Japanese importations was begun in December, 1908. A large number of egg masses, which showed by the exit holes (of Schedius) that they had been freely attacked by some parasite which had issued in the fall, were selected, then "sifted," and the eggs from each mass were then carefully examined and sorted into three lots, composed, respectively, of the healthy eggs, the eggs from which parasites had issued, and the eggs which were neither one nor the other. Those falling in this third division were scrutinized again with still more care. · Anastatus was quickly recognized, in most instances, and eggs containing its larvæ placed aside. In the majority of the remainder there was evidently no life, but in a considerable number minute, white larvæ could more or less plainly be seen, surrounded and more than half concealed by the remains of the embryonic caterpillars which had been destroyed. These eggs were isolated in small vials, in order that there could be no question concerning the identity of the particular host egg from which any particular parasite issued.

Long before this work was completed the necessity for all the care that was being expended to secure accurate results was made manifest by the emergence of no less than three species of parasites from isolated or partially isolated eggs. The first of these to appear was a species of Pachyneuron (determined by the senior author as *P. gifuensis* Ashm.), and on account of known habits of other members of the genus was placed as probably secondary. Nevertheless it was given an opportunity to prove itself a primary if it would, and the specimens as they issued were confined in vials with gipsy-moth eggs, some of which contained the healthy caterpillars, while others harbored the larvæ of Anastatus. The Pachyneuron paid not the slightest attention to either, but invariably died without attempting oviposition.

The next species to issue was *Tyndarichus navæ* How., and it was with considerable surprise that it was recognized as different from Schedius. On account of the strong superficial resemblance between the two it had been supposed up to that time that they were one and the same.

The third was *Perissopterus javensis* How., of which a single specimen only was reared. To date this record is unique, and the species has previously been reared only from scale insects.

There was other and pressing work to be done with the parasites of the hibernating brown-tail caterpillars, and a realization of the difficulties which were likely to attend the prosecution of the egg-parasite investigations, thus complicated by the discovery that five and possibly more parasites were involved of which only one was definitely proved to be primary, was the prime argument which finally resulted in the detachment of Mr. H. S. Smith from the cotton boll weevil investigations and his transfer to the laboratory staff. By the time he was prepared to undertake his new work a large number of eggs from which Anastatus, Tyndarichus, and Pachyneuron were positively known to have issued were ready for dissection and study, and to these were soon added a number from which Schedius was similarly known to have come, secured in the manner about to be described.

The first Schedius which was ever reared in a living condition issued from an isolated egg in the laboratory in December, 1908. It was a male, and it died before it could be furnished with a mate. The next individual issued on January 8 from an egg which had been isolated on December 19. It was a female, and she was immediately transferred to a large vial containing an egg mass freshly collected from the field. Within a few days after being thus confined she was observed in the act of oviposition, and parthenogenetic reproduction ensued. Her progeny began to issue February 16, and up to February 25 no less than 28 males were reared.

The experiment was tried of confining her with several of her asexually-produced progeny in the hope that she might thus be fertilized and produce females. The experiment did not succeed at that time, apparently because she was not able to deposit any more eggs. She remained alive until March 2, but was dead on March 6, after at least eight weeks of active life.

Fig. 18.—*Schedius kuvanæ:* Egg. Greatly enlarged. (Original)

The eggs from which these parthenogenetically-produced males issued were known beyond peradventure of a doubt to have produced Schedius, and never to have contained any other parasite, and together with those from which Anastatus, Tyndarichus, and Pachyneuron were known to have issued, made complete the series which was to be dissected.

The dissection work was mostly done by Mr. Smith, but he was not alone when it came to puzzling over the problems in parasite anatomy and parasitic interrelations which this work produced in abundance. The contents of the individual eggshells were scrutinized with the utmost care, and slowly the various anatomical remains found therein were associated with one parasite or another.

In the course of these studies it was discovered that Schedius deposits a large egg (fig. 18), which is supplied with a very long stalk. The egg is placed within the body of the unhatched but fully formed caterpillar, with the end of the stalk projecting outside.

Sometimes, and apparently usually, the end of this stalk passed through the shell of the egg as well as through the body of the cater-pillar, as indicated in the figures (fig. 19, Pl. XI, fig. 3). When the egg hatches, the larva does not entirely leave the shell, but remains with

its anal end thrust into it, and the stalk, which is hollow, becomes functional and acts like a lifeline attached to a submarine diver in supplying a connection with the outer air. As the larva grows the stalk increases in thick-ness, and the last anal segment of the larva becomes covered with a thick chitinized shield, which is unaffected by the action of strong caustic potash. There are two larval molts, and consequently three larval stages. During the entire course of both the first and second the young parasite remains quite firmly at-tached to its anal shield and lifeline and the cast skins are not entirely sloughed off, but

Fig. 19.—*Schedius kuvanæ:* Third-stage larva still retaining attachment to egg-stalk, and anal shield. Greatly enlarged. (Original)

are merely pushed backward. After the third ecdysis it retains this connection for awhile, and grows rapidly, but about the time when it reaches maturity the connection with the shield is broken, thus proving that it is not part and parcel of the integument. It would appear rather that this shield, including a tube within the egg-stalk (which, as stated, grows in thickness after the egg itself hatches), is actually part of the integument of the first-stage larva, and that the second and third stages merely continue to use what is in effect the skin of the first larval molt.

The host caterpillar is completely destroyed except for the harder chitinous parts, head, tarsal claws, hooks of the prolegs, etc., and the hair, which is left in a sort of hank, more or less completely surrounding and con-cealing the parasite larva. It is impossible to distin-guish between the larvæ of Schedius and those of its secondaries from an external examination of the eggs.

Fig. 20.—*Schedius kuvanæ:* Pupa. Greatly en-larged. (Origi-nal.)

After the larva reaches its full growth and casts off its anal shield, it quickly pupates (fig. 20) and very shortly thereafter issues as an adult. There is no indication of a desire to hibernate during any part of the preliminary stages, in which respect Schedius differs from nearly every other chalcidid which has been studied at the laboratory.

Schedius is to all purposes, if not to all intents, a secondary para-site upon occasion. In the spring of 1909 a generation was carried through to maturity within the larvæ of Anastatus, and at that time there was no difficulty experienced in induc-ing the Schedius females to oviposit in such. In the course of later experi-ments which were designed to deter-mine whether there was any preference shown between the eggs containing healthy caterpillars and those with the larvæ of Anastatus, only the healthy eggs were selected for oviposition by the parent females. What was more, although several later attempts were made to force Schedius to oviposit in eggs containing Anastatus larvæ, none but the first was successful.

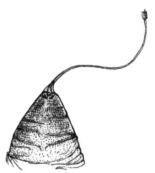

FIG. 21.—*Schedius kuvanæ:* Egg-stalk and anal shield of larva as found in host eggs of gipsy moth from which the adult Schedius has emerged, or in which the Schedius larva has been attacked by a secondary parasite. Greatly enlarged. (Original.)

Oftentimes two or more eggs are deposited in one host. Numerous in-stances have been found in which second-stage larvæ were feeding peaceably side by side as the result of such superparasitism, and still more have been observed in which the former presence of more than one individual was positively indicated by the presence of more than one egg-stalk and anal shield, but never, out of many thousands of examples under observation, has more than one adult parasite issued from one egg. What happens to the su-pernumerary individuals is not indicated further than that they disappear, and that their substance goes to nourish the sole survivor. Whether there is an actual struggle for supremacy in which victory comes to the strongest, or whether the struggle takes the form of a contest to deter-mine which shall quickest consume the available food supply, the loser calmly surrendering his body to the winner by way of forfeit, has never been revealed.

FIG. 22.—*Schedius kuvanæ:* Larval mandibles. Greatly enlarged. (Original.)

FIG. 23.—*Tyndarichus navæ:* Larval man-dibles. Greatly enlarged. (Original.)

The story of a triple tragedy is told in Plate XI, figure 3, which is drawn from a slide prepared by Mr. Smith. It represents a single gipsy-moth egg, which had been attacked by Anastatus before the embryonic caterpillar had developed sufficiently to leave perceptible remains. The Anastatus, after consuming the entire contents of the eggshell had reached the hibernating stage, and settled down to some 10 months of inactivity, when it was attacked by Schedius. No less than three Schedius eggs were deposited in fairly rapid succession (but probably by different parents) since the three larvæ, the outlines of which are shown, are practically equal in size. All are apparently about ready to molt for the second time, and after this molt, if they had been allowed to live, one would most certainly have gained the mastery and devoured the others.

Fig. 24 —*Pachyneuron gifu-ensis.* Egg. Greatly enlarged. (Original.)

But this conflict for supremacy, sanguinary as it is, is only the beginning of what might occur in the open in Japan. Tyndarichus and Pachyneuron are both habitually and essentially secondary parasites, and both prey not only upon Schedius, but upon each other with perfect impartiality. Either might attack the surviving Schedius, and be in turn the victim of the other, and there is no apparent reason why Schedius should not return to the fray and, by destroying its own secondary, start the battle all over again.

Fig. 25.—*Pachyneuron gifuensis:* Larval mandibles. Greatly enlarged. (Original.)

Such a long-drawn-out contest is hardly likely to occur very often, but in many instances tales scarcely less sanguinary have been told by the relics which strewed the field of battle. Among these relics the anal shield with egg stalk and the characteristic mandibles (figs. 21 and 22, respectively) have served as positive indication of the former presence of Schedius. Tyndarichus is betrayed by its mandibles (fig. 23), which, like those of Schedius, retain their characteristic form through all three stages. The former presence of Pachyneuron, curiously enough, is quite easily recognizable by its characteristic eggshell (fig.

Fig. 26.—*Anastatus bifasciatus:* Larval mandibles. Greatly enlarged. (Original)

24), which is of a substance which defies the action of hot concentrated caustic potash sufficently prolonged to result in the complete solution of the gipsy-moth eggshell. It may also be recognized by its mandibles (fig. 25), which are rather small and inconspicuous in any but the last stage. Anastatus, when its former presence can be proved at all, may be recognized by its mandibles also (fig. 26), but these are so small as to be very difficult to find, and it is altogether probable that there have been eggs dissected in which Anastatus was the original primary parasite, but of which fact no proof remained.

In order that some idea may be had of the conditions which actually prevail in the open in Japan, results of the dissection of 43 eggs from Japanese importations are given below. Many other eggs were dissected, in some of which the tale was too complicated to be unraveled, and it is, of course, necessary to leave out of consideration here the results of those dissections which were made before the significance of that which was found was fully recognized.

In the formulæ which follow the symbols are to be read as follows: $\times$ = Parasitized by; + = Superparasitized by.

Thus the conditions represented in the figure to which attention has already been drawn would be expressed:

Porthetria dispar $\times$ Anastatus $\times$ Schedius.
                                    + Schedius.
                                    + Schedius.

The host relations revealed by dissections of eggs from which Pachyneuron emerged are similarly indicated as follows:

Dispar $\times$ Schedius $\times$ Pachyneuron (20 times).
Dispar $\times$ Schedius $\times$ Pachyneuron.
                        + Pachyneuron (1 time).
Dispar $\times$ Schedius.
        + Schedius $\times$ Pachyneuron (3 times).
Dispar $\times$ Schedius.
        + Schedius.
        + Schedius.
        + Schedius.
        + Schedius $\times$ Pachyneuron (1 time).
Dispar $\times$ Anastatus $\times$ Pachyneuron (1 time).
Dispar $\times$ Anastatus.
        + Schedius $\times$ Pachyneuron 1 time).

Dissections of eggs from which Tyndarichus emerged resulted as follows:

Dispar $\times$ Schedius $\times$ Tyndarichus (11 times).
Dispar $\times$ Schedius.
        + Schedius $\times$ Tyndarichus (2 times).
Dispar $\times$ Schedius.
        + Schedius $\times$ Tyndarichus.
                        + Tyndarichus (1 time).
Dispar $\times$ Schedius $\times$ Pachyneuron.
                        + Tyndarichus (1 time).
Dispar $\times$ Anastatus $\times$ Tyndarichus (1 time).

Mention has already been made of the parthenogenesis of Schedius, and the fact that only males were produced in the first attempt of successful reproduction experiments in which only a single female was available. Numerous subsequent experiments have demonstrated beyond question that thelyotoky is the rule and that exceptions are rare if they ever occur.

In the course of the first unavoidable experiment in parthenogenesis the attempt was made to secure the fertilization of the female
through union with her own asexually produced offspring, but,
although she lived after they had completed their transformations, no
results were secured. It seemed to be within the bounds of possibility that success would follow if the experiment were differently
conducted, and accordingly in the fall of 1909 Mr. Smith repeated
it, with this variation, that the females, after they had deposited a
few eggs, were rendered dormant by exposure to moderate cold,
awaiting the issuance of their progeny. This time no difficulty was
experienced. The parthenogenetically produced males mated freely
with their respective parents, and the subsequent progeny in each
of several instances consisted of both sexes.

Females thus reared were mated with their brothers (which were
at the same time their nephews), reproduced with the ordinary
freedom, and their progeny were of both sexes in the usual proportions.
Still another generation showed no signs of weakness or any sort of
abnormality, and the experiment was discontinued.

In sexual reproduction the males appear always to be largely outnumbered by the females. Nothing like the diversity in this respect
which has been noted in the case of other chalcidids has been observed
in the case of Schedius.

REARING AND COLONIZATION.

When the first individuals of Schedius were secured from the
imported Japanese egg masses in April, 1909, there was no difficulty
in securing reproduction upon gipsy-moth eggs collected in the open,
but by the time the second generation was secured those which had
remained in the open were about to hatch, and would hatch almost
immediately they were brought indoors. A large quantity of eggs
had been placed in cold storage in anticipation of this, and it was
found that these would hatch nearly as quickly when they were
removed. Oviposition at any time within a few hours of the time
when the eggs would otherwise hatch was generally successful, but
when the eggs hatched within 36 hours after being exposed to the
degree of warmth necessary to secure oviposition of the parasite, it
soon became evident that not very much increase was to be expected.
Accordingly, the experiment was made of killing the host eggs through
exposure to just enough heat to bring this about. The parasites oviposited in these dead eggs with the same freedom that they would
attack the living, and reproduction ensued. The progeny, however,
were small and weak, and not as prolific as those secured earlier in
the spring.

Thus, in one way and another the species was carried through the
summer, and with the deposition of fresh gipsy-moth eggs early in
July much better results were secured, and the parasites immediately

began to increase rapidly in numbers with each succeeding generation. By August there were enough to make a small colony in the open possible without depleting the laboratory stock to a serious extent, and first one and later several small colonies were established in various localities in the moth-infested area.

At the same time reproduction work was continued on an ever-increasing scale at the laboratory, and by the first of the next year no less than 1,000,000 individuals, at a conservative estimate, were present in our rearing cages. Further attempts to increase this number were not successful, on account of the difficulties attending the handling of such an immense number at a time when the hatching of the host eggs followed too soon after their removal to high temperature.

The numbers in the laboratory suffered no decrease, however, and by the end of March colonization work on an extensive scale was begun. The parasitized eggs were divided into 100 lots, each of which contained approximately 10,000 of the parasite, and these were distributed to agents of the State forester's office, who placed them in the field in the hope and expectation that the parasites issuing from them would reproduce immediately upon the gipsy-moth eggs before the latter hatched.

There was also a large quantity of parasitized eggs remaining, and these were placed in cold storage in the hope that the emergence of the brood might be retarded until the fresh eggs of the gipsy moth should be available for attack in the latter part of the summer. This hope was not justified, because when the time came and the eggs were taken from cold storage not a single living parasite remained.

In Table IX are summarized the results of the reproduction work, as conducted in the laboratory from April, 1909, to the winter of 1909–10, and the dates when the first colonies were planted in the late summer and fall are therein indicated.

TABLE IX.—*Results of reproduction work with Schedius.*

Generation.	Number and source of parents.	Reproduction work begun.	Emergence of progeny.		Total number of progeny.	Colonized.
			Began.	Ended		
First. .	11 from imported egg masses.	Apr. 19............	May 19	June 14	114	....
Second...	114 from first generation...	May 19............	June 23	July 16	645	....
Third....	645 from second generation....	June 23............	July 16	Aug. 10	1,350	....
Fourth...	1,350 from third generation....	July 16............	Aug. 16	Sept. 7	11,999	10,980
Fifth- ...	1,019 from fourth generation...	Aug. 16............	Sept. 11	Sept. 29	6,286	3,280
Sixth....	3,006 from fifth generation ..	Sept. 11............	Oct. 5	Oct. 25	12,723	5,368
Seventh..	7,355 from sixth generation..	Sept 30......	Oct. 29	Nov. 15	35,423	5,639
Eighth..	29,784 from seventh generation.	Oct. 25....	....	....	[1] 219,627	20,115
Ninth....	199,512 from eighth generation.	Nov 21–Dec.21....	....	....	[1] 1,028,361	733,967
Tenth...	294,400 from ninth generation	Jan 5.	....	....	[1] 284,779	280,762

[1] Estimated.

The reproduction of the parasite in the field as a result of these early attempts at colonization was far in excess of expectations. The rate of reproduction in the laboratory (as indicated in the table) was greatly exceeded in the open, and hundreds of thousands of eggs in the immediate vicinity of the colony sites were known to be parasitized when the coming of cold weather put a stop to insect activity. In the one colony which was most closely watched, the parasitized eggs averaged some 30 to the mass (fig. 27), while everywhere within 50 yards of the center egg masses were so thick in spots as to hide the bark on the trees. Beyond the distance mentioned the number per mass fell off very rapidly, but some were found several hundred yards away from the point of liberation, in striking contrast to the results following the colonization of Anastatus.

FIG. 27.—Gipsy-moth egg mass, showing exit holes of *Schedius kuvanæ.* Enlarged about four times (Original.)

In October adults of what appeared to be the second generation were not uncommon in the field, and on any warm day they could be found, apparently ovipositing for a third generation. At the same time larvæ and pupæ were in abundance, and only a few days' exposure to the warmth of the laboratory was needed to bring them out from eggs collected in the field. Collections of eggs were made from time to time during the fall, in order that assurance might thus be had of the continued well-being of the parasite, and until December nothing untoward occurred. The first real winter weather came at the end of that month, and a few days later a lot of eggs was collected and brought in. Not a single parasite issued. The experiment was repeated and with the same results, and although many hundreds of masses have since been collected (some of them in the spring, after the caterpillars had issued for the purpose of determining whether there might not be reproduction of hibernated adults at that time, and the rest of them in the fall to see if by any chance the parasite had escaped detection in the spring), no trace of its existence could be found. In every instance, so far as the above-mentioned colony was concerned, the results were the same, and there seems to be no doubt that, in this particular locality at least, the species has become extinct.

In the spring one large colony of the Schedius was planted coincidently with the distribution of the 100 lots of parasitized eggs for colonization by the State forester's agents, and for two months following weekly collections of eggs were made with the expectation that a partial spring generation would follow. None of these collected egg masses produced the parasite, and again it failed to come up to that which was expected of it.

In the fall, as has already been mentioned, very large collections of eggs made in the vicinity of that which was considered to be the best and most promising of the colonies of 1909 failed to produce Schedius, and at the same time numerous smaller collections were made in each of the other colonies of 1909, as well as in a considerable number of the spring colonies of 1910. In only one of the colonies of 1909 was the Schedius recovered, and this, curiously enough, from that in which every attempt had been made to secure evidence of spring reproduction. Here it was found in one direction from the center of the colony only, and over a rather limited area. In the immediate vicinity of the colony site (within 100 yards) none could be found.

The collections which were made in each of the other colonies of 1909 were followed by curiously similar results. The parasite was recovered in one of them, and in one only, and although collections of eggs were made in all directions from the center and at varying distances, parasitized egg masses were only found in a limited area to one side and some distance away.

It was pretty conclusively demonstrated that the larvæ and pupæ of Schedius could not survive the rigor of the winter, and it is very difficult to say whether the recovery of the parasite in this last-mentioned instance is indicative of its ability to survive the winter as an adult. In 1909 a quantity of the adults was placed in a small cage in the open before the beginning of severe weather, and, although mortality was heavy, some of them lived for a long time after all of the younger stages were destroyed. None of them lived through until spring, but there is nothing to prove that they would not have done so had they had their choice of situations in which to hibernate.

It may be that females successfully hibernated in the instance of this colony, which appears to have lived throughout one year in the open. It may also be that the recovery of the species under these conditions is the result of dispersion of the individuals from some of the many spring colonies, several of which were located within a not unreasonable distance of this spot. It will require another year to demonstrate the truth of the matter.

The recovery of Schedius under any conditions at all was considered as sufficient to justify the repetition of the rearing work of the winter before, and accordingly, using a few individuals secured from the field early in the fall, a series of generations has been reared in the laboratory until the number now on hand (Jan. 1, 1911) runs into the hundreds of thousands. In the spring it is planned to establish one or two exceedingly large new colonies, sufficiently far distant from any of the others to make the recovery of the parasite elsewhere a certain indication that it is able to pass the winter in New England and thereby justify the labors which have been expended in its behalf.

THE PARASITES OF THE GIPSY-MOTH CATERPILLARS.

APPARENTLY UNIMPORTANT HYMENOPTEROUS PARASITES.

It would be presumption to state without qualification that the parasites which are here brought together as unimportant are in reality that. It may well be that among them are some which will be of sufficient promise to make advisable the trouble and expense incident to an attempt to transplant them to America, and which will serve to fill in the gap in the sequence which the apparent failure of *Apanteles fulvipes* has left. To determine more definitely their relative importance abroad is one of the objects of the work for the season of 1911, as at present planned, and something more than is known now is certain to be known a year from now unless the plans for the season go wrong from the beginning.

The various species coming in this category are called unimportant because they have never been received in imported material in numbers sufficient to make colonization in America possible, and only upon very rare occasions and in the instance of a few amongst them only, in numbers sufficient to indicate that they were of any importance whatever in effecting the control of their host abroad.

The investigations into the parasites and parasitism of various native insects more or less similar in one respect or another to the gipsy moth have served to throw considerable light upon the status of such parasites as these. It has been shown, in the instance of the tussock moth, that a parasite may be entirely absent in localities where the host is abundant, or else very rare under such circumstances and yet be sufficiently common to effect an appreciable amount of control in localities where the host is very rare. It is thus possible that some among these species may play a very important rôle in keeping its host, when already reduced to relatively small numbers, from increasing sufficiently to become of economic importance, and that at the same time they may play no part at all in reducing that insect from a state of or approaching noxious abundance to within its ordinary limits.

On the other hand, studies with the parasites of native insects have revealed the existence of what may be called accidental or incidental parasites. These may be important parasites of one insect and of no importance whatever in connection with another, nearly allied. Sometimes this is due to the fact that the one species of host may excite in the mother parasite the desire to oviposit, which is not excited by the other, and occasionally, as has more than once been observed, the presence of the favored host in the immediate vicinity will induce the parasite to oviposit in another species which under otherwise identical circumstances would be entirely ignored. At other times an insect may be acceptable to the mother parasite, but for some reason unacceptable to her progeny, so that only a very few out of the many eggs which are deposited will go through to maturity, and the species will be of necessity considered as rare and unimportant.

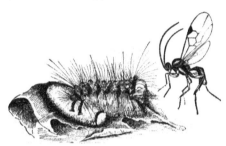

FIG. 28.—*Apanteles solitarius.* Adult female and cocoon Enlarged (Original)

The fact that there are included in every list of the parasites of a given host a few species which are thus to be considered as incidental or accidental lends force to the contention that among the recorded parasites of the gipsy moth are several at least which come into the same category. Just which these are is not altogether plain at this time.

APANTELES SOLITARIUS RATZ.

Cocoons of a solitary species of Apanteles (fig. 28) which attacks the very young to half-grown caterpillars of the gipsy moth throughout the greater part if not the whole of Europe have occasionally been received in shipments in which the caterpillars were not all in the last or next to the last stage. In those shipments which consisted of caterpillars in the third, fourth, and fifth stages at the time of collection, the cocoons of this species have been the most common. In no instance has a sufficient number been received to make possible anything like a satisfactory colony of this species, and in all scarcely more than 100 have been received since the beginning of the work.

The parasite undoubtedly attacks the first-stage caterpillars as well as those of the later stages up to the fourth at least, and perhaps the fifth. The host probably molts at least once, subsequent to attack, and remains alive after the emergence of the parasite larva, clinging to and seeming to brood over the cocoon of its mortal enemy. Numerous experiments have been made with other, similarly living caterpillers from which parasites have emerged, in an attempt to make them feed, and invariably these attempts have been unsuccessful.

Of all of the gipsy-moth parasites in Europe of which there is no present prospect of introduction into America, this species is the most promising, and yet, if dependence is placed upon the results of the rearing records, it is so scarce as to be wholly inconsequential as a parasite of this host.

METEORUS VERSICOLOR WESM.

Very occasionally cocoons of this common brown-tail moth parasite have been found in boxes of gipsy-moth caterpillars received from European sources, but never in any numbers. Altogether not nearly so many have been received as of the cocoons of *Apanteles solitarius.*

It is apparently an incidental parasite of no consequence, and were it not an enemy of the brown-tail moth as well, it is very improbable that any attempt would be made to introduce it into America.

As a parasite of the brown-tail moth it is of considerable promise, and as a brown-tail moth parasite it has been introduced and is apparently at this time thoroughly established over a considerable territory. Upon several occasions it has been reared from gipsy moth caterpillars collected in the field localities where it was particularly common as a parasite of the brown-tail moth, but, as in Europe, it expresses a strong preference for the last-mentioned host.

METEORUS PULCHRICORNIS WESM.

Quite a number of this species has been reared from cocoons found in the boxes of gipsy-moth caterpillars received from southern France, and a very few have also been received from Italy. None of the Meteorus which have been reared from the brown-tail moth in any part of Europe have been anything else than *M. versicolor*, so far as known. It is, of course, possible that two species similar in appearance might easily have been confused, and no attempt has been made to determine the specific identity of every specimen which has been reared for liberation.

There is nothing to indicate that *M. pulchricornis* is ever of more consequence as a parasite of the gipsy moth than is *M. versicolor*, and until evidence to the contrary is forthcoming it will not be considered as of importance or promise.

METEORUS JAPONICUS ASHM.

Specimens of this species were secured from boxes of young gipsy-moth caterpillars from Japan in very small numbers in 1908 and 1909, but so far as could be determined it was of no more importance in that country than were either of the two species already mentioned in Europe. In the winter of 1909–10 a few specimens were received from Mr. Kuwana, together with the statement that it was common as a parasite of the gipsy moth in Nagaoka, but not in Tokyo in 1908. Attempts to import it in 1910 were unsuccessful, and it is with the hope of confirming its importance, at least locally, and discovering some method of transplanting it to America, should such confirmation come about, that the investigations are undertaken in Japan in the year 1911.

LIMNERIUM DISPARIS VIER.

FIG. 29.—*Limnerium disparis:* Cocoon. Enlarged. (Original)

This interesting parasite was first received in June, 1907, in a shipment of small gipsy-moth caterpillars from Kief, Russia. A total of 18 of its peculiar cocoons (fig. 29) was received in June and July of that year in boxes which contained only a relatively small number of caterpillars when sent, and its importance as a parasite appeared to be considerable. It seemed probable that the larvæ spinning them had issued from caterpillars in the fourth and fifth stages.

The cocoons usually approach more nearly the spherical than that used as the type for the drawing. The walls are thin, but so dense as not only to be impervious to moisture, but to prevent the drying of the meconial discharge for months.

FIG. 30.—*Limnerium disparis:* Adult male. Much enlarged. (Original)

None of the cocoons was hatched on receipt. A single male adult (fig. 30) issued from one of these cocoons in August. No more adults appearing, some of the cocoons were opened from time to time during the fall and found to contain still living adults. Since it is known that several of the native species which spin similar cocoons actually do hibernate as adults within the cocoon, it is reasonable to suppose that the same is true of this.

No adults were reared, however, and their failure to emerge appears to be due to the drying during the winter of the semiliquid meconial discharge which effectually glued the adults to the sides of the cocoon and prevented their further movement.

The several native species spinning similar cocoons attack a variety of hosts, and one of them, *L. clisiocampæ*, is sometimes common and quite effective as a parasite of the host indicated by its specific name. The larvæ, after spinning the cocoon and before discharging their meconium, are very active for a period of about 24 hours, convulsively wriggling the body in such a manner as to make the spherical cocoon move about in an extraordinary manner. It is altogether probable that the gipsy-moth parasite has the same characteristic, and that the cocoons so spun in the trees are quickly dislodged, fall to the ground, and become hidden beneath leaves and débris.

Prof. Kincaid was especially instructed to seek for evidences of parasitism by this species in Russia on the occasion of his trip to that country in 1909. He did not find it at all abundant, however, and only secured three or four cocoons. In 1910 the junior author sought diligently for these cocoons in the forest about Kharkof, where the caterpillars had been very abundant the season before, but he was entirely unsuccessful and as a result thoroughly convinced that it was not an important parasite in any of the several forests visited. There was no opportunity at Kief to make a similar search, because the caterpillars had not been sufficiently abundant within recent years in any of the localities visited to make likely the discovery of these cocoons, even though the species had been of importance as an enemy of the gipsy moth.

In 1909 and again in 1910 it, or another practically indistinguishable species, was received in very small numbers from Japan, but at the same time under circumstances which were in a way as suggestive of the possible importance of the species as were those under which it was first received from Russia, as detailed above. As in the case of the Japanese Meteorus, it is hoped to be able to determine definitely whether it is to be considered as of more than technical interest in the connection in which it is here considered.

LIMNERIUM (ANILASTUS) TRICOLORIPES VIER.

From time to time several specimens of Limnerium cocoons, all of them oblong in shape, and most of them partially concealed by the skin of the host caterpillar, have been received from Europe. In no instance have they been in sufficiently large numbers to make the species appear promising as a parasite.

Less than a dozen specimens have been received, all told, and were it not for the fact that the remains of the host accompanied the cocoon, it would not be possible thus definitely to associate the parasite with its host.

APANTELES FULVIPES HAL.

The one among the hymenopterous parasites attacking the cater-pillars of the gipsy moth which has ever been received under circum-stances indicative of its unquestioned importance as an enemy of that host is at present known as *Apanteles fulvipes* (fig. 31). The name Glyptapanteles, as generically applied to it, has been regret-fully dropped, the more so since this name has already become familiar to many whose interest in parasites begins and ends with those which are included among the enemies of the gipsy and brown-tail moths. It was accepted, in the first place, on account of the immediate dis-tinction which it offered to Apanteles, as applied to *A. solitarius* and *A. lacteicolor* Vier., and because it seemed preferable to make the technical name the common name as well. Now, with an enforced change in the spe-cific name vaguely in prospect, it would seem advisable to adopt an arbitrary common name rather than to at-tempt to popularize the technical name, and should it again become desirable to write of it in a pop-ular way, this will probably be done.

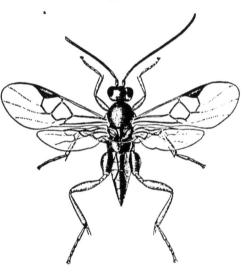

FIG. 31 —*Apanteles fulvipes* Adult Greatly enlarged (From Howard)

That a change in its specific designation will become necessary when it shall have been thoroughly well studied abroad seems probable, although there is no basis upon which to make such a change at the present time. If, as European taxonomists have agreed, it is synony-mous with *A. nemorum*, described by Ratzeburg as a parasite of *Lasiocampa pini* L. and is at the same time specifically identical with the form so determined by Marshall as a common species in England, there seems to be no reason why it should not be introduced success-fully into Massachusetts. A parasite with anything like the wide range of hosts accredited to this species abroad should find no diffi-culty in existing in America, and if the species which attacks the gipsy moth is proved to be identical with that which goes under the same name and attacks one or another of such a variety of hosts, no expense ought to be spared in attempting its introduction; always provided, of course, that the attempts already made prove not to be successful.

The story of these attempts, as told in the popular bulletin by the junior author, issued from the Massachusetts State forester's office in the spring of 1910, may well be quoted here, since there is little to be added to it.

Although this was almost the first parasite of the gipsy moth which attracted any attention in Massachusetts, and the first which it was attempted to import after the beginning of active work, it was one of the last to be liberated under satisfactory conditions, and its establishment in America is not yet certain. Extraordinary methods were necessary to bring it to America living and healthy, and it was not until Prof. Trevor Kincaid, who was selected by Dr. Howard as the best available man for the purpose, visited Japan and personally superintended the collection and shipment of the cocoons, that success was achieved. The story of Prof. Kincaid's experiences and of the difficulties which he met and overcame is interesting. He was accorded great and material assistance by the Japanese entomologists, and the work inaugurated by him in 1908 was continued with even greater success in 1909.

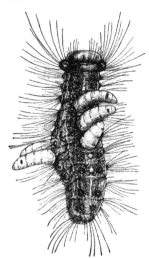

FIG. 32.—*Apanteles fulvipes:* Larvæ eaving gipsy-moth caterpillar. Enlarged. (Original.)

The adult parasite [fig 31] deposits a number of eggs beneath the skin of the active caterpillars, and any stage, from the first to and possibly including the last, may be attacked. The larvæ, hatching from the eggs, become full grown in from two to three weeks, and then work their way out through the skin of the still living caterpillar [fig 32], within the body of which they fed. Each spins for itself immediately afterward. for its better protection during its later stages, a small white cocoon. The number of parasites nourished by a single host varies in accordance with its size. There may be as few as 2 or 3 in very small caterpillars, or 100 or more in those which are nearly full grown.

The unfortunate victim of attack does not, as a rule, die immediately after the emergence of the parasite larvæ and the spinning of their cocoons, but it never voluntarily moves from the spot. Its appearance, both before and after death, surrounded by and seeming to brood over the cocoons, is peculiar and characteristic, and once seen can never be mistaken [fig. 33].

There is ample opportunity for two generations of the parasite annually upon the caterpillars of one generation of the gipsy moth. This is the rule in the countries to which it is native, and is to be expected in America.

The parasite was described from Europe more than seventy-five years ago, and has been known to be a parasite of the gipsy moth for a long time. Later it was described under a different name from Japan, and the Japanese parasite was for a time considered to be different from the European. Absolutely no differences in life and habit which can serve to separate the two are known, and, as the adults are also indistinguishable in appearance, they are considered to be identical.

It has been the subject of frequent mention under the name of Apanteles, as well as of Glyptapanteles, in the various reports of the superintendent of moth work, from the first to the fourth; and Dr. Howard, in the account of his first trip to Europe in the interests of parasite introduction, tells of its occurrence in the suburbs of Vienna. Largely on account of the fact that it is much more conspicuous than many of the other parasites, it has attracted more general attention. The Rev. H. A. Loomis, a missionary, and resident of Yokohama, was the first to call attention to its importance in Japan, and made several unsuccessful attempts to send it to America. Dr. G P. Clinton, mycologist of the Connecticut Agricultural Experiment Station, who visited Japan in 1909, observed the parasite at work, and reported most favorably upon its efficiency as a check to the moth. Numerous other attempts on the part of European and Japanese entomologists, including one elaborate experiment which involved the shipment of a large wire-screened cage containing a living tree with gipsy caterpillars and the parasite, were made, but with uniformly ill success. Upon every occasion the parasites all emerged from their cocoons and died en route.

When every other means failed, Prof. Kincaid, as already stated, was deputed to visit Japan, and to make all necessary arrangements for the transportation of the parasite cocoons in cold storage to America. The arrangements which· he perfected provided for continuous cold storage, not only en route across the Pacific, but during practically every moment from the time the cocoons were collected in the field in Japan until they were received at the laboratory in Melrose. Events justified the adoption of every precaution, and, with all the care, only a small part of the very large quantity of cocoons which he collected reached their destination in good condition. Hundreds of thousands were collected and shipped, and less than 50,000 were received alive—nearly all in one shipment in July.

The season in Massachusetts was early, and nearly all of the gipsy caterpillars had

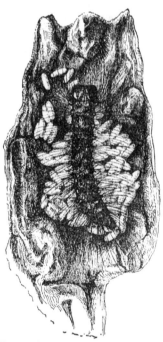

FIG. 33.—*Apanteles fulvipes* Cocoons surrounding dead gipsy-moth caterpillar. Slightly enlarged. (Original.)

pupated by that time, so that there was no opportunity for the parasite to increase in the field·upon this host that season. In 1909 the sites of the colonies were frequently visited, but not a single parasitized caterpillar was found which could be traced to colonizations of the year before. Keen disappointment was at first felt, but later developments have tended to throw a more encouraging light upon the situation.

In 1909 importations were continued, through the magnificent efforts of Prof S I. Kuwana, of the Imperial Agricultural Experiment Station at Tokio, with much more satisfactory results. In 1908 the season in Japan was very late, and it was not practicable t ι send any of the cocoons of the parasite until June and July; while in America the season was early, and by that time all of the caterpillars, as has already been stated, had pupated. In 1909 the season was rather early in Japan and correspondingly late in America; and, besides, through special effort, Prof. Kuwana was enabled to send a few thousands of the cocoons of the first generation, which reached the laboratory early in June   About 1,000 adults emerged from these cocoons after receipt, and the most of them were placed in one colony in a cold situation on the North Shore, where the caterpillars were greatly retarded, and where there were still some in the first stage. The remainder were colonized in warmer localities, where the caterpillars were one stage farther advanced.

Immediate success followed the planting of these colonies. Within three weeks cocoons were found in each, and the number of parasitized caterpillars was gratifyingly large. A very careful investigation was conducted, to determine the proportion which was attacked by native secondary parasites; and, while this was so large in one instance as seriously to jeopardize the success of the experiment, it was not so large in the others.

There were several thousands of this first generation known to have developed in the open upon American soil, which issued from the cocoons some four or five weeks after the colonies were established, but in only that one on the North Shore, where the caterpillars were in the first and second stages when the parasites were liberated, was there a full second generation. Here the larger caterpillars were again attacked, and an abundant second generation of the parasite followed.

Meanwhile, additional shipments of cocoons of the second Japanese generation were received early enough to permit of a generation in the open upon the native caterpillars, and several other colonies were successfully established. It is known that there were many thousands of the parasite issuing in at least five different localities during August, but immediately thereafter they were completely lost to sight, and it is futile to hope to recover traces of them before another spring.

Until the late summer of 1909 nothing occurred to indicate that this parasite would be likely to fly for any great distance from the point of its liberation; and, as has been already stated, it was looked for in vain in the summer of 1909 in the immediate vicinity of the colonies of the year before. In July, 1909, a strong colony was planted in an isolated woodland colony of gipsy moths in the town of Milton. It was rather confidently expected that it would attack these caterpillars so extensively as to destroy the major portion, but it was the cause of some surprise, when the locality was visited after the parasites of the new generation had mostly issued from the affected caterpillars, to find a smaller number of cocoons than there were individuals liberated in the first place, and only about one-fourth, perhaps less, of the caterpillars attacked. The circumstance was as discouraging as anything which had gone before, and for a few days nothing happened to change its complexion. Then, to the intense surprise of the writer, Mr. Charles W. Minott, field agent of the central division, sent to the laboratory a *bona fide* example of the parasite, which had been collected in the Blue Hills reservation, upwards of a mile away. There was no possible source except the Milton colony, and a spread of upwards of a mile in something under a week was indicated beyond dispute   At almost the same time the brood of Monodontomerus was found for the first time in pupæ of the gipsy moth in the field; and when the history of this species is considered, in the connection which it bears toward the circumstances surrounding the recovery of the Glyptapanteles so far from the point where it was liberated, the whole situation is altered.

Granted that the parasite disperses at the rate of one mile in each week of activity, and that it is able to adapt its life and habits to the climate and conditions in America, the chances are, that, instead of looking for it in the immediate vicinity of the points of colonization, it is quite as likely to be found almost anywhere in the infested area within 25 miles of Boston. If it is thus generally distributed, very large numbers in the aggregate may exist. and it may increase at a rate as rapid as that of Monodontomerus, and at the same time escape detection until the summer of 1911 or 1912.[1]

There is not very much to add to the account given above, further than the statement that all attempts to recover the species in the field in 1910 from the vicinity of colonies of the year before failed. It hardly seems likely that so conspicuous an object as the cocoon mass of this parasite should escape the notice of the many field men who are familiar with its appearance, and who know of the great interest and importance which would attach to its discovery. In consequence the failure to recover the species is of more significance than the failure in the instance of any other parasite which could be mentioned.

At the same time all hope has not been given up, especially in consideration of the curious circumstances which will shortly be described, surrounding the recovery of *Pteromalus egregius* as a parasite of the brown-tail moth. If, as can no longer be doubted, a minute and to all appearances an inactive insect like Pteromalus has dispersed over a territory of approximately 10,000 square miles within five years as the extreme limit, and if during that period it remained so rare as to defy all of our efforts to recover it, it is not impossible that *Apanteles fulvipes* will do the same. Should this come about, the year 1911 or 1912 would probably witness its sudden and simultaneous appearance throughout the greater part of the territory infested by the gipsy moth.

It must be confessed, however, that hope rather than faith has dictated these last lines. It is believed, and not without some foundation, that the failure of *Apanteles fulvipes* to exist here is due to the absence of an absolutely necessary alternate host, and that further attempts to introduce it will be unavailing. That is the reason why the most will be made of every opportunity to determine the truth or fallacy of the European records which accredit it with attacking a variety of insects representing half a dozen families, and two or three times that number of genera, many of which are represented by closely allied and sometimes by the same species in America. If investigations uphold the truth of these records, no expense ought to be spared in further attempts to establish the parasite in America, because of all those which attack the gipsy moth it is the one which was not only the most promising at the beginning, but which remains the most desired at the present time.

---

[1] The occurrence of the cocoons in the near vicinity of the colony sites immediately following the liberation is most natural, and in perfect harmony with the wide dispersion The female parasites as soon as they emerge are ready to deposit a small part of the eggs which they will eventually deposit if they live and have opportunity. After the deposition of this part, it is necessary for them to wait an appreciable time before they are ready to deposit any more.

In 1910 additional importations were made from Japan, and a large number of healthy adults was liberated sufficiently early in the season to allow for one generation upon the gipsy moth.   As in 1909, cocoon masses were found in the vicinity of these colonies about three weeks after their establishment.

An attempt will be made in 1911 to import enough cocoons from Russia to make possible a strong colony of the European race.   It is possible that it would succeed here when the Japanese would fail, and on the chance the experiment is undertaken.

SECONDARY PARASITES ATTACKING APANTELES FULVIPES.

It is safe to say that a better opportunity for an intensive study of the parasites of any one host which was itself a parasite has never been afforded than has come about at the laboratory in the case of the parasites of *Apanteles fulvipes*.

Hundreds of thousands of the cocoons of the primary parasite were collected in Japan after they had been exposed to attack by the secondaries, and, so far as can be judged, the latter stood the ordeal of the journey to America better than did the primary.   Even in those shipments which were just a few days too long en route and in which the Apanteles themselves had all issued and died before their receipt the secondaries had hardly begun to issue.   These, as well as the numerous shipments which were received in better condition, in so far as Apanteles was concerned, have produced many thousands of secondary parasites, which have all been carefully preserved, but not, as yet, carefully studied.   It is not even known how many species are represented in the assortment, which includes a considerable number of undescribed forms, but apparently there are at least 30, and probably more, from Japan alone.   Some are very rare, and are represented by but a few individuals among the thousands which have been reared.   Others are common at times, and rare or absent at others.   Some few are generally common, and practically always present.

The considerable shipments of cocoons which were collected in Russia by Prof. Kincaid and forwarded to the laboratory in 1909 were invariably so long en route as to permit the Apanteles to issue and die, but, as in the case of the Japanese shipments, the secondary parasites did not suffer.   Not nearly so much material of this sort has been received from European sources, and probably on that account alone the variety of secondary parasites reared has not been so large.   Nevertheless, more than 20 species have been recognized and probably at least 25 have been reared in varying abundance.

A good many of these secondary parasites have a very close resemblance to those which have been reared from the Japanese material.

In some instances they appear to be identical. In others it may be possible to find minor structural characters which, together with the difference in their habitat, will make it worth while to designate them by different names. Some very distinct species are peculiar to Europe or to Japan, and remain unrepresented by any nearly resembling them in the other country.

In 1909, as the immediate result of colonization work carried out under the happy auspices already described, it was possible to collect large numbers of *Apanteles fulvipes* cocoons under perfectly natural conditions in the open in America. This was accordingly done, with the result that no less than 18 additional hyperparasites were added to the list of those which attacked this host. Some of these were rare, others very common in this connection. A few appear to be undescribed.

The most interesting thing about them taken as a group, is the general resemblance which they bear to the similar groups of European and Japanese parasitic Hymenoptera having identical habits. Apparently there are about as many points in common between the American parasites of *Apanteles fulvipes* and the Japanese or the European as there are between the European and the Japanese.

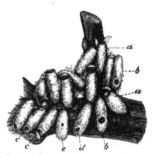

FIG. 34.—*Apanteles fulvipes:* Cocoons from which Apanteles and its secondaries have issued, as follows *a*, *Apanteles fulvipes; b*, Hypopteromalus; *c*, *Hemiteles* sp; *d*, Dibrachys; *e*, Asecodes. Enlarged. (Original.)

In the course of the work a total of 5,456 cocoons of *Apanteles fulvipes* was collected from several of the recently established colonies, but principally from two, representing the first among those planted in 1909 and in both of which a second generation occurred. Of this total, 1,531, or 28 per cent, had produced the Apanteles at the time of collection; 2,373, or 44 per cent, were attacked by secondaries (fig. 34); 634, or 12 per cent, were destroyed by various predatory insects, ants, etc.; and 918, or 17 per cent, remained unhatched in October, 1909. Among the unhatched cocoons was a considerable proportion which contained the hibernating larvæ of Asecodes, Elasmus, and Dimmockia. In more than one instance, too, hatching was prevented by superparasitism, and in others death probably resulted through the attack of predatory bugs. On at least one occasion *Podisus* sp. was found with its proboscis thrust through the wall of the cocoon and feeding upon the parasite larva or pupa within.

An idea of the variety of secondary parasites reared is conveyed by the tabulated list following.

List of secondary parasites reared from American cocoons of Apanteles in the order of relative abundance.

[In this list the number of individuals of Apanteles killed, not the gross number of the secondaries reared, is given. In case of tie, the species which was relatively the more important in the particular lot or lots from which it was reared is given preference]

Hypopteromalus	1,276	Pezomachus No. 65	15
Dibrachys	583	Pteromalid No. 68	6
Asecodes	[1] 161	Hemiteles No. 63	5
Hemiteles No. 60	[2] 58	Pteromalid No. 70	2
Hemiteles No. 61	52	Eupelmus	2
Hemiteles No. 75	[2] 49	Hemiteles No. 66	2
Pezomachus	64	Anastatus	1
Eulophid	71	**Total**	**2,288**
Hemiteles No. 62	18		

Local conditions as affecting the control of this parasite through hyperparasites were well represented in 1909 by a comparison between the relative abundance of secondary parasites in cocoons from two colonies, the "Reading-Wilmington," and the "West Manchester," which were planted at about the same time. In both reproduction was abundant, and a large number of cocoons was collected from each. Only those which were left in the field until all of the Apanteles which remained healthy had issued are counted in the following:

	Reading-Wilmington colony.		West Manchester colony.	
	Cocoons.	Per cent.	Cocoons.	Per cent.
Apanteles	70	8	543	66.5
Hyperparasites	624	68	162	20
Predators	8	1	22	2.5
Unhatched Oct. 20	218	23	89	11
Total	920	....	816	....

The West Manchester colony was located in rather dense forest, with a swamp, partly overgrown with brush and partly with thick forest on one side. The trees were large, and cocoon masses were frequently far beyond reach. Only those which could be reached from the ground were collected. There were more cocoons in this colony than in the other, but they were not quite so easily collected. It is of course possible that the larger number of cocoons explains in part the smaller percentage of hyperparasitism.

The increase in hyperparasitism in the cocoons of the second generation over the first can only be demonstrated in the case of the West Manchester colony, which was the only one where there was a second generation in sufficient abundance to permit of adequate field collections. In this it is or appears to be very striking, when the fact is taken into consideration that a considerable number of parasites hibernated in the cocoons of the second, while none were found in those of the first which failed to hatch after the 1st of September.

---

[1] Many Asecodes remaining unhatched within the cocoons will doubtless attempt to hibernate.

[2] Hemiteles No 60 and Hemiteles No. 75 may possibly be one and the same species. It is possible, too, that further study will cause a change in the relative position of the two species.

Two lots of each generation were collected after the healthy parasites had issued, and the results follow:

	First generation.		Second generation.	
	Cocoons.	Per cent.	Cocoons.	Per cent.
Glyptapanteles....................................	543	66.5	636	26
Hyperparasites.....................................	162	20	1,207	50
Predators...	22	2.5	102	4
Unhatched...	89	11	1 469	1 20
Total...	816	100	2,414	100

1 Oct. 20.

Other collections of cocoons from colonies where the Apanteles was liberated too late in the season to permit of two generations showed a high rate of hyperparasitism in the single generation, actually the first but corresponding to the second. Comparison in this instance is valueless, as local conditions enter in which can not be gauged.

This rather lengthy summary of a study in hyperparasitism has been prepared and is here presented with the object of illustrating the somewhat modified stand which it has been necessary to take concerning the subject in its relation to the project of parasite introduction. Were it within the bounds of possibility to introduce into America the parasites of the gipsy moth (*Apanteles fulvipes*, for example) without introducing the secondary parasites which preyed upon them abroad, it would unquestionably be possible to secure a greater meed of efficiency in America than that which the same parasites were capable of attaining in their native countries. This is on the supposition that the parasites themselves are no more likely to be attacked by the American hyperparasites than their hosts are likely to be attacked by the American primary parasites.

That the assumptions are fallacious, to a certain extent, is well proved by the results following the temporary establishment here of *Apanteles fulvipes*, as recounted above, and that the same results as those which followed the exposure of this parasite to American hyperparasites will result in the instance of others among the imported parasites is more than likely. In the case of *Compsilura concinnata* and *Apanteles lacteicolor* Vier. it is proved.

The truth of the matter is that the secondary parasites are very far from being as closely restricted to one or two species of hosts as are the primary parasites. This is in part due to the fact that they represent for the most part a much more degraded form of parasitism. Species like *Dibrachys boucheanus*, which is perhaps as generally abundant and omnivorous as any of the parasitic Hymenoptera, will attack anything which is dipterous or hymenopterous, provided it is physically suitable as food for its larvæ. *Apanteles fulvipes* and caterpillar parasites generally are governed in their host relations by physiological rather than by physical limitations, and the difference is as great as that which separates the true predator from the true parasite.

Did time permit, and were this the proper place, a lengthy digression might be made, in which several of the parasites typical of both groups, and which have been somewhat carefully studied at the laboratory, could be compared, the better to give strength to the statement just made. A little later attention will again be called to the matter.

At this time it is merely desired to define the modified stand which it has been necessary to take upon the question of hyperparasites. It is no longer possible, on account of the absence of their secondaries, to expect a much if any greater degree of efficiency from the imported parasites in America than the same species possess abroad. Since the foreign hyperparasites of the gipsy moth are generally the counterpart of the American species, which will become hyperparasitic upon the gipsy moth just so soon as there are any primary parasites, their introduction could not possibly do more than result in the existence in America of a somewhat greater variety of hyperparasites, which as a group would play exactly the same rôle as the lesser variety now existent here. Consequently the only secondary parasites which we have to fear are those which have no counterparts in America.

That such exist is beyond question; that they are in the minority is equally true. The only species which have been recognized as possibly or probably falling into that group are the hyperparasites reared from the gipsy-moth eggs from Japan, the Melittobia parasite of tachinids; the eulophid parasite of *Pteromalus egregius*, *Perilampus cuprinus*, and *Chalcis fiskei;* and, most unfortunately, two primary parasites already introduced, which are also secondary, *Pteromalus egregius* and *Monodontomerus æreus*. The two last mentioned are probably both beneficial rather than noxious in the final analysis, but nevertheless both are peculiarly adapted to act as secondary parasites of the brown-tail moth better than as secondary parasites of any other primary host.

It is not intended to ignore the secondary parasites in the future any more than in the past, but the same fears which have been expressed concerning their introduction are no longer felt in the same manner, and the benefits which were formerly expected to accrue through their exclusion are not so great as hoped.

TACHINID PARASITES OF THE GIPSY MOTH.

In proportion as one after another of the previously mentioned hymenopterous parasites of the gipsy moth have been eliminated from the lists as of no more than incidental or technical interest, and as the prospects for successfully introducing the one species which has been proved to be of preeminent importance abroad have grown less bright, the tachinid parasites have gained in the favor accorded to them, and from being considered as of secondary importance they have become of primary importance.

This change in attitude toward them would have come about in another way, even though it had not been forced through the comparative failure of the hymenopterous parasites to make good as yet. In nearly every instance in which the parasites of a native defoliating caterpillar have been studied, the tachinids have been found to play a part which was at least the equivalent of the part taken by the Hymenoptera, while in more than half the instances the tachinids have displayed superior efficiency. This is probably not true of the parasites of any other order than the Lepidoptera, and of only a portion of the larger representatives of that order.

For the most part the tachinids are restricted in their choice of hosts through purely physiological limitations, but to a material extent they are restricted through purely physical causes as well. The fall webworm offers a striking example of both. Literally thousands of tachinid parasites have been reared from it in the course of the past few years, and with the exception of an insignificant number of the imported *Compsilura concinnata*, only a single species has been found amongst them all. This species, at present known as *Varichæta aldrichi*, through its habit of depositing living larvæ upon the food plant instead of depositing eggs or larvæ upon the caterpillars, possesses a very distinct and powerful advantage in its attack upon this particular host. If the leaves or stems near a colony of young caterpillars are selected for larviposition, it is practically a certainty that the caterpillars will enlarge their nest to include these leaves; will thereby come in contact with the parasite larvæ, and thus complete the chain of circumstances through which parasitism comes about. A parasite having a similar habit would stand an infinitesimal show of providing for the future of its young if the webworm should suddenly change from a gregarious and nest building to a solitary and wandering insect. At the same time that the host escaped attack by Varichæta, it would lay itself open to attack by a variety of other species which are now only prevented from attacking it on account of the protection which its web affords.

But even though it were freely exposed to attack by all the species of tachinids which deposit eggs or larvæ directly upon or in their host, it would be immune to such attack by all but a small percentage of the species which might conceivably select it as a host. This is proved through the occasional ·occurrence of caterpillars bearing tachinid eggs, but with no evidences of internal parasitism showing on dissection. It is not merely necessary that the host be exposed to attack and acceptable to the instincts of the mother parasite; it is necessary that it possess certain physiological characteristics which force it to react in certain ways and no others to the stimulus of the parasite's presence. Unless the host does react in the manner to which the parasite is accustomed, the parasite which is unable to accommodate itself to circumstances beyond a certain extent will find itself in a position which would be comparable to that of a man suddenly thrust into a world where all the commonest laws of nature worked in an unfamiliar manner.

To say that many of the tachinids are physiologically restricted in their host relationships is equivalent to saying that they are restricted to a limited number of hosts, and this is true; probably more true than of the hymenopterous parasites taken as a whole, or of any large group of the hymenopterous parasites if the Microgasterinæ and a few similar groups of genera are excepted. It is probably true also that among those parasites which are the most closely restricted in their host relationships are to be found those which are the most effective in bringing about the control of their respective hosts. This is primarily due to the fact that a correlation usually exists between the life and seasonal history of such a parasite and some one or more hosts which it is particularly fitted to attack. The existence of a correlation between parasite and host of such intimate character makes possible the continued existence of the parasite independently of alternate hosts, and it is thus enabled to keep pace with the one species upon which it is peculiarly fitted to prey when other circumstances are favorable to its increase.

Some of the most interesting examples of correlation of this sort which have yet come to attention are to be found among the tachinid parasites of the gipsy moth or the brown-tail moth, and on this account as well as on a purely empirical basis they are now considered much more likely to become important enemies of these hosts than before their characteristics were so well understood.

THE REARING AND COLONIZATION OF TACHINID FLIES; LARGE CAGES VERSUS SMALL CAGES.

In more ways than can be recalled without taking up and discussing each species in turn has the necessity for a more complete knowledge of the tachinid parasites impressed itself upon those most concerned with their economical handling. The difficulties attending the successful hibernating of the puparia of Blepharipa and the mysterious disappearance of *Parexorista cheloniæ*, after it was considered to be thoroughly established in America, may be mentioned as conspicuous examples among the many oftentimes curious and sometimes apparently inexplicable problems which have come up for solution. Just at the present time there is pressing need of more and accurate

OUTDOOR PARASITE CAGE.

The tree is infested by gipsy-moth caterpillars, while the parasites are confined by the wire-gauze covering. Saugus, Mass., July, 1905. (From Kirkland.)

OUTDOOR CAGES COVERED WITH CLOTH AND INCLOSING INFESTED TREES: USED IN REARING PARASITES.
(FROM KIRKLAND.)

VIEW OF LARGE CAGE USED IN 1908 FOR TACHINID REARING WORK. (FROM TOWNSEND.)

PLATE XVI.

VIEW OF OUTDOOR INSECTARY USED FOR REARING PREDACEOUS BEETLES IN 1910. (ORIGINAL.)

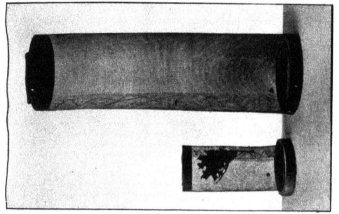

FIG. 2.—CYLINDRICAL WIRE-SCREEN CAGES USED IN TACHINID REPRODUCTION WORK IN 1910. (ORIGINAL.)

FIG. 1.—WIRE-SCREEN CAGES USED IN TACHINID REPRODUCTION WORK IN 1909. (ORIGINAL.)

information concerning the rapidity of dispersion of certain among these flies and concerning the host relations of certain others.

In an account of the methods used in conducting investigations into the lives and habits of the tachinids, published by Mr. C. H. T. Townsend three years ago, a good outline of the beginning of this work is given. It has been found necessary to modify to a certain extent the methods which seemed best at the time when this account was written, and in one particular at least it seems advisable to correct the statements therein made concerning the use of the large out-of-door rearing cage for tachinid reproduction work and investigation.

In the beginning the use of the large cages, consisting of a wooden frame covered with cloth or wire screen and inclosing a living tree, was attempted upon a considerable scale. Cages of this character had been so successfully employed in various somewhat similar lines of work as to justify their consideration in this, and accordingly a dozen or more were constructed and used for the confinement of all sorts of introduced enemies of the gipsy moth or the brown-tail moth, from *Pteromalus egregius* to *Calosoma sycophanta*, including the tachinid parasites.

The first of them, covered with wire gauze, was constructed in 1905 (Pl. XIII) and has been figured several times in various reports upon and accounts of the work, but it was never given a thorough test on account of the failure to secure parasites in any amount that first year. In 1906 cheesecloth coverings were substituted for wire and a number of cages, the general pattern of that figured herewith (Pl. XIV), was constructed and used that year and in 1907, but with pretty generally unfavorable results. It was found that only a very small number of caterpillars could be supported by the foliage of the inclosed trees or shrubs, and that it was necessary to feed them artificially exactly as was necessary in the smaller cages. The impossibility of keeping a variety of native insects out, as well as of keeping the foreign insects in, was another and only too apparent fault. In an experiment with tachinid reproduction in one of these cages in 1907, the number of flies introduced in the beginning grew steadily less day by day, with no adequate explanation for the disappearance of the missing individuals.

Another disadvantage accrued through the fact that when a caterpillar was in any way dislodged from the inclosed tree upon which it was expected to remain and feed, the chances were infinitely greater that it would find its way to the side, and then to the roof of the cage, than that it would, unassisted, regain its former position. The parasites, also, instead of staying about the tree where their business was supposed to demand their attention, would persistently remain in the uppermost recesses of the cage and refuse to come down.

All in all, the disadvantages were so many, in proportion to the advantages, and these latter were so largely imaginary in point of fact, as to result in the decision to discontinue the use of the large cages entirely in 1908.

The cage figured by Mr. Townsend (Pl. XV) was, however, an innovation in several respects. It was built independently of any tree which should serve as food for the inclosed caterpillars, but these caterpillars were confined within certain restricted limits and exposed to the attack of the tachinid flies at one and the same time by the use of the open "tanglefooted" tray. Here a most distinct advantage was gained. The floor of trodden earth (subsequently replaced by cement) effectually prevented the entrance of numerous insects which were formerly uninvited guests and thereby removed another serious disadvantage. An arrangement of double doors and wire-screened vestibule prevented the untimely liberation of the flies, and there were no longer so many inexplicable disappearances. The fact that the top of the cage was flat instead of being extended into the gable tended to keep the flies somewhere more nearly where they were wanted. In short, there were a great many advantages possessed by the new cage which were not possessed by the old, and there was some justification for considering it good.

In the meantime Mr. Burgess, who had taken over the Calosoma work in the fall of 1907, had developed the out-of-door cage along totally different lines, making it into nothing more than an out-of-door insectary (Pl. XVI), in which were conducted practically all of his numerous and varied investigations. It had seemed in 1907 as though the only one among the numerous imported insects which had done at all well in the out-of-door cages as then used had been the Calosoma, but the success attending their use for the rearing of this insect was so soon and so overwhelmingly eclipsed by the success which attended the use of small individual cages for single pairs of the beetles or individual larvæ as to render the advisability of their discontinuation for this purpose emphatic.

Some attempt was made to use the tachinid cage in 1909, but not to the extent to which it had been used the previous year. Late in the summer of 1909 reproduction experiments with small numbers of various species of tachinids were undertaken by Mr. W. R. Thompson, who used cages constructed after the familiar Riley type, but covered entirely with coarse fly screen. (Pl. XVII, fig. 1.) He succeeded in much of that which he undertook to do, and in 1910 continued the use of this type of cage, for a part of a quite extensive series of most interesting and successful experiments, but he also used a much smaller cage consisting of a wire-screen cylinder (Pl.

XVII, fig. 2, at right), about 8 inches in diameter and 12 inches high, with wooden top and bottom. His best results were secured through the use of this cylinder, and the reason appeared to be that the flies were less likely to fly and acquire sufficient momentum to injure themselves in small than in large cages.

In elaboration of the principle apparently involved, a still smaller cylinder (Pl. XVII, fig. 2, at left), scarcely 3 inches in diameter and shorter than that formerly used, was experimented with. Better results than ever before were secured upon the single occasion upon which this cage was used, and unless further experimentation results in additional modifications or in a reversal of the results first obtained, the cylinder cage figured herewith will be used almost exclusively in 1911.

As a basis for comparison of the utility of the large versus the small cages, the results attending the investigations into the biology of Blepharipa may be taken as an example.

Between 300 and 400 flies were used in an attempt to secure oviposition in the large cage in 1908, and no care that could be given them under these conditions was lacking. Not a single female completed her sexual development to the point at which she was capable of depositing fertile eggs, and no eggs of any sort were secured. Scores instead of hundreds of flies were used for the experiments in the spring of 1910, and many of the females lived throughout the period allotted for the incubation of their eggs and deposited them at the rate of several hundred daily, and abundant opportunity was thus afforded for the continuation of the studies into the lives and habits of the young larvæ under different conditions and in different hosts.

In short, after the most thorough tests, the use of the large out-of-door cages has been definitely abandoned for all phases of the work at the gipsy-moth parasite laboratory. It is not, however, intended to state thus dogmatically that similar large cages would not be adaptable to work with parasites of any other host.

HYPERPARASITES ATTACKING THE TACHINIDÆ.

Undoubtedly there is abroad an important group of secondary parasites of the gipsy moth and the brown-tail moth, included in which are some which attack the various species of tachinids to such an extent as indirectly to affect the welfare of the primary host. Very little is known of this hyperparasitic fauna, because practically all of the tachinids received have been from host caterpillars which were living at the time of collection. That it exists is well indicated by the tentative studies of the American parasites of *Compsilura concinnata*, which were made in 1910, and which will be the subject of mention at another place.

Occasionally, however, a few secondary parasites have been reared from puparia from abroad either because these puparia were collected in part in the open or because the parasites were of species which attacked the primary parasite during the life of the primary host. The number of secondary parasites having such habit is apparently very limited, and it has been definitely proved of but two genera, namely, Perilampus among the chalcidids and Mesochorus among the ichneumonids. The latter has never been reared as a parasite of any tachinid.

Because of the rather extraordinary precautions which were taken to avoid introducing into America the secondary, together with the primary, parasites of the gipsy moth and the brown-tail moth, the whole question of secondary parasitism is worthy of considerable attention in anything which purports to be a history, however abbreviated, of the operations conducted at the parasite laboratory. In the case of those attacking the tachinids it is better that they be briefly considered en masse, since there are very few among them with host relations restricted other than physically.

PERILAMPUS CUPRINUS FÖRST.

Actually, only a very little is known of this species from first-hand investigations further than that it is occasionally reared from puparia of any species of tachinid parasitic upon the brown-tail moth or gipsy moth in Europe, and under circumstamces which strongly indicate a habit of making its attack before the death of the primary host. At the same time it is felt that much is known of the probable habits of this species through analogy as the results of Mr. Smith's studies of the early history of the allied American species, *Perilampus hyalinus* Say, which attacks the parasites, both hymenopterous and dipterous, of the fall webworm. Presumably, like the American species, its minute first-stage larva, or "planidium," gains access to the host in some manner not quite clear, and after wandering about in its body for a time enters the bodies of such parasites as it chances to encounter.

That a secondary parasite having such habits might be expected to be peculiarly a parasite of the parasites of one particular host rather than of the same or similar parasites of another host, coupled with the fact that extraordinary precautions were obviously necessary to provide against its accidental importation, made *Perilampus cuprinus* appear peculiarly abhorrent, and for a time following the discovery of the early habits of *P. hyalinus* precautions against the importation of its congener were redoubled. In the course of time it was determined that it was never present in sufficient abundance to make it at all probable that it was a parasite of the gipsy moth or the brown-tail moth parasites to anything like the extent to which

P. hyalinus was thus peculiarly an enemy of fall webworm parasites, and thus a friend of the fall webworm. Neither, when it was present (which it was not, as a rule), was it ever known to emerge from infested puparia of the "summer issuing species" until long after the flies had ceased to emerge. From the puparia of species which hibernated as pupæ it never emerged until the spring and then appeared *before* the flies themselves. It was thus possible to provide against its escape with little trouble, and it is now considered as distinctly less menacing than the species which follows.

MELITTOBIA ACASTA WALK.

Another most extraordinary parasite of tachinids in Europe is *Melittobia acasta*, according to a determination furnished some years ago by Dr. Ashmead. It is thought probable that a careful comparison between the parasite of the tachinids and *M. acasta* will reveal specific differences, but at the time of writing such comparison has not been made. Of all of the secondaries which have been imported with the parasite material this has proved the most annoying.

Its most annoying characteristic is its minuteness, which enables it to pass through 50-mesh wire screen at will, and this, coupled with an extreme hardiness and an insidious inquisitiveness which seems to know no bounds, has resulted upon two occasions in an infestation of the laboratory which is comparable to a similar infestation, which will receive further mention, by the mite Pediculoides.

No one knows where it came from upon either occasion or how it first succeeded in gaining a foothold in the laboratory. Its first appearance was in 1906, when Mr. Titus encountered it in several lots of puparia of different species of tachinids from several European localities. Mr. Titus evidently thought, judging from his notes, that it had been imported in each instance with the material from those localities. He studied its habits that first year, and found that it would oviposit freely in confinement and that such oviposition was successful. He did not give it full credit for its insidiousness, and as a result it succeeded in eluding his vigilance and gaining access to a number of the lots of hibernating puparia of Blepharipa, upon which it reproduced with great freedom.

In the spring of 1907 this circumstance became evident through its emergence in some numbers from several of the lots of hibernating puparia early in June, after most of the flies had issued. An examination of the remaining puparia was thereupon undertaken and a vast number of larvæ, pupæ, and unissued adults destroyed.

At that time it was supposed that each of the lots of puparia were infested at the time of their receipt, but when an even larger amount of similar material was received from an even greater number of localities in 1907, and a smaller, but still a considerable amount in the course of the year following, and no trace of Melittobia was encountered, it began to become apparent that the quite general infestation of the puparia in 1906 had taken place after their receipt at the laboratory.

Vigilance unrewarded during the two years slackened somewhat in 1909, and late in the summer a new infestation of Melittobia suddenly developed. Where it originated was and remains wholly a mystery. Possibly the first individuals were received in a large shipment of sarcophagid puparia which had been collected in Russia and forwarded to the laboratory by Mr. Kincaid, who considered them to be gipsy-moth parasites. This lot of several thousand puparia was thoroughly infested, and a very large proportion contained either the exit holes or the brood of Melittobia when their condition was discovered.

But the infestation did not stop here. Various small lots of puparia of various sorts, inclosed in small pasteboard boxes, in cloth-covered vials, or in other receptacles were found to have been attacked by the parasite. It seemed suddenly to have come from nowhere and to have attacked everything at once.

A very general cleaning up was immediately instituted, but again, it was felt, after the damage had been done. The sarcophagid puparia, which would otherwise have served as the basis for a very necessary and desirable series of investigations into the true character of these flies, had to be destroyed. A large percentage of them was attacked by the parasite, and the rearing of the healthy remainder involved the isolation of each and all of them in a series of tightly stoppered vials. The Melittobia were issuing daily and immediately attacking the healthy remainder and there was no method short of breaking open each puparium which sufficed to determine its condition.

After the cleaning up had been accomplished, Mr. Smith began a series of investigations into the life and habits of the parasite, the results of which he intended to have prepared for publication before leaving the laboratory. Since he did not do this, and since the species is one which is likely to become a cause of annoyance should similar work to the present be undertaken, the following brief summary of the results of his studies may be given.

The minute females, after having been fertilized by the still more minute, blind, and wingless males, issue from the puparium in which they have passed their early transformations and go in quest of others which they may attack. They will also attack hymenopterous cocoons, but with less success, apparently, than in the case of the more favored host. In the course of this search they will enter the damp earth for a distance of several inches in quest of puparia which have been buried therein, and since they can pass through well-nigh invisible cracks and are in possession of an acute maternal instinct, they are able to enter receptacles of all sorts by means of openings far too small to permit the passage of any other among the secondary parasites which have been studied, not excepting those from the gipsy-moth eggs.

Having located their prey, oviposition follows, the eggs are deposited upon the surface of the nymphs in an irregular circle surrounding a wound made by the ovipositor. They are very small but appear to swell somewhat before hatching, and if the puparium is broken open so that they are freely exposed to the air, they will not hatch at all. Contrary to expectations the larvæ and their mode of life presented nothing abnormal. The number of larvæ or pupæ which had been found in the hibernated Blepharipa in the spring of 1907 was so extraordinarily large in comparison to the size of the mother insect that it was considered likely that some form of polyembryony or pædogenesis would be found upon further study.

Becoming full fed, they will pupate immediately if the temperature is uniformly high, but will hibernate if it is allowed to fall below a point which was not determined. As soon as pupation has taken place the sexes are easily separable, through the absence of wings and eyes in the males. The male pupæ develop much more rapidly than the females and the adults issue in advance of their mates. They are invariably in the great minority, and their relative numerical strength is still further reduced through the terrific duels which follow their emergence. Notwithstanding their physical defects in the matter of sight and powers of flight, their seeming weakness otherwise, and their small size, even when compared to their mates, they possess a courage and a vigor that is most surprising. In the instance of a colony which had been removed, from the puparium in which it was reared through its early stages, to a small glass cell, the several males which issued well in advance of the females engaged forthwith in conflict, in the course of which a considerable number was killed. The survivors of this Lilliputian battle royal calmly awaited the issuance of the members of their harems and proceeded to mate with one and all with an ardor which seemed to know no limit.

Mr. Smith also conducted an experiment in parthenogenesis, the results of which were and remain unique in the annals of the laboratory. As in every other instance in which an attempt has been made to secure parthenogenetic reproduction with the hymenopterous parasites, it was successful, but in this case to a limited degree only, in that the females positively refused to deposit more eggs than they would normally have produced males had they been properly fertilized. Instead of depositing sufficient to provide for the complete consumption of the host, only four or five would be deposited at a time, and notwithstanding that after the depositing of what probably amounted to barely 5 per cent of those which filled their abdomens fairly to bursting, they ceased, and nothing short of impregnation served to arouse their maternal instincts again.   As virgins they displayed a longevity lacking in the case of the fertilized individuals, and in those instances in which they were properly cared for easily outlived the time necessary for their scanty progeny to complete its transformation.

This progeny, as was expected, was exclusively of the male sex, which, when afforded opportunity, promptly united with their virgin mothers, who thereupon displayed the normal desire to deposit their eggs.   As in the instance of Schedius, the fruit of such unnatural union consisted of both sexes.

Nothing approaching this characteristic of Melittobia has been encountered in any similar studies which have been made of the parthenogenetic reproduction of the parasitic Hymenoptera.   In every instance either one sex or the other has been the result, and oviposition by virgin mothers, in so far as any observations to the contrary have been made, is perfectly normal and as free as by mated females.

It formed a strong argument in favor of the sex of the egg, in this particular species, having been determined before fertilization took place, a characteristic which is certainly not possessed by the majority of the parasites studied.

CHALCIS FISKEI CRAWF.

This large and fine representative of its genus has been received from Japan each year since the first large shipments came from that country in 1908 as a parasite of Crossocosmia and Tachina.   It is of interest in that it is fairly common, and worthy of consideration on that account, but more on account of its having been reared under circumstances which tend to indicate that it somehow gains access to the tachinid larva before the latter leaves its host.   This evidence is not sufficiently complete to justify an outright statement to the same effect, but it is sufficiently convincing to make its possibility worthy of mention.   On this account the species acquires an importance which it would otherwise lack, and as a possible specific enemy of the parasites of the gipsy moth it is worthy of special endeavors looking toward its exclusion.

MONODONTOMERUS ÆREUS WALK.

As will be mentioned again under the discussion of this species as a primary parasite of the gipsy moth and the brown-tail moth, Monodontomerus is commonly reared as a secondary as well as a primary parasite.   Its occurrence as a secondary is altogether too frequent and under such conditions as to make its recognition as such too plain to permit excuses in its behalf similar to those which have been put forward in the case of *Theronia fulvescens* Cress. It is therefore considered as a secondary just as much and as habitually as it is a primary and the question as to whether it is of enough more importance in one rôle than it is in the other to render it more than neutral remains to be decided. Apparently its value as a primary is sufficient to render void its noxiousness as a secondary, and to leave a considerable margin to its good, but this margin does not seem quite as wide now as it did a year ago, and it will require a year or two more to determine the true status of the parasite.

MISCELLANEOUS PARASITES

There are quite a number of small chalcidids, the most of them being *Dibrachys boucheanus* Ratz, which are occasionally received with shipments of tachinids from abroad. None of them is of any importance whatever in this connection, from the point of view gained through the study of the material collected and sent under the conditions which have prevailed in the past. Sometimes when lots of loose puparia have been shipped as such, loosely packed, two or three among them have produced a colony of Dibrachys or some other parasite of similar size and habits, and these individuals have immediately set about the propagation of

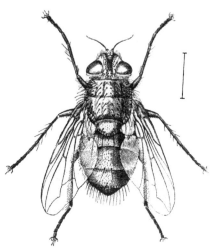

FIG. 35—*Blepharipa scutellata·* Adult female   Enlarged.
(Original)

their species with such good effect as to bring about the destruction of the larger part of the remaining puparia.

No serious effort has as yet been made to sort the Chalcididæ thus reared to species, much less to determine their specific identity.

BLEPHARIPA SCUTELLATA DESV.

Among the tachinid parasites of the gipsy moth caterpillars or the brown-tail moth caterpillars, *Blepharipa scutellata* (fig. 35) is the most conspicuous representative of the group characterized by the habit of depositing eggs (figs. 36 and 37) upon the foliage of trees or other plants frequented by its host with the deliberate intention that they shall be devoured. It is also an exceedingly close ally to the

Japanese *Crossocosmia sericariæ*, which was the subject of the original investigations by Dr. Sasaki through which this peculiar habit was discovered. The full life of the fly from the deposition of the eggs to the issuance of the adult, some 10 or 11 months later, has been the subject of a special series of investigations by Mr. W. R. Thompson, who, it is expected, will shortly publish the results of his studies.

FIG. 36 — *Blepharipa scutellata:* Eggs *in situ* on fragment of leaf. Enlarged (Original.)

It is worthy of note that the results of Dr. Sasaki's observations have been abundantly confirmed in very nearly every respect in which there is not an actual difference between the habits of Blepharipa and those of Crossocosmia. Each female fly is capable of depositing several thousands of eggs upon the foliage of trees frequented by the caterpillars of the chosen host, but it is not known to what extent she employs discretionary powers in the selection of these trees. Presumably she is attracted to those upon which the host caterpillars are most abundant. Whether one sort of tree is more attractive to them than another is not known. The young larvæ hatching from the eggs which have escaped maceration by the mandibles of the caterpillars pass through the wall of the alimentary canal and immediately proceed to take full advantage of the physiological changes brought about in the host organism as the direct result of their presence. There are two larval ecdyses and three larval stages (as is the case with every other parasite of which the transformations are sufficiently well known to make any statement possible), and the manner of life undergoes a change with each ecdysis.

The first-stage larva embeds itself in the tissues of the host, which apparently react in a manner somewhat suggestive of the reaction which results in the growth of a vegetable gall following attack by a gall-making insect. The drawings of these

FIG. 37.—*a*, Egg of *Blepharipa scutellata*, showing characteristic sculpture and markings; *b*, egg of *Pales pavida*. Greatly enlarged. (Original.)

gall-like bodies containing the larvæ (fig. 38), as well as the drawings of the egg and of the second-stage larval "funnel" were prepared under the direction of Mr. Thompson as illustrations for his forthcoming paper.

The second-stage larva undergoes a complete change in its manner of life, and its activities result in the formation of a tracheal "funnel," as illustrated in figures 39 and 40. In this stage the larva breathes through the spiracle of its host, to which the "funnel," which is apparently formed by the adventitious growth of a main branch of the trachea, is directly attached.

But few of the parasites, the early stages of which have been studied at the laboratory, exhibit a more clearly defined physiological relationship with their host than does Blepharipa. This relationship is comparable in many ways to that between the cynipid gall-makers and the oak tree which serves as their host. As is well known, many species of cynipids are closely restricted to one species of oak, or, at least, to several nearly allied species, and the same is to be expected of parasites like Blepharipa and others here spoken of as physiological, and thus limited in their host relationships.

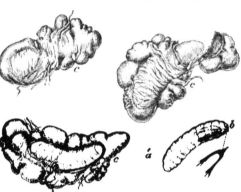

Fig. 38.—*Blepharipa scutellata.* First-stage larvæ: *a*, Natural size; *b*, greatly enlarged, *c, c, c,* greatly enlarged *in situ* in atrophied tissue of host. (Original.)

The gipsy moth itself is comparable to the parasites in which the host relations are determined by physical rather than by physiological conditions. In its choice of food, although it prefers oak to almost any other of the native trees, it can and does attack all or nearly all varieties

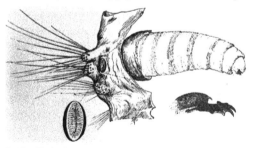

Fig. 39.—*Blepharipa scutellata:* Second-stage larva *in situ* in a portion of its tracheal "funnel." Greatly enlarged. (Original.)

of deciduous trees, and even conifers and herbaceous plants when necessity demands.

The development of the Blepharipa is directly correlated to the development of the host, and as a parasite of the gipsy moth, its larva awaits the pupation of the host before assuming the aggressive, and destroying it (Pl. XVIII, fig. 1). Its own pupation is accomplished in the earth (Pl. XVIII, fig. 2), and the pupa develops adult

characters in the fall. The space between the pupa or nymph and
the shell of the puparium is filled by a small quantity of liquid, and
the complete drying up of this liquid is very prejudical to the health
of the individual, and is usually sufficient to prevent its emergence.

The difficulties which have stood in the way of a successful introduction of *Blepharipa scutellata* into America have differed in many
respects from those which have accompanied the work with any of
the other species, saving only the closely allied Crossocosmia. The
first importations of full-grown caterpillars or freshly-formed pupæ
of the gipsy moth in 1905 resulted in the securing of a considerable
number of hibernating puparia. There were several hundred at least,
but although they were kept under conditions which would be satisfactory in the case of most of the tachinids, not a single Blepharipa
issued in the spring of 1906. The death of the insect did not take
place until after the fly was
fully formed and apparently
nearly ready to issue from the
puparium.

A great many different
methods of hibernating these
puparia have been experi-
mented with at the laboratory
with variable, and until the
winter of 1909 with poor, results. During the winter of
1907–8 the puparia were kept
in moist earth and a 10 per

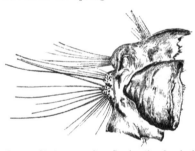

Fig. 40 —*Blepharipa scutellata* Basal portion of tracheal
"funnel" Greatly enlarged. (Original)

cent emergence from a total of 5,000 was secured. The year before
they were also hibernated in earth, but the emergence was less,
amounting to only 3 per cent of the total, and the year following
still less, being only about 1 per cent.

In 1909 for the first time since the inception of the work large numbers of living gipsy-moth pupæ containing the immature maggots of
Blepharipa were received at the laboratory from Hyères, France,
through the magnificent efforts of M. René Oberthür, of Rennes, and
as a direct result of the senior author's trip earlier the same year.
Some idea of the size of these shipments may be gained by reference
to Plate XIX, figures 1 and 2, which show a small proportion of the
total number of packages at the time of their receipt at the laboratory.
For the first time it was possible to allow the formation of the puparia
under natural conditions in the earth. During each of the preceding
years the caterpillars and pupæ had been received from abroad by
means of the ordinary methods of transportation and puparia had
been formed in the boxes on receipt. They were often injured and
always thoroughly dried when received. This year provision had

FIG. 1.—BLEPHARIPA SCUTELLATA:
FULL-GROWN LARVA FROM GIPSY-
MOTH PUPA. ENLARGED ABOUT
SIX TIMES. (ORIGINAL.)

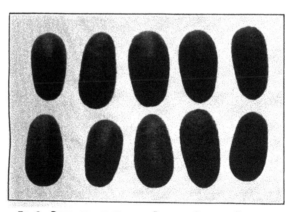

FIG. 2.—BLEPHARIPA SCUTELLATA: PUPARIA. SLIGHTLY ENLARGED.
(ORIGINAL.)

FIG. 1.—IMPORTATION OF GIPSY-MOTH CATERPILLARS FROM FRANCE IN 1909, EN ROUTE TO LABORATORY AT MELROSE HIGHLANDS, MASS. (ORIGINAL.)

FIG. 2.—IMPORTATION OF GIPSY-MOTH CATERPILLARS FROM FRANCE IN 1909; RECEIPT AT LABORATORY, MELROSE HIGHLANDS, MASS. (ORIGINAL.)

been made for cold storage in transit, with the results as mentioned above.

A very large number of the parasites were secured in this manner, and several thousands of the maggots were allowed to enter the earth in the open in forests infested by the gipsy moth. Others were allowed to pupate in a natural manner in forest soil or in a mixture of garden loam and sand in a variety of containers in the laboratory grounds.

An examination of these puparia was made from time to time during the winter and they were found to be uniformly in a much more satisfactory condition than the hibernating puparia had ever before been at that season of the year. So far as could be determined even up to within a few weeks before the emergence of the flies would naturally take place, there was no difference in the condition of the puparia hibernated in different kinds of soil or under slightly different environment.

Beginning quite early in the spring and continuing through a considerable period, flies emerged in very variable proportions from the different lots of puparia. The emergence in a few instances was well up toward 100 per cent. In others it was much lower, and in a few none of the flies completed their transformations. The reasons for these differences were not obvious in every instance, but it was obvious that unless conditions are practically identical with those which prevail in the open, the flies will fail to issue in the spring. Moisture is an essential, but is by no means the only essential to success. Nor can failure be attributed to unduly high or low temperatures, or unnatural and abrupt changes in the temperature during the period of hibernation.

The average percentage of emergence from all of the different lots of pupæ has not been as yet accurately calculated, but it was far in excess of any that was secured before, and three colonies which were considered to be satisfactorily large and strong were established in different parts of the infested area. It was not really expected that any of the new generation would be recovered from the field during the course of the first season, and it was therefore considered a particularly good omen when a few were recovered, without difficulty, and under conditions which indicated that dispersion at a quite rapid rate had accompanied a rapid rate of increase. The species has not yet been placed on the list of those considered as thoroughly established, since it is not certain that it will pass through the complete seasonal cycle in the field, but it is confidently expected that it will live through successfully and that it will be recovered in 1911 in larger numbers. If these expectations are realized there is every reason to believe that it will become a parasite of consequence within the next five years.

Curiously enough, among the imported gipsy-moth enemies that which most nearly resembles Blepharipa (if the practically identical Crossocosmia be excepted) in the part which it will probably take in the control of the gipsy moth is *Calosoma sycophanta*. No two of the imported enemies differ more radically in their method of attack than do these, the extent of their differences being fairly well exemplified by the fact that the gipsy moth eats Blepharipa, while the Calosoma eats the gipsy moth, which is literally true.

In one very important respect they are similar in that both are able to exist continuously upon the gipsy moth without being forced to have recourse to any other insect so long as the gipsy moth retains a certain degree of abundance. Both work to their best advantage and multiply most rapidly at the expense of the moth when the latter is superabundant.

It is only necessary to consider the powers of reproduction (potentially several thousandfold) possessed by the tachinid to see what an enormous rate of increase is likely to prevail in localities where practically complete defoliation occurs without becoming so complete as to bring about wholesale destruction of the gipsy moth through disease. Under such circumstances a very large proportion of the eggs deposited upon the foliage would perforce be eaten, as compared with the proportion eaten were the caterpillars present in small numbers. The percentage of parasitism would remain practically the same in both instances, but the gross number of parasites completing their transformations would be tremendously increased with a resulting increase in the percentage of parasitism the following generation, whenever the gipsy moth becomes unduly abundant. In like manner the Calosoma, which works at a disadvantage when the caterpillars are scarce, finds the conditions resulting through superabundance exceptionally favorable for its rapid increase.

Theoretically, therefore, Blepharipa ought to act as an agent in the reduction in the prevailing numbers of the gipsy moth whenever it exceeds a certain degree of abundance, and this is the rôle which it is expected to play. Theoretically, Calosoma will play practically the same rôle. Together their activities ought to result in the breaking up of dangerous colonies of the gipsy moth, and thereby render the work of the other parasites and of such native enemies as birds, predatory bugs. etc., doubly effective.

COMPSILURA CONCINNATA MEIG.

Quite a good many of the parasites of the gipsy moth attack the brown-tail moth also, but there is only one among them, *Compsilura concinnata* (fig. 41), which is equally important as a parasite of both. The remainder, if they attack both hosts, are more or less partial to one or the other.

In its method of attack Compsilura is the opposite of Blepharipa. Its eggs hatch in the uterus of the mother, and the tiny magots are deposited beneath the skin of the host caterpillar by means of a sharp, curved "larvipositor," which is situated beneath the abdomen. They usually seek the alimentary canal, in the walls of which they establish themselves during the first stage of their larval existence.

Growth is rapid, and in the summer is in no way correlated with the growth and development of the host. About two weeks are required for the complete development of the maggot, irrespective of the stage of the host at the time of attack, and at the end of that period it issues, and usually drops to the ground for pupation. The puparia from maggots which issue from caterpillars which have spun for pupation are not infrequently found in the cocoons in the case of the brown-tail moth; and even in the case of the gipsy moth, which does not spin cocoons worthy of the name, the puparia are often found immediately associated with the host remains.

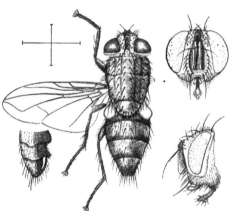

FIG. 41.—*Compsilura concinnata.* Adult female and details. Much enlarged. (Original.)

It requires a surprisingly short time for the females to attain full sexual maturity after their emergence, three or four days apparently being sufficient. This, with two weeks for the growth of the larva, and one week, or perhaps a little more, for the pupal period, makes possible a generation every four weeks during the warmer months of the year.

The position of the larva in the alimentary canal, together with certain structural characteristics, consisting of minute anal hooks, which are only known amongst other first-stage tachinids in the very similar genus Dexodes, makes possible the quite accurate determination of *Compsilura concinnata* from its first-stage larva alone, and only from observations which have been made upon these larvæ is it possible to say anything definite and at first hand concerning its habits of hibernation. Larvæ, which are almost certainly *Compsilura concinnata*, have been occasionally found in living brown-tail moth caterpillars during the winter months. It is presumed if these larvæ were able to mature under these circumstances, that they would have been reared before now from some among the hundreds of thousands of brown-tail caterpillars which have been carried through their first three or four spring stages in the laboratory. None having been reared under these circumstances, the only logical conclusion is that they start into activity so early and develop so rapidly as to cause the death of the host before they are sufficiently advanced to pupate successfully. This is not necessarily the true explanation of the failure to rear the species from hibernating brown-tail caterpillars fed in confinement, but it appears to be the best.

Ordinarily in the summer the larvæ do not pass over into the pupa of the host, but occasionally they do so. In the late summer and fall, when the host caterpillar is of a species which hibernates as a pupa, the parasite appears to be aware of that fact in some subtle manner, and likewise prepares for hibernation. Its larvæ (or what are without much doubt its larvæ) have several times been found in hibernating pupæ of several species. The adult has never yet been reared from pupæ under these circumstances, and the record is on that account open to some question.

The larger part of the Compsilura which were imported from 1906 to 1908, inclusive, issued from puparia (Pl. XX, fig. 1) found free in the boxes of brown-tail caterpillars from abroad. A companion species, *Dexodes nigripes*, which is indistinguishable from Compsilura in any of its preparatory stages, has also been reared under exactly similar circumstances, but curiously enough, if Compsilura was common in material from the same locality, Dexodes was apt to be rare, or vice versa. Some few were reared from gipsy-moth importations during this same period, but not in anything like the numbers which were secured from the brown-tail moth material, and it was not considered as of particular importance as a gipsy-moth parasite until 1909, when it was found to be very common among the tachinid parasites secured from shipments of gipsy-moth caterpillars from southern France.

The first colonies of Compsilura were planted in various localities within the gipsy-moth infested area in 1906, and in 1907, according to the records of the laboratory, a single fly was reared from gipsy-moth caterpillars collected in the immediate vicinity of one of these colonies. There is some reason to doubt the truth of this record, since every attempt at recovery made in 1908 failed.

In 1907 a much larger colony than any ever liberated before was located in the town of Saugus, in the near vicinity of one of those of the previous season. In 1908 none was colonized. In 1909 several very large and satisfactory colonies were planted in several places within the infested area, and for the first time it was felt that the species had been given a fair opportunity to prove its effectiveness as an enemy of the gipsy moth and brown-tail moth in America.

FIG. 2.—TACHINA LARVARUM: PUPARIA. ABOUT TWICE NATURAL SIZE. (ORIGINAL.)

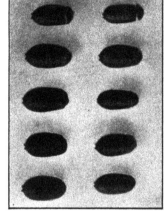

FIG. 4.—PAREXORISTA CHELONIÆ: PUPARIA. ABOUT TWICE NATURAL SIZE. (ORIGINAL.)

FIG. 1.—COMPSILURA CONCINNATA: PUPARIA. ABOUT TWICE NATURAL SIZE. (ORIGINAL.)

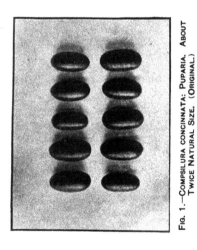

FIG. 3.—SARCOPHAGA SP.: PUPARIA. SOMEWHAT ENLARGED. (ORIGINAL.)

Hardly was the final establishment of what was for a few days considered to be the first satisfactory colony of *Compsilura concinnata* accomplished than the necessity for the expenditure of further labor on its account was obviated by the discovery that it could be recovered from the field in small but at the same time in very satisfactory numbers. Only an insignificant number was reared from the collections of gipsy-moth caterpillars made in 1909, but later in the fall of that year field men who were scouting for evidences of the spread of Calosoma and searching under burlap bands for its molted larval skins began to bring into the laboratory bona fide puparia of Compsilura found under the same circumstances. It was thus possible to delimit its range with some accuracy, and it was found to extend over a considerable territory, with the 1906–7 colony in Saugus much nearer to its center than any other more recently located colony. (See fig. 42.) There could be no doubt that the species was well established and spreading and multiplying at a rapid rate.

The results of the season of 1910 were awaited with very great interest, in expectation that they would confirm those of the year before. That these were confirmed, and most conclusively and satisfactorily, is evidenced by the results of the rearing work as summarized in Tables IV and V (pp. 141, 142), which give the results of rearing work for that year. The total number of the parasites reared or otherwise recovered from the field as indicated by these tables is very far short of the total secured.

Compsilura concinnata is recorded as a parasite of a large number of hosts in Europe, and will doubtless be found to attack an equally large number in America when it shall have become thoroughly established and abundant over a wide territory. Already some half dozen native hosts are known, and it would easily be possible to double or treble this list in the course of another season's work, should it be conducted with that end in view.

Few subjects for speculation are so overcrowded with possibilities as that of the effect which the importation of new parasites having a wide range of hosts will have upon native parasites and their hosts. The increasing abundance of Compsilura offers a most excellent opportunity to answer numerous questions which naturally arise when this subject is considered, and it is hoped that it may be made the most of. Already several highly significant observations have been made.

One of the most interesting of these resulted from a series of collections of tussock-moth caterpillars made by Mr. Wooldridge in the summer of 1910 for the purpose of determining the prevalence of parasitism in various localities and under slightly different conditions. All of these collections were of necessity made under urban conditions, since the tussock moth is rare in the country in eastern Massachusetts, and while it was expected that Compsilura would eventually be recovered as a parasite of this host, it was hardly expected that it would become of importance as a parasite so soon as 1910, or, for that matter, that it would become of importance as a parasite in cities at any time.

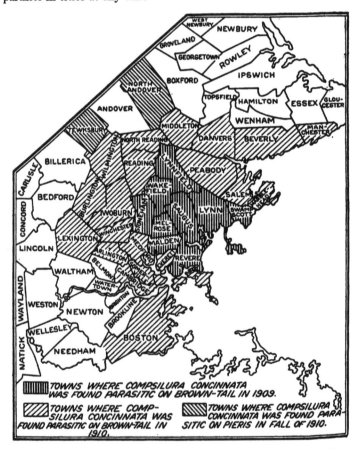

Fig. 42.—Map showing distribution of *Compsilura concinnata* in Massachusetts. (Original.)

The only one of the localities chosen for the tussock-moth collections which was within the limits of Compsilura's distribution so far as known when the work was instituted was in the city of Lynn, Mass., and from this a total of 110 caterpillars was collected on July 18, 1910. On July 29 the tray in which they were contained was carefully examined. Thirteen of the tussock-moth caterpillars had pupated and remained alive. The remainder had died, principally as the result of parasitism.

In all 96 tachinid puparia and 1 cocoon of Meteorus were found. Of these puparia 95 were *Compsilura concinnata* and the other apparently *Tachina mella*. The Meteorus, incidentally, proved to be of the introduced species, *Meteorus versicolor*.

Parasitism by native tachinids was probably considerably higher than would be indicated by the fact that only a single puparium was secured as against 95 of the imported species, but because the latter completes its larval development much more rapidly than does *Tachina mella*, it would almost certainly be the victor in case of a conflict.

Later collections of tussock-moth caterpillars made for the express purpose of determining the limits to the distribution of Compsilura resulted in its discovery throughout practically all of greater Boston, and it may be that it will have some effect in reducing the importance of this insect as a pest in that city and its suburbs. With the end of experimenting further along this line the puparia secured from the Lynn collection, together with several hundred more from gipsy-moth caterpillars, were sent to Washington, where they were liberated upon the grounds of the Department of Agriculture, where the tussock moth is periodically a pest.

Another indication of good which may possibly result from the introduction of this tachinid resulted from an investigation, begun in September, 1910, by Mr. J. D. Tothill, into the parasites of the imported cabbage butterfly ([*Pieris*] *Pontia rapæ* L.). He found that in localities where Compsilura was known to be common the summer before, it was actually abundant as a parasite of this pest, and as high as 40 per cent had been attacked in some instances.

There is no native tachinid known to have quite the same habits as Compsilura, neither is there any with quite so varied a list of hosts. Both the cabbage butterfly and the tussock moth are commonly considered as pests, the one generally and the other in cities, and both can probably sustain additional parasitism without much difficulty. But in the case of the other native insects liable to attack by the imported parasite, and already thoroughly well controlled by various agencies, of which parasitism is one, the outcome of the struggle which is likely to ensue is probably going to be different. In the case of such an one it is reasonably safe to predict that one of two things will happen. Either the prevailing abundance of the host will be reduced through the introduction of a new factor into its natural control, or the host will maintain its present relative abundance, and its parasites will suffer directly in the struggle into which they will be forced by the advent of the tachinid.

It is yet too soon to begin to speculate upon what the actual out-come in specific instances will be. An investigation of the parasites of the fall webworm was undertaken in the fall of 1910 on the suppo-sition that Compsilura would find it an acceptable host, but although it is freely attacked when outside of its web in rearing cages in the laboratory, it was not at all commonly attacked in the open, as will be seen by reference to the brief summary of the results of the work in the concluding pages of this bulletin.

There has never been a good opportunity to study the parasites of the Tachinidæ, owing to the fact that some of the species pupate upon or beneath the surface of the soil, and are therefore difficult to find in sufficient quantities to make a comprehensive study possible. So abundant was Compsilura, however, as to make it possible to collect its puparia in considerable abundance and with comparatively little trouble at the base of trees upon which the gipsy-moth cater-pillars were common, and accordingly a number was so collected in the late summer of 1910. Not enough attention was given to the work to make the results as definite as is desired, but these were sufficient to indicate that secondary parasitism was undoubtedly of very common occurrence, and that it might be a factor of some con-sequence in limiting the effectiveness of the parasite. No less than six species of secondaries were reared, including *Monodontomerus æreus*, which was common, Dibrachys, another small chalcidid, a species of Chalcis, a proctotrypid, and a Phygadeuon. It is hoped that circumstances will permit of a more thorough study of this subject in 1911, and should the parasite show an increase propor-tionate to that which was indicated by its abundance in 1910 over that of 1909, the project should be very easy of accomplishment.

It is fortunate that, under the present circumstances, with the gipsy moth and the brown-tail moth both exceedingly abundant and uncontrolled, there should be at least one parasite which was equally drawn toward both. It is easily possible that the first individuals which are reared upon the brown-tail moth as a host may attack the full-fed caterpillars for a partial second generation the same season, and then, together with the bulk of the brood coming from this host, turn their undivided attention to the gipsy-moth caterpillars. In a similar manner the first individuals to go through their transforma-tions upon the gipsy moth, together with the partial second genera-tion upon the brown-tail moth, may attack the less advanced gipsy-moth caterpillars for a partial third brood before the necessity for an alternate host becomes apparent.

There is thus possible uninterrupted increase for two complete gen-erations at least, and probably for a partial third, but unfortunately the necessity for an alternate host, though delayed until no more than one such host is necessary in order that the seasonal cycle may be rounded out, is not done away with. Unless the parasite hibernates in the brown-tail caterpillars such a host must be found among the native Lepidoptera, and while the number of species available probably runs into the hundreds, they are, with few exceptions, already controlled by their native parasites. Compsilura, if it continues to increase, will have to overcome these parasites in the competitive struggle for possession, and, as already stated, the outcome of this struggle is awaited with interest. Upon it will very largely depend the effectiveness of Compsilura as a parasite of the gipsy moth and the brown-tail moth in America.

TACHINA LARVARUM L.

This rather important parasite of the gipsy moth (fig. 43), and to a more limited extent of the brown-tail moth in Europe, is so similar to the American *Tachina mella* Walk. as to make the separation of the two by structural characters alone difficult at best, and in some instances impossible. It is similarly closely allied to *Tachina japonica* Towns., and the three species or races appear to occupy about the same position in the natural order

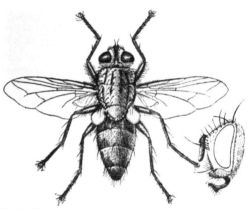

FIG. 43.—*Tachina larvarum:* Adult female and head in profile. Enlarged. (Original.)

of things in the several countries which they inhabit. The European and the American are both quite catholic in their host relations, and while the same can not be said of the Japanese in the present state of our knowledge, it will doubtless be found true when this knowledge shall be more extensive.

From an economic standpoint *Tachina mella* and *Tachina larvarum* are distinct enough specifically, if we are to consider their parasitism from an economic aspect, since the one is habitually and commonly a parasite of the gipsy moth, while the other is not. It would appear that *Tachina mella* attacks the gipsy moth quite as freely in America as *Tachina larvarum* does in Europe, but, as has already been mentioned, the attack is not successful from either the economist's or the parasite's point of view.

This is one of the less striking of several examples of a species which differs from another in biological rather than in structural characteristics. Others are to be found in the European race of Trichogramma, in the Japanese Apanteles parasitic upon *Euproctis conspersa*, which so resembles the brown-tail Apanteles of Europe, or in *Parexorista cheloniæ*, examples which will be again referred to on subsequent pages.

The large, flattened, and conspicuous eggs characteristic of Tachina and its allies are the most, and in fact the only familiar type of tachinid eggs, and they are deposited before embryological development has taken place in at least a part of the instances which have come under direct observation. The larva issues through an irregular hole in one end, and immediately forces an entrance through the skin of its host. The life cycle is longer than in the case of Compsilura, but just how much longer is not known. Sometimes the larva is carried over into the pupa of its host, but not very often. Very frequently it kills the host after it has prepared for pupation. Nearly always it leaves the host remains before pupating, on its own account, but occasionally puparia within the caterpillar skin or pupal shell are found.

The puparium (Pl. XX, fig. 2), unfortunately, is practically inseparable from that of *Tricholyga grandis* or *Parasetigena segregata* in its structural details, so that it is necessary to rear the fly before the species can be determined.

As a parasite of the gipsy moth, *Tachina larvarum* may and sometimes does take preeminent rank. Caterpillars from Holland have been received from which more puparia were secured than there were hosts, and the same has occurred on at least one other occasion in the instance of a box of caterpillars from Italy. When *Tricholyga grandis* is common, *Tachina larvarum* is rare, or at least has been rare in each instance in which the two species have been specifically determined as they issued from the imported material. It has also been entirely absent from some lots of caterpillars which did not produce Tricholyga.

It was about the first, if not the very first, parasite to be received alive in the course of the parasite-introduction work, and mention will be found in Mr. Kirkland's first report as superintendent of moth work, of its having been reared from Italian material in 1905. It was not secured in sufficient abundance to make colonization possible until 1906, but in that year quite a number of small colonies was planted in various localities in the infested territory. In 1907 it was received, but in not such large numbers, and still smaller numbers were secured and colonized in 1908. In 1909, for the first time, really satisfactory colonies were planted, and one of these colonies was strengthened by the liberation of more individuals in 1910.

So far as it has been possible to determine, no results followed these several attempts at colonization. Of all of the tachinids liberated in

1906, this was colonized the most satisfactorily, or so it is believed (having been confused with *Tricholyga grandis* it is impossible to state definitely which of the two was the more abundantly reared and liberated that year), and it ought to have been recovered by 1910 if it is ever to be recovered as a result of early colonizations. That it has not been recovered as a result of the 1909 colonization work is not at all surprising, because there is every prospect of two or three years elapsing between the liberation and the recovery of any species, and more particularly of those which, like Tachina and many others of the tachinid parasites of both the gipsy moth and the brown-tail moth, are not received from abroad until after the season is so far advanced as to make immediate reproduction upon either of the hosts mentioned impossible.

It is unfortunately true that it would be impossible to distinguish it from *Tachina mella*, should it be reared, since *T. mella* is occasionally reared as a parasite of the gipsy moth or the brown-tail moth, but it is still more unfortunate that no adults of any species which could by any possibility be referred to either were reared in 1910 from the gipsy moth.

In 1910 Messrs. Thompson and Tothill conducted an experiment to determine whether *T. mella* and *T. larvarum* would hybridize. The results were negative, and not of sufficient strength to be at all decisive. If it could be proved that hybridization took place freely, the fact in itself would probably be sufficient to render the European species of no account as an enemy of the gipsy moth in America. Interbreeding with a vastly superior number of another race, the principal and only economically important distinguishing characteristic of which was inability to breed upon a certain host, would undoubtedly result in the sinking of the racial characteristic, and *T. larvarum* as a race would almost immediately cease to exist. This is the more probable in the light of the experiences, yet to be related, which attended the attempted introduction of the brown-tail parasite *Parexorista cheloniæ*.

On this account, and on no other, *Tachina larvarum* has been tentatively eliminated from the list of promising parasites of the brown-tail moth and the gipsy moth. It may not establish itself here in America, and under the peculiar circumstances, proof to the contrary being lacking, its possible hybridization may make further attempts to import it useless.

TACHINA JAPONICA TOWNS.

Pretty nearly all that has been said of *Tachina larvarum* may be said with equal truth of *Tachina iaponica*, in so far as its value in America is concerned. It may possibly be that it is sufficiently distinct as a species to make possible its successful establishment, even presuming that *T. larvarum* should not be established, but the chances are not particularly in favor of such an outcome. It has not been so long nor so satisfactorily colonized, and there is yet a chance that it will be recovered as a result of the colonies which have been planted, or which are likely to be planted in the future. No especial attempt will be made to test its ability to exist as a race apart from *T. mella*, but it is expected that its puparia will be imported in some numbers in 1911 or in 1912, in connection with work involving the importation of other Japanese parasites.

TRICHOLYGA GRANDIS ZETT.

Although generically distinct from Tachina, according to the at present accepted and as is increasingly evident artificial classification of the Tachinidæ, *Tricholyga grandis* is so similar to *Tachina mella* and *T. larvarum* as sometimes to be separated with difficulty from those species. In Europe Tachina and Tricholyga attack the gipsy moth with nearly equal freedom, but relatively a very few Tricholyga have been reared from the brown-tail moth. The fact that Tachina and Tricholyga do not usually occur in the same locality the same year has already been the subject of comment. If may be that a careful review of the rearing records of the two will show that Tricholyga is increasingly important as a parasite in the more southerly localities, but such review has not been made with this point in view.

It was not until 1909 that it was definitely separated from Tachina in the records of the rearing and liberation of the tachinid parasites, and up to 1908 the two species were so inextricably mixed as to make it very difficult to state with any approach to accuracy the relative proportions of the two among the number colonized. There was only a single specimen of Tricholyga among the several Tachina which were preserved for museum specimens from among those imported in 1906 and 1907, and on this account it is probable that Tachina was in the considerable majority.

In habits Tricholyga differs from Tachina in only a single conspicuous respect. It deposits the same sort of eggs, similarly placed; its larvæ appear to have the same feeding habits, and about the same length of life cycle; but unlike Tachina it seems habitually rather than occasionally to pupate within the caterpillar skin or pupal shell of its victim. On this account some of its puparia have been difficult to find in the boxes of imported caterpillars, and it has been found advisable when they are present at all, to keep the dead caterpillars inclosed until such flies as are present have emerged.

Like Tachina, it probably hibernates in the puparium, but neither of the two has ever attempted to hibernate when reared from imported European gipsy-moth caterpillars. The introduction and establishment of Tricholyga will depend upon the existence of an alternate host, and its effectiveness as a parasite upon its ability to make a place for itself in the established American fauna. Several attempts to secure its reproduction in the laboratory on other hosts than the gipsy moth have been measurably successful, and there is good reason to believe that it will find conditions suitable to its continued existence here.

Notwithstanding its similarity to Tachina it appears to be a perfectly good and distinct species, and since it is not known to be represented by any very close ally in America, the objections which have been raised against the probable establishment of Tachina do not apply.

It is unfortunately impossible to say more concerning the likelihood of its becoming established here, since there is much doubt concerning its colonization. If, as is possible, it formed the bulk of the so-called Tachina liberated in 1906 and 1907, it ought to have been recovered before now; if, on the contrary, it was sparingly present among the tachinids reared and liberated during those years, there is no reason to expect its recovery before 1911, and perhaps not until 1912, as the direct result of the large colonies which were liberated in 1909, and which would represent the first satisfactory colonization of the species in America.

In the popular bulletin issued in the spring of 1910, through the office of the State forester of Massachusetts, it was stated that in the fall of 1909 it had already been recovered upon several occasions as a parasite of the gipsy moth, and under such circumstances as to make it possible that it was already established and dispersing rapidly. This statement was in part at least based upon erroneous identification, but at the present date it is expected that 1911, or at the latest 1912, will see its recovery under bona fide circumstances as an established and promising parasite of the gipsy moth.

PARASETIGENA SEGREGATA ROND.

A third species of the group which includes Tachina and Tricholyga, and which deposits similar large, flattened eggs, is to be found in *Parasetigena segregata*, which occurs throughout Europe in very variable abundance as a parasite of the gipsy moth, but not of the brown-tail moth. It is the one species of gipsy-moth parasite which appears to be more common toward the northern limits of the range of this particular host, a fact which may be explained in part by the fact that it is a common parasite of the nun moth (*Liparis monacha* L.) as well. It differs from either Tachina or Tricholyga in that it has but a single generation a year. It hibernates in the puparium, and the flies issue coincidently with, or perhaps if anything a little in advance of, those of Blepharipa in the spring.

Consequently it possesses the material advantage of being independent of an alternate host, and theoretically there is nothing to prevent its rapid increase whenever conditions favor the increase of the gipsy moth. The most that can be said against it is its inability to effect the control of its other and apparently more favored European host, the nun moth, which to a greater extent than the gipsy moth is a pest in the forests of northern and central Europe. Perhaps it may find conditions in America more favorable than in Europe, and thereby be able to do more toward effecting the control of its host here than abroad.

So far as known its larval habits agree very exactly with those of Tachina in all of their essential particulars. It leaves the host caterpillar before pupation, and only upon rare occasions is carried over into the pupa.

The first specimens which were reared in connection with the work of parasite introduction were found mingled with those of Blepharipa, which issued from hibernated puparia in the spring of 1908. There were only a very few of them, but there were enough to make it possible for Mr. Townsend to determine the salient features in its life history and to create a desire to secure more for colonization purposes.

Relatively very few puparia were secured in importations of 1908, and it remained for those of 1909 to produce the number which was necessary to make a satisfactory colony of the species possible. Its puparia being indistinguishable from those of Tachina and Tricholyga, it was necessary to await the emergence of those species before attempting to count upon Parasetigena, but after the others of the Tachina group had ceased to issue, it was found that a very satisfactory number of unhatched and healthy puparia remained. This number was subsequently increased by the importation of several hundred which had been reared from the nun moth, and which subsequently proved to be specifically identical with, or at least indistinguishable from those from the gipsy moth.

For the most part these puparia successfully hibernated, and in excess of 1,000 of the flies were reared in the spring and colonized in one locality where there was every opportunity for them to multiply to the limit of their powers upon the gipsy moth. An attempt to recover the species in the locality later in the season failed, but since it was not expected that it would be recovered so soon the disappointment was not very keen. It would undoubtedly be more encouraging from a practical standpoint if it were positively known that the species was reproducing freely, but the failure to recover it is in no way so significant as would have been the failure in the case of Blepharipa.

Blepharipa, which was colonized in the same locality and under the same circumstances and subsequently recovered, does not leave its host until after the latter has pupated, and since the collection of pupæ is very much less difficult than the collection of caterpillars, its recovery was that much more easy and certain. Although upward of a dozen Blepharipa were secured from collections made in this colony, all of them were from pupæ and none from the caterpillars, which had to be depended upon for Parasetigena.

A determined effort will be made to recover both species in 1911, and the results of the season are anticipated with much interest.

CARCELIA GNAVA MEIG.

This is probably the least understood of the tachinid parasites of the gipsy moth. It appears to be not at all well distributed throughout Europe and has never appeared in sufficient abundance to give it rank as among the important parasites except in the material from southern France. From that region it has been secured in sufficient numbers to make its colonization possible on a scale that is quite as satisfactory as the colonization of Tricholyga, or, for that matter, of Compsilura until after Compsilura was found to be established.

It was received in gipsy-moth caterpillar importations as early as 1906, but in very small numbers in that year, and in still smaller numbers in 1907 and 1908. In 1909 the very large and until then unprecedented importations from the Hyères region produced several thousands of flies, and more were received in 1910, which went to strengthen colonies of the previous year. Curiously enough, in 1910 it was almost the only tachinid parasite secured from this region, on account of which the gross number colonized is in excess of any other species. Like Tachina, Tricholyga, and Compsilura, it is practically certain that an alternate host will be a requisite if it is to complete its seasonal cycle in America. If this disadvantage can be overcome, there is every reason to expect its recovery in 1911 or 1912. That it was not recovered in 1910, in spite of the fact that some 10,000 caterpillars of the gipsy moth were collected in the immediate vicinity of the most satisfactorily liberated colony of the summer before, entirely loses its significance when it is taken into account that neither was Compsilura recovered from these 10,000 possible hosts, and Compsilura was also colonized at the same time and in the same place and under circumstances very much more favorable to its establishment than those which accompanied its original and effective colonization two or three years before. Better than Compsilura has done is expected of none of the tachinids, and neither Tricholyga, Carcelia, nor Parasetigena, nor Zygobothria, which is the next parasite to be considered, has had the opportunity which Compsilura has demanded in each instance in which it has been colonized, to prove itself of value.

ZYGOBOTHRIA NIDICOLA TOWNS.

Pretty much everything which has been said of Carcelia may be said of Zygobothria, not so much because it is similar in its habits as because we have very little first-hand knowledge of its habits. It probably deposits living maggots upon the body of its host or else very thin-shelled eggs containing maggots ready to hatch; but this is not certainly known. It always leaves its host before pupation and forms a free and characteristic puparium with roughened surface and protruding stigmata very unlike that of any of the other tachinid parasites of the same host.

It is not quite so common as a parasite of the gipsy moth as is Carcelia and not so many have been colonized, but the colonies have been very satisfactory notwithstanding, and there is about as much reason to expect the establishment of this species as in the case of any of the others. Like several of the others, it was not colonized until 1909, and its recovery is hardly to be expected until 1911 or 1912, and as in the case of these others its establishment and value as a parasite will very largely depend upon its ability to find a sufficient supply of acceptable hosts.

CROSSOCOSMIA SERICARIÆ CORN.

Many years have passed since Dr. Sasaki published the most interesting and surprising results of his investigations into the life and habits of the so-called "uji" parasite of the silkworm in Japan, and his account of the manner in which this serious enemy of that insect gained access to its host was so extraordinary in the light of that which was known concerning the oviposition of tachinids in general as to cause the truth of his discovery to be questioned by several eminent entomologists.

His work has been most carefully reviewed in connection with the · investigations which have been carried on at the laboratory into the life and habits of the allied species, *Blepharipa scutellata,* and it was with much satisfaction that his account of the biology of Crossocosmia was found to apply almost equally well in nearly all of its details to the biology of the European parasite of the gipsy moth. There was one important point of difference, however, in that the first-stage Blepharipa was never found ensconced in the ganglion of its host, while Crossocosmia, according to Dr. Sasaki, habitually chooses this position.

In 1908 quite a number of the puparia of a Japanese parasite of the gipsy moth was received from that country, which, so far as external characteristics were concerned, were indistinguishable from those of Blepharipa from Europe. None of the flies issued the following spring owing to the bad conditions under which the puparia were received, but an examination of the pupæ, which like those of Blepharipa developed adult characters in the fall, was sufficient to convince Mr. Townsend that the species was nothing else than *Crossocosmia sericariæ* itself.

Mr. Townsend's determination of the species was partially confirmed in the spring of 1910 when several hundred of the flies were reared from puparia received the previous summer. Later the same year, through the kindness of Dr. Kuwana, specimens of the bona fide "uji" parasites, reared from silkworms, were received at the laboratory. No differences whatever were discernible and the confirmation appears complete.

There was an opportunity, during the summer of 1910, to dissect a few of the caterpillars of *dispar* from Japan, and among those so dissected by Messrs. Thompson and Timberlake were found several which contained the young larvæ of Crossocosmia in the ganglia, exactly as described by Dr. Sasaki. Thus it was that his account of the life of the "uji" was confirmed in its every particular in which his remarks were based upon actual observation and not in part upon speculation as to the significance of certain obscure phenomena. To Mr. Townsend, and perhaps more particularly to Mr. Thompson, who has devoted considerable time and performed a vast amount of tedious and in some instances unremunerative dissection work, is the credit due for thus removing all reflection upon the accuracy of Dr. Sasaki's remarkable observations.

In practically every respect, except in the location of the first-stage maggots in the body of their host, the life and habits of Crossocosmia as a parasite of the gipsy moth agree with those of Blepharipa. In Japan it is of about the same relative importance as a parasite as Blepharipa in Europe. Its habits of pupation and the difficulties experienced in providing for its successful hibernation are identical.

Its value as a parasite of the gipsy moth in America depends very largely upon the success which attends the attempts to import and establish the European parasite. Should this be accomplished, as now appears probable, any special efforts to import Crossocosmia might well be deemed unnecessary. It is highly improbable that two species having habits so exactly similar would be any more effective than one.

But it is pretty evident that in one other and very important respect the habits of Blepharipa are different from those of Crossocosmia. It is apparently quite as abundant in Europe as is Crossocosmia in Japan, but even in the most important silk-producing regions it is yet to be recorded as an enemy of the silkworm. It would appear that in their respective host relations the two species possess a difference, and it is probable that it will be found to extend to other hosts than the silkworm when all the hosts of both species are known.   In consequence it is not only well to have Crossocosmia to fall back upon in case Blepharipa fails to come up to expectations, but it is well that it be given a trial in order that the relative value of the two species may be determined.

CROSSOCOSMIA FLAVOSCUTELLATA SCHINER (?).

It was with considerable surprise, accompanied with no small degree of doubt as to the accuracy of our records, that the presence of a species of Crossocosmia was recognized among the flies issuing from European puparia in the spring of 1910.   At first it was thought that there must have been some Japanese puparia mingled with them, and when reference was made to the notes it was found that something like 15 or 20 larvæ of *Crossocosmia sericariæ* had been received the summer before, and that their disposition was not indicated. Accordingly, for a time it was supposed that the Crossocosmia issuing were from these, but it was not long until more adults had issued than could possibly be accounted for in that manner.   There were as many Japanese Crossocosmia puparia producing Crossocosmia as the notes called for, with never a Blepharipa among them, and when after a time it became apparent that the number of European Crossocosmia would run into the hundreds and that they came from a variety of lots of puparia under several numbers and received at different times, it was finally decided that the existence of what has every appearance of being an European race of *C. sericariæ* could no longer be doubted.

Its occurrence in Europe is the more surprising because, like Blepharipa, it has never been recorded from the silkworm in any of the silk-producing districts.   In its distribution it also exhibited peculiarities, practically all that issued having come from a lot of puparia received in gipsy-moth caterpillar importations from the vicinity of Charroux, a town in western central France, and one which would hardly be expected to differ particularly in its fauna from other localities from which material was received.

Only a very few specimens of this European Crossocosmia were pinned for the collection, but so far as the closest scrutiny manifests there is not the slightest structural difference between the bona fide "uji" parasites reared from the silkworm—that which is consequently believed to be the same species reared from the gipsy moth in Japan—and the species under present consideration from France, which is seemingly not present, or, if present, not common in other parts of Europe from which parasite material has been received.

The specimens reared, to the number of several hundred, with several hundred of the Japanese Crossocosmia, were colonized together, and under favorable circumstances, as indicated by the recovery of Blepharipa from the immediate vicinity as the result of coincidental colonization. Should the two species be in very truth the same, they will probably hybridize, and enough have been liberated to make one good colony. Should they refuse to intermingle, there is not a sufficient number to make what past experience has indicated as a "satisfactory" colony of either.

UNIMPORTANT TACHINID PARASITES OF THE GIPSY MOTH.

There are not as many unimportant dipterous as there are unimportant hymenopterous parasites of the gipsy moth in Europe, and there are other reasons why they need not be considered at so much length. One of them, *Pales pavida* Meig., which is occasionally present in shipments of gipsy-moth caterpillars, is much more commonly received as a parasite of the brown-tail moth, and *Dexodes nigripes* Fall., which is very rarely associated with the gipsy moth, is a very common parasite of the other host. Both of these species will be discussed later, and something will be said of their life and habits and of what has been done toward securing their establishment in America.

Of the remaining tachinids which have been reared from imported material from Europe, none has been positively associated with the gipsy moth itself. There is always the chance that one or two caterpillars of some other species may have been accidentally included amongst those of the gipsy moth, and while the number of such has always been very small, the chance that a strange parasite should be reared from them rather than from the gipsy-moth caterpillars is large.

To date at least 98 per cent of the tachinid puparia which have been received from Japan as parasitic upon the gipsy moth have been either of Tachina or Crossocosmia. The remaining 1 or 2 per cent have been of various species, among which was one that resembled *Pales pavida* and another has been described as "Compsilura-like." There have been so few of these strange forms as to make impossible a definite statement as to their host relations. It seems rather curious that against the 8 European tachinids, all of which are of at least local importance as parasites of the gipsy moth, Japan should be able to produce only two. It may be that the tachinid fauna of Japan is much less extensive than that of Europe or of America. It may also be that a more thorough survey of the Japanese situation will reveal the presence of species which have not been received hitherto on account of the inadequacy of the methods of collection and shipment. It is rather expected that the latter may be the true explanation and that the apparent scarcity of tachinids in the parasite fauna of the gipsy moth in Japan may not prove to be real.

PARASITES OF THE GIPSY-MOTH PUPÆ.

THE GENUS THERONIA.

The discussion of the pupal parasites of the gipsy moth may well begin with mention of the most generally distributed of all—Theronia. The genus has already been the subject of brief comment in the account of the American parasites, and something was said of the habits of *Theronia fulvescens* in its relation to this host in America, and of its unimportance. The form which by courtesy is thus specifically designated is very imperfectly differentiated from *T. atalantæ* Poda, which prevails throughout Europe in relatively about the same abundance in relation to the gipsy moth. It is readily distinguished from the American form by its habitat and to a less satisfactory extent by color.

In Japan occurs still another, indistinguishable biologically (so far as its biology is known) or morphologically, but differing in color from either the American, from which it is most distinct, or from the European. It has been described as *Theronia japonica* Ashm.

The rôle played by these so-called species in the countries to which they are severally native is nearly identical and at the same time unimportant, when economically considered. The likelihood that either the European or the Japanese would become relatively more effective in America than the American itself seems so very remote as to make unworthy of consideration any serious attempts to introduce and colonize either. Quite a good many of the European have been liberated in America from time to time, but in a purely incidental way. More will probably be received in the future and similarly liberated.

It was in the winter of 1907–8 that the late Mr. Douglas Clemons, of the laboratory, found a large number of the females of *T. fulvescens* congregated beneath old burlap bands in a tract of woodland in which the gipsy moth was actively being fought. Some of these females were dissected some days later and found to be without fully developed eggs, and on the basis of these inadequately conducted dissections it is supposed that, as in Monodontomerus, the males die in the fall, leaving the females to hibernate. It would, in other words, mean that the species is single-brooded.

The subject ought to have been still further investigated, but the unimportance of the species from an economic standpoint has robbed it of interest other than that which has attached to the remarkable and suggestive vagaries which it has exhibited in its host relations.

If it is, in truth, single-brooded, like its host, it ought to multiply much more rapidly than it has done, in view of the superlative opportunities which the past 10 years have afforded.

THE GENUS PIMPLA.

The several forms of the genus Pimpla which have been reared from gipsy-moth pupæ received from Europe and Japan are not, like the forms of Theronia, confusing and indefinitely separable, but good and distinct species. There are 3 European, and a like number of Japanese, making together, with the 2 American, a total of 8 of the genus known to attack this host. Notwithstanding their variety, all the species acting together in any one locality have never effected the degree of parasitism resulting from the attack by Theronia in the same locality. Being collectively of so little importance it is unnecessary to say more concerning their relative importance individually.

Quite a little has been learned at first hand concerning the two European species most frequently encountered, *Pimpla instigator* Fab. and *Pimpla examinator* Fab. Both have been received in considerable numbers in shipments of brown-tail moth pupæ, and have been liberated to the number of several hundred each in 1906, 1907, and 1909. Neither has since been recovered from the field.

Both have been carried through all of their transformations in the laboratory upon the gipsy moth, the brown-tail moth, or the white-marked tussock moth, and in the case of *P. instigator* upon all three above-mentioned hosts. The early stages of the larvæ have not been seen. In nearly every respect, so far as observed, they resemble each other in habit and biology and also *P. (Hoplectis) conquisitor* Say. and *P. pedalis* Cress., their American congenors. The one point of difference between them is the tendency of *Pimpla instigator* to hibernate within the pupa of the brown-tail moth. A very few have been reared each spring since 1908 from cocoon masses received the summer before. The proportion thus hibernated is very small.

Pimpla instigator, like the American *P. (Hoplectis) conquisitor*, may become hyperparasitic on occasion. On August 7, 1907, five female specimens of *P. instigator* were confined with several tussock-moth cocoons which contained the cocoons of *Pimpla (Epiurus) inquisitoriella* Dalla Torre, from some of which adults were emerging, and all of which had been spun for several days. Oviposition was immediately attempted. It was certainly successful, for on August 29, at least two weeks after the Epiurus had ceased to issue, a greatly dwarfed male *P. instigator* appeared and it was followed by another similarly small male on September 3. There is not the slightest doubt that the European parasite attacked the native and that its larvæ fed to maturity. At the same time it is not likely that it would have done so had the cocoons of the native not been associated with the proper host of the other.

The third species, *Pimpla brassicariæ* Poda, is much less commonly reared from either the gipsy moth or the brown-tail moth than the other two. Apparently its habits are identical.

Hardly enough have been received of the three Japanese species to indicate their relative abundance. The most striking of them, *Pimpla pluto*, appears to be the only one of the trio which has been described, and to the others Mr. Viereck has given the names *P. disparis* and *P. porthetriæ*. It is possible that they are just a trifle more common in connection with the gipsy moth in Japan than are the corresponding species in either Europe or America. At the same time Theronia has outnumbered all three together in the Japanese material studied at the laboratory.

Hardly anything is known about them. Not enough have been received to make colonization possible, and only upon one occasion to permit of laboratory reproduction with fertilized females, and upon this occasion there was no time to devote to their further study. .

Presumably, except for minor differences, all of the Japanese Pimpla will be found to conform very exactly in biology and habit to the American and European. All will probably be found to attack a very large variety of hosts, and all will defer their attack until their host has entered the prepupal or pupal state. The females of all will probably be ready to oviposit for a new generation almost immediately following their emergence, and the length of life cycle, dependent upon temperature, will be about three or four weeks. There will necessarily be more than one generation each year unless the hibernating individuals should live long enough to' deposit eggs for another hibernating generation, as might easily be possible in the case of *Pimpla instigator*, and conceivably possible in the case of each of the others.

Pimpla conquisitor and *Pimpla pedalis* are among the most generally effective of the pupal parasites of the medium-sized cocoon-spinning Lepidoptera in the Northeastern States. The first named is perhaps the most common and effective of all the parasites of the tent caterpillar and about as effective as any other one as a parasite of the tussock moth. It does not vary much in relative abundance from one year to the next, and appears to play a part which is rather to be compared to that taken by the birds than to that taken by most of the parasites. It is, like Theronia, so impartial in its attentions to all of the different species of its hosts as scarcely to be affected by an unusual abundance or unusual scarcity of any one among them in particular.

The same is very likely to be true of the European and Japanese species. The part played by each in the localities where it is native is probably similar to that taken by *P. conquisitor* or *P. pedalis* in America. On this account it is not considered as probable that either *P. examinator* or *P. instigator* will ever become established in America as a result of the not very satisfactory colonies which have been liberated. They will, of necessity, enter into direct conflict with the American species for a share in the business of being parasites upon a certain section of the insect community, including a large number of species of which the gipsy moth is but one. Competition may result in cut rates and more and cheaper parasitism for a time, but eventually, if the newcomers ever secure a foothold at all, they will either drive the natives out of the business or else share and share alike with them in accordance with an amicable and natural agreement.

In consequence, no assistance is expected from the various foreign species of the genus Pimpla as parasites of the gipsy moth or of the brown-tail moth. They are merely liberated when received, under the best conditions which can be afforded looking for their establishment, and if they are ever recovered from the field, the most that is expected of them is that the circumstances surrounding such recovery will exemplify the truth of the above remarks.

ICHNEUMON DISPARIS PODA.

One of the most distinctive of the gipsy-moth parasites, and one of the first, if not the very first, described as attacking that host, *Ichneumon disparis* is at the same time one of the less common, if dependence is to be placed upon the rearing records at the laboratory. It may be that it is never common, or it may be that it is eastern and southern in its distribution in Europe, rather than central and western; some few incidents in connection with its importation have indicated that perhaps its scarcity in European imported material was due to such material having been collected outside of its natural range. In any event not more than two score of individuals have been reared in the course of the five years since the work was begun.

Very little is known of its life and habits, other than that it probably attacks the pupæ or perhaps the prepupæ, and never the active caterpillars. It is thought possible that it hibernates as an adult, and if this is true, it might conceivably be a parasite of importance could enough be secured to make possible a sufficiently strong colony. To date there never has been a single mated pair available for liberation at any one time.

THE GENUS CHALCIS.

The first few boxes of parasite material which were received in 1905 produced among other things quite a large number of Chalcis, a part of which issued from the pupæ of the gipsy moth and a part from dipterous puparia, supposed at that time to be those of tachinid parasites of the same host. All of them appeared to be of one species, *Chalcis flavipes*, and on the supposition that those which appeared to issue from the gipsy-moth pupæ might actually have come from tachinids which were inside, all were destroyed.

In 1906 and 1907 very few Chalcis were received from any source and there was no opportunity to determine the true host relations of the European species. In 1908 a considerable shipment of gipsy-moth pupæ from Italy arrived in good condition for the first time since 1905, and another shipment from Japan, also in good condition, reached the laboratory almost coincidently. Both were soon found to contain Chalcis in some numbers, and, as it soon developed, in considerable variety.

It is not necessary to go into any details as to the steps through which it was finally decided that no less than six species of Chalcis were present in these two shipments, of which two were easily separable by conspicuous structural and color characters. The others were more or less confusing to one who had only a few doubtfully identified specimens in the collection, and little knowledge of what were the characteristics of a species in the genus.

With the assistance of biological and geographical characters, the separation was finally effected, and the 6 have been since definitely identified by Mr. Crawford, as below. To the list are added 2 more, 1 of which is Japanese and the other American, making a total of 8 in all that have been definitely associated with the one host.

Chalcis flavipes Panz. Primary parasite of the gipsy moth in Europe.

Chalcis obscurata Walk. Primary parasite of the gipsy moth in Japan.

Chalcis callipus Kirby. Primary parasite of the gipsy moth in Japan, according to rearing note attached to a specimen forwarded to the laboratory through the kindness of Mr. Kuwana.

Chalcis fiskei Crawf. Parasite of the tachinids *Crossocosmia sericariæ* and *Tachina japonica* in Japan, and thereby a secondary parasite of the gipsy moth.

Chalcis compsiluræ Crawf. Parasite of *Compsilura concinnata* in America, and therefore a secondary parasite of the gipsy moth. (*Chalcis ovata* Say has never been reared as a parasite of the gipsy moth, although it is not improbable that it will be found to attack it when the moth shall extend its range southward into territory where the Chalcis is more common than it appears to be in eastern Massachusetts.)

Chalcis minuta L. Parasite upon sarcophagids associated with the gipsy moth in Europe. Since the status of the sarcophagids themselves remains to be determined, it is impossible to state that of the Chalcis. It is believed that the sarcophagids are scavengers, and neutral, in which case the Chalcis would also be neutral.

Chalcis fonscolombei Duf. Also a parasite of sarcophagids associated with the gipsy moth in Europe.

Chalcis paraplesia Crawf. Parasite upon sarcophagids associated with the gipsy moth in Japan.

It is thus seen that the genus Chalcis is a little of everything in its relations to the gipsy moth. Of the 8 species, 3 are enemies, 2 are friends, and 3 are undertaker's assistants. To round out the series, one may expect to find a species attacking tachinids in Europe, 1 attacking the gipsy-moth pupæ as a primary parasite, and another attacking sarcophagid puparia in America.

So far as known tachinids are never attacked by the species which prey upon the sarcophagids, although this statement presupposes a discriminating instinct which has rarely been encountered among the parasites of the Diptera generally. For the most part, and in fact with no other exception, so far as the experiences of the laboratory have gone, the parasites which will

FIG. 44.—*Chalcis flatipes:* Adult. Enlarged. (From Howard)

attack the one will attack the other family also. There are several records, including that already mentioned which was made in 1905, of the rearing of Chalcis from tachinid puparia, but these have either been made before a distinction was made between the puparia of the two families, or else there have been a large number of mixed tachinid puparia involved, and in such instances it is always possible and usually the case that a few sarcophagids are present.

As parasites of the gipsy-moth pupæ, *Chalcis flavipes* and *C. obscurata* are closely allied, and exceedingly similar in every respect. *Chalcis flavipes* (figs. 44, 45) appears to be rather restricted in its range in Europe and has never been received from any localities outside of the watershed of the Mediterranean, if an exception is made of the portion of southern France which drains into the Atlantic. The Japanese *C. obscurata* (fig. 46) has been present in every shipment of pupæ from Japan, but the exact localities from which these shipments came is not known.

Both are, or appear to be, invariably solitary, notwithstanding that there is an ample food supply in one pupa for several individuals. Invariably there is an abundance of unconsumed matter in the host pupa, and on this account the parasite has rarely been successfully reared from any of the imported pupæ except the small males, in which this matter is in such small amount as partially to dry before

FIG. 45.—*Chalcis flavipes:* Female. Hind femur and tibia, showing markings Greatly enlarged. (From Crawford)

receipt at the laboratory. In the large female pupæ the decomposing contents of the pupal shell form a semiliquid mass, which is shaken about while the material is in transit, and completely overwhelms the larva or pupa of the parasite. The parasite is able to withstand this condition to a remarkable extent, but not to the extent frequently brought about by the unnatural conditions incident to transshipment.

Partly on this account, but still more owing to the difficulties which have stood in the way of securing an adequate supply of gipsy-moth pupæ in good condition from localities where Chalcis occurs, it has not yet been possible to colonize either the European or the Japanese species satisfactorily, nor, so far as known, successfully. Only a few hundred have been received, all told, since their status as primary parasites was first established in 1908. Had they all been of one species, received at one time, and colonized in the same place, there would be some reason to expect that the colonization would be followed by establishment. There were two species, however, they were not all colonized in one place, and colonization has extended over three years. The best and largest colony was liberated in 1909 and strengthened by the addition of the small number received from abroad in 1910.

FIG. 46 — *Chalcis obscurata:* Female. Hind femur and tibia, showing markings. Greatly enlarged. (From Crawford.)

A single specimen was reared from a lot of gipsy-moth pupæ collected in the immediate vicinity of the colony shortly after it was founded in 1909, but none issued from similar collections made in 1910.

Both *Chalcis flavipes* and *C. obscurata* have been carried through all of their transformations in the laboratory on American pupæ. The females are able to oviposit very shortly after emergence, and will do so with considerable freedom in confinement, making possible the artificial multiplication of either species were it possible to secure a supply of host pupæ. In the act of oviposition the female firmly grasps the active host-pupa with her powerful hind legs and resists all of its efforts to dislodge her. The egg has not been observed nor

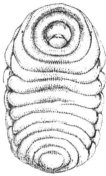  

FIG. 47.—*Chalcis flavipes:* Full-grown larva from gipsy-moth pupa. Much enlarged. (Original)

FIG. 48.—*Chalcis flavipes:* Pupa, side view. Much enlarged. (Original)

FIG. 49. — *Chalcis flavipes:* Pupa, ventral view. Much enlarged. (Original)

the early-stage larvæ. The full-fed larva is quite characteristic in appearance, and well represented in the accompanying illustration (fig. 47). The pupa (figs. 48, 49) is almost invariably located in the

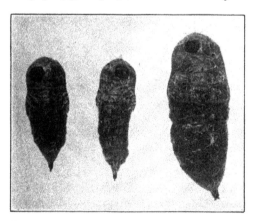

FIG. 50.—Gipsy-moth pupæ, showing exit holes of *Chalcis flavipes.* Enlarged. (Original)

anterior portion of the host pupa, and the exit hole (fig. 50) of the adult is characteristic, being smaller than that of Pimpla or Theronia, and rarely at the extreme end, as is the case with the ichneumonid parasites. The pupal exuvium is also characteristic and, curiously enough, yellow in the case of the Japanese, and black in that of the European species. Between three and four weeks of ordinary summer weather are necessary for the complete life cycle.

The adults are very long lived, and a few of both species were kept in confinement from early in August, 1910, until December of the same year. During this time they were offered numerous sorts of pupæ, but after it was no longer possible to secure those of the gipsy moth there was no further reproduction. It has always been supposed that it was the adults which hibernated, and the longevity of the individuals mentioned above lends strength to this supposition. If

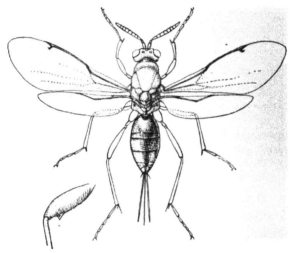

FIG. 51.—*Monodontomerus æreus:* Adult female. Greatly enlarged. (Original.)

correct, there need be only a single generation annually, and the species would therefore be independent of any other host. Neither is known to attack the brown-tail moth, but both have been reared through their transformations upon pupæ of the white-marked tussock moth.

Both of the species of Chalcis are of considerable importance as parasites of the gipsy-moth pupæ in their respective habits, and so far there has nothing occurred to destroy confidence in their ability to become of importance here provided a sufficiently large number may be secured to enable them to become established. It is confidently expected that they will disperse at a very rapid rate, and on this account it will be necessary that the colonies be large and strong, so that extinction through too great scarcity during the first or second season following colonization will not result. Renewed efforts to make this possible will be made this coming season, and at the same

FIG. 1.—VIEW OF LABORATORY INTERIOR, SHOWING CAGES IN USE FOR REARING PARASITES FROM HIBERNATING WEBS OF THE BROWN-TAIL MOTH IN 1910-1911. (ORIGINAL.)

FIG. 2.—"SIFTING" GIPSY-MOTH EGGS FOR EXAMINATION AS TO PERCENTAGE OF PARASITISM. (ORIGINAL.)

time special efforts will be made to recover the species from the field as a result of earlier colonization.

MONODONTOMERUS ÆREUS WALK.

Few among the parasites have been the cause of a larger variety of mingled feelings than this, and the history of its introduction into Massachusetts is in many respects unique and apart from similar histories of the other parasites.

The females (fig. 51) have the curious habit of hibernating in the winter webs of the brown-tail moth, and the species is rather a parasite of the brown-tail moth pupæ than of the gipsy-moth pupæ, although it is sometimes common in the latter connection as well. It was first received at the laboratory in the winter of 1905 in shipments of brown-tail-moth hibernating nests and was reared from these nests in the spring. It has been recorded as a parasite of the gipsy moth, and a colony was planted by Mr. Titus early in the spring of 1906. The records of this colony have apparently been lost and it will never be known exactly how many individuals were included in it.

Some 1,700 issued from the imported nests that first spring, but not all of them were liberated. Dr. W. H. Ashmead,[1] to whom the specimens were sent for determination, stated it as his opinion that it was a secondary parasite rather than a primary, since few or none of the group to which it belonged were definitely known to be primary parasites upon lepidopterous hosts. Accordingly the work of colonization was stopped almost as soon as it was begun, and for a period of more than two years Monodontomerus was treated as a secondary parasite, and destroyed whenever found. During this period, many thousands issued from importations of brown-tail-moth cocoons, and much doubt was felt as to its actually being a secondary, on account of the numbers alone, since it enormously outnumbered all other hymenopterous parasites (whether primary or secondary) reared. For reasons which would be obvious to anyone who has ever had any experience in handling the cocoons of the brown-tail moth, no serious effort was made to determine its host relations by the dissection of the brown-tail moth pupæ. A few pupæ were sought out from which it had issued, and no trace of any other host was found, but such was the state of our technical knowledge at that time as to render questionable such evidence. We were· not sufficiently familiar with the appearance of pupæ from which Monodontomerus as a secondary parasite had issued, and dared not give any more than negative weight to the fact that no remains of any other primary host than Monodontomerus could be found. Moreover, against this negative evidence indicative of primary parasitism, was much that

---

[1] Now deceased.

was positive, indicative of secondary parasitism, because every little while the Monodontomerus would issue from tachinid puparia which had been sorted out from the cocoons and pupæ. Still more frequently it was reared from puparia of sarcophagids.

In the summer of 1908 the shipment of gipsy-moth pupæ from Italy, which served the purpose of establishing the status of the European species of Chalcis in their relation to the gipsy-moth, served also in establishing the status of Monodontomerus as a primary parasite of this host. A large number of the pupæ which were examined was found filled with the larvæ (fig. 53) or pupæ (figs. 54, 55) of the parasite, and even when the larvæ were still immature and feeding there was absolutely no trace of any other parasite present in the majority of instances. There was such trace in a few, and it was found that the former presence of Pimpla, Theronia, or any tachinid was very easy of determination, no matter how completely it might have been destroyed.

It was felt that a mistake had been made in not liberating the very large number of Monodontomerus which had been secured through the earlier shipments, and it was resolved to colonize them as fast as they were secured in the future. Hardly anything was less expected than that the species should even then be established.

Each winter since that of 1906–7 (Pl. XXI) large numbers of the winter nests of the brown-tail moth had been collected in a vain endeavor to secure evidences of the establishment of *Pteromalus egregius*, but without results. In the winter of 1908–9 this work was undertaken anew, and almost the first lot which was brought into the laboratory was shortly productive of a number of Monodontomerus, exactly as lots collected in the open in Europe had been productive of the species each season since their importation had been begun. The circumstance, surprising and unexpected, was also gratifying, coming as it did so soon after the investigations which had served to demonstrate the primary parasitism of the species. The surprise and gratification was increased materially when it was discovered, through the collection of a large quantity of the winter webs, that the parasite was distributed over a considerable territory indicated by the area I on the accompanying map (Pl. XXII), and though the actual number recovered was small, the rapid rate of dispersion was sufficient to indicate a very rapid rate of increase. It was estimated, in fact, that at least a 25-fold per year increase and a 10-mile per year dispersion had followed the colonization three years before.

In 1909 an examination of the pupæ of the gipsy moth in the field revealed the presence of what was actually a small, but under the circumstances a gratifyingly large, number which contained the larvæ or pupæ of the parasite, and the results of the winter scouting work were awaited with confidence and interest. They were quite as satisfactory as could be expected. The collections of nests from areas II and III on the map (Pl. XXII) produced the parasite in abundance, and in area I, throughout which it was found the winter before, it was very much more abundant, as will be seen by reference to the tabulated summary of the results of the work for the winter in Table X.

TABLE X.—*Monodontomerus æreus as distributed over its area of dispersion.*[1]

Section.	Year.	Number of brown-tail nests collected.	Number of Monodontomerus recovered.	Monodontomerus per 1,000 brown-tail nests.	Section.	Year.	Number of brown-tail nests collected.	Number of Monodontomerus recovered.	Monodontomerus per 1,000 brown-tail nests.
1	1908	5,574	39	6	4	1910	1,698	234	137
	1909	2,200	708	376	5	1910	701	215	305
	1910	1,508	495	328	6	1910	2,836	521	183
2	1908	947	0	0	7	1910	1,050	260	246
	1909	1,107	124	112	8	1910	555	86	151
	1910	700	182	260	9	1910	825	538	652
3	1909	770	34	49	10	1910	500	1	2
	1910	260	13	50	11	1910	1,600	95	59

[1] Refer to the map (Pl. XXII) for the area included in each section.

TABLE REPRESENTING THE RAPID MULTIPLICATION OF MONODONTOMERUS IN THE FIELD

	Brown-tail nests collected.	Monodontomerus recovered	Monodontomerus per 1,000 brown-tail nests.
Average of sections 2 and 3 in 1909	1,927	168	87
Average of sections 2 and 3 in 1910	960	190	199

In 1910 a fairly satisfactory number of the parasites was reared from collections of brown-tail moth cocoons made in the field, but when the gipsy-moth pupæ were examined in the field as in 1909, scarcely if any more were found to be parasitized. This was anything but encouraging, because it had been expected that parasitism would amount to at least 1 per cent, if the rate of increase which had prevailed up to 1909 had continued. It appeared that the Monodontomerus was either inclined to pass over the gipsy-moth pupæ in favor of other hosts, or else that its rate of increase had received a sudden check before it was sufficiently abundant to become of aid in the control of the moth. As before, that winter's work was anticipated with interest since its results would be more directly comparable with those of the year before than was that summer's work.

The collections of winter webs were first made in the territory included within the range of the parasite in the winter of 1909–10 (areas I, II, and III), and the fact soon became manifest that instead of increasing it had actually decreased in abundance throughout that territory in the course of the year. It was inexplicable, in view of the unlimited opportunities for increase, and it was, to say the least, discouraging.

Following these preliminary collections, which were intended for no other purpose than to indicate the rate of increase, collections from towns and cities to the westward of its known distribution the previous winter and to the northward in southern New Hampshire and southernmost Maine were made. It was rather confidently expected that it would be found in Maine just over the New Hampshire line, and also that this would mark the limits of its distribution in that direction.

How far removed the expectations were from the reality is well indicated by the accompanying map[1] (Pl. XXIII), and still more by a study of the table. It will be seen that instead of stopping at the Maine State line, Monodontomerus has extended its range for a full hundred miles to the northeastward, and that to the north and west it has pretty nearly reached the limits of the present known distribution of the brown-tail moth itself. But what is more surprising, it is actually much more abundant in a large part of this new territory than it was in Massachusetts a year before.

It will also be observed that the distribution has been much more rapid toward the north and east than toward the west and south, which is true also of that of the gipsy moth and the brown-tail moth. Whether this will prove to be the rule with others of the parasites remains to be seen. It is not indicated in the instance of any other as yet.

Monodontomerus appears to pass through but a single generation annually. The females are sometimes, perhaps habitually, fertilized before they actually issue from the pupal shell of the host. The males invariably die before the winter, or at least out of many thousands of individuals which have been secured in the winter from brown-tail-moth nests at home and abroad, only females have been present. Dissection of a considerable number of hibernating females has failed to result in the finding of even partially developed eggs. Neither has it been found possible to keep females alive in the spring until eggs should develop, although some have remained in a state of activity in confinement for several months.

Beginning in 1906, and each year thereafter until 1909, numerous attempts were made to secure reproduction in confinement. Dipterous larvæ and puparia as well as pupæ of the gipsy moth and the brown-tail moth were supplied as hosts, and females from hibernating nests as well as those from gipsy-moth and brown-tail moth pupæ and other sources were used in these experiments. Failure resulted in every instance, due, apparently, to the impossibility of keeping the parents alive until eggs should be developed.

---

[1] The maps and tables have been prepared by Mr. H E. Smith, to whom the work of caring for the nests as they have been received at the laboratory has largely been intrusted.

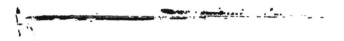

& IN NEW ENGLAND FROM WHICH MONODONTOMERUS ÆREUS HAS BEEN COLLECTED IN HIBERN

The egg which is figured (fig. 52) was dissected from a female which was imported in 1909 with cocoon masses of the brown-tail moth and which was evidently hibernated. She was given no opportunity to oviposit. In 1910 several females were collected in the open in June, and these, upon being supplied with fresh pupæ of the brown-tail moth, immediately oviposited.

The very characteristic larvæ (fig. 53) feed externally upon the pupæ of tachinids within the puparium, but internally within the pupæ of Lepidoptera. The pupæ (figs. 54, 55) are also characteristic, and the appearance of that of the female is indicated by the accompanying illustrations. The exit hole (fig. 56) left in the gipsy-moth

FIG. 52. – *Monodontomerus æreus.* Egg. Greatly enlarged. (Original.)

pupæ is invariably smaller than that left by Chalcis, and larger than that of Diglochis. It may be located anywhere, in which respect it differs from any of the larger of the pupal parasites.

As a secondary parasite, Monodontomerus has been reared from tachinid puparia upon numerous occasions both from those which have been received from abroad and from those collected in America. It was rather expected of it that its attack would be confined to those

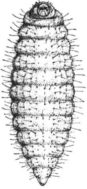

FIG. 53.—*Monodontomerus æreus:* Larva. Greatly enlarged. (Original.)

FIG. 54.- *Monodontomerus æreus:* Pupa, side view. Greatly enlarged. (Original.)

FIG. 55.— *Monodontomerus æreus:* Pupa, ventral view. Greatly enlarged. (Original.)

which were immediately associated with one or another of its chosen hosts, but as usual it did the unexpected, and it has been reared from Compsilura puparia which were collected at the base of trees upon which the caterpillars of the gipsy moth had been common. It has also been reared from tachinids parasitic upon the tussock moth (and from the tussock moth as a primary parasite), from the tent caterpillar, in which it was apparently parasitic upon Pimpla, and from the cocoons of *Apanteles lacteicolor* Vier., the imported brown-tail moth parasite. Like another anomalous species, *Pteromalus egregius,* it betrays a distinct partiality for anything that savors of the brown-tail moth. It is thereby led to seek out the molting webs of the brown-tail caterpillars in the spring and consequently comes into contact with the Apanteles cocoons.

THE SARCOPHAGIDS.

There has been considerable controversy in the past concerning the habits of the Sarcophagidæ, and a wide difference of opinion as to whether they were to be considered as truly parasitic, or whether they were merely scavengers, attacking and feeding upon insects which had died through some other cause. In the case of those species which are reared from grass-hoppers there seems to be no further question that they are to be classed as true parasites or at least that they are as truly parasitic as many of the more degraded among the hymenopterous parasites. This seems not to have been proved of any of the species which are found within the pupæ of the larger Lepidoptera.

Fig. 56 - - Gipsy-moth pupa, showing exit hole left by *Monodontomerus æreus*. Enlarged. (Original.)

If judgment were to be based upon the occurrence of sarcophagids in the shipments of gipsy-moth pupæ from abroad, it would certainly be judged that the sarcophagids were parasitic. Their puparia (Pl. XX, fig. 3), have frequently outnumbered the tachinid puparia, and even the tachinid puparia and hymenopterous parasites together. Unfortunately, there is nothing known of the circumstances under which this material was collected in any instance, and for all that is known to the contrary, the sarcophagids actually entered gipsy-moth pupæ which had been attacked and killed by another parasite, Chalcis for example, and by feeding, first upon the unconsumed contents of the pupal shell, and later upon the body of the true parasite, which might be destroyed either through accident or design on the part of the intruder, would become, in effect, secondary parasites.

If judgment were to be based upon the results of a quite elaborate series of investigations into the relations between the native sarcophagids and the gipsy moth in America, it would unavoidably be to the effect that these sarcophagids were scavengers and nothing more. We are confronted with conflicting evidence, presented by a much greater abundance of sarcophagids associated with the gipsy moth in Europe than is similarly associated with it in America, which is suggestive of two things: Either the sarcophagids are associated with the gipsy moth because they are parasitic upon it or because of the presence of its parasites, which is quite as reasonable an explanation. It will require much careful work in Europe before it will be possible to settle this point at all definitely. Meanwhile it does not seem to be advisable to attempt the introduction of the European sarcophagids until we know whether they are an aid in the control of the moth or a possible hindrance to the work of the parasites.

The special investigations which were conducted for the purpose of determining the exact status of the sarcophagids in America in relation to the gipsy moth were conducted by Mr. T. L. Patterson, and have been made the subject of a special report.[1]

Another series of investigations, conducted by Mr. P. H. Timberlake, upon the parasites of the pine "tussock moth" in northern Wisconsin, resulted in the accumulation of evidence which pointed quite convincingly to the parasitic character of certain sarcophagids which he encountered in abundance associated with this insect. Unfortunately it is not wholly convincing. If it could be accepted at its full face value it would mean that in these flies we have a group of dipterous parasites wholly distinct from the tachinids, and working in a wholly different manner. The tachinids are caterpillar parasites, and never, so far as has been recorded, attack the caterpillar after it has spun for pupation. The sarcophagids, like Pimpla, Theronia, etc., are pupal parasites and will be grouped together, and at the same time apart from the hymenopterous pupal parasites, even as the tachinids as a group stand beside but apart from the hymenopterous parasites of the caterpillars.

THE PREDACEOUS BEETLES.

It is very probable that further studies into the subject of natural predatory enemies of the gipsy moth will result in the addition of a considerable number of names to the list of predaceous beetles which attack it in one stage or another of its existence and with more or less freedom. The egg masses received from abroad have very frequently been infested with small dermestids, and in the forests in the vicinity of Kief, Russia, in September, 1910, large numbers of the larvæ of a species not yet determined were found feeding, to all appearances, upon the eggs of the moth as well as upon the covering of felted hair.

That these larvæ do actually eat the eggs was demonstrated by Mr. Burgess during his association with the moth work as conducted by the State board of agriculture in 1899 and later his observations were confirmed by a series of simple experiments conducted at the laboratory.

In the spring of 1908 a large number of cocoons of the tussock moth with egg masses attached was collected in East Cambridge, Mass., and from them in June a number of dermestid beetles issued, deter-

---

[1] U. S. Department of Agriculture, Bureau of Entomology, Technical Series 19, Part III, March 22, 1911.

mined by Mr. E. A. Schwarz as *Anthrenus varius* Fab. and *Trogoderma tarsale* Melsh. The Trogoderma was the more common of the two. Later, in the fall, another collection of old cocoons was made for the purpose of determining the status of these beetles. It was found that both of them fed, as larvæ, upon the eggs of the tussock moth, and when they were confined in vials with eggs of the gipsy moth they fed not only upon the hairy covering of the egg masses, but also upon the eggs themselves. Larvæ apparently of one of these species have several times been received at the laboratory associated with egg masses of the gipsy moth, which were in each instance collected upon the sides of buildings or in other situations different from those under which egg masses are most frequently encountered.

As soon as the gipsy-moth caterpillars hatch, if, as frequently happens, the egg mass is situated in some particularly well-sheltered spot, the young caterpillars are liable to attack by small carabid beetles, several species of which have been found under burlap bands in the spring apparently feeding upon the gipsy-moth caterpillars in this stage. Several of these species were made the subject of casual study in the summer of 1910, the results of which will be published later.

The elaterid genus Corymbites, though not generally recognized as predaceous, is undoubtedly more or less addicted to a diet of living insects. An adult of one species was once found feeding upon the cocoons of *Apanteles fulvipes;* and the larva of another, upon one occasion, at least, upon the pupæ of the gipsy moth. There are many species in the New England States. Some of them are nocturnal, and it is not at all beyond the limits of probability that they may be found listed among the predatory enemies of the gipsy moth and the brown-tail moth when these lists shall have been finally completed.

Among the coccinellids the large *Anatis 15-punctata* Oliv. has more than once been observed, as a larva, attacking the small caterpillars of the gipsy moth, and it is not at all unlikely that the species is actually of as much consequence as some of the minor parasites in assisting in the control of the pest.

The lampyrids, too, include amongst their numbers many species which are either occasionally or habitually predatory. One such which abounds in eastern Massachusetts in the spring flying about in the tops of the trees and crawling over the foliage was encountered in the spring of 1910 in the act of destroying a small gipsy-moth caterpillar. Probably one beetle would not destroy many caterpillars in the course of its life, but there are such swarms of the beetles as to make an average of even one caterpillar count materially in the end. Some of the lampyrids are nocturnal, as in fact are a great many of the proved or probably predatory Coleoptera, and their association with the gipsy moth is not likely to be established unless special effort toward that end is undertaken. Such studies require time and patience, but are none the less necessary if we are ever to know all that is to be known about the subject.

None of the beetles mentioned is likely to attack the later-stage caterpillars, but among the larger Carabidæ is to be found a variety of species which are not only able, but more than willing to destroy the full-fed caterpillars and pupæ whenever opportunity offers. There are many such in Europe which do not occur in America, and altogether a considerable number of different species has been received from abroad and tested as to ability to assist in the control of the gipsy moth in this country.

Three characteristics in addition to ability and willingness to attack the gipsy moth are necessary if the introduction of a beetle is to be seriously undertaken as an economic experiment. It must breed at the proper season of the year, so that its larvæ may receive the advantage of the practically unlimited food supply which the present superabundance of the gipsy moth gives; it must be able to withstand the rigors of the New England climate, and not only the adult beetles but their young must be arboreal in habit. An abundance of species both native and foreign will feed freely upon the gipsy moth in confinement, but of these only a few will seek out the caterpillars or pupæ in the situations in which they are to be found in America. The adults of a portion of this number do habitually climb into the trees in search of their prey, but not all such are similarly arboreal during their larval stages. Of those which are arboreal, or which appear to be arboreal, during all of their active life, a part appear to breed at the wrong season of the year and another part do not extend their range into a sufficiently high latitude to make them effective as enemies of the gipsy moth. There is not a single species native to America which meets all of the delicate requirements of the situation, but such a species has been found abroad in *Calosoma sycophanta* L. (See Pl. I, frontispiece, adult eggs, larvæ, and pupa.) This, of all of the numerous species of predaceous beetles which have been investigated at the laboratory, bids fair to be of real assistance in the fight which is being waged.

Like all the larger carabids inhabiting the temperate regions, this species is terrestrial during a considerable portion of its life cycle, but both adults and young, which are equally voracious, climb freely into the trees in search of their prey. The eggs are deposited in the earth, and the young larvæ upon emerging are possessed of a remarkable vitality and sufficient strength and cunning to enable them to seek out and successfully to attack, when found, the largest and most active of the gipsy-moth caterpillars. They also attack the pupæ with even greater freedom, and once ensconced within such a mass of pupæ as is frequently encountered in partially protected situations upon a badly infested tree, will rapidly complete their growth without leaving the spot. The full-fed larvæ seek the earth and, burrowing well below the surface, construct a vaulted pupal cell, within which the final transformation takes place during the late summer or fall. The adult beetles remain quiescent and as a rule do not issue until late in the succeeding spring.

The breeding season coincides almost exactly with the caterpillar season. The hibernated beetles begin egg deposition just a little before the caterpillars are large enough to be easily found and attacked by their young; the height of their activities in this direction is at a time when their young are best provided for, and they cease oviposition very shortly after the gipsy moths themselves begin to deposit eggs for a new brood. Very shortly thereafter, with summer still at its height, the adult beetles, both male and female, burrow deep into the soil and become dormant, awaiting the arrival of another spring.

The data as given above concerning the life and habits of *Calosoma sycophanta* have been accumulated by Mr. A. F. Burgess, who has had full charge of that part of the laboratory work which had to do with the predatory beetles since the late summer of 1907. Up to that time the pressure of other work was so great as to render impossible any systematic studies along that or similar lines. The first of the adult beetles of this species, together with a smaller quantity of another, *Calosoma inquisitor* L., were imported and in part liberated in the spring of 1906. A few were confined within the large out-of-door cages of the type already figured and briefly described (see Pl. XIV), and reproduction was secured in the instance of *Calosoma sycophanta*. Neither Mr. Titus nor Mr. Mosher was able to give this phase of the work the attention which it really deserved, and while their observations were sufficient to cover most of the salient points in the life history of the predator, there was still an abundance of opportunity for further studies.

In 1907 early and not very systematic surveys of the several field colonies established by Mr. Titus the year before failed to result in the recovery of the beetle. Accordingly, when similar experience with others among the introduced insects had indicated that larger colonies were likely to be required, it was determined to liberate all the adult Calosomas in one locality as they were received from abroad, and thus secure its establishment, if this were possible, before attempting further artificial dispersion. This was done, and several hundred had been received and thus liberated by the time Mr. Burgess was ready to take full charge of the work.

Although it was quite late in the season, Mr. Burgess, with the assistance of Mr. C. W. Collins, who has remained associated with him ever since, succeeded in securing the eggs and in carrying to maturity several larvæ of the species in close confinement in jars of earth, and effectually demonstrated the superiority of this method over that involving the use of the large out-of-door cages. The following spring the work was undertaken upon a considerably larger scale, and along still more specially developed lines. From the hibernated parent stock, and from newly imported beetles, he reared large numbers of larvæ, a part of which were allowed to complete their transformations in confinement, while others were colonized directly in the open when about half grown. These larval colonies promised to be successful, and accordingly the work of rearing the larvæ and distributing them throughout the gipsy-moth-infested area in eastern Massachusetts, with an occasional incursion into other parts of the infested area, was continued throughout 1909 and 1910.

Meanwhile, beginning in the late summer of 1907 and continuing uninterruptedly until the close of the season of 1910, Calosoma has been steadily gaining in the confidence of those who have watched its progress. Its larvæ were first recovered from the field at just about the time when Mr. Burgess first took over the beetle work, and its ability to complete its seasonal cycle in America unassisted was thus indicated. The large colony was also proved to be unnecessary.

Its progress in the field was slow at first, even in the instance of the large adult colony founded in 1907 before it was known to have become established. In 1908 its larvæ were found in abundance in the center of this colony, but not to any great distance away from the point where the beetles had first been liberated. In 1909 the spread was more rapid, but at the same time restricted in comparison with that which became evident in 1910. As will be seen by reference to the accompanying map (Pl. XXIV) which has been prepared by Mr. Burgess from the results of the scouting work of three years, its apparent or discernible dispersion has been at a rapidly increasing rate each year in the instance of colonies which, like these, chanced to be so happily located as to allow for unrestricted and uninterrupted increase from the start.

At the present time there is every prospect that a continued rapid increase for a few years more will result in an abundance of the beetles sufficient to render very efficient aid in the fight against the moth. It is not expected that they will be of very much assistance in localities in which the moth is reduced to such numbers as to make control through parasites such as Compsilura and others of its character possible, but it is expected that whenever the moth breaks out of bounds, and increases to such abundance as to afford the beetles and their larvæ an unlimited food supply, first migration and later rapid multiplication of the beetle will result. In this respect the rôle played by Calosoma is similar to that which is rather confidently expected of Blepharipa.

THE EGG PARASITES OF THE BROWN-TAIL MOTH.

THE GENUS TRICHOGRAMMA.

The parasites belonging to the genus Trichogramma, of which several have been reared from eggs of the brown-tail moth, are the most minute of any which have been handled at the laboratory, and are among the smallest of insects. The egg of the brown-tail moth is in form of a flattened spheroid, approximately as large in its greatest diameter as the printed period which ends this sentence. Normally two or three individuals of the parasite pass through all of their transformations from egg to adult upon the substance of a single host egg, and in exceptional instances as many as 10 perfect adults are known to have issued from one egg. This is the more remarkable when it is remembered that the female Trichogramma is sexually mature at the time of issuance, and ready to deposit a large number of eggs for a new generation.

FIG. 57 —*Trichogramma* sp. in act of oviposition in an egg of the brown-tail moth. Greatly enlarged. (Original)

The mother parasite exhibits little discretion in the selection of host eggs for attack (fig. 57), and if any dependence is to be placed upon observations which have been made in the laboratory, she is quite as likely to select eggs which contain caterpillars nearly ready to hatch as those which are freshly deposited. The feeding habits of her young are such as to permit a considerable latitude in this respect, but there is a certain limit, and after the embryological development has passed beyond a certain point in the host egg, the attack by the parasite is unsuccessful. It is much better that the host egg be dead than that it contain a living embryo in the later stages of its development.

The life cycle, from egg to adult, varies very considerably in length in accordance with the prevailing temperature. In the summer it may be completed in as short a period as nine days, while in the fall three weeks or more may be required. If the temperature falls below certain limits the young parasites will hibernate or attempt to hibernate, and thereafter their development may be delayed for several weeks, or even months, even though they are exposed to continuous high temperature during this period.

After about one-third of the time requisite for the completion of the life cycle has elapsed, the eggs begin to turn dark, and finally become shining, lustrous black (fig. 58). This change is brought about by the preparation of the larvæ for pupation.

Three races or species have been reared from the eggs of the brown-tail moth, two of them being European and the third American. The American, according to Mr. A. A. Girault, to whom the series of mounted individuals was submitted for determination, is the common and widely distributed *Trichogramma pretiosa* Riley. One of the European, which is here referred to as the *pretiosa*-like form, is or appears to be structurally identical with the American *pretiosa*. It differs in that the progeny of parthenogenetic or unfertilized females is either of both sexes, or else exclusively female, while the progeny of unfertilized females of the American species has always been exclusively male in the very considerable number of reproduction experiments with such females which have been carried on at the laboratory.

The other European species may at once be distinguished from either of its congeners by its dark color, as well as by other characters of taxonomic value. Like the American race of *T. pretiosa* which was studied at the laboratory, it produced males exclusively as the result of parthenogenetic reproduction.

FIG 58 —Eggs of the brown-tail moth, a portion of which has been parasitized by *Trichogramma* sp. (Original)

It seems to the writer that in the two morphologically identical but biologically distinct races of Trichogramma (*T. pretiosa*, American or European) we have what is nothing less than two species, quite as distinct as are the species of bacteria, for example, which are founded upon cultural characters. If the manner in which a bacterium reacts when cultivated upon a certain medium prepared after a fixed formula may be considered as sufficient to separate it specifically from an otherwise indistinguishable form which reacts in a different manner under identical circumstances, why may not the same distinctions be made to apply to insects? It may not appeal to the taxonomist and student in comparative insect morphology, but it certainly will appeal to the economic entomologist, who has, or ought to have, a greater interest in the biological than in the anatomical characteristics of the subjects of his investigations. The case of Trichogramma is by no means unique. That of *Tachina mella*, which is practically indistinguishable from *T. larvarum* but which reacts differently in its association with the gipsy moth, is another. Another is to be found in the American and European races of *Parexorista chelonix*. There are also others, which need not be mentioned here, but which will receive attention, it is hoped and intended, at some future time.

These statements concerning the behavior of the several forms of species of the genus Trichogramma are based upon the results of

something like 275 separate but similar experiments in their reproduction in confinement in the laboratory. It is of course possible,
since it was especially desired to continue the experiments as long
as possible with individuals of known parentage, that the results are
misleading. Possibly had American Trichogramma been collected
in the open from a variety of sources, a race might have been found
which was arrhenotokous, even as the similar search might have
resulted in the discovery of a thelyotokous race in Europe. As it
is, the American stock was once renewed. In 1907 a series of experiments was conducted with parent stock reared from brown-tail moth
eggs collected in Maine, and in 1908 a similar but more extensive
series with parent stock from eggs of the brown-tail moth collected
in Massachusetts. In each instance the results were the same.

The longest series of experiments with the arrhenotokous European race was with the progeny of individuals reared from one lot
of European eggs from the Province of Carniola, Austria. Similar
experiments with one other lot of females upon another shipment
of eggs from the same Austrian province and perhaps from the same
locality resulted similarly, but the series was not nearly so long. In
the first-mentioned series 13 generations were reared in the laboratory,
all but the first three being parthenogenetic. Males were secured
at one time, and for a limited number of generations, but soon disappeared, even from the progeny of mated females. The results
of these experiments will be published in detail later.

Importations of egg masses of the brown-tail moth which had been
collected in the open in Europe were first attempted in the summer of
1906, and from almost the first of those which were received at the
laboratory a few examples of the *pretiosa*-like European form were
reared. Mr. Titus attempted to secure reproduction in the laboratory that first season, but as he had no supply of host eggs in which
embryonic development was not considerably advanced, his attempts
met with failure.

In 1908 a larger number of egg masses of the brown-tail moth was
imported from a great variety of European localities, and as before,
the *pretiosa*-like Trichogramma was quickly secured. The failure
of the previous season and its cause had early been taken into account, and some time before a large quantity of fresh eggs of the
brown-tail moth had been collected and stored at a temperature
sufficiently low to prevent embryological development. When supplied with a quantity of these eggs the imported Trichogramma oviposited with the greatest freedom, and in the course of a few generations had increased enormously, so that many thousands were liberated later in the fall. It was conclusively demonstrated that
even though the host eggs were dead, abundant reproduction could
be easily obtained under laboratory conditions.

A large number of parasitized eggs, containing the brood in various stages of development, were placed in cold storage and kept until the following June and July, when, upon being removed, a few of the parasites completed their transformations. With these as parents large numbers were reared in the laboratory upon the fresh eggs of the brown-tail moth, at that time abundantly available, and the cold-storage experiment was repeated during the winter of 1908-9 with much better results than before. An abundant supply of parent females was available in the summer of 1909, and a great many thousands of the parasite were reared and liberated under the most favorable conditions which could possibly be desired or devised. Many thousands were known to have issued from parasitized eggs contained in small receptacles attached to the branches of the trees upon which the brown-tail moths were even then depositing eggs in abundance.

No false hopes were felt as to the probable success of this venture. It has been amply demonstrated in the laboratory that the females were unable to penetrate the egg mass for the purpose of oviposition, and the location in the mass of the few eggs parasitized by the American race of *pretiosa* indicated sufficiently well the inability of that species to do better in the open than either it or the European would do in confinement.

Accordingly no disappointment was felt, when it was found that the degree of parasitism effected by the European species in the immediate vicinity of the colony sites was hardly, if any, greater than that ordinarily effected by the native species. It is hardly a physical possibility for Trichogramma to effect more than a small percentage of parasitism in the egg mass of the brown-tail moth, and the value of the genus as represented by the three species or forms which have been studied at the laboratory is slight.

At the same time, it is not felt that the labor which has been expended in an attempt to give Trichogramma a fair test has been altogether lost. There are numerous other hosts upon which it is a very efficient parasite, and it is easily conceivable that at some future time it will be found possible to utilize it in some manner which the circumstances themselves will suggest.

As a possible example may be mentioned the tortricid *Archips rosaceana* Harris, which at times becomes a pest in greenhouses devoted to the growing of roses. In Volume II, No. 6, of the Journal of Economic Entomology, Prof. E. D. Sanderson describes such an outbreak in a large rose house in New Hampshire under the heading of "Parasites." Prof. Sanderson says:

The outbreak observed by us furnished a case of the most complete parasitism we have ever seen. When first observed, in late July, from one-third to one-half of the eggs were parasitized by a species of Trichogramma. Two weeks later it was difficult to find an egg mass in which over 95 per cent of the eggs did not contain the black pupæ of the parasite and in most cases 99 to 100 per cent were affected. So effective were the parasites that the control of the outbreak was undoubtedly due to them much more than to any remedial measures.

At about the time when the American parasite was reaching a state of efficiency, a large number of eggs of the brown-tail moth containing the brood of one of the European species was sent to Prof. Sanderson for liberation in the rose house. They were received too late for service, but had they been sent at an earlier date it might easily have been claimed, and with perfect confidence, that the final results were the direct outcome of the colonization experiment.

In such circumstances as these it would (or at least it seems from this distance as though it would) easily be possible and practicable to collect masses of the parasitized eggs and by keeping them in cold storage have ready at hand within the following twelvemonth a supply of the parasites which would be available should the natural stock perish through lack of food, and the destructive increase of the host follow. Parasitized eggs could be sent from one greenhouse to another, and stock could be kept in cold storage in one city to be drawn upon by a florist in any other part of the country when need arose.

Another possible use for the parasite is as an enemy of Heliothis, which is causing serious injury to tobacco in Sumatra. Dr. L. P. De Bussy, biologist of the tobacco growers' experiment station at Deli, has already undertaken its introduction there, and will attempt to handle it after somewhat the same manner as that above described.

TELENOMUS PHALÆNARUM NEES.

A small number of this species was reared from imported eggs of the brown-tail moth from several European localities in 1906, and an attempt was made to secure reproduction in the laboratory. Oviposition was secured, as in the instance of similar attempts with Trichogramma, but it did not result successfully, and apparently for the same reason.

In 1907 a somewhat larger number was reared, and an abundant supply of suitable host eggs having been provided, this number was soon increased several fold, and one large, and several smaller colonies of the parasite were liberated under very satisfactory conditions late in the summer. It was found that the reproduction could be secured upon host eggs which had been killed through exposure to cold, and the experiment was made of hibernating the brood in cold storage, but without success.

In 1908 the quantity of eggs of the brown-tail moth imported was smaller than during the previous year, and only a very small proportion of them proved to be attacked by the Telenomus. Not nearly enough for a satisfactory colony were reared, and again it was attempted to hold the brood over winter in cold storage, and again the attempts failed.

If judgment is based upon the percentage of parasitism by this species in the lots of egg masses of the brown-tail moth which have been received from abroad, it is an unimportant parasite in Europe. Partly on this account, and more, perhaps, because it was colonized so satisfactorily in 1907, no further attempts to secure its introduction into America have been made. Neither has a serious attempt to recover it from the field in the vicinity of the 1907 colony site been made, and it may have become established from this colony.

The plans for field work in 1911 include the collection of a large number of eggs of the brown-tail moth from the general vicinity of the larger colonies of 1907 and, if arrangements can be perfected, for a study of the extent to which the eggs of the brown-tail moth are attacked by parasites in Europe. As in the case of every other class of parasite material received at the laboratory, nothing is known of the circumstances under which those egg masses which were received from 1906 to 1908 were collected. It may easily be that they were collected too soon following their deposition to permit of their having been parasitized to anything like the extent which would have come about had they been allowed to remain in the open for a few days longer, and in at least one instance the receipt of the masses with a dead female moth accompanying each was sufficient to more than justify such doubts.

PARASITES WHICH HIBERNATE WITHIN THE WEBS OF THE BROWN-TAIL MOTH.

Partly because it has been practicable to import the gipsy moth and the brown-tail moth in the hibernating state in better condition than it has been possible to import their active summer stages, but equally because there has been ample time and opportunity to study them during the winter months when only a limited amount of field work could be done, it has been possible to learn more of the parasites which hibernate within the gipsy-moth eggs and the nests of the brown-tail moth than of those parasites which are only associated with the same hosts during a more or less limited time in the summer. The winter nests of the brown-tail moth have from the beginning been the subject of an increasingly intensive study, and as a result more is known of the parasites which hibernate within them than of any other group of brown-tail moth or the gipsy-moth parasites without excepting even the parasites of the gipsy-moth eggs.

Very large numbers of these nests, amounting in the aggregate to more than 300,000, have been imported each winter from that of 1905–6 to that of 1909–10, inclusive, Now that all of the primary

FIG. 59.—Larvæ of *Pteromalus egregius* feeding on hibernati caterpillars of the brown-tail moth. Much enlarged. (Orig nal.)

and the parasite larva soon issues and s
the molting web, which may or may not
There is no second generation upon the
tail moth the same season.

Meteorus versicolor Wesm. Habits es.
of Apanteles until after the caterpillars h
spring. The parasitized individuals usu
are overcome and destroyed away from
The cocoons, which are characteristic of
end of long threads. The adults issuin
attack the larger caterpillars of the br
generation.

Zygobothria nidicola Towns. Hibernat
of Apanteles and Meteorus. The affect
grown and spin for pupation before being
Sometimes they pupate. The parasite ad
when the moth would have issued had tl
transformations. There is but one genera
nate host is necessary.

Compsilura concinnata Meig. Hibernating larvæ are occasionally found, but apparently do not complete their transformations in the spring.

Mesochorus pallipes Brischke. Occasionally reared as a parasite of *Apanteles lacteicolor*. The Apanteles larva reaches full maturity and spins its cocoon, but is overcome before pupating. The Mesochorus adult issues from the cocoon a very few days later than would the Apanteles had it remained alive.

Entedon albitarsis Ashm. An internal parasite within the larvæ of *Pteromalus egregius*.

The appearance of the hibernating larvæ of the Pteromalus is indi-

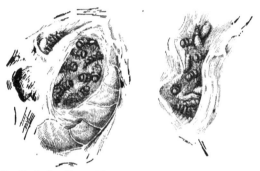

FIG. 60.—Portion of brown-tail moth nests, torn open, showing caterpillars attacked by larvæ of *Pteromalus egregius*. Enlarged. (Original.)

cated fairly well in the accompanying illustration (fig. 60), which represents a "pocket" of parasitized caterpillars torn open. Very little of interest is associated with the life and feeding habits of these larvæ. The female pierces the host caterpillar with her ovipositor preliminary to the deposition of her egg externally, and the caterpillar thus stung is frequently rendered quiescent, and may even die before the hatching of the parasite larvæ.

The hibernating larva of Apanteles is so small as to be very difficult of detection until after it has resumed activity in the spring and increased in size. Its

FIG. 61.—*Apanteles lacteicolor:* Immature larva from hibernating caterpillar of the brown-tail moth. Much enlarged. (Original.)

exact appearance during the hibernating stage can not be described, because nearly every specimen found has been injured more or less in the removal. The accompanying illustration (fig. 61) is from a sketch made by Mr. Timberlake of a half-grown larva from life. None of the preserved specimens shows the curious projection beneath the anal bladderlike appendage which latter is characteristic of the early stage larvæ of the subfamily to which Apanteles belongs. The head and mouthparts are strikingly dis-

similar from those of the first-stage Meteorus, and are so little differentiated as to be indescribable.

The hibernating stage of Meteorus is in remarkable contrast to
that of Apanteles. The accompanying drawing (fig. 62) is from a
balsam mount, and represents an individual which has resumed activity and grown very slightly larger and plumper than is character-

FIG. 62.—*Meteorus versicolor:* Immature larva from hibernating caterpillar of the
brown-tail moth. Much enlarged   (Original)

istic of its hibernating condition. These larvæ are curiously anomalous, in that though they are actually first-stage, the head alone is
considerably larger than the original egg as deposited by the mother.

An interesting series of dissections made by Mr.
Timberlake in the spring of 1910 served to explain this apparent anomaly. The eggs are very
small when first deposited and almost globular.
Apparently with the beginning of embryological
development they begin to grow and by the time
the inclosed embryo begins to assume the characteristics of the larva they have reached a diameter at least four times greater than that of the
newly deposited egg. The enormous chitinized
head, with strong, curved mandibles, is in strange
contrast to the undifferentiated cephalic segment
of Apanteles and is apparently closely analogous
to the large-headed, heavily mandibled larvæ of
the Platygasters, as described by Ganin, Marchal,
and others. There are many points of resemblance
between the two forms, and it would seem, without
going into the matter at all deeply, as though the
type of embryological and early larval development
characteristic of Meteorus were essentially the same
as that of the Platygasters and many ichneumonid

FIG 63 —*Zygobothria ni-*
dicola First-stage larvæ
in situ in walls of crop of
hibernating brown-tail
moth caterpillar. Great-
ly enlarged. (Original)

genera. while that of Apanteles would have to be
considered as of an essentially different type.

In both Apanteles and Meteorus the later larval stages are much
more conventionalized and more like the familiar type.

The position assumed by the Apanteles larva is not very definitely
known. The Meteorus larva usually lies superior to the alimentary

canal, its axis parallel to the axis of the body of the host caterpillar, and its head in the ultimate or penultimate body segment, and pointed toward the rear.

The larvæ of Zygobothria are similarly assigned to a definite position, and in otherwise healthy caterpillars have invariably been found embedded in the walls of the crop, as indicated by figure 63. In appearance they are typical of the tachinid first-stage larvæ generally, and with no extraordinary points of difference from most others of the group to which they belong. Those of Compsilura (fig. 64) may be found in similar positions, but they are easily distinguishable from Zygobothria by the presence of the three chitinous anal hooks or spines, as indicated in the accompanying figure.

Nothing is known of the hibernating stage of Mesochorus. It does not seem probable that it should resemble the planidium of Perilampus, which, like Mesochorus, is a secondary parasite which gains access to its host before the latter has left the body of the caterpillar which harbors both primary and secondary. It is presumed that it will be representative of a highly specialized type of development which fits it for the peculiar rôle which it plays, but that this development will have been along wholly different lines from that which has taken place in the case of Perilampus. The whole genus, apparently, possesses habits similar to those of *Mesochorus pallipes*. A very beautiful and, according to Mr. Viereck, an undescribed species has been reared from the cocoons of *Apanteles fiskei*, parasitic upon a species of Parorgyia, under circumstances which indicate positively that attack was made while the primary host was still alive. The same may be said of another undetermined species which has similarly been reared from *Apanteles hyphantriæ*.

FIG. 64.—*Compsilura concinnata:* First-stage larva. Greatly enlarged. (Original)

Mesochorus pallipes is not an uncommon parasite of *Apanteles lacteicolor* Vier., having been reared from only a few among the many localities from which its host has been secured in numbers, but the average proportion of parasitized individuals has been only about 2 per cent.

The interrelations of these several parasites thus closely associated with one stage of the same host, and consequently with each other, are interesting and peculiar. Pteromalus, of course, cares little whether the host caterpillar selected for attack is parasitized by one or more of the endoparasites which hibernate as first-stage larvæ. The female will undoubtedly attack parasitized as freely as it will attack unparasitized caterpillars and its larvæ develop as satisfactorily upon the one as upon the other, and at the expense of the other internal parasites as well as of the primary host. But the matter does not stop here. The adults issue from the hibernating nests at just about the time when the Apanteles are issuing from the young caterpillars and spinning their cocoons in the molting webs, which are very frequently in the outer interstices of the very same nests from which the Pteromalus are also issuing. The females of the latter are ready to oviposit almost immediately following their eclosion, and will oviposit with the greatest freedom in the cocoons of Apanteles or of Meteorus whenever they chance to encounter them. Thus it comes about that the Pteromalus, after passing one generation as a primary parasite of the brown-tail moth, immediately passes another as a secondary upon the same host. Undoubtedly it would thrive equally as well upon Mesochorus as upon Apanteles. Proof of this could unquestionably be secured through the careful dissection of the very large number of cocoons from which it had issued in the laboratory, some of which, it is certain, must have contained Mesochorus as well, but proof is really unnecessary. By doing so, it becomes tertiary upon the same host as that upon which it is habitually and regularly a primary and secondary parasite.

Entedon, were it to follow Pteromalus through its varied adventures, would in like manner (as it probably does) become successively secondary, tertiary, and quaternary.

Monodontomerus, commonly a primary parasite upon the pupa of the brown-tail moth or gipsy moth and only present as a regular guest in the winter nests, is none the less pretty intimately connected with them in other ways. It directly attacks the cocoons of the Apanteles, acting in all respects like a secondary parasite, and thereby comes into direct conflict with Pteromalus, one of the other of which must develop at the expense of its competitor. It also will become tertiary whenever it chances to attack a cocoon containing Mesochorus as a secondary parasite on Apanteles. It is also a parasite of tachinid puparia, and especially of tachinid puparia which it encounters associated with the gipsy moth, or the brown-tail moth, and thereby becomes a parasite of Zygobothria and in consequence a secondary parasite of the brown-tail moth.

Should Apanteles and Meteorus, or Apanteles and Zygobothria chance to become located in the same host, the Apanteles, because of its more rapid development in the spring, would certainly be the winner.

When Meteorus and Zygobothria enter into competition for possession of the same host individual, Meteorus is invariably the winner and is in no way affected by the presence of the other parasite. In fact, Zygobothria is twice apt to be the victim of Meteorus, which goes through two generations before the tachinid has entered its second stage.

Table XI has been prepared for the purpose of showing these inter-relations graphically. It is to be understood, of course, that not in every instance have the exact relations thus set forth been actually observed; but it is perfectly safe to say that they are not only within the bounds of probability, but that they actually occur in nature. The only point concerning which doubt is felt is in the hyperparasitism of Entedon upon Pteromalus, when Pteromalus itself is hyperparasitic upon Apanteles or Meteorus.

TABLE XI.—*Possible interrelations between parasites hibernating in brown-tail cater-pillars.*

Primary parasites.	Secondary super- or hyper-parasites.	Tertiary super- or hyper-parasites.	Quaternary super- or hyper-parasites	Quinquinary super- or hyper-parasites.
Pteromalus egregius Apanteles lacteicolor.	Entedon albitarsis [1] Pteromalus egregius [1] [2]	Entedon albitarsis [1]		
	Mesochorus pallipes.[2]	Pteromalus egregius.[2]	Entedon albitarsis.[1]	
	Monodontomerus æreus.[2]			
Meteorus versicolor.	Apanteles lacteicolor.[2]	Pteromalus egregius.[2]	Entedon albitarsis.[1]	
		Mesochorus pallipes.[1]	Pteromalus egregius.[2]	Entedon albitarsis.[1]
		Monodontomerus æreus.[1]		
	Pteromalus egregius.[1] [2]	Entedon albitarsis.[1]		
Zygobothria nidicola.	Apanteles lacteicolor.[2]	Pteromalus egregius.[1] [2]	Entedon albitarsis.[1]	
		Mesochorus pallipes.[1]	Pteromalus egregius.[2]	Do.[1]
		Monodontomerus æreus [1]		
	Meteorus versicolor [2]	Pteromalus egregius.[1] [2]	Entedon albitarsis.[1]	
	Monodontomerus æreus.[1]			
	Pteromalus egregius [2]	Entedon albitarsis.[1]		
Monodontomerus æreus				

[1] Hyperparasitic relations.   [2] Superparasitic relations.

PEDICULOIDES VENTRICOSUS NEWP.

During the winter of 1908–9 trouble was experienced in the work of breeding Pteromalus, the exact nature of which was not immediately apparent. There were numbers of the reproduction experiments in which the proportionate number of progeny to parents used was much below that which had hitherto been secured as the result of similar work in the previous spring. An examination of the nests of the brown-tail moth which had been used in these experiments disclosed the presence of vast numbers of the adults and young of a mite, determined by Mr. Nathan Banks as *Pediculoides ventricosus* Newp. The gravid females were attached to the caterpillars of the

brown-tail moth, or to the larvæ or pupæ of the parasite, indiscriminately, and in some of the reproduction cages practically every host
and parasite had been attacked.

It was not known where these mites came from, but it was presumed that they were brought in from the field upon nests of the
brown-tail moth. By the time that they had been discovered they
were in practically everything in the laboratory. Even tachinid
puparia were not immune to attack, and there were numerous
instances in which the wandering young had forced their way through
tight cotton plugs, which would ordinarily have prevented the passage of bacteria.

Much time and trouble was necessary before the laboratory was
finally cleared of the pest; but it was finally accomplished by the
rigid separation of every rearing cage containing life which had
been present before the invasion became apparent from those which
were begun afterwards. The general cleaning up and policy of seg-
regation proved effective, and by spring the last of the mites appeared
to have died; nor has a single specimen been observed since.

As parasites of the brown-tail moth the mites were singularly effective. If it were possible to bring about a general infestation of the
nests in the early fall, it would doubtless result in the destruction of
a very large proportion of the hibernating caterpillars; but unfortunately this seems to be not at all practicable. It is not even certain
that the parasite was actually brought into the laboratory in nests
of the brown-tail moth, though this would seem to be the most likely
explanation of its presence.

The fact that its presence has never once been detected in any of
the many thousands of similar nests which have been brought in at
other times indicates rather conclusively that it is not actually an
enemy of any consequence in the field.

PTEROMALUS EGREGIUS FÖRST.

It was quite late in the spring of 1905 before the senior author was
able to organize a corps of European collectors, and as a consequence
only a very small quantity of parasite material was imported during
the summer of that year; but during the fall and winter following,
well within a year after the work was first authorized by the Massachusetts Legislature, importation was begun in earnest. More than
100,000 hibernating nests of the brown-tail moth were received from
abroad that winter, and since scarcely anything was surely known
of the parasites which were likely to be reared from them, the early
discovery of the hibernating brood of *Pteromalus egregius* (fig. 60,
p. 263) was hailed with satisfaction. The circumstance has already
been the subject of comment in an earlier section.

In the spring of 1906 some 40 large tube cages (Pl. X, fig. 1), each capable of accommodating several thousand nests, were constructed after the model of a cage which had been successfully used for a somewhat similar purpose in California.   Hardly had the nests been placed in these newly constructed cages before the caterpillars began issuing in extraordinary numbers, and with them many thousands of adult parasites, representing a great variety of species.   *Monodontomerus æreus* was about the first to issue, and with it was a quantity of *Habrobracon brevicornis*.   A little later *Pteromalus egregius* (fig. 65) appeared in an abundance which exceeded that of all the other parasites taken together, and it was followed shortly afterwards by swarms of its own little parasite. determined by Dr. Ashmead as *Entedon albitarsis*.

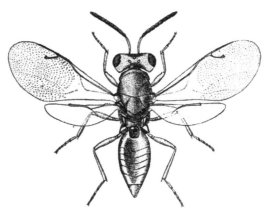

FIG. 65.—*Pteromalus egregius:* Adult female.   Greatly enlarged.   (Original.)

Mr. Titus at once recognized Entedon as hyperparasitic and proceeded as assiduously to destroy it as he was assiduous in saving the Pteromalus.   Of the myriad of other parasites issuing, the vast majority were represented by so few individuals as to render it very improbable that any among them were enemies of the caterpillars of the brown-tail moth.   Nearly all of the more common species, aside from Pteromalus and Entedon, were representative of genera or groups of genera well known to be parasitic upon Cynipidæ, of which large numbers issued from the galls on oak leaves that had been used by the caterpillars in the construction of their nests.   There remained as possible parasites of the caterpillars of the brown-tail moth only *Habrobracon brevicornis*, *Pteromalus egregius*, and *Monodontomerus æreus*.

It looked for a time as though the Habrobracon might be parasitic upon the hibernating caterpillars, and quite a large number of them was liberated in the spring of 1906, but it was later discovered that, like Monodontomerus, they merely sought the hibernating nests for the protection which was thus afforded during the winter. One colony of Monodontomerus was also established early in the spring of 1906, but almost immediately thereafter the action was regretted on account of the doubts which Dr. Ashmead expressed concerning the host relations of this species. He was certain that it was a parasite of Diptera, and that it could not be a primary parasite of the gipsy moth or of the brown-tail moth. As events have since abundantly proved, he was right and wrong at one and the same time.

The separation of the parasites from the exceedingly large number of caterpillars which issued coincidently, and the subsequent separation of Pteromalus from the remaining species, was a task of huge proportions, but eventually it was accomplished, and some 40,000 Pteromalus were liberated in several localities, as indicated on the accompanying map. At the same time an attempt was made to determine the habits of the species, and reproduction experiments were conducted, using the active caterpillars of the brown-tail moth as hosts.

The females were frequently observed to take peculiar interest in these active caterpillars of the brown-tail moth. They would frequently alight upon their backs and appear to oviposit, and since nothing was then known or suspected of the well-nigh total depravity of this species in so far as its habits of oviposition are concerned, it was only natural to suppose that it was really possible for successful oviposition to take place under these circumstances. Nothing less was expected than that there would prove to be a second generation of the parasite, developing within the active caterpillars, or perhaps in the pupæ.

Attempts to discover some trace of this generation were futile, but failure could not altogether be attributed to the fact that such a generation did not exist. As it happened, every one of the several colonies of the parasite was situated within a territory to the northward of Boston over which the brown-tail moth was exceedingly abundant. Late in the spring the host of caterpillars was suddenly destroyed by an epidemic of a fungous disease which was so complete and overwhelming as to leave very few survivors. Even now, four years later, the brown-tail moth has not reached its former abundance over a considerable portion of the territory affected, notwithstanding that there has been steady and fairly rapid annual increase throughout this period. It looked, in fact, as though the parasites had suffered to an even greater extent than their hosts (since they were not so thoroughly well established), and failure to recover Pteromalus from the field during the summer, or even during the winter following, was thought to be the result of the epidemic of disease.

Another importation of the hibernating nests consisting, like the first, of about 100,000 from various localities in Europe, was received the next winter and handled in the same manner as was the other, but affairs at the laboratory did not run as smoothly as they might in the spring of 1907 at about the time when the Pteromalus were issuing. Mr. Titus was absent on account of sickness which eventually forced him to resign from his position at the laboratory, and neither Mr. Crawford, who first took his place, nor the present incumbent, who finally assumed charge the latter part of May, was sufficiently familiar with the work to carry it on to as good advantage as Mr. Titus would have done had he retained his health. Partly on this account and partly on account of weather conditions which were very unfavorable to the issuance of the parasites, only about 40,000 of the Pteromalus were reared and liberated. As before, they were colonized in various localities within the infested area as soon after their emergence as was practicable, and as before attempts to secure laboratory reproduction were made.

All of these attempts to secure the reproduction of the parasite in 1906 or in 1907 failed, since only active caterpillars of the brown-tail moth or gipsy moth were used. All sorts of theories to explain this were formulated, but that which seemed the most reasonable at the time, namely, that the parasite did not actually reproduce upon active caterpillars or pupæ, but only upon inactive caterpillars after the construction of their nests in the fall, could not be given an actual test, since inactive caterpillars were not available. An attempt to carry the living Pteromalus adults through the summer did not succeed, and with the death of the individuals in confinement, and the almost immediate disappearance of those which were liberated in the field, the investigations were necessarily brought to a close.

Meanwhile, as will be detailed later on, a variety of other parasites was found to be present as minute larvæ which hibernated within the still living caterpillars, and for the purpose of securing these as well as an additional supply of the Pteromalus, further extensive importations of the nests of brown-tail moths were made during the winter of 1907–8. A radical modification in the policy of the laboratory was inaugurated at the same time, and instead of discontinuing its activities during the winter months, the experiment was made of keeping it open for the purpose of conducting a series of winter investigations, and the study of the hibernating caterpillars of the brown-tail moth and of their parasites was selected as the subject for the first winter's work.

The first lot of nests arrived from abroad in December, and instead of awaiting the coming of spring they were immediately brought into a warmed room in the hope that the parasites might thereby be forced into activity. The experiment was successful. The first of the

Pteromalus began to issue coincidently with the beginning of the new year, and they were at once supplied with a quantity of nests of the brown-tail moth collected in the open and containing living caterpillars. In most cases the females almost instantly entered these nests and oviposited upon the still dormant caterpillars (fig. 66) with the result that in three and four weeks large numbers of a second generation began to issue. This successful outcome to what was considered to be, until that time, an experiment of rather doubtful utility, was very encouraging, since it was at once evident that any desired number of Pteromalus might easily be reared in captivity. Accordingly the work of rearing it on a large scale was begun, with the result that by the end of March American nests which contained the progeny of some 100,000 individuals were available for colonization.

Meanwhile large numbers of nests of the brown-tail moth—several thousand, in fact—had been collected in the neighborhood of the colonies which had been planted in 1906 and 1907 and no Pteromalus issued from them. It was evident that the colonization experiments of the spring of 1907 were no more successful than those of the spring before, and it was no longer possible to consider the bad results as due to the unusual mortality of the brown-tail moth in the vicinity of the colonies. It was necessary to seek some other explanation for this apparent failure to establish the one parasite which had been imported and colonized in wholly satisfactory numbers, and it was thought that this might be found in the circumstances under which the parasites were reared and liberated.

In 1906 and 1907 the adults had been liberated in the field some two or three weeks sooner than they would normally have issued as adults on account of their development having been hastened by the storing of the nests of the brown-tail moth at an artificially high temperature during the time that they were in transit from Europe. This, it was believed, might be responsible for the fact that the species had failed to establish itself and it was planned to do things very differently in the spring of 1908.

In accordance with these plans the nests containing the brood (as well as quantities of healthy caterpillars) were placed in large tube cages, which were fitted with a "tanglefooted" shield within, intended to prevent the emergence of the caterpillars without hindering the egress of the winged parasites, and four colonies, each of which was estimated to consist at the very least of 50,000 of the parasite larvæ, were located in four widely separated localities in eastern Massachusetts. The cages were simply taken into the field and left, so that the parasites were free from the moment of their emergence.

Considerable trouble was experienced at first on account of the "tanglefooted" shields failing to do all that was expected of them in the matter of preventing the escape of the caterpillars, but aside from that, the experiment promised to be highly successful. Instead of losing track of the parasites immediately following their liberation, they were found to be present in abundance in and about these cages throughout May and June, and even in July Mr. Mosher (who conducted this work) observed a few alive and apparently waiting until the next generation of hibernating caterpillars would be open to their attack.

Not all of the Pteromalus brood was liberated in this manner, but a part of the artificially infested nests was placed in cold storage at a constant temperature of approximately 30° F. and kept during the summer and until the formation of the brown-tail moth nests in the fall. Then a part of them was removed as a check on the condition of the remainder, and when it was certain that many, if not most of the Pteromalus had survived, a considerable number of them was allowed to issue in the open in a locality where they would find an abundance of fresh nests of the brown-tail moth ready at hand. Others of the stored Pteromalus were held for the purposes of winter reproduction, in case the further colonization of the parasite seemed worth attempting.

At first it appeared that the colonies of 1908, both spring and fall, were successful. In the vicinity of each of them (but particularly of that which was planted in the fall) the larvæ of the parasite were found in the nests of the brown-tail moth, and for the first time it was known to have lived over summer out of doors. Extensive rearing work was organized in the laboratory, with the intention of securing at least 1,000,000 for colonization in 1909, and certain technical investigations into the life of the parasite, which were begun in the spring of 1908, were continued.

The results of these biological investigations soon became startling in their nature. Gradually, as they were continued, and the results of one experiment after another became apparent, a tale of insect duplicity·was unfolded the like of which has never been quite equaled in any similar investigation. It is not possible to give the story in anything like complete detail, but a brief summary ought to be presented, if for no other purpose than to illustrate the degradation to which a parasite may sink.

It was found that the instinct of the female Pteromalus was first to seek the immediate vicinity of the feeding caterpillars, or of the nests or molting webs which they had deserted, and second to oviposit upon nearly anything which she encountered, providing it resembled in the slightest degree a dormant caterpillar of the brown-tail moth inclosed in its hibernating web (fig. 66). Attempted oviposition upon active caterpillars was only one of innumerable

teromalus egregius: Female in the act of oviposition through t
hibernating caterpillars of the brown-tail moth.   Greatly enla

t hesitation.   The cocoons of small Hy₁
les and Limncrium, were especially attra
.d whether associated with the brown-tail
results of this indiscriminate oviposition
eposited upon dead caterpillars of the br
perished except in one instance, in which t
killed and "pasteurized."   Upon this oc
of the larvæ lived, and at least one went t
ctive caterpillars of the brown-tail moth an
llars removed from the silken envelope.
ıd themselves within their nests, oviposi
if the caterpillar moved to any extent aft
ıting caterpillars, removed from their nes

a very little as a rule, and the least motion was sufficient to dislodge the egg or young larva of the parasite.

Oviposition upon any other host was equally unsuccessful, provided that the host was free to move about to any extent, but whenever it was confined within the limits of a cocoon, and was not too large, it usually fell a victim to the parasite. Especially was this true of the hibernating larvæ of hymenopterous parasites within their cocoons, and from these the largest and finest Pteromalus were reared.

If the parasite or hymenopterous larva was very small, as in the instance of the larva of Apanteles, it was very likely to be killed by the Pteromalus in the process of oviposition and, as a common result, her progeny would perish also.

Evidence to indicate that the female parasite possesses discriminative powers which enable her intelligently to select suitable hosts for her young is wholly lacking, and in consequence, when several individuals are given access to a single nest of the brown-tail moth, the chances are that all of them will concentrate their attack upon the few caterpillars which chance to be most readily accessible, to the exclusion of all others. The outcome is one of the manifold phases of superparasitism. The larvæ hatching from the superabundance of eggs are unable to reach their full development. They ma. complete their transformations but the adults produced are small, weak, and in extreme instances wholly unfit for further reproduction.

In the work of rearing the parasite for colonization purposes, no matter how many parent Pteromalus were used, the number of caterpillars which were parasitized by them would be a small percentage of those in the nests exposed to their attack, and invariably when more than a few females were used as parents the nests had to be torn open, so as to expose a large number of caterpillars equally. Otherwise the progeny would be so small as to be practically worthless for further reproduction, colonization, or anything else. This in itself was sufficient to render Pteromalus of very much less value from an economic standpoint, and the extraordinary avidity with which it attacked the cocoons of other hymenopterous parasites was anything but a point in its favor. Most especially was this true when the life and habits of *Apanteles lacteicolor* Vier. were taken into consideration. It soon became evident that Pteromalus was peculiarly fitted to act as its most dangerous enemy, and since, between the two, Apanteles was much the more promising parasite, it was decided to abandon all further effort toward the introduction of Pteromalus, and the work of rearing was discontinued.

Some 250,000 larvæ and pupæ were on hand at the time when this decision was reached, and these were placed in cold storage. It was considered probable that the species was already introduced, if it were possible to do so, as the result of the elaborate colonization work already described, and that any harm which might result was probably already done, so it was determined to use these larvæ and pupæ for the purpose of giving the parasite one more opportunity to retrieve a lost reputation. The brood lived through the summer in cold storage without much loss, and in the fall one tremendous colony of some 200,000 individuals was established in the midst of a tract of small oak, well infested with nests of the brown-tail moth. The adults issued at a time when there was nothing to prevent their entering these nests and ovipositing immediately, and there were enough of them to destroy all of the caterpillars of the brown-tail moth within a considerable radius. There were many larvæ to be found in the nests that winter, but, as was the case in the laboratory, only a few of the more exposed caterpillars were attacked.

A rather elaborate series of nest collections was made within a radius of a mile of the center of the colony, but the data obtained were of little consequence. From only a part of the many lots of nests did any of the parasite issue, and its probable rate of dispersion was not definitely indicated. One lot of nests collected a little over a mile away produced a few individuals, and this was the only instance in which it could be shown to have traveled so far.

At the same time large collections were made in the vicinity of the 1908 colonies, from which, it will be remembered, some few parasites had been recovered the winter before. In no instance was it again recovered, and there was everything to indicate that it had failed to establish itself.

No attempt whatever was made to rear it for colonization in 1910, and until the beginning of the winter of that year it was considered that the story of Pteromalus in America was complete. It is the unexpected which usually happens in the gipsy-moth parasite laboratory, however, and even as the rough manuscript for the last few pages was being prepared, Chapter II of the history of *Pteromalus egregius* in America was about to begin.

Every winter since that of 1906–7, to and including the present, an increasingly large number of the hibernating nests of the brown-tail moth have been collected from various localities throughout eastern Massachusetts and confined in tube cages in the laboratory. In the first two winters this was done for the express purpose of recovering Pteromalus and, as has been already stated, without result. In the winter of 1908–9, it was found that Monodontomerus was to be recovered in this manner over a considerable territory and under conditions which were both interesting and instructive. Accordingly, beginning with that winter, the collections have been made general throughout the territory in which it was thought likely that Monodontomerus would occur, and with less reference to the

MAP

localities in which Pteromalus had been colonized. Several thousands were thus collected in 1909–10 (as may be seen by reference to Table X) and a much larger series of collections was planned for the winter of 1910–11.

On the face of the results of this work during the two previous winters, nothing was much less likely than that Pteromalus should be recovered from any of these collections of nests. When a few specimens of a pteromalid which looked very much like it did issue early in December, they were accorded a rather cool reception, and made to identify themselves by reproducing upon hibernating caterpillers of the brown-tail moth in confinement. Before such identification was complete, it was rendered unnecessary through the issuance of considerable numbers of what could no longer be questioned as the true *Pteromalus egregius* from no less than 10 lots of nests collected in different towns scattered all the way from Milford, Mass., down near the Rhode Island line, to Dover and Portsmouth, N. H., just across the Piscataqua River from Maine. At the time of writing they are still emerging from the collected nests, and the extent of their dispersion is not yet known, but Mr. H. E. Smith, who is attending to the rearing cages, has prepared a map (Plate XXV) showing the location of the original colonies as well as the towns from which recovery has been made the present winter.

Sufficient data have already been accumulated to make certain the astounding fact, that as a result of the colonization work conducted between 1906 and 1908, the parasite is now thoroughly established over a territory which undoubtedly includes portions of four States, and during the period of its dispersion it spread itself out so thin as to make its recovery impossible except in the immediate vicinity of the colony sites, and for a short period immediately following colonization. Until this time Monodontomerus has held the record for rapid dissemination, but this record is now eclipsed.

It is impossible to determine whether the first of the colonies were after all successful, or whether they actually died, as was supposed, and success finally resulted from the very much larger colonies in 1908. If the early colonies lived, it means that no less than four years elapsed before any evidence to that effect was forthcoming. This fact, in its relation to circumstances attending the colonization of another parasite, *Apanteles fulvipes*, which seems not to have succeeded in establishing itself any more than Pteromalus appeared to have established itself as a result of those early colonizations, will sustain some hope for the ultimate recovery of this parasite until 1912 or 1913.

If, on the other hand, the establishment of Pteromalus resulted from the very much larger and in every way satisfactory colonizations of 1908, it may mean, in its reference to *Apanteles fulvipes*, that very much larger colonies will be necessary before we can hope to see that species established in America. To colonize it under more satisfactory conditions than those which prevailed in 1909 would be well-nigh impossible except at a very heavy expenditure, because the favorable conditions in 1909 were primarily due to the unusual coincidental circumstance of an early season in Japan, and a late season in Massachusetts. Such coincidences can not be depended upon, and without them, a tenfold expenditure over that of 1909 would be insufficient to secure equally favorable conditions for the establishment of the species, and a proportionately larger expenditure to better them.

The story of Pteromalus has been given at length because of the bearing which it has upon the question of what constitutes a satisfactory colony of any species of parasite. Except in a few instances, of which Calosoma and Anastatus are conspicuous, we frankly do not know the answer, and it is only through the study of such phenomena as those which have accompanied the recovery of Pteromalus that we are able to judge the probable character of the answer in the instance of those parasites which for some obscure reason or another have failed to make good their establishment in America.

APANTELES LACTEICOLOR VIER.

The story has already been told of how, during the winter of 1905–6, some 100,000 hibernating nests of the brown-tail moth were imported, placed in large tube cages in the laboratory at North Saugus, and how some 60,000 Pteromalus and countless thousands of caterpillars of the brown-tail moth issued into the attached tubes, and were sorted with difficulty. There is not a single published record outside of those emanating from the laboratory, so far as was then known, or is known now, which suggested the possibility of this particular sort of caterpillar harboring other parasites than those which issued as adults from its nests. Mr. Titus recognized that this might well be possible, however, and rather with the purpose of determining the fact than with the expectation of securing such parasites in any quantity for liberation, he caused some of the caterpillars to be fed in confinement and under observation. His foresight was well rewarded when, in the course of time, a number of cocoons of an Apanteles (fig. 67) was found in these cages, and the fact that at least one parasite hibernates within the living caterpillars was demonstrated.

The following spring he laid his plans for the wholesale rearing of this parasite and whatever other parasites might chance to be present. A considerable number of wood and wire-screen cages (Pl. XXVI, fig. 1), modifications of the familiar Riley type, was procured, and as the caterpillars issued from the cages containing the second large importation of hibernating nests in the spring of 1907 they were placed in these cages and fed. As a result of his enforced absence from the laboratory at a critical period these cages lacked the proper attention, and things went wrong with many of them. A few, however, were measurably successful, and eventually about 1,000 of the Apanteles were reared and colonized. Meteorus was discovered to have similar hibernating habits, and Zygobothria was also reared under circumstances which were sufficient to indicate its hibernating habits to the satisfaction of the junior author of this bulletin, but not to that of the senior. The Apanteles, in accordance with what was then the policy of the laboratory with regard to parasite colonization, were liberated in no less than three widely separated localities. None of the colonies, so far as known, was successful.

As anyone who was unfortunate enough to be associated with the laboratory during the spring and summer of 1907 will undoubtedly be willing to testify, the discomfort caused by handling quantities of caterpillars and cocoons of the brown-tail moth was literally dreadful. The poisonous spines upon the young caterpillars are neither so abundant nor so virulent as those upon the older caterpillars, but they are bad enough, and the task of feeding the inmates of the numerous cages which contained some thousands was a task of no little magnitude and one involving much physical discomfort.

FIG. 67 —*Apanteles lactescolor:* Adult female and cocoon. Much enlarged. (Original.)

The instant the door of one of these cages was opened, if the day was warm and its occupants active, a variable, but usually a large number would crawl outside, and to attempt to brush them back was but to afford opportunity for more to escape. Consequently thousands did escape and had to be brushed up and destroyed after each day's feeding. To keep the cages clear of débris was well-nigh out of the question, and every time that some attempt was made to clean them out more thousands of caterpillars escaped and had to be destroyed.

When the Apanteles and the Meteorus cocoons were discovered to be present in variable abundance in several of the cages trouble began in earnest, because they were for the most part firmly attached to the sides, or cunningly concealed in the midst of an accumulation of unconsumed food, so that much time was required to find and remove

them.   During this operation the caterpillars, stirred into unusual
activity, were crawling over everything in the immediate vicinity, but
more particularly over the outside of the cage and the person of the
operator.

If a sufficient number of these caterpillar parasites were to be reared
to make possible satisfactory colonies another year, it was obviously
exceedingly desirable to devise some other means of feeding the
caterpillars than that afforded by the closed cage, and accordingly,
in the winter of 1907–8, when the first active caterpillars began to
emerge from the nests which had been kept in the warmed part of
the laboratory for the purpose of securing Pteromalus, all sorts of
experiments were made in the hope of discovering some method
whereby the disadvantages above recounted might, at least in part,
be obviated.   The feeding tray illustrated herewith was the result of
these experiments, and as soon as it was found to be practicable,
enough to accommodate several thousand caterpillars were constructed, and one wing of the laboratory "annex," illustrated in
Plate XXVI, figure 2, and Plate XXVII, was fitted for their accom-
modation.

In all respects these trays were a success.   There was occasionally
some trouble caused by the caterpillars finding or constructing a
"bridge," by which they passed from the interior of the tray directly
to the frame above the concealed band of "tanglefoot," but when
sufficient care was used in feeding and in searching for bridges before
they were completed this was almost completely done away with.

It was manifestly impossible to feed more than a very small part
of the caterpillars from the many thousands of nests which had been
imported during this winter, and accordingly the caterpillars from a
few nests in each lot were fed in small trays in the laboratory during
the late winter and early spring, and the extent to which they were
parasitized by Apanteles was thus determined.   The most highly
parasitized nests were saved, and the larger part of those less highly
parasitized were destroyed forthwith, since it was no longer desired
to save the Pteromalus which might be reared from them.

A good many Apanteles were reared in the course of this work, and
since they issued long before the resumption of insect activities out
of doors they were used in a series of reproduction experiments upon
active caterpillars of the brown-tail moth feeding upon lettuce indoors.
It was found to be easy to secure reproduction when caterpillars
which had not molted since leaving the nest were used as hosts, but
if they had molted once successful reproduction was secured with
great difficulty or not at all.   The adult Apanteles were very far from
being as strong and hardy as the adult Pteromalus and could not be
kept alive to deposit more than a small part of their eggs.   There
were other reasons, too, why reproduction upon caterpillars in confine-

FIG. 1.—RILEY REARING CAGES AS USED AT GIPSY-
MOTH PARASITE LABORATORY. (ORIGINAL.)

FIG. 2.—INTERIOR OF ONE OF THE LABORATORY STRUCTURES, SHOWING TRAYS USED IN
REARING APANTELES LACTEICOLOR IN THE SPRING OF 1909. (ORIGINAL.)

VIEW OF THE LABORATORY INTERIOR, SHOWING CAGES IN USE FOR REARING PARASITES FROM HIBERNATING WEBS OF THE BROWN-TAIL MOTH IN THE SPRING OF 1908. (ORIGINAL.)

ment could not be looked upon as a feasible method for obtaining the parasites for liberation, and all ideas of laboratory reproduction work on a large scale were regretfully abandoned before the spring was far advanced.

A program for the colonization of Apanteles during the year 1908 was definitely formulated as the direct result of this experimentation, by which it was hoped to afford the parasite the best possible opportunity for speedy establishment. In accordance with this plan the nests which had been found to contain the more highly parasitized caterpillars were divided into three lots. The larger of these was placed in the same form of tube cage which was used for the Pteromalus-rearing work in 1906 and 1907; the next larger was placed in cold storage, and the nests remaining were brought into the laboratory toward the end of March and the caterpillars forced into premature activity. There was not very much room available for this indoors, so that the number of nests thus treated was decidedly limited, but from the caterpillars issuing from them no less than 2,000 Apanteles cocoons were secured during the latter part of April, and the adult Apanteles, to the number of about 1,300, which issued from them were liberated in the one colony in the field just as the caterpillars of the brown-tail moth were issuing from their nests and beginning to feed out of doors.

It was known that under natural conditions the parasite never issued as an adult at this season of the year, but it was reasonably certain that it would immediately reproduce upon the small caterpillars which in a week or two more would be so large as to make reproduction impossible. It was hoped in this manner to give the individuals liberated in this colony a certain advantage over those liberated later by allowing them superior opportunities for immediate reproduction and incidentally an opportunity for one more generation during the year than would be possible in colonies established at a later date.

To a certain extent these expectations were realized. It was positively ascertained that the parasite did take advantage of the opportunity offered and that it did actually pass one generation upon the newly active caterpillars of its chosen host. The experiment is not known to be a practical success, however, because all subsequent attempts to recover Apanteles from nests of the brown-tail moth collected in the vicinity of this colony have failed.

As soon as the caterpillars in the nests which had been placed in the large cages became active they were transferred to the larger trays which had been provided especially for them (Pl. XXVI, fig. 2), and fed first with lettuce and later with fresh foliage collected in the field. They did remarkably well at first, and about May 20 the cocoons of Apanteles began to appear in the trays in large numbers. The collection of these cocoons and their removal to small cages for the rearing of the Apanteles was simplicity itself, compared with the similar process the year before.   A large sheet of paper, thickly perforated with small holes,[1] or what was equally suitable, a strip of ordinary mosquito netting, would be spread over the pile of débris in each tray and fresh food placed on it.   In the course of 24 hours the great majority of the caterpillars would have crawled upon this paper or netting, and could be removed instantly, and with scarcely any disturbance, to a fresh tray.   The sorting over of the contents of the tray in which they had been feeding, for the cocoons of their parasites, could then be conducted without the annoyance of their presence, and with a minimum of discomfort.   This is, of course, the same method used in feeding the silkworm of commerce.

In all some 15,000 cocoons were secured in this manner, but only about 10,000 of the adults were reared and liberated.   Some 100 of the cocoons produced the secondary parasite *Mesochorus pallipes*, and a considerably larger number a small pteromalid, which was vaguely familiar in appearance, but which was not at that time recognized as identical with *Pteromalus egregius*, concerning which so much has already been written.

The Apanteles were carefully separated from their enemies and three colonies were established in the field.   Two of these were rather small, but one of them was made very large, and to comprise more than two-thirds of the total number reared.   It was no longer a question that the small colony was sometimes a mistake, and that it was invariably safer to liberate large colonies and to establish the species first of all, and to bring about dispersion later, if artificial dispersion should appear to be necessary.

It is interesting to note, in this connection, that neither in 1909 nor 1910 was it possible to find any trace of the Apanteles in the neighborhood of either of the two smaller colonies mentioned above, while from the larger it was recovered in 1909, and by 1910 had spread to a distance of several miles at least.

The third lot of nests, which was placed in cold storage before the caterpillars became active in the spring, was left there until early in July, when it was removed.   A part of the caterpillars immediately became active, but it was at once evident that many of them had died as a result of the unnatural conditions.   The weather was exceedingly hot immediately following and suitable food for the young caterpillars could not be obtained.   In consequence, a great many of them died from one cause or another, or from a combination of several; the larger part of the caterpillars died soon after having become active, and it seemed as though those containing the Apanteles suffered much greater proportionate mortality than the others; in any event, only about 250 of the cocoons were secured, when it was

[1] This paper was originally imported from France for use in a similar manner in rearing silkworms.

hoped to secure 10 times that number at the very least. The few that were reared were placed in the field in accordance with the program mapped out the winter before, just about the time when the new generation of brown-tail caterpillars was beginning to construct winter nests, and when there was no possible excuse for failure on the part of the Apanteles to reproduce to the full extent of its powers. In so far as the puny colony thus planted could possibly be expected to succeed, this one was a success. Quite a number of cocoons was found the next spring in the molting webs of the caterpillars from the near-by nests, and it was evident that if the Apanteles had been reared as successfully at this season of the year as it had been hoped would be the case, no better plan for the rearing and colonization of the species could be devised.

There was no possibility of judging the success or failure of the Apanteles colonies of 1908 until the following spring at the earliest, and whether they succeeded or failed it was obviously desirable to continue the work at least one year more. Accordingly, more nests were imported in the winter of 1908–9, and from among them those which were the most highly parasitized were selected for rearing the Apanteles. No attempts to establish colonies out of season were made this time.

Partly as a result of experience gained the year before, partly because more caterpillars were fed, and partly because several among the lots of nests received this year were very heavily parasitized, the number of Apanteles reared and colonized was about 23,000, or twice the number of the year before. They were distributed in three colonies, one of which was near the site of the only successful late spring planting of 1908. There was no apparent necessity for this, but the accuracy of the theory of the large colony and establishment at any cost was becoming more and more evident, and it was resolved to let no opportunity slip by which a possible advantage might be lost. The two remaining outlying colonies were each as large as the successful colony of 1908, and Apanteles was recovered in the spring of 1910 in the vicinity of both.

By the spring of 1909 the pteromalid, which had commonly been reared from the Apanteles cocoons, was identified beyond question as *Pteromalus egregius*. It was found that the females persistently haunted the trays in which the caterpillars were feeding, and that they were very free in ovipositing in the cocoons of Apanteles whenever they encountered them. It was discovered, furthermore, that if the weather was hot and humidity low, the Apanteles larva or pupa in the cocoon attacked would die and dry up before the Pteromalus was full-fed, so that nothing would emerge. A few of the unhatched cocoons, of which there were more than 25 per cent in the summer of 1908, were saved and examined after these facts were known, and in several of them what were almost certainly the eggs and very young larvæ of Pteromalus were found. It thus became evident that at least a portion of this unfortunate mortality was due to hyperparasitism by Pteromalus, a considerable number of which had been free in the compartment where the caterpillars of the brown-tail moth had been feeding. In 1909 pains were taken to prevent a recurrence of these circumstances, and as a result only a very few of the cocoons were lost through attack by Pteromalus. That Pteromalus was to be considered as an aggressive enemy of Apanteles could no longer be doubted, and when it was remembered that the adults naturally emerged from the nests of the brown-tail moth in the open at almost the precise time (Pl. XXVIII, fig. 1) when the Apanteles larvæ were emerging and spinning their cocoons, more often than otherwise in the outer interstices of these same nests, and that, furthermore, the Pteromalus was prone to linger in the vicinity of these nests in preference to any other place, its true duplicity was at last realized.

A much smaller number of over-wintering nests of the brown-tail moth was imported during the winter of 1909–10 than during any other since the beginning of the work, more for the purpose of securing Zygobothria, if possible, than for the rearing of additional Apanteles. The large trays were used as before (Pl. XXVI, fig. 2) for the rearing of a number of the caterpillars in the spring, and 10,000 or more Apanteles were reared and liberated in one colony at some distance from any of the others.

Until late in the summer of 1910 considerable doubt was felt as to the ability of this Apanteles to pass through the summer months successfully in large numbers. That it was able to live from June to August or September at all was rather more than was expected when it was first liberated. When, in 1908 and 1909, it proved its ability to do that much it remained to be determined whether it was going to be dependent upon an alternate host during that period or not, and if dependent whether a sufficient abundance of such hosts would be found in America to support as many of the parasites as would needs be carried through the summer, if it were to become an aggressive enemy of the brown-tail moth when this insect is in abundance.

It was with much satisfaction, therefore, that *Apanteles lacteicolor* Vier. was recovered as a parasite of Datana and Hyphantria late in the summer of 1910. Both hosts are common at that season of the year in Massachusetts, and both are parasitized to a considerable extent by tachinids. It is certain that the Apanteles will develop at the expense of these parasites as well as that of their hosts, and the chances are good that it will replace them to a certain extent, without bringing about a serious reduction in the prevailing abundance of these hosts; in short, that it will find a permanent place for itself in the American fauna.

FIG. 1.—COCOONS OF APANTELES LACTEICOLOR IN MOLTING WEB OF THE BROWN-TAIL
MOTH. (ORIGINAL.)

FIG. 2.—VIEW OF LABORATORY YARD, SHOWING VARIOUS TEMPORARY STRUCTURES,
REARING CAGES, ETC. A, A, OUT-OF-DOOR INSECTARY USED FOR REARING
PREDACEOUS BEETLES. (ORIGINAL.)

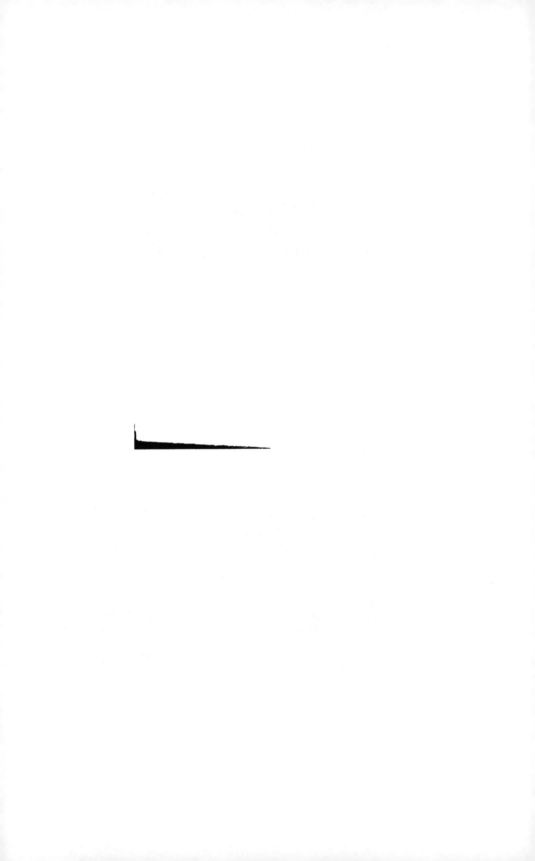

The diversity in the host relations of the parasite thus indicated is also encouraging. If it is capable of attacking arctiid as well as notodontid caterpillars with as much apparent freedom as it does liparids, there ought always to be plenty of available hosts to carry the species over the two or three months which must elapse after its emergence from the hibernating brown-tail moth before it can attack the young caterpillars of the same species for a second generation. It was hoped for a time that it would succeed in passing one generation upon the gipsy moth, but although it has been forced to oviposit in gipsy-moth caterpillars, and its larvæ have upon a single occasion attained their full development upon this host, there is no indication that it ever attacks it voluntarily in the field.

It is interesting and perhaps significant that in its relations with Datana it affects the host caterpillars exactly as in its relations with the brown-tail moth. The caterpillars died before the emergence of the parasite larvæ, and were left as nothing more than mere skins containing a small quantity of a clear liquid. In this respect, *Apanteles lacteicolor* Vier. differs materially from *A. solitarius*, or from many other among its congeners, which leave the host in a living condition but so seriously affected as to be unable to feed again.

Just as the proof of this bulletin is being read (June 12, 1911) word is received from the laboratory at Melrose Highlands that 4,000 cocoons of this parasite have been secured from brown-tail moth webs taken in the field in Malden and other towns.

APANTELES CONSPERSÆ FISKE.

In the summer of 1910 several boxes of the cocoons of an Apanteles parasitic upon the Japanese brown-tail moth, *Euproctis conspersæ* Butl., were received at the laboratory through the kindness of Prof. S. I. Kuwana. All of them had hatched at the time of receipt, and the circumstance would hardly be worthy of mention were it not for the fact that the adults which were dead in the boxes proved upon examination by Mr. Viereck to be identical in all structural characteristics with *Apanteles lacteicolor* Vier. It would appear that here was still another example of that phenomenon which has several times been mentioned without having been particularly designated, but which is, in effect, the existence of what has been termed "physiological" or "biological" species.

It is not so difficult to conceive as to find proof of the existence of two species which are so nearly alike structurally as to be indistinguishable by any taxonomic characters commonly recognized, but which are, at the same time, different. This difference may be exemplified by the sex of the parthenogenetically produced offspring, as in the instance of the European and American races of *Trichogramma pretiosa*. It may lie in the instincts of the female, which lead her to select certain hosts in preference to certain others, as in the instance of the American and European races of *Parexorista cheloniæ*. Again, it may be in the ability of the young larvæ to complete their development upon a certain host, as in the case of *Tachina mella* and *Tachina larvarum*. Or again, it may be that the difference lies as between *Apanteles lacteicolor* Vier. and *A. conspersæ* Fiske in the methods of attacking the host.

As has already been recounted, no less than 45,000 adults of *Apanteles lacteicolor* Vier. have been reared at the laboratory and liberated in the field. In addition a very large number has been reared under close observation during the winter or spring, and there has been a large number of more or less successful reproduction experiments conducted, in most instances with great care. In all this time there has not been a single exception to the rule, that the larva of *Apanteles lacteicolor* Vier. is solitary, and kills its host before issuing from its body. Nothing whatever, either in the field, or in the many experiments in reproduction, or in the occurrence of the parasite in shipments of larger caterpillars from Europe, has indicated in any way that it may ever attack the large caterpillars successfully, or that it is ever anything else than solitary.

Had *Apanteles conspersæ* Fiske been received as a parasite of the Japanese brown-tail moth without other data than the mere rearing record it would undoubtedly have been considered as identical with *Apanteles lacteicolor* Vier., but it is impossible so to consider it in view of the fact that it is not solitary but gregarious; that it attacks, not the small but the large caterpillars, and, if appearances of the material from Mr. Kuwana were not deceiving, that the host is left alive instead of being killed before the emergence of the parasite larva. These differences are, or ought to be, sufficient to make of it another species.

It is not at all improbable that if it were given the opportunity it would attack the caterpillars of the European brown-tail moth, and it is hoped that enough can be collected in Japan and forwarded to America to make the experiment possible.

METEORUS VERSICOLOR WESM.

A very few specimens of this parasite were imported in 1906 with caterpillars of the brown-tail moth and the gipsy moth from several European localities. In 1907, as already stated in the account of *Apanteles lacteicolor* Vier., a few specimens of Meteorus (fig. 68) were reared from caterpillars imported in hibernating nests the winter before. There were very few, less than 100 all told, and not enough to colonize with any likelihood of success. It was therefore decided to use them in a series of reproduction experiments, on the chance that a much larger number might be reared for colonization.

There were plenty of caterpillars of the brown-tail moth available and the smallest that could be found were confined in a cage with the first of the parasites that were reared.   Oviposition was not observed, and the parent adults did not live very long, but the caterpillars did very well for about 10 days, after which the cocoons of Meteorus began to be found in the cages in most gratifying numbers.   It seemed as though success was assured, and other similar experiments were immediately begun in the hope that some method would be found for prolonging the life of the adult parasites in confinement and securing more abundant reproduction.

The days of rejoicing over this, the first successful reproduction experiment with any of the parasites imported in 1907, were very few. In about a week the adults began to issue from the cocoons, and all proved to be males.   It looked like a curious coincidence at first, but when one after another of the various lots of cocoons hatched and out of the total of 156 every single individual was of the one sex, it was evident that something serious was the matter.   Where the trouble lay was not ascertained at that time, nor has it been determined as the result of other experiments similarly conducted in later years.   In all, 244 adult Meteorus have been reared in confinement, and among them there have been just 5 females, not one of which was secured until the late summer of 1908.

Fig. 68 — *Meteorus versicolor:* Adult female and cocoons. Much enlarged. (Original.)

Breeding Meteorus on a large scale for colonization purposes under circumstances like these can not be considered as an economically profitable venture.

The numbers of Meteorus reared from the caterpillars imported in the hibernating nests were increased by the addition of some few more secured from importations of full-fed and pupating caterpillars later in the season, and a small colony was planted in 1907, but it was so small as to make its success more than doubtful, and it was determined to rear enough for at least one good colony in the spring of 1908.

A description has already been given of the methods which were perfected during the winter of 1907-8 for the rearing of *Apanteles lacteicolor* Vier. in large numbers from the caterpillars of the brown-tail moth imported in hibernating nests, and these methods applied equally well to Meteorus. It was not nearly so common as the Apanteles, and only about 1,000 adults were secured for colonization. These were all liberated in one colony at a convenient place from the laboratory, and in order that they might have an opportunity for immediate reproduction a very large number of retarded caterpillars of the brown-tail moth from nests which had been placed in cold storage during the winter were liberated upon trees in the immediate vicinity.

These caterpillars seemed to be not at all injured as a result of their abnormal experience, but immediately began to feed voraciously and to grow apace. That they were injured soon became evident, but it could not be determined whether such injury was due to the enforced lengthening of their period of hibernation, to the hot weather which then prevailed, or, possibly, to the fact that the foliage was much more advanced than that upon which caterpillars newly emerged from hibernation usually fed. They began to die at an alarming rate inside of two weeks, and when it was time to make a collection for the purpose of determining whether the Meteorus had found them or not hardly more than 300 could be found out of the thousands which had been liberated. These were removed to a tray in the laboratory, and from June 23 to July 15 no less than 76 Meteorus cocoons were removed. From these 43 adults, of which 16 were females, were reared.

These were the first females of the second generation which had been secured at the laboratory, and a part of them was used in a reproduction experiment similar to those which had resulted in the production of males the previous year. Curiously enough, the adults of the third generation reared from these parents, under circumstances identical with those which had been used in earlier reproduction experiments, consisted of both sexes, there being 5 females out of a total of 40. These were the first females of the species ever reared from adults in confinement.

In the spring of 1909 the caterpillars from a few nests which had been collected the winter before in the vicinity of this first satisfactory field colony were fed in the laboratory, and from them a few cocoons of Meteorus were secured. It was certain that the species had completed the cycle of the seasons in the open, but it was also rather evident that it was not very common. If this were due to widespread dispersion, as might easily be the case, it might possibly result in the species spreading out so thin as to be lost, and it was resolved to place the Meteorus reared in 1909 in the same general vicinity, on the theory that by spreading over the same territory the colony might be materially strengthened throughout. This was done, and about 2,000 individuals were liberated during that spring and summer, the most of which came from hibernating caterpillars, but a part of which was imported as parasites of the full-fed and pupating caterpillars. The experiment of colonizing large numbers of retarded caterpillars in the vicinity was repeated, and with similar results to those secured in the previous season.

In 1910 a larger number of the cocoons was found, but at the same time a very few were secured from the caterpillars which had been collected in the vicinity of the colony. It did not look as though much was to be expected from the parasite at first, but when, toward the end of June, collections of full-fed caterpillars were made from various localities for the purpose of determining the status of the tachinid parasites, the results were much more encouraging. Cocoons of the second generation of Meteorus were soon found in some numbers and to a distance of a mile or more from the original colony center. Within a rather limited area near the colony center they could almost be said to be abundant, so abundant that 50 were collected in the course of about two hours' work. They are far from being conspicuous objects, being wholly disassociated from the caterpillar which served as host, and on this account the number collected was considered to indicate a very satisfactory abundance.

Its rate of dispersion, so far as indicated by the results of the summer work upon the caterpillars of the brown-tail moth, was too slow to be satisfactory, but in the early fall a single specimen, definitely determined by Mr. Viereck as of this species, was secured from a lot of caterpillars of the white-marked tussock moth collected in the city of Lynn, some 7 miles from the colony site. This would indicate a rapidity of dispersion in excess of that of Compsilura, and one which is distinctly satisfactory.

Another specimen was reared in the fall of 1910 from a caterpillar of the fall webworm collected in the open, and this was also considered as satisfactory evidence of its ability to exist here. At the present time there seems to be every reason to expect that it will be found in 1911 over a more considerable territory and in a much greater abundance than in 1910.

ZYGOBOTHRIA NIDICOLA TOWNS.

The few caterpillars which Mr. Titus saved from among those emerging from the hibernating nests in the spring of 1906 all died before pupation, and no other parasite than Apanteles and a single specimen of the Apanteles parasite, *Mesochorus pallipes*, was reared from them. In 1907 trouble was again experienced in carrying the caterpillars from imported nests through to maturity, but among the thousands which were fed in the cages at the North Saugus laboratory, as described in the account of the introduction of Apanteles, a few did reach the point of pupation, and from them a very few Zygobothria adults (fig. 69) were reared. There was no ground for doubting that the tachinids actually issued from the imported caterpillars of the brown-tail moth and that they had actually been present as hibernating larvæ within the caterpillars when they were received from Europe, but at the same time the circumstance seemed so improbable as to be refused immediate credence. Confirmation of the records was accordingly sought in 1908, and preparations were made to carry large numbers of the caterpillars from imported nests through to maturity in the large trays, already mentioned in the discussion of Apanteles.

For a time everything went well, and the caterpillars passed through three of the spring stages and assumed the colors characteristic of the last with scarcely any mortality. Then, for some reason, they ceased to feed freely, and began to die, and even those which did feed ceased to grow. Eventually practically all of them died, but of the few which survived to pupate, a very few contained the parasite, and although only about half a dozen of the adult Zygobothria were reared, they were sufficient to prove beyond question the validity of the earlier conclusions. The death of the caterpillars from imported nests in 1906 was supposed to be due to the epidemic of fungous disease which affected those in confinement quite as generally as those in the open, and in 1907 death was presumed to be the result of the unsanitary con-

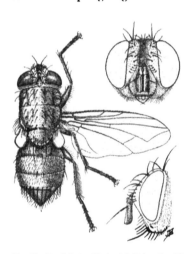

Fig. 69.—*Zygobothria nidicola:* Adult female, with front view of head above and side view below. Much enlarged. (Original.)

ditions which resulted from the use of the closed cages. In casting about for a cause in 1908, the drying of the food in the open trays before the caterpillars fed upon it was deemed to be sufficient, and consequently, in 1909, it was determined to use extraordinary precautions and to rear a large number of the tachinids if it were possible.

In the early spring of 1909 a considerable number of the imported caterpillars was dissected before they began to feed, and in some lots a high percentage was found to contain the hibernating larvæ of the Zygobothria (fig. 63, p. 264). These lots were to be given especial care, and little doubt was felt as to the success of the outcome, because no particular difficulty had been experienced in feeding small numbers of the caterpillars from native nests through all of their spring stages.

As was the case in 1907, the caterpillars passed through the first three spring stages with scarcely any mortality, and, as before, trouble was finally encountered. In the first place a considerable proportion of the trays was infected with the fungous disease, which had been accidentally brought in from the field, and these had to be destroyed summarily. There were still a number of the trays unaffected, however, and these were given the very best care which previous success with native caterpillars and failure with imported caterpillars suggested. In spite of all the results were exactly as before, and, as before, only an insignificant number of the Zygobothria completed their transformations. It was all the more surprising because there were several of the smaller and choicer lots which were kept in a cool, airy place, side by side with trays of native caterpillars, fed upon the same food and given identically the same attention, and yet every single individual of the one lot died, while nearly every individual of the others went through to maturity.

It began to look as though there was something wrong which was outside of the power of anyone at the laboratory to remedy, and it was resolved to test the matter thoroughly in 1910.

The caterpillar-dissection work which was begun in the spring of 1909 was carried on quite extensively in the winter of 1909–10, and among the several lots of hibernating nests imported that winter those which came from Italy and France were found to contain a very large percentage of caterpillars bearing the larvæ of Zygobothria (fig. 63, p. 264). These caterpillars, as soon as they emerged from these nests in the spring, were separated into two lots. A part of them was fed in trays, as before, and another part was immediately placed in the open, upon small oak trees which had previously been cleared of native nests of the brown-tail moth with this end in view.

The caterpillars, as usual, did remarkably well in both cases, and as usual the three spring stages were passed in the normal manner. At the end of that time those which had been fed in trays began to die, and those in the open to disappear. Mr. Timberlake, who was assiduously trying to follow the development of the Zygobothria maggots throughout their later stages, found it increasingly difficult to find the caterpillars in very large numbers in the field where they had been colonized, and finally of the thousands originally present only about 150 could be found. These had reached their last stage by this time, and they were collected and brought into the laboratory. Within a few days all but a very small number had died, and as there was a good chance that a few native caterpillars were present, there was nothing to indicate that all of the survivors were not native instead of imported.

At the same time that the imported caterpillars were liberated, as above described, a number of experiments in the similar colonization of native caterpillars was begun, and in every instance in which they were not overtaken with some well-defined calamity—fire in one instance, starvation in others—they went through to maturity in large numbers and in a perfectly normal manner.

It is no longer to be doubted that in the case of tne imported caterpillars some element other than any which is operative during the feeding period of the caterpillars in the spring is to be held responsible for their wholesale demise. The uniform ill success which has invariably attended the attempts to feed the brown-tail caterpillars from imported nests through to maturity can no longer be considered as either coincidental or the result of inexperience in this sort of work. Something else is responsible, and in looking about for parallel instances the results which have attended all attempts to feed caterpillars of the brown-tail moth, no matter from what source, out of season, are possibly to be considered as comparable.

Hundreds of experiments involving the feeding of native and imported caterpillars upon lettuce during the late winter and early spring have invariably resulted in carrying the caterpillars through their first three spring stages and in their death before pupation. This may be due to the character of the food.

A smaller number of experiments in feeding caterpillars of the brown-tail moth, which had been retarded in their emergence from the winter nests, have always resulted in a manner not altogether incomparable. Many thousands of these caterpillars have been kept in cold storage for about one month after they would normally have issued and then placed upon their favored food plants in the open. Upon several occasions when this has been done the caterpillars have fed very freely at first, grown rapidly, and appeared to be perfectly healthy. Then they would begin to die, almost exactly as the imported caterpillars would begin to die in the fourth spring stage, and it does not appear that any of them have ever completed their transformations. This may be due to the weather conditions and unsuitable food. It is believed that it is indirectly due in this instance, and in the instances of the caterpillars from imported nests, to the fact that both the one and the other have been subjected to abnormal conditions during hibernation. The imported nests are always exposed for a considerable period during the winter to an unduly high temperature. The caterpillars are almost upon the point of becoming active—sometimes they are beginning to become active—when the nests are received at the laboratory. As soon as possible after their receipt they are placed under out-of-door temperature again, with the result that the caterpillars become inactive and remain so until the time when they would normally have issued from the nests had they not been exposed to undue warmth during the winter.

It makes little difference whether the nests are exposed to one temperature or another during the winter so long as the caterpillars are not actually stirred into activity; the date of final emergence in the spring remains practically unchanged. Roughly speaking, if brown-tail nests are exposed to a constant high temperature beginning at any time during October the caterpillars will die without becoming active; during November they will die if kept too warm, but become active in a little over a month if kept warm and humid; in December they will sometimes become active by the 1st of January if they are kept fairly humid, and during January they will nearly always become active in a little less than a month, no matter what the conditions of humidity; after the 1st of February activity is resumed in something like two weeks; after the 1st of March in about one week, and later in a few days. If kept at a high temperature for three weeks in December or two weeks in January and then placed under natural conditions for the rest of the winter, their emergence will not be appreciably hastened in the spring, but if the attempts to rear Zygobothria from imported caterpillars which have been handled in much this manner are to be properly interpreted, subjection to such abnormal conditions results in a subtle disarrangement of the vital processes, and the insect is metabolistically unbalanced.

It is hardly necessary (to return to the story of Zygobothria) to state that these successions of almost total failures were not only puzzling, but decidedly exasperating. In 1910, for example, we estimated the number of apparently healthy Zygobothria larvæ on hand in apparently equally healthy caterpillars to be something like 40,000, of which something like 10,000 or 15,000 were in the caterpillars which were feeding and growing in a perfectly natural manner in the open. Long before it was time for these caterpillars to pupate we had given up all hope of more than an insignificant number of these parasites going through to maturity, and, as a matter of fact, there is no record of a single one among them going through. Every resource had been exhausted the winter before in attempting to secure a shipment of nests of the brown-tail moth in good condition from some locality where there was a likelihood of Zygobothria occurring in abundance as a parasite, and the failure was even more complete than usual. There remained only the alternative of importing large numbers of full-fed and pupating caterpillars of the brown-tail moth, collected in the same localities, and the prospect that this would be successfully accomplished was far from brilliant. The senior author was in Europe at the time when these conclusions were formed and was putting forth his utmost endeavors to bring about this very thing, but June passed, and with the advent of July it became certain that no shipments of any consequence would be received.

It was known that the parasite could be secured in this manner because small numbers had been reared from the imported quantities of full-fed and pupating caterpillars which were received at the laboratory in 1906 and several hundred from similar shipments in 1907. This latter year no accurate records had been made of the number of each species of tachinids emerging from the importations of brown-tail moth material, but it was known that somewhere between 300 and 500 individuals had been reared, the most of which were colonized at North Saugus. This was the only lot of adult flies of any consequence which had been reared and liberated, and since special efforts which had been made to recover this and other species liberated at the same time and place had failed in both 1908 and 1909, it was not considered to be at all likely that the attempted colonization was successful.

The situation, in so far as Zygobothria was concerned, could hardly have appeared worse than it was at the beginning of July, 1910. No one species of anything like equal importance had been quite so difficult to secure in adequate numbers and, moreover, there was no immediate prospect of finding a way to overcome the difficulties attending its importation. Consequently no similar circumstance, except perhaps the recovery of the gipsy-moth parasite, *Apanteles fulvipes*, could have caused a livelier satisfaction than was felt when several bona fide specimens of Zygobothria were reared from a lot of cocoons of the brown-tail moth which had been collected in the field some time before. The first specimen to issue was a male and it was followed by several more of the same sex. The males are markedly different from the females in appearance and not quite so distinctive, and we did not feel absolutely sure of their identity at first, but when after a few days a female was secured in the same manner from American cocoons there was no possible doubt that the species was not only established in America as firmly as three generations from a small beginning would permit, but dispersing with considerable rapidity, since of the seven specimens reared none was from less than 1 mile of the original colony site and one was from at least 3 miles distant. It is certain that the species must have spread over at least 30 square miles since its colonization three years ago, and when the millions of brown-tail moth caterpillars which are present in that territory are compared with the few thousands which produced the seven Zygobothria reared in 1910, it is equally certain that its increase has been at the same time enormous.

It bids fair, judging from this, to do exceedingly well in America. Unlike *Compsilura concinnata*, *Pales pavida*, and other tachinids, which rank of some importance as parasites of the brown-tail moth and gipsy moth in the Old World, it is wholly independent of any host other

than the brown-tail moth, and its rate of multiplication, being unquestionably more rapid than that of the brown-tail moth, ought not
to be checked until it has become a factor in the control of its own
particular host.

It may be added as a postscript that a few days after writing the
above a few hundred caterpillars of the brown-tail moth collected in
the field from hibernating nests were dissected in the laboratory.   In
them were found several of the characteristic first-stage Zygobothria
larvæ (fig. 63, p. 264) embedded in the walls of the gullet.   The evidence presented by this small number of dissections is less satisfactory
than though the number were larger, but if it is to be accepted the
rate of increase of Zygobothria in 1910 is considerably better than was
expected.

PARASITES ATTACKING THE LARGER CATERPILLARS OF THE BROWN-TAIL MOTH.

HYMENOPTEROUS PARASITES.

In Europe after the caterpillars of the brown-tail moth resume
activity in the spring they become subject to attack by a variety of
tachinid parasites, but so far as has been determined by rearing work
with imported material the only hymenopterous parasite of any consequence is Meteorus, which passes the winter as a first-stage larva in
the hibernating caterpillars.

In fact, only a single other parasite has ever been reared from
imported caterpillars which may not have come from some other accidentally included host, and this is the *Limnerium disparis*, which has
already received attention as a minor parasite of the gipsy moth.   It
would certainly seem as though there were likely to be others attacking the caterpillars of the brown-tail moth in Europe in spite of the
fact that none has been secured, and this supposition is upheld by the
published results of a study in the parasites of the brown-tail moth
which was made a few years ago by a Russian entomologist, Mr. T. W.
Emelyanoff.   He mentions a number of parasites which have not been
reared at the laboratory from imported material, and among them one,
Apanteles vitripennis Hal., which is so common, according to his account, that the "cocoons are sometimes accumulated together in great
numbers."   Any suspicions that the Apanteles thus observed by him
is identical with *A. lacteicolor* Vier, as reared at the laboratory and
which is the only representative of the genus that has been reared
from caterpillars collected in Russia or elsewhere, is at once dispelled
by his detailed account of the early life and habits of the species which
he had under observation and which differ in all essential particulars from the life and habits of *A. lacteicolor*.   The caterpillars
are attacked soon after they leave the nests.   Instead of dying in

their molting webs they crawl down the trunks of the trees, and the cocoons of the parasite are found in splits and holes in the bark, rarely higher than from 1 to 1½ yards from the ground.   The host caterpillar is left alive and remains for some time clinging to the cocoons of its parasite, something which has never been observed in the case of *A. lacteicolor*.

The plans for the coming season, if they materialize, call for a thorough study of the Russian parasitic fauna of the brown-tail moth, and it is sincerely hoped that the observations of Mr. Emelyanoff may be confirmed.

TACHINID PARASITES.

Several of the tachinids which attack the brown-tail moth have already been mentioned in the course of the discussion of the gipsy-moth parasites.   Among them *Compsilura concinnata* is the only species which is of real importance in connection with both hosts.

Tachina larvarum is not uncommonly encountered as a brown-tail moth parasite, but never so commonly as it frequently is in its other connection.   *Tricholyga grandis* has also been reared in small numbers from cocoon masses of the brown-tail moth.

The tachinid parasites of the brown-tail moth, which are either unknown as parasites of the gipsy moth or which are rarely encountered in that connection, include a considerable variety of species, several of which appear to be of little or no real importance.   As will be seen, they include amongst their number species which represent the extreme of diversity in habit.

DEXODES NIGRIPES FALL.

Another example of the artificiality of the present accepted scheme of classification of the tachinid flies is to be found in the separated positions therein occupied by the two exceedingly similar species *Compsilura concinnata* and *Dexodes nigripes*.   So similar are these two that if a few hairs and bristles were to be rubbed from the head of one it would be practically impossible to distinguish it from the other, even though everything in connection with the early stages and life of each was known.   The one point of difference of any consequence from an economist's standpoint is the more restricted host relationship of Dexodes, which, though equally common with Compsilura as a parasite of the brown-tail moth in Europe, is exceedingly rare as a parasite of the gipsy-moth caterpillars.   In every other respect, except host relationship, the habits of the two are identical, and so far as known their earlier stages are absolutely indistinguishable.

Dexodes was first received and liberated as a parasite of the brown-tail moth in 1906, and it was the first of the tachinid parasites to be carried through all of its transformations in the laboratory upon American hosts.   This was accomplished by Mr. Titus in one of the large out-of-door cages in 1906, and again with somewhat more success in one of the smaller indoor cages in 1907.   As with Compsilura, only two weeks are required for the larval development, a week or ten days for the pupal stage, and three or four days for the female to reach her full sexual maturity.   As is also true with Compsilura the larvæ are deposited by the female beneath the skin of the host caterpillar.

Several small colonies were planted by Mr. Titus in 1906, followed by several more small ones and one larger one in 1907.   None was liberated in 1908, but in 1909 one very large and satisfactory colony was put out.   In 1910 only a single specimen of the parasite was received from abroad and this, curiously enough, in a shipment of gipsy-moth caterpillars.

It was confidently expected that in 1910 at least a few specimens would be recovered from the field as a result of the earlier colonization work, but these expectations were not realized.   Of all of the tachinid parasites of the brown-tail moth, not excepting *Compsilura concinnata*, it was the one most satisfactorily colonized in 1906 and 1907, and on this account it was expected to find it established in the field.

It is considered as one of the most likely of the as yet unrecovered parasites to be recovered from the field in 1911 or 1912.

PAREXORISTA CHELONIÆ ROND.

No brown-tail moth material was received from abroad during the summer of 1905, and consequently nothing was known of the hibernating tachinids which attack this host until the spring of 1907, when they began to issue from the puparia of the previous summer's importations.   All that were reared that spring were of the one species, which has since been determined as *Parexorista cheloniæ* Rond., and to date no other species hibernating as a puparium and with but one annual generation has been reared from this host. Nothing to compare with the difficulties which attended the hibernation of the principal gipsy-moth parasite having similar habits was encountered in the case of Parexorista.   Its puparia (Pl. XX, fig. 4) were carefully covered with earth the first winter and the second, but it was then found that this precaution was unnecessary and that the percentage of emergence was quite as large when the puparia were kept dry as when they were damp.   The difference appears to be associated with the state in which the pupæ themselves hibernate.   Those of Blepharipa and Crossocosmia develop adult characters in the fall, and it is in reality the unissued adults which hibernate.   Those of Parexorista do not develop adult characteristics until spring, and besides in Parexorista the space between the pupa or nymph and the shell of the puparium is dry and does not, as in Blepharipa, contain a small quantity of colorless liquid.

Accordingly the percentage of emergence of *Parexorista chelonix* has aggregated nearly or quite 90 per cent each year as against the relatively small percentage of Blepharipa which has been carried through its transformations.

A very few of the flies were liberated in 1907, but there were too few puparia of the species received the summer before to make anything like a satisfactory colony of the species possible. In 1908 in excess of 2,000 of the flies issued from the previous season's importations, and of these about 1,500 were liberated in one colony under circumstances which were the most favorable that could be imagined. The remaining ones were used by Mr. Townsend in a successful series of experiments which have already been summarized in an account of his first year's work, published as Part VI of Technical Series 12 of the Bureau of Entomology. Other equally satisfactory colonies were established later, but of these nothing more need be said at this time.

The large colony liberated by Mr. Townsend in the spring of 1908 consisted of flies of both sexes, very many of which had mated before they were given their freedom. This circumstance, which was not considered as particularly of interest at the time, has acquired significance more recently, as will be shown.

Later in the spring and early in the summer caterpillars collected from the immediate colony site were found to contain the larvæ of the parasite, and a calculation involving the number of caterpillars within a limited area immediately surrounding the point of liberation, the number of flies liberated, and the percentage of parasitism prevailing in this area indicated a very satisfactory rate of increase. It will be remembered that the flies were in part ready or nearly ready to oviposit when they were given their freedom, so that dispersion did not have to be taken into consideration to the extent which is necessary when a long period elapses between the time of liberation and the time of recovery.

In 1909 similar collections of caterpillars and cocoons were made in the same and in nearby localities, and the number of Parexorista which was secured from them was gratifyingly large. These collections had not been made with the view of determining the rate of dispersion, but it was apparent that the increase had been accompanied by a rate of dispersion that was, at the very least, satisfactory and which, for all evidence to the contrary, might be phenomenal.

Accordingly in 1910 a series of collections was planned, some of which were to be made in exactly the same localities as those from which the flies were recovered the year before and which were designed to be indicative of the prevailing rate of increase, while others at varying distances and in different directions from the colony center were designed to show the rate of dispersion. No doubt whatever was felt concerning the recovery of the parasite, which was considered to be as firmly established as the Calosoma or *Compsilura concinnata*. The results afforded another example of the obtrusiveness of the unexpected. Not a single Parexorista puparium was secured from any of the material included in this series of collections.

This was, all things considered, the most serious setback of any which the parasite work has experienced since its inception. It was never doubted from the first that some among the parasites would be unable to exist in America, and no species was really credited with having demonstrated its ability to do so until it had lived over at least one complete year out of doors. Parexorista had done this and more, having gone through two complete generations, unless, what was not at all likely, its puparia had all been killed some time during the fall or winter.

Without indulging in unnecessary speculation as to the reason for its disappearance, the following facts are presented for consideration:

There is in America a tachinid known as *Parexorista cheloniæ*, which is morphologically identical with the European race so far as may be determined through a painstaking comparison of the two. It is a common parasite of the tent caterpillars *Malacosoma americana* Fab. and *M. disstria* Hübn. The adult flies issue at the same time in the spring as do those of the European parasite of the brown-tail moth. The same type of egg is deposited; the larvæ are indistinguishable in any of their stages or habits during their several stages; the third-stage larvæ issue at the same time and form puparia which are apparently the exact copies of the European, and the hibernating habits are the same. The one and only difference is that the American *Parexorista cheloniæ* does not attack the caterpillars of the brown-tail moth, while the European *Parexorista cheloniæ* is perhaps the most important of the tachinid parasites of this host.

Mr. W. R. Thompson, whose excellent and painstaking work makes possible the above comparison between the two races, went a step further in his investigations. He found by actual experiment that in confinement, at least, the European males would unite with the American females with as much freedom as with those of their own species. Granted that similar intermingling of the races takes place in the open, and the reason for the nonrecovery of *Parexorista cheloniæ* as a parasite of the brown-tail moth in the summer of 1910 is no longer a mystery.

It was stated a few paragraphs back that the flies which were colonized in the spring of 1908 were largely mated at the time of liberation. Their progeny, which issued in the spring of 1909, would therefore be of the pure-blooded European stock. Issuing at the same time were a vastly larger number of the American race, because as it happened there was an incipient outbreak of *Malacosoma disstria* in that very locality, which was quite heavily parasitized by Parexorista. There were easily 50 or 100 of the American flies to one of the European race present in that general vicinity in the spring of 1909. The chances that the pure-blooded European females were fertilized by American males were therefore a good 50 or 100 to 1 at the most conservative estimate.

Being of the European race, their instincts led them to attack the caterpillars of the brown-tail moth, and the attack was successful, as witnessed by the number of puparia which were secured from the collected caterpillars and pupæ in the summer of 1909, but these puparia, instead of representing the pure-blooded European stock, as was then supposed, represented the half-breed stock resulting from the promiscuous mating of their mothers. Evidently the females issuing from them in the spring of 1910 lost the cunning which is characteristic of the European race, which makes possible the deposition of the soft-shelled eggs amongst the bristling poisonous spines of the host without injury.

Mr. Thompson, in his experiments with the American female which had been fertilized by an European male, found that she was neither anxious to oviposit upon the caterpillars of the brown-tail moth nor able to do this successfully. A proportionately large number of the eggs deposited upon this host were either pierced by the poison spines or else the young larvæ came in contact with these and died before entering. A few larvæ did succeed in gaining entrance, and one or two passed through their transformations, but when the natural disinclination to attack the caterpillars of the brown-tail moth was associated with a heavy mortality following occasional attack the percentage of parasitism is reduced to the minimum.

In consequence of these observations in field and laboratory, the name of *Parexorista cheloniæ* has been erased from the list of promising European parasites of the brown-tail moth and placed at the head of the list of the imported parasites which are proved unfit.

It is a pity, too, as has incidentally been stated, because it is about the most common of any of the tachinid parasites in Europe, and, moreover, is one which is entirely independent of any alternate host.

PALES PAVIDA MEIG.

There is a very considerable group of tachinid parasites of the brown-tail moth which appears to be more commonly encountered in material from southern European localities than from those in the north. One of these, *Zygobothria nidicola*, has already been the subject of lengthy discussion. The fact that though apparently southern in its distribution in Europe, it has manifested a strong tendency to become thoroughly acclimatized here, has lent encouragement to the attempts which have been made, and which will be renewed, looking toward the establishment here of others having a somewhat similar distribution.

Pales pavida (fig. 70) is perhaps as promising as any among these, although it is possible that it appears so on account of a somewhat larger knowledge which we possess concerning its life and habits. It was first imported in not very large numbers in 1906. In 1907 about as many were secured and colonized as of the successfully introduced Zygobothria, and more were colonized in 1909. The fact that it has not been recovered is by no means to be taken as positive asssurance that it is not established, and it is well within the bounds of possibility that it will be recovered in 1911 or 1912.

It is one of the species which deposits its eggs upon the leaves to be eaten by its host (fig. 36, p. 214) and was the first species having this habit to be carried through all of its transformations in the laboratory. In 1908 Mr. Townsend succeeded in carrying some of the flies through the period allotted for the incubation of their eggs, but he did not succeed in securing oviposition. In 1909 Mr. Thompson had better fortune, and not only secured eggs in abundance,

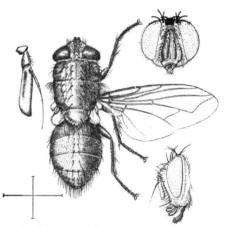

Fig. 70.—*Pales pavida:* Adult female, with front view of head above and side view below, and antenna at left. Much enlarged. (Original.)

but fed these eggs to a variety of caterpillars and secured either the puparium or the fly in nearly every instance. He also secured much interesting data upon the early stages, and upon the life and habits of the early stages, a story of which is left for him to tell. The accompanying illustrations of the eggs and larvæ were prepared under his direction. That of the egg (fig. 37, *b.* p. 214) is of interest in comparison with that of the egg of *Blepharipa scutellata* (fig. 37, *a,* p. 214), as showing the difference in the characteristic microscopic markings. That of the larva will give a good idea of the integumental "funnel" (figs. 71, 72), formed by the ingrowing epidermis, as differing from the tracheal "funnel" characteristic of the larva of Blepharipa, as figured on pages 215 and 216.

Not very much that is definite can be said of the seasonal history of Pales. It undoubtedly will require another host than the cater-pillar of the brown-tail moth in order that it may complete its seasonal cycle, but that it will find such a host is pretty certain. It would rather appear, from what has been observed, that it will attempt to hibernate as an adult. Whether or not it will be able successfully to do this in New England remains to be proved.

FIG. 71.—*Pales parida:* Second-stage larva *in situ* in basal portion of integumental "funnel." Much enlarged (Original.)

It has occasionally been reared as a parasite of the gipsy moth, and if successfully introduced into America it ought to be of some assistance in this rôle also. Unfortunately, as a parasite of the caterpillar of the brown-tail moth, it does not issue until about the time when the moth would have issued had the individual remained healthy. It requires some little time for the females to develop their eggs, and it is not at all likely that, like Compsilura, it will be found to pass one generation upon the caterpillars of the brown-tail moth, and the next upon the gipsy-moth caterpillars.

ZENILLIA LIBATRIX PANZ.

This parasite, like Pales, deposits its eggs upon foliage to be eaten by its host, but, unlike Pales, it has not been reared through its stages in the laboratory. Like Pales, it is southern in its distribution, and in relative importance they are about equal, judging from the numbers of each which have been reared at the laboratory. It was colonized in small

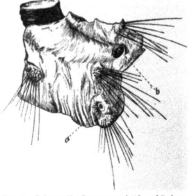

FIG. 72.—*Pales parida:* Integumental "funnel," showing orifice in skin of host caterpillar. Much enlarged. (Original.)

numbers in 1906, in larger numbers in 1907, and in very small numbers subsequently. The circumstances attending its colonization are as satisfactory, so far as known, as those attending the colonization of Pales and Zygobothria, and it is hoped that it may be recovered in the course of 1911 or 1912. It is also hoped that a large number will be imported in 1911.

MASICERA SYLVATICA FALL.

This tachinid appears not to be uncommon as a parasite of the brown-tail moth in Italy, but has not been received from other countries in more than the most insignificant numbers.

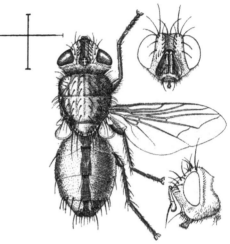

FIG. 73.—*Eudoromyia magnicornis:* Adult female, with front and side views of head at right Much enlarged. (Original)

Not enough have been received to make anything like colonization possible, and it is one of the species which it is hoped to receive in 1911.

EUDOROMYIA MAGNICORNIS ZETT.

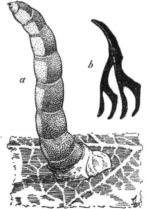

FIG. 74.—*Eudoromyia magnicornis: a,* First-stage maggot attached to leaf, awaiting approach of a caterpillar; *b,* mouth-hook of maggot. *a,* Greatly enlarged; *b,* highly magnified. (From Townsend.)

This (see fig. 73) is the most distinctive of the tachinid flies parasitic upon the brown-tail moth, and the only one among the parasites of either the gipsy moth or the brown-tail moth which has the habit of depositing its active larvæ upon the food-plant of its host. This habit was first discovered by Mr. Townsend, who gives an account of the manner of the discovery in Technical Series VI, part 12, of this bureau, from which the accompanying figure (fig. 74) was taken.

It is another of the group of tachinid parasites which appear to be southern rather than northern in distribution, on account of which it has been found impossible to secure a sufficient number to make adequate colonies practicable. It was colonized together with *Zygobothria nidicola*, *Pales pavida*, and *Zenillia libatrix*, in about the same numbers in 1906 and 1907, and, like the two last named, it is hoped to recover it in 1911. It is also hoped to import and liberate a much larger number than hitherto during that year.

CYCLOTOPHRYS ANSER TOWNS.

Mr. Townsend described this species as new from specimens reared in 1908 from brown-tail moth material received from the Crimea. It has not been detected in shipments of similar character from any other locality in sufficient numbers to indicate it as being an important parasite, nor have enough been received from the Crimea to make possible its colonization. It is hoped that this may be done in the course of the year 1911.

It is one of the relatively few species of tachinids attacking the larvæ of the brown-tail moth which deposit large, flattened eggs upon the body of the host caterpillars.

BLEPHARIDEA VULGARIS FALL.

This is almost the only tachinid parasite of either the gipsy moth or the brown-tail moth which is of no apparent importance in connection with either host and which at the same time has been reared a sufficient number of times to make its host relationship reasonably certain. The few specimens which have been received have mostly come from various parts of the German Empire. Very little is known of its life and habits, and it is not considered as being of sufficient importance to warrant further investigation.

PARASITES OF THE PUPÆ OF THE BROWN-TAIL MOTH.

By far the most important of the parasites of the pupæ of the brown-tail moth in Europe appears to be Monodontomerus, an account of which has already been given in the discussion of the parasites of the gipsy moth. It is more frequently reared in connection with the brown-tail moth than with the gipsy moth, and some of the shipments of cocoons have produced it in extraordinary numbers.

Theronia, also mentioned as a parasite of the gipsy moth, is about the next in importance, but the European *T. atalantæ* Poda is no more frequently reared than the American *T. fulvescens* Cress.

The same species of Pimpla already mentioned as parasites of the gipsy moth in Europe attack the brown-tail moth as well. Like Theronia and Monodontomerus, they are more frequently encountered in this connection than in the other.

No species of Chalcis has been reared from any European material received to date, and in this respect the parasitism of the pupa of the brown-tail moth differs from that of the gipsy moth. It further differs in that two small gregarious chalcidids, both of them closely allied or identical with American species of the same respective genera, have occasionally been reared from imported cocoon masses. Neither of these is common. One, *Diglochis omnivora* Walk, appears to be specifically indistinguishable from the form which goes under the same name in America, where it has occasionally been reared from the gipsy moth and abundantly from the brown-tail moth. The other is a species of Pteromalus, which, according to Mr. Crawford, is hardly to be distinguished from the tussock-moth parasite, *Pteromalus cuproideus* How.

Enough of the latter species have been reared to make small colonies possible, but these colonies have been so very small as to make its establishment improbable. It is hoped that a larger number will be imported in 1911, but since it appears to be of very slight importance in Europe no great enthusiasm is felt over the prospect.

SUMMARY AND CONCLUSIONS.

The work of introducing into America the parasites and other natural enemies of the gipsy moth and the brown-tail moth has been more arduous than was anticipated when it was begun. It was soon found that the published information concerning these enemies was deficient and unreliable, and that much original research was necessary in order that they might be intelligently handled. Later it developed that the rate of dispersion of the introduced species was so very rapid as to necessitate larger and stronger colonies than had been contemplated.

The policy originally adopted of employing foreign entomologists to collect the eggs, caterpillars, and pupæ of these pests abroad for shipment to the Massachusetts laboratory, where the parasites which they contained might be reared, has resulted in the successful importation and colonization of a considerable number of the parasites which a study of this material, after its receipt at the laboratory, has indicated as being of importance. Numerous others successfully imported have been colonized, but so recently as to render the success of the experiment uncertain. On account of the rapidity of dispersion, which results in the parasites being very rare over a large territory instead of being common over a restricted territory, as long a period as four years may elapse before it is possible to recover them after colonization. It has been found impossible to secure certain of the parasites in adequate numbers for colonization under satisfactory conditions. The proportion of such is very small, it is true, but at the same time it may easily be that ultimate success or failure may depend upon the establishment. not of the most important among the parasites and other natural enemies, but of a group or sequence of species which will work together harmoniously toward the common end. Viewed in this light, the importance of parasites which otherwise might be considered as of minor interest is greatly enhanced.

It is impracticable to determine certain facts in the life and habits of those parasites which have been colonized under conditions believed but not known to be satisfactory. Further detailed knowledge is necessary before we can judge whether the circumstances surrounding colonization were in truth the best that could be devised. Furthermore, so long as original research is confined to the study of material collected by foreign agents, some of whom are technically untrained, it is practically impossible to secure the evidence necessary to refute published statements concerning the importance of certain parasites abroad which the results of first-hand investigations have not served to confirm. It is believed that these statements are largely based upon false premises, but should this belief prove ungrounded it would mean that there are important parasites abroad of which little or nothing is known first-hand.

A determined effort was made in 1910 to better the deficiencies in the foreign service without going to the lengths of adopting a radically changed policy, but the results were not satisfactory. Lack of assurance that a continuation of the work in 1911 along similar lines would bring more favorable results made its continuation inadvisable. It therefore became a question of adopting new and radically different methods in so far as the foreign service was concerned.

In favor of a policy of inactivity was the prospect of an immediate reduction, as opposed to an increase, in expenditures should renewed activity be decided upon. There was the chance that the parasites already introduced and colonized would be sufficient to meet the demands of the situation.

On the other hand, the vast majority of defoliating native insects, which rarely or never become so abundant as to be considered injurious, prove upon investigation to support a parasitic fauna similar in all its essential characteristics to that supported by the gipsy moth in countries where it is similarly a pest at very rare intervals or not at all

Parasitism appears to be unique among the many factors of control, in that no other agency similarly increases in efficiency in direct proportion as the efficiency of other agencies, such as climatic conditions, miscellaneous predators, etc., diminishes. In short, the apparent importance of parasitism as a factor in the natural control of defoliating insects has been decidedly enhanced as a result of these more or less technical and intensive studies. It can be said with the utmost assurance that if a sufficient number and variety of parasites and other natural enemies of the gipsy moth which act in a manner comparable to the true facultative parasites, as above described, can be introduced into America, the automatic control of the gipsy moth will be permanently effected.

During the past four years the "wilt" disease has been increasing somewhat in efficiency, but notwithstanding that it is and has been prevalent in every locality in which the gipsy moth has been allowed to increase unchecked, the gipsy moth still continues to be a menace to the life and health of valuable trees which have been protected during this time at a considerable cost. In parts of Russia, where parasitic control is obviously inefficient, control through disease is not sufficient to keep the gipsy moth from increasing until defoliation of large areas results. Similarly, in America, the destruction of very large numbers of caterpillars of several sorts of the larger Lepidoptera has been observed, but in no instance until after the caterpillars involved had increased to such numbers as to become a pest.

It may be that the parasites already introduced and established, or likely to become established, will prove to be sufficient for the purposes intended. Only events themselves can be depended upon to answer this question, and from five to six years must pass before the answer is known. During this period the gipsy moth will continue to disperse and multiply, and large expenditures will be necessary to prevent much more rapid dispersion and multiplication than has prevailed in the past.

Expenditures amounting to a very small percentage of the total will suffice to carry on the parasite work. If the parasites already introduced are sufficient to meet the needs of the situation, the expenditure projected will have been needless and unnecessary. If the parasites already introduced are not sufficient, it may be that this deficiency can be made up in time to avoid much if any delay in the day of final triumph.

THE PRESENT STATUS OF THE INTRODUCED PARASITES.

PARASITES OF THE GIPSY MOTH

EGG PARASITES.

ANASTATUS BIFASCIATUS Fonsc.

Received first in 1908. Colonized unsuccessfully in 1908 and successfully in 1909. First recovered in immediate vicinity of colony in 1909. Increased notably in 1910, but indicated dispersion is only about 250 feet per year. Artificial dispersion necessary. Apparently well established.

SCHEDIUS KUVANÆ How.

Received first in 1907, dead, and in 1909, living. Successfully colonized in 1909. Recovered in immediate vicinity of colony site in 1909. Doubtfully recovered in 1910. Establishment very doubtful on account of climatic conditions

HYMENOPTEROUS PARASITES OF CATERPILLARS.

APANTELES FULVIPES Hal.

Received first in 1905, dead, and in 1908, living. Colonized unsatisfactorily in 1908 and under exceptionally favorable conditions in 1909. Two generations recovered in immediate vicinity of colony site in 1909. Not recovered in 1910 except from recent colony. Establishment doubtful on account of lack of proper alternate hosts.

TACHINID PARASITES.

COMPSILURA CONCINNATA Meig.

First received in 1906 and colonized same year. Colony strengthened in 1907. Recovered doubtfully in 1907 from immediate vicinity of a colony site. Certainly recovered and found to be generally distributed over considerable territory in 1909. Marked increase in 1910. Apparently established.

CARCELIA GNAVA Meig.

Doubtfully colonized in 1906. Satisfactorily colonized in 1909. Not recovered from field. Establishment hoped for.

ZYGOBOTHRIA GILVA Hartig.

Doubtfully colonized in 1906. Satisfactorily colonized in 1909. Not recovered from field. Establishment hoped for.

TACHINA LARVARUM L.

First received in 1905 and colonized in 1906. Much more satisfactorily in 1909. Not recovered. Establishment doubtful on account of hybridization with similar American species.

TACHINA JAPONICA Towns.

First received and poorly colonized in 1908. A better colony put out in 1910. Recovery doubtful on same account as above.

TRICHOLYGA GRANDIS Zett.

Doubtfully received and colonized in 1906. Satisfactorily colonized in 1909. Recovered from immediate vicinity of colony site in 1909. Not recovered in 1910, but establishment hoped for.

PARASETIGENA SEGREGATA Rond.

First received in 1907 and colonized in 1910. Not recovered. Establishment hoped for and expected.

BLEPHARIPA SCUTELLATA R. D.

First received in 1905. Colonized under very unsatisfactory conditions in 1907. Satisfactory colonization for first time in 1909. Recovered from immediate vicinity of colony site in 1910   Establishment confidently expected.

CROSSOCOSMIA spp.

First received in 1908 and colonized in 1910 under fairly satisfactory conditions. Not recovered. Establishment rather doubtful on account of unsatisfactory colony.

PARASITES OF THE PUPA.

MONODONTOMERUS ÆREUS Walk.

First received in 1906. Colonized in 1906. Recovered, generally distributed over considerable area, in winter of 1908-9. Firmly established and dispersing at a very rapid rate.

PIMPLA spp. (See Parasites of the brown-tail moth.)

CHALCIS OBSCURATA Walk.

First received in 1908. Colonized in 1908 and 1909, but not satisfactorily. Establishment doubtful on account of small size of colony.

CHALCIS FLAVIPES Panz.

First received in 1905. Colonized in 1908 and 1909 but in unsatisfactory numbers. Recovered from immediate vicinity of colony site in 1909. Not recovered in 1910. Establishment doubtful on account of small colony.

PREDACEOUS BEETLES.

CAOSOMA SYCOPHANTA L.

First received in 1906. Colonized same year. Recovered from immediate vicinity of colony site in 1907. Found generally distributed over limited area in 1909 Firmly established and increasing and dispersing rapidly.

PARASITES OF THE BROWN-TAIL MOTH.

PARASITES OF THE EGG.

TRICHOGRAMMA spp.

First received in 1906. Colonized in 1907; more satisfactorily in 1909. Recovered from immediate vicinity of colony site in 1909. Not recovered in 1910. Establishment probable, but the species of no importance as a parasite.

TELENOMUS PHALÆNARUM Nees.

First received in 1906 and colonized satisfactorily in 1907. No attempts toward recovery since made. Establishment hoped for. Not an important parasite.

PARASITES ATTACKING HIBERNATING CATERPILLARS.

PTEROMALUS EGREGIUS Först.

Received first in 1906. Colonized in large numbers that year and in 1907 but much more satisfactorily in 1908. Recovered from immediate vicinity of colony site in 1909. Not recovered in 1910. Found to be generally distributed over very extended territory in 1911.[1] Apparently well established.

APANTELES LACTEICOLOR Vier.

Received first in 1906. Colonized in 1907. Satisfactorily colonized in 1908 Recovered in immediate vicinity of colony site in 1909. Generally distributed over considerable area in 1910. Apparently firmly established.

METEORUS VERSICOLOR Wesm

First received in 1906. Colonized satisfactorily in 1908. Recovered in immediate vicinity of colony site in 1909. Generally distributed over limited area in 1910. Apparently firmly established.

ZYGOBOTHRIA NIDICOLA Towns.

First received in 1906. Colonized unsatisfactorily only in 1906-7, but notwithstanding was recovered in 1910 over a considerable territory. Apparently firmly established.

TACHINID PARASITES OF LARGER CATERPILLARS.

COMPSILURA CONCINNATA Meig. (See Gipsy-moth Parasites.)

TACHINA LARVARUM L. (See Gipsy-moth Parasites)

DEXODES NIGRIPES Fall.

First received in 1906 and colonized satisfactorily in 1906 and 1907. Still more satisfactorily colonized in 1909. Not recovered. Establishment hoped for.

EUDOROMYIA MAGNICORNIS Zett.

First received in 1906. Colonized in about the same numbers as *Zygobothria nidicola* in 1906 and 1907. Not recovered. Establishment doubtful on account of small size of colonies.

---

[1] Winter of 1910-11.

PALES PAVIDA Meig.
    Status same as that of Eudoromyia.
PAREXORISTA CHELONIÆ Rond
    Received first in 1906. Colonized very unsatisfactorily in 1907 and satisfactorily in
1908 Recovered in immediate vicinity of colony site in 1908 and in larger numbers
in 1909. Not recovered in 1910. Establishment very doubtful on account of hybridization with American race.

PARASITES OF THE PUPÆ.

PIMPLA EXAMINATOR Fab.
PIMPLA INSTIGATOR Fab.
    First received in 1906 and colonized in 1906 and 1907 unsatisfactorily. Establish-
ment doubtful. Of better promise as parasites on account of great similarity to
American species
MONODONTOMERUS ÆREUS Walk. (See Gipsy-moth Parasites)

The gross number of each of the various species which have been
colonized since the beginning of the work up to and including the
season of 1910 is given in the accompanying tabulated statement:

Hymenopterous Parasites		Tachinid Parasite.—Con.	
Schedius kuvanæ How....	1,061,111	Parasetigena segregata Rond..................	1,187
[3] Pteromalus egregius Först..................	354,300	[4] Crossocosmia sericariæ Corn....................	699
Anastatus bifasciatus Fonsc.	177,210	[2] Pales pavida Meig.......	476
Trichogramma spp.......	76,000	Tachina japonica Towns..	471
Apanteles fulvipes Hal....	57,700	[2] Zenillia libatrix Panz....	161
Apanteles lacteicolor Vier.	44,310	[2] Zygobothria nidicola Towns..................	109
Monodontomerus æreus Walk..................	15,325	Masicera silvatica Fa'l....	23
Telenomus phalænarum Nees..................	4,650	[2] Eudoromyia magnicornis Zett....................	
Meteorus versicolor Wesm.	3,113	[1] Unclassified tachinids (1906–07)..............	9,420
Pimpla spp..............	583		
Chalcis spp.............	338		
	1,794,640		68,343
Tachinid Parasites.		**Predatory beetles.**	
Carcelia gnava Meig.......	15,581	Calosoma sycophanta L...	17,742
[2] Tricholyga grandis Zett..	8,721	Carabus auratus L.........	478
[2] Zygobothria gilva Hartig.	7,502	Calosoma inquisitor L.....	262
[2] Compsilura concinnata Meig..................	6,777	Carabus arvensis Hbst....	108
[2] Dexodes nigripes Fall....	5,040	Carabus nemoralis Müll...	100
[2] Blepharipa scutellata R. D..................	5,109	Calosoma reticulatum Fab.	83
Parexorista cheloniæ Rond.	5,026	Carabus violaceus L.......	62
[2] Tachina larvarum L.....	2,036		18,835
		Total..................	1,881,818

[1] Including species marked (2)
[2] Species which are also included under "unclassified tachinids."
[3] Does not include progeny of 114,000 individuals liberated in 1908.
[4] Including also C. flavoscutellata.

THE DEVELOPMENTS OF THE YEAR 1910.

At the beginning of the year 1910 the statement was made that if the parasites maintained the rate of progress which was then indicated by the results of the recent field work, the year 1916 would see the triumphant conclusion of the experiment and the automatic control of the gipsy moth through parasitism. This prophecy was also dependent upon the measurable success of the importation work which was planned for 1910.

The importations of 1910 were disappointing, and did not result in the colonization of the few parasites which have not yet been liberated in America under satisfactory conditions. Neither has the progress of the parasites in the field been quite as satisfactory as was hoped and expected.

The failure of Schedius to demonstrate as clearly as might be wished its ability to survive the winter was the first unfavorable development in 1910. Recovery of *Apanteles fulvipes*, while not expected, was hoped for, and although its nonrecovery can not be considered as surely indicative of its inability to establish itself here, it is none the less disquieting. Discovery of the error in identity which had resulted in misapprehensions concerning the status of *Tricholyga grandis* was a serious blow to expectations concerning the future of this species. Most serious of all was the nonrecovery of the important brown-tail moth parasite, *Parexorista cheloniæ*, which was considered to be thoroughly well established at the close of the season of 1909. Similarly, the failure of Monodotomerus to increase in efficiency to the extent which was expected, was viewed with apprehension, as possibly indicative of what might result with others of the imported species.

To offset these several and various reverses was the unexpectedly satisfactory increase in abundance and dispersion of Calosoma. Anastatus did better than was expected in the matter of increase in numbers and in effectiveness and slightly better in dispersion. Blepharipa was recovered, when recovery was not expected so soon following its liberation, and Compsilura was considerably more abundant, and promised more efficient assistance than had been hoped for. Among the brown-tail moth parasites, Apanteles gave evidences of a more rapid increase and wider dispersion than was expected, and Meteorus was also unexpectedly abundant over a limited area, and later showed evidence of rapid dispersion. The recovery of *Zygobothria nidicola*, after its disappearance for two or three years, was the most satisfactory and unexpected of the favorable results of the season's field work until the recovery of Pteromalus in the fall and during the winter. Although this latter is not an important parasite, its nonestablishment was practically conceded, and the circumstances surrounding its recovery are considered to be highly gratifying and significant.

It is by no means easy to draw a balance which should fairly represent the status of the work as a whole in 1910 as compared with 1909, but after long consideration it was definitely decided that the present status of the parasites was perhaps less favorable to ultimate success than was the apparent status of the work one year before. Recognition of this fact had much to do with the formulation of the policy for the continuation of the work in 1911. It is hoped that by putting forth an especial effort the small amount of lost ground may be regained, and that by 1912 it will be possible to state with assurance that the progress hoped for at the close of 1909 has been more than equaled, and that the chances are still favorable to the successful outcome of the work and to the establishment of an efficient and automatic control of the gipsy moth by the year 1916.

It should be understood that the manuscript of this bulletin was completed in the first week in January, 1911, and that no more recent developments of the situation have been considered in it, except for an incidental mention of the progress of *Apanteles lacteicolor*.

O

INDEX TO BULLETIN NO. 91, BUREAU OF ENTOMOLOGY.

7362°—11——1

A

Stilpnotia salicis, host of Carcelia excisa.................................... 89
gnava...................................... 88
Compsilura concinnata.............................. 89
Parexorista cheloniæ.............................. 92
Tachina larvarum.............................. 90
Sugar-cane borer (Rhabdocnemis obscurus), quest of parasites for introduction
into Hawaiian Islands.. 36
food plant of Perkinsiella saccharicida............................. 35
plant-lice 35
leafhopper. (See Perkinsiella saccharicida.)
Sycamore, food plant of white-marked tussock moth (Hemerocampa leucostigma) 101
Syntomosphyrum esurus, parasite of brown-tail moth in America.... 139, 144, 147–149
gipsy moth in America................... 139
parasite of ladybird larvæ................................ 30
Tabanid flies, prey of Monedula carolina................................ 45
transmitters of trypanosomiasis of dromedaries................. 45
Tachina, biological character separating the species mella and larvarum...... 257
japonica, gross number colonized................................ 310
host of Chalcis fiskei........................... 240
parasite of gipsy moth in Japan, position in "sequence"... 121
introduction into United States... 227–228
reared at laboratory................. 88
status in United States in 1910...... 308
recorded host...................... 88
larvarum, gross number colonized................................ 310
parasite of brown-tail moth in   Europe,   position   in
"sequence".............. 136
introduction into United States 296
reared at laboratory........... 91
status in United States in 1910 308,
309
gipsy moth in Europe, position in "sequence".. 132
introduction into United States,
habits.......................... 225–227
reared at laboratory............... 88
recorded in literature............. 88
status in United States in 1910..... 308
recorded hosts... 90
latifrons, parasite of brown-tail moth, recorded in literature........ 91
recorded hosts... 92
"Tachina-like" parasite of gipsy moth, results of rearing work of 1910........ 142
Tachina mella and Tachina larvarum, biological differences................... 286
parasite of brown-tail moth in America................. 93, 145, 147–149
gipsy moth in America....................... 90, 139–140
white-marked tussock moth (Hemerocampa leucostigma)... 221–223
noctuarum, parasite of gipsy moth, recorded in literature........... 88
recorded hosts... 90
parasite of gipsy moth, relative abundance in Massachusetts and
Russia... 127
Tachinidæ, hyperparasites.. 207–213
Tachinid flies, rearing and colonization, large cages versus small cages........ 204–207
parasites of the brown-tail moth............................... 296–304
gipsy moth.. 202–236

ERRATA TO BULLETIN NO. 91, BUREAU OF ENTOMOLOGY.

Page 6, line 15 from bottom, indent *Limnerium (Anilastus) tricoloripes Vier.* 2 ems more.
Page 9, line 3, for *The Calosoma beetles* read *Calosoma sycophanta.*
Page 9, line 18, for *The gipsy moth* read *Different stages of the gipsy moth.*
Page 9, line 19, for *The brown-tail moth* read *Different stages of the brown-tail moth*
Page 10, line 18 from bottom, for *webs* read *web.*
Page 24, line 20, after *Muls* insert period.
Page 35, line 7, for *Dactylopius* read *Pseudococcus.*
Page 41, line 2 from bottom, for *trypanosome* read *protozoan.*
Page 42, line 6, for *trypanosomiasis* read *piroplasmosis.*
Page 88, line 11 from bottom, for *polychlorus* read *polychloros.*
Page 89, last line, for *Haloisus* read *Hyloicus.*
Page 90, line 13 from bottom, for *Lymantria* read *Porthetria.*
Page 91, left-hand column, omit *Digonichæta setipennis Fall., Digonichæta spinipennis Meig., Nemorilla sp.,* and *Nemorilla notabilis Meig.,* and close up the column.
Page 91, omit last three lines.
Page 92, omit lines 2, 3, 11, 12, and 13.
Page 93, lines 4, 5, and 6, omit *Blepharipeza leucophrys Wied., Sturmia discalis Coq.,* and *Exorista grisescans V. de Wulp.*
Page 93, line 4, left-hand column, insert *Phorocera saundersii Will.*
Page 93, line 6, left-hand column, insert *Exorista boarmiæ Coq.*
Page 93, right-hand column, move *Tachina mella Walk.* from line 6 to line 5.
Page 95, line 14 from bottom, for *pathogenetically* read *parthenogenetically.*
Page 116, line 10 from bottom, for *III* read *111.*
Page 145, line 2, for *Chalcis sp.* read *Chalcis compsiluræ Crawf.*
Page 145, lines 5 to 9, inclusive, read as follows: *has been examined and compared with other species of the genus by Mr. J. C. Crawford, who has found it to be distinct and has described it under the name Chalcis compsiluræ.*
Page 145, line 14 from bottom for *" Native Parasite of chrysorrhœa"* read *(?) Phorocera leucaniæ Coq.*
Plate VI (facing p. 156), line 2, for *below, at* read *Japanese variety, lower.*
Plate VI (facing p. 156), line 3, omit *enlarged* and *somewhat reduced.*
Plate VI (facing p. 156), line 4, omit *somewhat reduced.*
Plate VI (facing p. 156), line 5, after *center* insert *All slightly reduced.*
Plate VII (facing p. 160), line 2, for *another winter nest* read *cocoon in leaf.*
Plate VII (facing p. 160), line 3, omit *somewhat reduced.*
Plate VII (facing p. 160), line 4, for *the eggs hatching* read *torn open, showing eggs.*
Plate VII (facing p. 160), line 5, after *leaf* insert *All somewhat reduced.*
Page 170, line 5 from bottom, after *Staud* insert period.
Page 183, line 2 from bottom, for *thelyotoky* read *arrhenotoky.*
Page 193, line 2, *Apanteles fulvipes Hal.* should be printed in 10-point small capitals.
Page 198, line 11, the words *secondary parasites attacking apanteles* should begin with capitals.
Page 199, figure 34, the upper *b* in cut should be *d.*
Page 199, figure 34, the lower *c* in cut should be *a.*
Page 213, line 12, for *are* read *is.*
Page 229, line 5 from bottom, for *Liparis* read *Porthetria.*
Page 237, line 12 from bottom, for *Hoplectis* read *Itoplectis.*
Page 237, line 17 from bottom, after *Say* omit period.
Page 237, line 18 from bottom, for *Hoplectis* read *Itoplectis.*
Page 258, line 8, for *arrhenotokous* read *thelyotokous.*
Page 258, line 9, for *thelyotokous* read *arrhenotokous.*
Page 258, line 15, for *arrhenotokous* read *thelyotokous.*
Page 267, Table XI, column 2, line 6, after *Mesochorus pallipes,* for [2] read [1].
Page 267, Table XI, column 2, line 8, after *æreus,* for [2] read [1].
Page 285, line 18 from bottom, for *conspersæ* read *conspersa.*
Page 309, line 7, for *Cuosoma* read *Calosoma.*
Page 310, left-hand column, line 5, between *Anastatus* and *bifasciatus* insert space.
Page 310, right-hand column, line 11, for *silvatica* read *sylvatica.*

○

FIG 1 —ORANGE COVERED WITH SOOTY MOLD

FIG 2.—LEAF OF ORANGE COATED WITH SOOTY MOLD

U. S. DEPARTMENT OF AGRICULTURE,

BUREAU OF ENTOMOLOGY—BULLETIN No. 92.

L. O. HOWARD, Entomologist and Chief of Bureau.

WHITE FLIES INJURIOUS TO CITRUS IN FLORIDA.

BY

A. W. MORRILL, Ph. D.,

AND

E. A. BACK, Ph. D.

ISSUED JULY 12, 1911.

WASHINGTON:
GOVERNMENT PRINTING OFFICE.
1911.

LETTER OF TRANSMITTAL.

UNITED STATES DEPARTMENT OF AGRICULTURE,
BUREAU OF ENTOMOLOGY,
Washington, D. C., March 2, 1911.

SIR: I have the honor to transmit herewith, for publication as Bulletin 92 of the Bureau of Entomology, a manuscript prepared by Drs. A. W. Morrill and E. A. Back, dealing with the life history of the white flies injurious to citrus trees in Florida.

The investigation of the citrus white flies in Florida, under the general direction of the assistant chief of this bureau, Mr. C. L. Marlatt, was begun in 1906, and is now approaching completion. There has already been published a bulletin (No. 76) dealing fully with the general subject of fumigation with hydrocyanic-acid gas for the white fly. A circular (No. 111) has also been issued, giving brief directions for winter fumigation.

The present publication is a general account of the two species of white flies which are of special economic importance to the citrus grower in Florida. The publication includes the history of these insects in the United States, their distribution and food plants, and a very detailed study of the habits and life cycle of the two species. A great deal of painstaking and minute work has been done, and the information secured furnishes an accurate foundation for the developing of the best means of control.

Supplementing this publication, which deals largely with life history and habits, it is proposed to publish a bulletin on control by sprays, fungi, and other enemies, and to supplement or reissue in revised form the bulletin dealing with fumigation.

Respectfully,

L. O. HOWARD,
Entomologist and Chief of Bureau.

Hon. JAMES WILSON,
Secretary of Agriculture.

3

CONTENTS.

ILLUSTRATIONS.

WHITE FLIES INJURIOUS TO CITRUS IN FLORIDA.

INTRODUCTION.

The present bulletin includes the principal results of studies of the two species of white flies most destructive to Citrus in the United States, commonly known as the citrus white fly (*Aleyrodes citri* R. & H.) and the cloudy-winged white fly (*Aleyrodes nubifera* Berger). With these pests successful control measures must be based on a complete understanding of the insects themselves. On this account the study of the insects, their life history, seasonal history, habits, food-plant relationships, and related topics has occupied an unusually important position in the white-fly investigations.

The authors have concluded that unless natural enemies capable of controlling the two white-fly pests are existent and are secured, control measures will require permanent expert supervision for the most satisfactory and economical results—not supervision of work in individual citrus groves, but supervision aimed principally to properly correlate individual efforts and to take full advantage of favoring local conditions. For supervision of this nature, a good foundation of extensive and reliable studies of the insects is necessary. While the portion of the white-fly investigations herein reported is comparatively extensive, it is necessarily not exhaustive and in the course of time certain features of this work can undoubtedly be continued with profit as an aid to the future improvement of control measures.

The white-fly investigations now in progress were begun in July, 1906, by the senior author, who was in field charge up to the time of his resignation from the bureau in August, 1909, The junior author's connection with these investigations dated from June, 1907. The life-history studies of the first two years have been largely superseded by the more extensive work of the third year. Practically all of the data presented under the subjects of the life history and habits and the seasonal history of each species are based on studies by the junior author and were written by him. The remainder of the bulletin was written by the senior author.

SPECIES OF WHITE FLIES AFFECTING CITRUS.

Twelve species and one subspecies of the family Aleyrodidæ are known to breed upon citrus. The list of these insects, the authority for the original description, the recorded distribution, and the food-plant records are given in Table I:

TABLE I.—*Aleyrodidæ that breed upon citrus.*

Species.	Described by—	Occurrence.	Food plants other than citrus.
Aleyrodes citri (syn. aurantii).[1]	Riley and Howard.	North and South America, Asia, Japan.	See list, p. 29.
Aleyrodes floccosa	Maskell	Mexico, Jamaica	Gualacum officinale.
Aleyrodes floridensis	Quaintance	United States (Florida)	Persea gratissima (alligator pear), Psidium guajava (guava).
Aleyrodes giffardi	Kotinsky	Hawaii	None recorded.
Aleyrodes howardi	Quaintance	Cuba, United States (Florida).	Do.
Aleyrodes marlatti	do	Japan	Do.
Aleyrodes mori	do	United States (Florida)	Mulberry, sweet gum, etc.
Aleyrodes mori arizonensis	Cockerell	United States (Arizona)	None recorded.
Aleyrodes nubifera	Berger	United States (Florida, Louisiana), Cuba.	Ficus nitida.
Aleyrodes spinifera	Quaintance	Java	Rosa spp.
Aleyrodes struthanthi	Hempel	Brasil	Michelia flava, Loranthus (struthanthus) flexicaulis.
Aleyrodes vitrinellus [2]	Cockerell	Mexico	None recorded.
Paraleyrodes perseæ	Quaintance	United States (Florida)	Do.

[1] Mr. A. L. Quaintance, after careful comparison of material from Maskell's collection, evidently type material, with *A. citri*, concluded that Maskell's *aurantii* was the same as Riley and Howard's *citri*. Through the kindness of Mr. Quaintance the authors have had an opportunity to examine the material referred to and agree with him in considering *aurantii* a synonym of *citri*.

[2] There seems to be some doubt as to the identity of the food plant of this species, for in connection with the description the authority for it gives the following food plant record: "On the under side of leaves which appear to be those of orange."

Of the Aleyrodidæ referred to above, *A. citri*, *A. giffardi*, *A. howardi* (Pl. II, figs. 2, 4), and *A. nubifera* are known to be orange pests or capable of becoming orange pests. *A. floridensis*, *A. mori* (Pl. II, fig. 1), *A. mori arizonensis*, and *Paraleyrodes perseæ* (Pl. II, fig. 3) apparently are not likely to cause injury to citrus, while the remainder of those listed are doubtful in this respect.

Paraleyrodes perseæ is found in all sections of Florida and is frequently quite abundant, but in only one instance has it been known to cause blackening of the foliage of citrus trees. This was in the winter of 1906–7 and occurred in a pinery where in one section citrus nursery trees were being grown. In the course of two or three months after being first noticed the insects were reduced to the point of scarcity through parasitism by a new species of Encarsia, which Dr. L. O. Howard has described under the name of *Encarsia variegata*. Observations extending over three years indicate that this parasite will effectively control *P. perseæ* and that it is unlikely that this aleyrodid will ever cause noticeable injury under ordinary conditions. It is, however, possible that the appearance of a prolific hyperparasite of *Encarsia variegata* might seriously interfere with the present equilibrium in nature.

of Entomology, U. S. Dept. of Agriculture.

The woolly white fly (*Aleyrodes howardi* Quaintance (Pl. II, figs. 2, 4)) was first discovered in this country at Tampa, Fla., by the junior author in November, 1909. The insect appears to be of recent introduction, since the infested area has been under observation at intervals during the past three and a half years by the several men connected with these investigations.[1]

Of the four species known to be destructive to citrus, *Aleyrodes citri* and *A. nubifera* are included in the investigations herein reported.

THE CITRUS WHITE FLY.

(*Aleyrodes citri* R. & H.)

HISTORICAL REVIEW.

ORIGIN.

The origin of the citrus white fly is by circumstances quite definitely indicated to be Asiatic. The present known occurrence of it in Japan, China, and India will be referred to under the subject of distribution. The list of food plants, showing as it does the natural adaptations in this respect, indicates in itself that the fly is not native to either North or South America, but to Asia. Moreover, if the citrus white fly were a species native to the Gulf coast region of North America, or if it had been introduced before 1850, it would almost certainly have become a pest worthy of mention by Townend Glover in his reports on the orange insects of Florida published in the United States Agricultural reports for 1855 and 1858. According to these reports orange growing was very extensive in proportion to the population and very profitable in spite of the temporary check due to the freeze of 1835. The principal orange-growing district in Florida was, at the time of the reports of Glover, already mentioned, the northeastern section of the State, along the St. Johns River and at St. Augustine. Orange growing on a large scale gradually spread to the south and southwest, the center of production being correspondingly moved. To-day citrus fruits are generally grown in all the counties of the peninsula of Florida, yet, according to the authors' estimates, only about 40 per cent of the orange groves of the State are infested by *A. citri*.[2] These infestations in the different sections are almost without exception readily traceable to the ordinary sources of dissemination, with all the evidence strongly against the fly having been a native species infesting uncultivated food plants. The same may be said in regard to the occurrence of the citrus white fly in orange-growing regions in Mississippi, Alabama, Louisiana, and Texas.

[1] The Woolly White Fly, a New Enemy of the Florida Orange. Bulletin 64, Part VIII, Bureau of Entomology, U. S. Department of Agriculture, 1910.

[2] *A. nubifera* alone occurs in not more than 5 per cent of the groves. In 15 of the 40 per cent above mentioned both *A. citri* and *A. nubifera* occur.

EARLY HISTORY IN THE UNITED STATES.

Riley and Howard give the following account of the status of the citrus white fly previous to 1893:

For many years an important and interesting species of the type genus has been known to infest orange trees in Florida and in more northern greenhouses, and more recently the same form has appeared in injurious numbers in the orange groves of Louisiana. In the Florida Dispatch, new series, volume 11, November, 1885, this species received the name of *Aleyrodes citri* at the hands of Mr. Ashmead. The Florida Dispatch, however, is a local newspaper of no scientific pretensions, and the description accompanying the name was entirely insufficient to enable recognition aside from the food plant. We adopt the name in connection with a full description, not with a view of encouraging such mode of publication, which is not sanctioned by the canons of nomenclature formulated and generally accepted, but as a manuscript name, satisfactory in itself, the authority to be recognized for it being comparatively immaterial.

Our first acquaintance with the species was in June, 1878, when we found it occurring in profuse abundance on the leaves of the citrus trees in the orangery of this department. Some observations were made upon its life history during that summer, and all of its stages were observed. During the following years we observed it in Florida, and it was studied by two of our agents, Mr. H. G. Hubbard, at Crescent City, and the late Joseph Voyle, at Gainesville. The species was not treated in Mr. Hubbard's report on the insects affecting the orange, as we wished to give it a fuller consideration than could then have been given, and other duties prevented doing so in time. Moreover, at the time when Mr. Hubbard's report was prepared the insect had not become of especial economic importance.

Since that time many further notes have been made in Washington, and we have received the species from Pass Christian, Miss.; New Orleans, La.; Baton Rouge, La.; Raleigh, N. C.; and many Florida localities; and during the past year or two it has become so multiplied in parts of Louisiana and Florida as to deserve immediate attention.

The authors quoted above specifically recorded the occurrence of the white fly in Florida only at Gainesville (Alachua County), Crescent City (Putnam County),[1] and Manatee (Manatee County). Dr. H. J. Webber in 1897 (basing his statement on records in 1893 and 1894) referred to the occurrence of the white fly at the following additional points: Evinston (Alachua County), Ocala and Citra (Marion County), Ormond (Volusia County), Panasoffkee (Sumter County), Orlando (Orange County), Bartow (Polk County), and Fort Myers (Lee County). Prof. H. A. Gossard in 1903 mentioned only the following additional localities specifically: Tallahassee (Leon County), Lake City (Columbia County), Jacksonville (Duval County), and Candler (Marion County). In the same publication the following additional

[1] Examination of the specimens of white flies in the collection of the Bureau of Entomology, collected by Mr. H. G. Hubbard in 1895 and bearing the locality label "Crescent City," indicate that this record with little doubt refers to *Aleyrodes nubifera*. Circumstances known to the authors, but which need not be discussed here, show that with little doubt the citrus white fly was the species present at Crescent City before the freeze of the winter of 1894–5. The specimens collected by Mr. Hubbard probably came from the Hubbard grove at Haw Creek, several miles southeast of Crescent City.

counties were reported more or less infested without reference to definite localities: Baker, Jefferson, Leon, and Brevard.

Messrs. Riley and Howard and Dr. H. J. Webber advance no theories in regard to the original Florida infestations. Prof. Gossard, however, has the following to say in regard to the matter:

The fly seems to have been first known throughout the region comprised in Volusia, Marion, Lake, Alachua, and Orange Counties, from which I have little or no doubt it was transferred to the Manatee country and to local centers along the northern borders of the State.

According to reliable information received from Mr. M. S. Moreman, of Switzerland, Fla.; Mr. A. M. Terwilliger, of Mims, Fla., and Mr. T. V. Moore, of Miami, Fla., the citrus white fly appeared in the northern part of St. Johns County at a date which indicates that this section was one of the first or possibly the first to be infested in the State of Florida. Mr. Terwilliger informs us that he first observed the white fly at Fruit Cove on the St. Johns River in 1879 in a grove of large seedling trees owned by Col. McGill. The McGill grove adjoined the grove of the Rev. T. W. Moore, whose son, Mr. T. V. Moore, corroborates Mr. Terwilliger on the point of the occurrence of the white fly in this section prior to 1880. According to Mr. Moreman the white fly was known in the vicinity of Switzerland on the St. Johns River in 1882, and was first discovered in his own grove in 1888. The species concerned is with little doubt the citrus white fly, *A. citri*, for the authors and Mr. W. W. Yothers have been unable to find specimens of any other species at Switzerland or St. Augustine, the two points visited in the northern part of St. Johns County, or at Green Cove Springs, located a few miles below Switzerland on the west side of the St. Johns River in Clay County. These early reports of the citrus white fly in this section of the State are supported by the fact that the earliest collected specimens of this species in the collection of the Bureau of Entomology bear the date 1888 and the locality label "St. Nicholas," a point located in Duval County about 15 miles north of Fruit Cove.

Interesting information concerning the early history of white-fly infestations in Florida has been obtained from Messrs. Borland and Kells, citrus growers at Buckingham, Lee County, Fla., formerly of Citra, Marion County. According to these gentlemen, the presence at or near Panasoffkee, in Sumter County, Fla., of a small white insect which caused blackening of the foliage of orange trees became known among orange growers around Citra, at that time in the heart of the orange-growing district of Florida, in 1881 or 1882. The grove of Bishop Young, of Panasoffkee, was one of the first reported infested. It is believed that Bishop Young, after traveling in Asia (Palestine?), brought back with him plants which he set out, and in a year or two thereafter blackening of the foliage of near-by

citrus trees in association with a new insect pest first became noticeable. The white fly affecting citrus trees at Panasoffkee was exterminated by the freeze of 1894–1895 and, so far as the authors can
learn, has not reappeared. There seems to be at present no means of
determining whether the report given above refers to the citrus
white fly or to the cloudy-winged white fly.

Mr. A. J. Pettigrew, of Manatee, Fla., a reliable observer who has
been in the citrus nursery and orange-growing business in Manatee
County since 1884 and who has been familiar with the white fly since
its first discovery in that country, has furnished the authors with a
statement concerning the early history of the pest in that section of
Florida. According to Mr. Pettigrew, Messrs. C. H. Foster and F. N.
Horton each received from Washington, D. C., 6 tangerine trees in
1886 or 1887—as near as can be determined at this time, although
possibly earlier by a year or two. A year or two after the trees were
received and planted, the fly was noted by Mr. Pettigrew as abundant
on a rough lemon near one of these tangerines, and the following year
it was first noted as abundant in a seedling orange grove near by.
At Mr. Pettigrew's suggestion specimens were sent to the Department
of Agriculture at Washington and identified as a white fly. These
specimens were probably sent to Washington in 1891, for a letter
from Mr. Foster, dated January 8 of that year, was published in
Insect Life[1] with the reply. The oldest specimens of the citrus
white fly now in the collection of this bureau, which were collected
in Manatee County, Fla., bear the date of March 5, 1891, with "Man-
tee" as the locality record. These were probably sent in by Mr.
Foster in connection with later correspondence than that referred
to above.

Concerning the history of the citrus white fly in Louisiana, Prof.
H. A. Morgan in 1893 made the following statement:

> This pest, common from Baton Rouge to the Gulf, is known as the white fly. Orange
> growers claim that it has been recently introduced—that is, within the last ten years—
> and it is supposed to have come in upon plants brought to the New Orleans exposition
> in the year 1885. The present wide distribution of the white fly in the southeastern
> United States is due to the lack of restrictions, until very recently, against shipments
> of infested nursery stock and of privets and the Cape jessamine.

LITERATURE.

The citrus white fly was first given a valid scientific name and
adequately described by Riley and Howard in an article published
in Insect Life[2] in April, 1893. Following the account of the early
history heretofore quoted, these authors describe the different stages
of the insect in detail, give an account of the habits and life history,
and give records with discussion of results obtained by a correspondent

---

[1] Insect Life, vol. 4, p. 274.   [2] Id., vol. 5, no. 4, pp. 219–226, 1893.

in Manatee County, Fla., who had undertaken some cooperative experiments in spraying.

During the same year (1893) Prof. H. A. Morgan, then entomologist of the Louisiana Agricultural Experiment Station, gave an account of the citrus white fly in Louisiana in a bulletin of that station.[1]

The Division of Vegetable Physiology and Pathology of the United States Department of Agriculture began investigations of citrus diseases in Florida in 1893. These included investigations of the "sooty mold" resulting from white-fly infestation, and the first report on the subject was published by Swingle and Webber in 1896[2] and a more extended report by Dr. H. J. Webber in 1897.[3] Conclusions from a series of spraying experiments are included in this publication and many important observations are recorded, particularly in connection with the two most useful fungous enemies of the white fly which were discovered by Dr. Webber in the course of his work.

Prof. H. A. Gossard, then entomologist of the Florida Agricultural Experiment Station, published, in 1903,[4] an account of the white fly situation up to that time, with his conclusions from observations extending over several years.

In a volume entitled "Citrus Fruits," published in 1904 by Prof. H. H. Hume, four chapters are devoted to citrus insect pests and methods of control, the white fly receiving due attention.

Since the present investigations by the Bureau of Entomology have been in progress, Dr. Berger, entomologist of the Florida Experiment Station, has published two bulletins[5] which present a summary of white-fly conditions with recommendations for control, particularly with reference to the use of fungous enemies. In the later published of the two mentioned, the specific distinctions are pointed out and illustrated, separating from the common *A. citri* the form which Dr. Berger has named *A. nubifera*.

Messrs. P. H. Rolfs and H. S. Fawcett, in a bulletin issued in July, 1908,[6] discuss in a general way the use of fungous parasites of the white fly in Florida and give recommendations for the introduction of the three most common species. The most important contribution to our knowledge of the fungous parasites of the citrus white fly is contained in a paper by Prof. H. S. Fawcett, published in 1909.[7]

---

[1] The Orange and Other Citrus Fruits. By W. C. Stubbs and H. A. Morgan. Spec. Bul. La. Agr. Exp. Sta., pp. 71–73, 1893.

[2] The Principal Diseases of Citrus Fruits in Florida. By W. T. Swingle and H. J. Webber. Bul. 8, Division of Vegetable Physiology and Pathology, pp. 25–28, 1896.

[3] The Sooty Mold of the Orange and its Treatment. Bul. 13, Division of Vegetable Physiology and Pathology, U. S. Department of Agriculture, 1897.

[4] White Fly. Bul. 67, Fla. Agr. Exp. Sta., June, 1903.

[5] White Fly Conditions in 1906, the Use of Fungi. Bul. 88, Fla. Agr. Exp. Sta., January, 1907; White Fly Studies in 1908, Bul. 97, Fla. Agr. Exp. Sta., February, 1909.

[6] Bul. 94, Fla. Agr. Exp. Sta., July, 1908.

[7] Special Studies No. 1, University of State of Florida, 1909.

In Louisiana the demand for information concerning the citrus white fly has resulted in a publication on this subject by Mr. A. H. Rosenfeld in 1907.[1] The discovery of the white fly in California in the same year led to the publication, by Prof. C. W. Woodworth, of a circular of general information,[2] and of a second circular[3] dealing with the methods of eradication that were being employed in that State. A very complete account of the white-fly infestation in California was given by Mr. C. L. Marlatt, assistant entomologist of the Bureau of Entomology, before the Entomological Society of Washington.[4]

The foregoing paragraphs refer to the principal publications in which the citrus white fly is treated, exclusive of short papers in horticultural periodicals, press bulletins, experiment station reports, and transactions of the Florida State Horticultural Society. Numerous press bulletins have been issued by the Florida State Experiment Station dealing with several phases of white-fly control and written from time to time as the occasion demanded by Prof. Gossard, Dr. Sellards, Dr. Berger, and Prof. Fawcett.

Reviews of the white-fly situation for the year, with notes on new observations, have been included in their annual reports by each of the first three named, who have served successively as entomologist at the Florida Experiment Station. Many important papers and discussions on the white fly have been published in the Transactions of the State Horticultural Society, but for the most part these have been incorporated or the ground covered more fully in the regular bulletins referred to.

Taken as a whole, the literature on the citrus white fly is quite extensive, giving a fairly good idea of the status of the white fly and progress in methods of control from year to year since the publication of the paper by Riley and Howard referred to in the opening paragraph.

The description of the different stages and the account of the life history and habits of the citrus white fly by Riley and Howard have been followed quite closely by subsequent writers, few additional records having been made up to the beginning of the present investigations. Records of food plants, miscellaneous life-history records, general results of field experiments, and conclusions from general observations on the efficiency of spraying, fumigating, and natural control by fungous diseases have been published by Messrs. H. J. Webber, H. A. Gossard, E. H. Sellards, E. W. Berger, and H. S. Fawcett. Comparatively little real data has been published so far in

---

[1] Circular 18, State Crop Pest Commission of Louisiana, 1907.
[2] Circular 30, California Agricultural Experiment Station, 1907.
[3] Circular 32, California Agricultural Experiment Station, 1907.
[4] Proceedings of the Entomological Society of Washington, vol. 9, pp. 121-123, 1908.

connection with experimental work with the white fly. A review of all the literature to date shows that data have been published on the effect of kerosene emulsion on white-fly eggs, by Riley and Howard; on the subject of effects of cold upon white-fly larvæ and pupæ, by Prof. Gossard; on the percentage of trees infected by the spore-spraying method of introducing the fungous parasites, and on the amount of honeydew secreted by the larvæ of the insect, by Dr. E. W. Berger; upon subjects related to fumigation,[1] by the senior author of the present bulletin; and on laboratory experiments with the fungous parasites, by Prof. H. S. Fawcett. Aside from the above, practically no data have been heretofore published.

INJURY.

NATURE OF INJURY.

The direct injury by the citrus white fly may be included under two main heads: (1) Injury by removal of sap from foliage, and (2) injury from fungous growth known as sooty mold (Meliola), which develops upon foliage and fruit on the excretions of the insects.

The direct injury is principally included as loss in value of trees, extra expenses of maintenance, and losses from scale insects and diseases, which more seriously affect white-fly infested trees.

LOSS OF SAP.

The amount of sap extracted by the insects is not generally considered an item of great importance compared with the injury from sooty mold. While the extraction of sap by itself probably would not cause sufficient injury to make the white fly rank as an important citrus pest, it is doubtless of considerable importance when combined with the lowered assimilative powers of the foliage due to the sooty mold. As mentioned more in detail under the subject of feeding habits, it has been estimated that the loss of sap per day amounts to about one-half of a pound for 1,000,000 larvæ and pupæ.

SOOTY MOLD.

Sooty mold is the principal evidence of white-fly injury, and is the most important element of damage, affecting both the foliage and fruit. (See Pl. I, frontispiece.) No special attention has been given by the authors to its botanical aspects, but the following notes concerning it are taken mainly from Dr. H. J. Webber's report on this subject:[2]

---

[1] Fumigation for the Citrus White Fly as adapted to Florida Conditions. Bulletin 76, Bureau of Entomology, U. S. Department of Agriculture, Oct. 31, 1908.

[2] Bulletin 13, Division of Vegetable Physiology and Pathology, U. S. Department of Agriculture, pp. 5–11, 1897.

The sooty-mold fungus is a species of the genus Meliola [1] of the order Pyrenomycetes. Dr. Webber states that in Florida and Louisiana it is quite generally known as smut or black smut, but as the fungus concerned is not a smut fungus these terms are erroneous, and their use should be discontinued. When abundant on leaves and fruit of citrus, this fungus forms a dark-brown or black membranous coating composed of densely interwoven branched mycelial filaments. At first this coating covers only limited spots or is not thick enough to form a distinct membrane, but later, if the honeydew-secreting insects are abundant, the coating becomes thick enough to be entirely removed from the leaf and torn like paper. (Pl. III, figs. 1, 2.)

Frequently the fungus membrane becomes detached at some point and is caught by the wind and large fragments torn off. These fungus fragments are found scattered about in badly infested groves in the fall, being especially noticeable during the winter after a high wind or after the trees have been sprayed.

Dr. Webber recognized several forms of reproductive agents, which are easily distributed by various means, but principally by winds. The fungus is entirely saprophytic in so far as known, deriving its nourishment from the honeydew secreted by certain insects. As such honeydew falls mostly on the upper surface of the leaves and on the upper half or stem end of the fruit, the sooty mold develops most densely in these places, but it is usually present to a greater or less extent on the lower surface of the leaves, sometimes developing in tufts on drops of honeydew which diseased insects fail to expel in a normal manner. Sooty mold also develops on the twigs and in some cases on the sides of buildings when heavily infested trees are growing near by.

Seasonal history of sooty mold.—The sooty mold resulting from the attacks of the citrus white fly is most abundant late in the season. Very little sooty mold develops during the winter months, while the films of blackish mycelium gradually become removed from the leaves by winds and rains and much is knocked off in picking the fruit, in spraying, pruning, fumigating, etc. The thicker the coating of sooty mold, the more readily and thoroughly it is removed. By the time of the appearance of the new spring growth the greater part of the sooty mold on the old leaves has disappeared and from this time to the 1st of May there is very little, if any, evidence of a new growth of this fungus. Slight blackening of spring growth has been noted as far north as Island Grove in Alachua County, Fla., as early as May 20, the average number of live larvæ and pupæ per leaf being estimated as about 50, not including old leaves which were practically uninfested. By June 20, leaves from McIntosh, in the same county,

---

[1] Generally referred to *M. camelliæ* (Catt.) Sacc., but perhaps including more than one species.

SOOTY MOLD.

Fig. 1.—Sooty mold on orange leaf following white-fly attack; broken and falling from leaf.
Fig. 2.—Sooty mold on cinnamon tree following attacks by cinnamon scale. (Original.)

with an average of about 11 live larvæ and pupa cases[1] per leaf, were slightly blackened. In general, heavy coats of sooty mold on leaves are common in Florida by the 1st of June in groves heavily infested by the citrus white fly.

Effect of sooty mold on leaf functions.—Dr. Webber has discussed the effect of sooty mold on leaf functions in the report already referred to, and as there is nothing to add at this time, the following paragraph (pp. 10–11) is quoted:

When it is remembered that various investigations have shown that the process of phytosyntax[2] is almost entirely checked in a plant placed in the back part of a living room, opposite a window, where the light is fairly bright, but diffused, it can readily be judged that the effect of the dark, compact mycelial membrane of the sooty mold covering the leaves would be to almost wholly check the process of phytosyntax in the orange tree. Quite bright or direct sunlight is necessary for the best results. The injurious effects of sooty mold on the phytosyntax was clearly demonstrated by Busgen. He removed the fungus membrane from a small portion of a leaf and exposed the leaf to the sun. In the evening, after a sunny day, the leaf was plucked and the chlorophyll extracted with alcohol. After this leaf was treated with iodine, the parts from which the membrane had been removed in every case stained a dense blue, indicating the formation of an abundance of starch, while the surrounding portions of the leaf, which were protected from the sun by the fungus membrane, remained entirely uncolored, showing that no starch was formed. The stomata, or breathing pores, are also to some extent closed by the sooty mold, and in this way the passage of gas is more or less hindered. In the orange leaf, however, the stomata are confined to the lower surface, where generally there is but little sooty mold. In plants where the stomata are on the upper surface of the leaf also, the damage resulting from the obstruction of the passage of gases would probably be considerably greater.

EXTENT OF INJURY.

In the following discussion the statements concerning injury and the estimates of the extent of this injury by the citrus white fly refer to groves in which the fly has become well established and in which no remedial measures have been practiced.

INJURY TO FRUIT.

Unless otherwise stated, oranges and tangerines are referred to. These constitute more than 88 per cent of the citrus fruit crop of Florida. The total injury to grapefruit by the citrus white fly is rarely over 15 per cent and is frequently inappreciable.

Ripening retarded.—Ripening of fruit on heavily infested citrus trees is greatly retarded, and in case of the formation of a very heavy coating of sooty mold on the upper half of the orange the rind underneath it may remain green indefinitely while the lower half of the fruit is

---

[1] Some of the first generation had matured, but are properly included with the insects responsible for the sooty mold present.

[2] " Phytosyntax " refers to the process of the formation of complex carbon compounds out of simple ones under the influence of light; "photosynthesis" is a more common term for this process of assimilation.

well colored. The retardation of ripening, delaying as it does in some cases the time when the fruit is marketable and materially increasing the percentage of culls, causes injury which is very conservatively estimated to range from 2 to 5 per cent of the value of the crop. The injury to grapefruit in retardation of ripening by the citrus white fly is much less, varying from none at all to 2 or 3 per cent.

Number and size.—The greatest injury by the white fly is in the reduction of the salable crop of fruit. Dr. Webber on this point makes the following statement:[1]

The effect of the sooty mold on the orange is very noticeable, the growth being usually greatly retarded and the blooming and fruiting light. In serious cases growth is frequently entirely checked, and blooming and fruiting wholly suppressed until relief is obtained.

Prof. Gossard has estimated [2] that during a six-year period the reduction in yield due to the citrus white fly is from 25 to 40 per cent.

Replies to a circular letter of inquiry addressed to orange growers and the observation of the authors in Florida indicate that the reduction in yield due to the citrus white fly amounts to 50 per cent, on the average, when no artificial methods of control are practiced.

From information received from many growers and from personal observation, the authors would estimate that with continued good care and with the additional fertilizer usually given infested trees the -reduction in yield in different groves in a series of years amounts to an average between 20 and 50 per cent.

The decrease in yield due to white-fly infestation ordinarily consists of a decrease in the actual number of fruit produced and also in the packing size. From information obtained it seems a conservative estimate to consider that oranges and tangerines are reduced either one or two packing sizes as a result of white-fly attack. For each packing size, the number of reduced fruit remaining the same, the reduction in the crop would average about 12.5 per cent.

Expense of cleaning.—Fruit noticeably affected with sooty mold requires cleaning before marketing. One of the most economical machines for washing fruit used in Florida is a California washer used by Mr. F. D. Waite, of Palmetto, and Mr. F. L. Wills, of Sutherland. The cost of washing with these machines ranges from 1.4 to 2.5 cents per box. The cost of cleaning with the simplest machines is about 5 cents per box. Mr. E. H. Walker, of Orlando, Fla., estimates the cost of hand cleaning oranges at 10 cents per box as a minimum and 7 cents a box for cleaning grapefruit. In consideration of the foregoing it is estimated that the range in cost of cleaning the sooty mold from fruit to be from 1 to 10 per cent of the value of the crop.

Shipping and keeping quality.—The sooty mold produced by the white fly and other citrus pests does not, so far as known, affect the

---

[1] Loc. cit., p. 9.   [2] Bul. 67, Fla. Agr. Exp. Sta., p. 617.

shipping quality of the fruit directly, but the processes of cleaning have been proved to be of considerable importance in this respect. The subject of the deterioration in shipping quality of citrus fruits has been thoroughly investigated in California by agents of the Bureau of Plant Industry under the direction of Mr. G. H. Powell.[1] Their report shows in a conclusive manner that the amount of decay in shipment is very materially increased by brushing or washing the fruit to remove the sooty mold. Table II, arranged from data published in the report referred to, shows the effect of dry brushing and washing fruit on the percentage of decay.

TABLE II.—*Effect, on decay, of cleaning sooty mold from fruit.*

Record No.	Unbrushed fruit apparently sound.	Dry brushed fruit apparently sound.	Washed fruit apparently sound.
	Per cent.	*Per cent.*	*Per cent.*
1.....	2.7	6.6	17.8
2.....	1.9	4.2	10.0
3.....	....	1.8	2.6

It will be observed that dry brushing increased the amount of decay to about two and one-half times the decay in the unbrushed in record No. 1, and to about two and one-fifth times in record No. 2. Washing increased the amount of decay to about six and two-thirds times in record No. 1, and to about five and one-fifth times in record No. 2.

The injury from cleaning the fruit is due to the increased opportunities for infection with spores of the blue mold and to mechanical injuries in the process of cleaning. The chances of decay are still further increased whenever the fruit is not thoroughly dried before packing. Washing in constantly running water or by running the fruits through brushes with water constantly sprayed over them is considered much less objectionable than the ordinary systems.

Flavor.—The attack of the white fly is generally supposed to affect the quality of the fruit in a marked degree. Dr. Webber and Prof. Gossard describe the flavor as insipid as a result of heavy infestations. The latter presents the results of chemical analyses of samples of the fruit of tangerine trees in two adjoining groves. In one grove the white fly was completely controlled by spraying; in the other the fly was unchecked. The analyses showed that there was, in the samples from the latter grove, 15 per cent less reducing sugar, 15 per cent less sugar dextrose, and 5 per cent less citric acid. While oranges and tangerines are frequently much affected in flavor, thoroughly blackened groves in many cases produce as well flavored fruit as can

---

[1] Bulletin 123, Bureau of Plant Industry, U. S. Department of Agriculture.

be found in the market. When trees are supplied with as much fertilizer as they can use to advantage the white fly does not ordinarily affect the flavor of the fruit to such a noticeable extent as is commonly believed. It is suspected that a well-grounded prejudice against the white fly rather than a discriminating taste is responsible for a large part of the supposed effect on the, flavor of the fruit in infested groves.

Increased injury from scale insects and from plant diseases.—The number of culls is in some cases very much increased by diseases and insect pests which thrive after the trees have been weakened by the white fly. There are no data available showing the usual increase in percentage of fruit injured by scales and by diseases of the trees as a result of white-fly infestation, but this is a consequence observed by many citrus growers and is properly considered a factor of white-fly injury. As such it is conservatively estimated to vary from 1 to 5 per cent in groves thoroughly infested, although an instance of a valuable crop being completely ruined by secondary scale attack has come under the authors' observation.

Market value.—Imperfections in fruit rind due to diseases and insect pests as followers of the white fly and to failure of fruit rind to color up normally, in addition to the direct effect on the size of the marketable crop as heretofore discussed, usually lower the average grade even after the fruit is cleaned by the most approved methods. A few growers claim that after being cleaned their oranges and tangerines bring as good prices as any, and leaving out of consideration instances where it is claimed that most or all of the fruit is rendered absolutely unsalable under any conditions, we may conservatively estimate the depreciation in market value to range from none at all to 10 per cent.

Sooty-mold-blackened oranges shipped without cleaning have a market value ordinarily from 25 to 50 cents less per box than the same fruit would have cleaned.[1] Certain Florida brands of oranges well advertised, carefully graded, and packed, would fail to bring within a dollar a box of their average value if they appeared on the market blackened by sooty mold.

Losses to growers estimated on basis of prices paid by orange buyers.— The authors are indebted to Mr. E. H. Walker, of Orlando, for the information that during the season of 1907–8 orange buyers in Florida paid from $0.75 to $1.45 for oranges free from white-fly effects, and from $0.50 to $1 per box for fruit blackened by white fly; during the season of 1908–9 the price paid for clean fruit varied from $0.60 to $1 per box, and from $0.50 to $0.75 for fruit blackened by sooty mold. The loss to the growers is not entirely represented by these

---

[1] Statement based on information from Mr. E. H. Walker, Orlando, Fla.

figures, since, according to Mr. Walker, the best prices were not paid for sooty-mold-blackened fruit until late in the season after the clean fruit had nearly all been shipped or disposed of. Clean fruit at this time would have been proportionally more profitable.

INJURY TO TREES.

Weakening of vitality.—It is doubtful if the white fly is ever the direct cause of the killing of trees, limbs, or twigs in well-fertilized groves. It does, however, seriously stunt the growth of all heavily infested trees, and may temporarily entirely check the growth of young trees. Its greatest effect on the vitality of the tree is an indirect one. Infestation by the white fly appears to weaken the resistance of orange and tangerine trees to foot rot, die back, melanose, wither tip, and drought, and favors the multiplication of the purple and long scales which are second to the citrus white fly as citrus pests in the Gulf coast regions.

Depreciation in value.—The selling values of citrus groves are greatly reduced by white-fly infestation, and citrus nurseries have their territory for sales much restricted and values reduced. Concerning the reduction in value of groves of bearing trees one of the most experienced dealers in orange groves in the State estimates that it is in general about one-third. For years California has been closed to Florida nurserymen as a field for the sale of citrus nursery trees, and a similar quarantine regulation has recently gone into effect in Arizona. In Florida and in citrus-growing sections of other Gulf coast States a guaranty of freedom from white fly is generally required, especially when the purchaser contemplates planting a more or less isolated grove.

SUMMARY OF LOSSES.

The estimates in the foregoing pages refer to ordinary losses where the white fly is unchecked by natural enemies or by artificial methods of control and not to exceptional or occasional losses. These estimates, as applying to the fruit, are summarized in Table III.

TABLE III.—*Estimates of losses to orange crops by white fly in uncontrolled condition.*

	Maximum.	Minimum.	Mean.
	Per cent.	Per cent.	Per cent.
Ripening retarded	5	2	3½
Number and size of fruits	50	20	35
Cost of cleaning	10	1	5½
Deterioration in shipping quality	6	2	4
Indirect injury: Increased scale and disease effects on fruit	5	1	3
Loss in market value	10	0	5
Total	86	26	56

The mean of the total percentage of estimated loss is considered by the authors to represent about the normal loss which the citrus white fly is capable of causing in orange groves. It is estimated that the condition is reduced to about 45 per cent loss in the average infested grove as a result of net profits from spraying with contact insecticides and of the natural efficiency of fungous diseases.

From extensive records obtained in the course of their investigations the writers estimate that the citrus white fly infests at present 45 per cent of the citrus groves in Florida. Of this, 5 per cent is a sufficient allowance to represent the groves so recently infested that normal abundance of the pest has not been reached. An injury of 45 per cent in 40 per cent of the groves is equal to about 18 per cent of the entire value of the crop as it presumably would have been if the white fly were not present.

The latest Florida citrus crop concerning which statistics are available is that of 1907-8.[1] The orange crop for that season is valued at $3,835,000. With an estimated total loss of about 15 per cent this represents 85 per cent of the value of the crop if not affected by the white fly. Accordingly, the estimated loss in Florida is calculated to have been about $680,000 for oranges and similarly on the basis of 10 per cent loss to grapefruit on a valuation of $469,700, the percentage of infested groves the same as in the case of the orange groves, a loss of $16,700 is estimated, making the total loss in valuation of fruit about $696,700 for the crop of 1907-8. The crop of 1908-9 was doubtless affected to the extent of $750,000 by the citrus white fly.

At present the spread of the fly into uninfested groves is undoubtedly faster than at the rate of 5 per cent new infestations per year. Even on this basis, however, the annual increase in depreciation in the value of Florida citrus groves due to white-fly infestation is more than $200,000 per year.[2] In addition, the citrus nursery business in Florida is affected to an extent hard to estimate, but which would be only nominally represented by $50,000 per year.

Figures are not available which would allow approximate estimates to be made of the damage by the citrus white fly in the Gulf coast citrus-growing sections outside of Florida, but the widespread occurrence of the white fly in those States indicates that the losses are heavy.

INCREASED COST OF MAINTENANCE.

The items of expense of maintenance principally affected by the white fly are fertilization, spraying, and fumigation. In Florida

---

[1] Tenth Biennial Report of the Commissioner of Agriculture of the State of Florida.

[2] This is not shown by actual depreciation, for the number of groves coming into bearing for the first time each year more than covers the loss.

ordinarily the expense of the fertilizer necessary to maintain orange trees in good productive condition varies from 10 to 20 cents per box of fruit produced. The wide range given is largely due to differences in soil conditions. Mr. E. O. Painter, in response to an inquiry on the subject, writes that citrus trees infested with white fly in his opinion require at least 15 per cent increase in fertilizer for best results under the circumstances. On the basis of cost of fertilizer amounting to 10 to 20 cents per box and an increase of 15 per cent due to white fly infestation, the extra expense which may be charged as white fly injury amounts to 1.5 to 3 cents per box.

Cost of control measures properly chargeable to increased cost of maintenance.—Estimates based on the experience of the writers in fumigating and in spraying for the white fly give the range in expense of the former method of control as 5.5 to 14 cents per box of oranges produced, and of the latter method 12.5 to 20 cents per box. These estimates refer to thorough control, with the result that production is entirely unaffected by the white fly. The minimum estimate on the expense of fumigation refers to groves so located that the migrations of adults from outside groves does not make treatment necessary more than once in two years. The maximum estimate refers to conditions where treatment every year is required to prevent loss. Increase in production, due to destruction of scale-insect pests, is not taken into consideration. In the estimates of expense of control by spraying the minimum estimate refers to cases where three applications of insecticide per year have resulted in satisfactory control. This result can be attained only after the insect has been brought into complete subjection, such as referred to in the introductory paragraph of the subject of artificial control. Insecticides costing more than 1½ cents per gallon when mixed ready for application have not been taken into consideration.

DISTRIBUTION.

As has been shown in the historical review, the citrus white-fly at present is generally distributed in North America. In the northern part of the United States it occurs in greenhouses, and in the southern part, and in limited districts in California, it occurs on citrus, China trees, privet, cape jessamines, and other food plants. In the present publication we are concerned only with the distribution of the species in the citrus fruit-growing regions of the United States.

IN THE UNITED STATES.

According to the statistics of the Florida commissioner of agriculture, in 1905 there were 17 counties in the State reporting more than 5,000 bearing citrus fruit trees. In all but two of these, Dade and St. Lucie, the citrus white fly (*Aleyrodes citri*) occurs to a greater or less extent. (See fig. 1.) The 17 counties referred to, arranged in order of the number of bearing citrus trees, is as follows: Orange, Lake, Volusia, Polk, Putnam, Brevard, Hillsboro, De Soto, Lee, Manatee, Dade, Marion, St. Lucie, Osceola, Sumter, St. John, and Alachua. Palm Beach as well as Dade and Monroe Counties are infested with the cloudy-winged white fly, as hereafter noted, but so far as known the citrus white fly does not occur there. In order of the percentage of groves infested the foregoing counties which are known to be infested would be arranged about in the following order, so far as our observations and records show: Marion, Alachua, St. John, Manatee, Orange, Lee, Volusia, Polk, Putnam, Lake, Hillsboro, Sumter, De Soto, Osceola, and Brevard. If the groves infested by the cloudy-winged white-fly only were also taken into consideration, Hillsboro and Lake Counties would be transposed in the list, as would Osceola and Brevard, but aside from this there would be no change. The arrangement is only approximate, being based on observations made by the various men connected with the white-fly investigations upon information and samples of infested leaves received from correspondents and upon

Fig. 1.—Map showing distribution of the citrus white fly (*Aleyrodes citri*) in Florida. (Original.)

nearly 250 replies received in response to circular letters sent out in the spring of 1907.

At the present time the writers estimate that throughout the State of Florida about 40 per cent of the citrus groves are infested by the citrus white fly, and that an additional 5 (or 10) per cent are infested by the cloudy-winged white fly alone.

The citrus white fly occurs in nearly all the larger towns in northern Florida, infesting the various food plants which are grown as ornamentals as well as the citrus fruit trees which are grown to a limited extent. The insect is of common occurrence, principally on China trees, cape jessamines, and on privet and hedges of *Citrus trifoliata* in South Carolina and in southern Georgia, Alabama, Mississippi, Louisiana, and Texas. In the last two States citrus fruits are being grown quite extensively, and a large percentage of the citrus-growing localities are infested.

Aside from the Gulf coast States, citrus fruits in the United States are grown only in California and Arizona. The citrus white fly does not occur in Arizona. In California the pest was first discovered in May, 1907. Mr. C. L. Marlatt has given the following account of the distribution of the white fly in that State in 1907:[1]

Marysville is situated a few miles north of Sacramento, and the first infestation seemed limited to this town, but toward the end of the summer the white fly was discovered well established at Oroville, in Butte County, some 26 miles to the north of Marysville. The Marysville infestation was confined to the town and to yard trees or small garden orchards. Oroville lies in a considerable orange district, and the white fly had been carried from the town into several of the adjacent orchards and had become rather widely scattered. Shortly after the discovery of the fly at Marysville it was found also to have established itself locally near Bakersfield,[2] in the southern end of the San Joaquin Valley, and separated only by a mountain range from the citrus districts of southern California.

IN FOREIGN COUNTRIES.

For years the citrus white fly has been supposed to be an introduced species, and much interest has been attached to its occurrence elsewhere than in North America. Prof. H. A. Gossard in 1903 stated that Mr. Alexander Craw, of the California State commission of horticulture, had received this species on plants from Chile, where it was reported to be a great pest. Mr. G. W. Kirkaldy, in his catalogue of the Aleyrodidæ, in 1907, gives "Mexico, Brazil, and Chile (?)" as the known habitats of the citrus white fly outside of the United States. The writers are informed by Prof. A. L. Quaintance that he was told in person by the late Prof. Rivera, of Santiago, Chile, that the citrus white fly was abundant in that country. Prof. Carlos Camacho, chief vegetable pathologist at Santiago, Chile, is also, according to Prof. Quaintance, authority for the statement that it occurs there.

The Bureau of Entomology received, in 1906, specimens of an aleyrodid on orange leaves from China which Prof. Quaintance determined as *Aleyrodes citri*,[3] and still more recently it received, through Mr. August Mayer, in charge of plant-introduction garden, and through the California state commission of horticulture, specimens of orange leaves infested with what Prof. Quaintance has identified as this species from different parts of China and Japan.

The occurrence of the citrus white fly in India (northwestern Himalayas) has recently been established by Prof. Quaintance, who has compared Maskell's *A. aurantii*, collected in the region mentioned

---

[1] Proceedings of the Entomological Society of Washington, vol. 9, pp. 121–122, 1908.

[2] Specimens of the species present at Bakersfield were examined by the senior author at the California State Insectary at Sacramento and found to be the cloudy-winged white fly (*A. nubifera*).

[3] Proceedings of the Entomological Society of Washington, vol. 8, Nos. 3–4, p. 107.

above, with *A. citri*, and failing to find any differences in the egg and pupal stages found it necessary to regard the name given by Maskell as a synonym of that given by Riley and Howard.

The citrus white fly does not occur in Cuba, so far as known, although it is not unlikely to be found there, since there have been heavy shipments of nursery stock from infested citrus nurseries in Florida to that country during the last few years.

FOOD PLANTS.

AUTHENTIC AND QUESTIONABLE RECORDS.

The separation as distinct species of two forms formerly considered as belonging to the species *Aleyrodes citri* makes it necessary that all of the reported food plants of the citrus white fly be verified. Nearly 60 species of the genus Aleyrodes have been recorded for North America. Of these less than 20 have been described in the first larval stage in a manner which distinguishes them, although when carefully studied this stage has been found to have striking specific characters. The second and third larval stages rarely possess distinguishing characters. The fourth or pupal stage, or the empty pupa case, is used as the basis of specific descriptions in the Aleyrodidæ, but even in this stage a careful microscopic examination is usually necessary to positively determine the species. Good specific distinctions in the adult stage have been found only in a few species, and even those entomologists who have made a specialty of the Aleyrodidæ do not attempt to distinguish the different species in this stage. It is obvious, therefore, that a list of food plants should properly include only those verified by entomologists, with determinations of the species made since the status of the two most abundant citrus-infesting species of Aleyrodes has been fully recognized. Dr. E. W. Berger has recently arranged the full list of food plants and reported food plants in a graphic manner, separating the list into two classes according to the degree of preference, and each class is subdivided into native and introduced species. This method of grouping the food plants is here adopted (see Table IV) with the transposition of the lilac and coffee from class II to class I and omitting certain reported food plants in order to restrict the list to include only positive records, leaving the others for a separate discussion. Dr. Berger has recently discovered the citrus white fly on wild olive, and has also verified Prof. Gossard's report of the citrus white fly on *Viburnum nudum*. Both of these food plants, together with the green ash, probably will eventually be found to be subject to heavy infestation and be placed in class I.

TABLE IV.—*Definitely known food plants of the citrus white fly (Aleyrodes citri).*

CLASS I. PREFERRED.

Introduced:

1. Citrus (all species cultivated in America).
2. China tree (*Melia azedarach*).
3. Umbrella China tree (*Melia azedarach umbraculifera*).
4. Cape jessamine (*Gardenia jasminoides*).
5. Privets (*Ligustrum* spp.).
6. Japan persimmon (*Diospyros kaki*).
7. Lilac (*Syringa* sp.).
8. Coffee (*Coffea arabica*).

Native:

9. Prickly ash (*Xanthoxylum clava-herculis*).
10. Wild persimmon (*Diospyros virginiana*).

CLASS II. OCCASIONALLY INFESTED.

Introduced:

11. Allamanda (*Allamanda neriifolia*).
12. Cultivated pear (*Pyrus* spp.).
13. Banana shrub (*Magnolia fuscatum*).
14. Pomegranate (*Punica granatum*).

Native:

15. Smilax (*Smilax* sp.).
16. Cherry laurel (*Prunus laurocerasus*).
17. Wild olive or devilwood (*Osmanthus americanus*).
18. Viburnum (*Viburnum nudum*).
19. Green ash (*Fraxinus lanceolata*).

In addition to those in the foregoing list [1] there are several plants reported as food plants of the citrus white fly which, while probably true food plants, can not consistently be included in the recognized list until the observations have been repeated and the infesting species positively identified. In some instances where eggs or larvæ have been found there is doubt as to whether the white fly could develop to maturity on the plants in question. Plants upon which the insect is unable to develop to maturity can not properly be considered true food plants. The following is the list of plants reported as food plants, but which in each case require further observations either as regards the ability of the insect to reach maturity thereon or as regards the species of white fly concerned, in view of the recent separation of *A. citri* and *A. nubifera:* Water oak, reported by Prof. A. L. Quaintance; *Ficus altissima, Ficus* sp. (from Costa Rica), and scrub palmetto, reported by Prof. H. A. Gossard; honeysuckle and blackberry, reported by Dr. E. H. Sellards; oleander, reported by

---

[1] In addition to those already mentioned as being food plants in Florida, the following plants are on record at the State insectary at Sacramento, Cal., as food plants of the citrus white fly observed at Marysville and Oroville by agents of the State commission of horticulture: English ivy (*Hedera helix*), yellow jessamine (*Jasminum odoratissimum*), *Ficus macrophylla*, bay (*Laurus nobilis*), tree of Heaven (*Ailanthus glandulosa*), and crape myrtle (*Myrtus lagerstrœmia*). Information concerning the authorities for the plants listed is not available.

the senior author of the present publication; camellia, reported by Dr. E. W. Berger. In the case of the last two plants mentioned the uncertainty as to their proper standing is on the possibility of the insect reaching maturity thereon and not on the identity of the infesting species.

The present status of the plants which have heretofore been listed by entomologists as food plants of the citrus white fly is shown in the foregoing paragraphs. There are doubtless numerous additional introduced species and a few additional native species of plants occurring in the United States which serve or are capable of serving as food plants of the citrus white fly, but for the reasons connected with the identification of the insects, stated in the opening paragraph under the subject of food plants, reports of food plants other than those included in classes I and II should never be credited unless verified by or made by an entomologist. There are no important food plants occurring in the Gulf coast region omitted from this list, and future additions to the list probably will be of little significance economically as affecting the control of the pest. There is a widespread belief that many other common trees, shrubs, and vines in Florida are food plants of the citrus white fly, but the correctness or falsity of this belief can be readily ascertained in the case of the individual plants suspected by submitting specimens of the foliage and of the infesting insect to the Bureau of Entomology or to the State experiment station.

There are three common causes for erroneous reports concerning citrus white-fly food plants. The first is the presence of sooty mold on many plants, due to other honeydew-secreting insects, such as aphides, scale insects, and mealy bugs. The insects themselves are not seen in this case and the mistaken idea is due to ignorance of the fact that other insects than the citrus white fly excrete honeydew on which the same species of sooty mold fungus thrives. The second cause for erroneous reports in this respect is the misidentification of the insect concerned. The necessity for the identification of the infesting insect by an entomologist has been discussed. The third cause is the frequent occurrence of the adult citrus white fly on the foliage of plants upon which it does not breed and upon which it seldom or never deposits an egg. In the course of the present investigation by the Bureau of Entomology several trees and shrubs have been thoroughly tested as possible food plants by cage experiments, and observations have been made on these and other plants, showing that if it is possible for the citrus white fly to develop on one of them, it is, at the most, of too rare occurrence to be of any significance. Cage tests have been made with oak (*Quercus brevifolia*), magnolia (*Magnolia foetida*), blackberry (*Rubus* spp.), laurel cherry or mock orange (*Prunus caroliniana*), and cultivated figs (*Ficus carica*) and crape myrtle (*Myrtus lagertraemia*). In each case a rearing cage (Pl. VII) was attached to the end of a branch covering new growth and from 50 to 100 adults of *A. citri* were confined therein. Except in the case of the blackberry, in which no observation was made on the point, the adults were noted as resting contentedly and apparently feeding on the leaves for one or two days after being confined. In every case, however, all the adults were dead on the fourth day after confinement on the plants noted, although check lots of adults collected at the same time but confined on branches of citrus trees lived for a normal period. No eggs were deposited in any of the tests, although the check lots deposited eggs on the citrus leaves in a normal manner.

Each of the five plants tested with the cage experiments have in addition been subjected to very careful examinations by the writers under such circumstances that the opportunities for infestation by the citrus white fly were at their best. In addition, particular attention has been given to examinations of species of oaks (*Quercus* spp.) and bays (*Persea* spp.), guavas (*Psidium* spp.) and mulberries (*Morus* spp.), when located near, and in some cases with branches intermingling with infested citrus or other favorite food plants.

ECONOMIC SIGNIFICANCE OF FOOD PLANTS, AND INTERRELATIONSHIP BETWEEN FOOD PLANTS AND INSECTS.

Entomologists familiar with the present white-fly situation agree in their conclusion that a requisite for satisfactory control of this pest is proper attention to food plants other than citrus fruit trees. Mr. H. G. Hubbard, who was a well-known authority on orange insects, being a special agent of the Bureau of Entomology, was a strong advocate of destroying food plants of the white fly that were of no value. Dr. Sellards, formerly entomologist at the Florida Experiment Station, Dr. Berger, the present entomologist, Prof. P. H. Rolfs, director of the Florida Experiment Station, and the authors have each emphasized the importance of the relation of the various food plants to white-fly control.

The following paragraph from the senior author's bulletin on the subject of fumigation for the citrus white fly [1] states in a general way the situation in this respect as viewed by entomologists who have investigated the white fly:

The presence of food plants of the white fly other than citrus trees, in citrus fruit-growing sections, constitutes a serious menace and in itself often prevents successful results from remedial work. Fortunately the list of food plants is limited, and the greater number of those thus far recorded is subject to infestation only when located near or in the midst of heavily infested citrus groves. The food plants which are of most importance in connection with the white-fly control are the chinaberry trees, privets, and cape jessamine, and these—except for the last, in certain sections where grown for commercial purposes—can be eradicated readily, or their infestation may be prevented where community interests precede those of the individual in controlling

[1] Bulletin 76, Bureau of Entomology, U. S. Department of Agriculture, pp. 9–10.

public sentiment. These food plants favor the rapid dissemination of the white fly from centers of infestation and their successful establishment in uninfested localities. They seriously interfere with the success of fumigation, as well as of all other remedial measures, by furnishing a favored breeding place where the white fly can regain its usual abundance in a much shorter time than would be the case if it were entirely dependent upon citrus fruit trees for its food supply. The plants mentioned, together with *Citrus trifoliata* (except where used in nurseries), and all abandoned and useless citrus trees should be condemned as public nuisances and destroyed in all communities where citrus fruit growing is an important industry.

Not only is a knowledge of the relation of the various noncitrus food plants to white-fly injury of great importance, but it is also of considerable importance to growers to know the capability of the insect for multiplying on the different citrus fruit trees in order that advantage may be taken of it in the arrangement of new groves and the improvement of old groves.

CITRUS.

It is a matter of common observation that injury from the white fly is most marked on citrus fruits of the Mandarin group. This group includes the Tangerine, Satsuma, and King of Siam. The sweet oranges are next to the mandarins in this respect, followed by the kumquats and grapefruits.

The relatively less injury to grapefruit by the citrus white fly (*A. citri*) is sometimes obscured by the presence of *A. nubifera.* Blackening of foliage and fruit by the citrus white fly is more noticeable on grapefruit trees when they are surrounded by or are otherwise unfavorably located in respect to oranges or tangerines. Solid blocks of grapefruit trees rarely show more than slight effects of white-fly infestation when only the citrus white fly is present. An example of this is the Manavista Grove at Manavista, Manatee County, Fla. This grove consists of 22,000 grapefruit trees, and appreciable blackening of the foliage is rarely seen except occasionally where orange groves adjoin. Only one record, based on actual examination of leaves, illustrating the difference in the degree of infestation of adjoining blocks of grapefruit and orange trees is available. The grapefruit block consisted of about 400 trees located immediately north of a block of 200 or 300 orange trees and separated on the west by a public road from a grove of about 800 orange trees. On April 23, 1909, after practically all the overwintering pupæ had matured, an examination of 100 or more leaves collected at random from each grove, counting the pupa cases, showed an average of 8 insects that had reached maturity on the grapefruit leaves, 27 on the orange leaves of the block south, and 56 on the orange leaves of the block west. No studies have been made to determine the different degrees of susceptibility to white-fly injury among the different varieties of grapefruit, but the Royal variety appears to be more nearly immune than any other of those commonly grown. This was first pointed out by Mr. F. D. Waite, of Palmetto, Fla. In this connection it should be noted that the Royal variety in its general characteristics is not a typical grapefruit.

The reason for the partial immunity of grapefruit trees to white-fly injury is as yet obscure. Several observations on grapefruit and orange trees growing side by side give no basis for the supposition that it is a matter of food-plant preferences of the adult flies. In some cases the differences in the amount of new growth must be taken into consideration. Counts of adults, pupa cases, and hatched eggs of the citrus white fly on alternating grapefruit and orange trees, six in all, located on the laboratory grounds at Orlando, were made on June 4, 1909, when no new growth was present on the trees. The leaves were selected at random and, with the exception of a few upon which adults were counted, they represented the spring growth of 1909. The difference between the number of the adults on 500 grapefruit and 500 orange leaves, 87 and 104, respectively, is not as great as would be expected, considering the much greater number of insects that had matured on the orange up to the time of the examination. There were about twenty times as many pupa cases on the 100 orange leaves as on the 100 grapefruit leaves, or 6 and 120, respectively. This was offset by the presence of about three times as many live pupæ on 10 grapefruit as on 10 orange leaves, 41 and 14, respectively, making the sum of the pupa cases and live pupæ 4.16 per leaf in the case of the grapefruit and 2.6 per leaf in the case of the orange. This is about the same proportion as the number of hatched eggs on the two food plants. The condition of the leaves, as shown by this data, fails to indicate any cause for the partial immunity of grapefruit trees.

The examinations by Mr. W. W. Yothers of two leaves picked at random from each tree in a small isolated grove consisting of 41 grapefruit and 28 tangerine trees gave rather striking figures, showing more rapid multiplication of the citrus white fly on the latter than on the former. The first examination was made on November 4, 1908, and the second on June 8, 1909. On the former date the average number of live and dead white-fly larvæ and pupæ per leaf was 31.9 on the grapefruit and 96.2 on the tangerine, 16.6 and 80.9, respectively, being alive. During the winter a series of fumigating experiments reduced the numbers of the white fly so that at the second examination the number per leaf was 1.1 on the grapefruit and 2.25 on the tangerine. The arrangement of the two kinds of trees in the grove was such that they had equal chances of becoming reinfested by the insects which escaped the effects of the experimental tests.

The difference in the degree of injury between orange and tangerine trees is less marked than between tangerine and grapefruit or orange and grapefruit, but the difference is nevertheless usually quite noticeable. The practical application of this difference in the degree of adaptation of the citrus white fly to the various citrus food plants will be discussed in a forthcoming bulletin dealing with the artificial control of the white fly.

CHINA TREES AND UMBRELLA TREES.

While China trees (Pl. IV) and umbrella China trees (Pl. V), when grown for shade and ornamental purposes, are, as has been pointed out, very injurious to citrus fruit-growing interests, the investigation of the utility of these plants as trap foods gives an increased importance to a definite knowledge concerning them as citrus white-fly food plants. Their injuriousness to citrus growers is very clear to professional entomologists, but not as generally appreciated by the citrus growers themselves as is desirable.

The umbrella tree is recognized by botanists as a variety of the China tree. This variety is the one most commonly grown except in a few localities, and observations reported herein specifically refer to it and not to the China tree. The latter tree has, however, been under observation by the authors, and no noticeable difference has been observed between the two trees in their relation to the citrus white fly, and the data and observations are in the main fully as applicable to the one as to the other.

The numbers of the white fly which mature on individual umbrella trees have been estimated in three instances and found to range between 25,000,000 and 50,000,000 where trees are favorably located with respect to nondeciduous food plants. Examinations were made by selecting 10 or more leaves at random and from each selecting a leaflet which appeared to represent the average condition of all the leaflets composing the leaf. In two instances it was found that the infestation was fully as great toward the top of the tree as on the lower parts. In one instance an extensive examination of different parts of an infested umbrella tree showed a decrease from lower branches to top branches of 50 per cent. In order to be fully conservative, this percentage has been used as the basis of the calculations, making the average infestation throughout the tree 75 per cent of the infestation of the leaves of the lower branches. Full-grown leaves were found to consist of about eighty-two leaflets. Complete records were made of eggs and of live and dead larvæ and pupæ, but only a part of this data will be presented. The estimates and counts of both leaves and insects in the case of the first tree were made by the senior author; in the cases of the second and third trees the estimates of the number of leaves per tree represent the average

THE CHINA TREE.

Fig. 1.—China tree defoliated during winter. Fig 2 —Same tree in full foliage in summer.
(Original)

THE UMBRELLA CHINA TREE.

Fig. 1.—Leaflet of umbrella China tree, showing infestation by *A. citri.* Fig. 2.—Umbrella China tree with orange tree in rear of house. (Original.)

of three estimates—one by the senior author, one by the junior author, and one by Mr. W. W. Yothers. The counts and estimates of insects were made by the senior author in the second instance and by the junior author in the third. The data obtained in these examinations bearing on the number of insects the umbrella trees are capable of maturing are given in Table V.

TABLE V.—*Number of citrus white flies developing on umbrella China trees.*

Tree.	Date of examination.	Estimated number of leaves on tree.	Number of pupa cases per leaf.	Number of live pupæ per leaf.	Estimated number of insects matured on tree.	Estimated number of larvæ and pupæ alive at time of examination.
1	Oct. 28, 1906	20,000	2,478	4.1	49,560,000	82,000
2	Aug. 19, 1908	25,000	1,910	13.2	47,702,000	330,000
3	Aug. 25, 1908	12,000	2,230	4.0	26,760,000	58,800

The three trees examined are not in any way exceptional as regards the degree of infestation, but may be considered as representative of the condition of China trees and umbrella China trees in localities where the citrus white fly is established. Tree No. 1 was located by the roadside near a 5-acre grove of newly bearing budded orange and grapefruit trees which were lightly infested by the white fly and on which it is estimated that not over 100,000 insects could have matured on any one tree during the season. Tree No. 2 was located most unfavorably for a heavy infestation, standing in a vacant lot in the business section of Orlando and having its source of citrus white-fly infestation in the spring almost entirely restricted to two neglected and worthless orange trees of small size growing within a radius of 100 feet. Tree No. 3 was located in front of the laboratory at Orlando, with 36 orange and grapefruit trees on the grounds. The least conservative of the authors' estimates would place the number of white flies which matured on any one of these citrus trees during the year 1908 as not over 500,000, with the average of the 36 trees at about one-half this number. It is estimated, therefore, that the one umbrella tree produced upward of three times as many adult citrus white flies during the year 1908 as the 36 citrus trees on the laboratory grounds combined. The important relation of the remarkable multiplication of the citrus white fly on China and umbrella trees to the spread of the pest will be discussed under the heading "Spread."

Two new points of importance have been established by the present investigations in regard to umbrella China trees as citrus white fly food plants. First, this insect shows in one respect a greater degree of adaptation to this food plant than to citrus plants, as shown by the very low rate of mortality in the immature stages. Table VI gives the data obtained by five counts made at Orlando, Fla., during these investigations.

TABLE VI.—*Mortality of citrus white fly on umbrella China tree leaves.*

Date.	Pupa cases.	Live larvæ and pupæ.	Dead larvæ and pupæ.	Mortality.
				Per cent.
Oct. 28, 1906....	806	1	111	12.0
July 8, 1908....	497	338	192	18.7
July 21, 1908...	113	256	51	12.1
Aug. 19, 1908..	232	49	152	35.1
Aug. 25, 1908..	312	169	70	13.4
Total...	1,960	813	596	18.2

The record made on August 19, 1908, showing the highest percentage of dead stages of the white fly, was based upon 10 leaflets selected from a single leaf and is not considered so typical of the condition throughout the tree examined as is the case in the other records. In contrast to the low mortality records as shown by the insect forms present on the leaves of the umbrella trees, 26 records of mortality in citrus groves gave an average of 57.9 per cent dead on the leaves. These records were based on the examination of about 2,000 leaves and over 100,000 white-fly forms. It should be noted that the mortality in the above records is based on the number of live and dead larvæ and pupæ, and of pupa cases present on the leaves at the times of the examinations. The actual mortality would be represented by the difference between the total live larvæ, live pupæ, and pupa cases and the number of hatched eggs. On umbrella China tree leaves this difference is slight and represented for the most part by the number of dead larvæ and pupæ found on the leaves. In the case of the citrus trees, on the other hand, the number of citrus white-fly forms on the leaves ordinarily represents only from 25 to 30 per cent of the total number of eggs deposited. This disappearance from the leaves is discussed elsewhere. Its significance in this connection is that the actual mortality on citrus leaves is much higher than the average per leaf of 57.9 would indicate. The citrus white fly forms on the leaves show a mortality on the umbrella tree amounting to only one-third of the mortality on citrus trees. The consideration of the number of hatched eggs as a basis for mortality estimates would reduce this to about one-fifth. The figures refer to citrus groves where the citrus white fly is well established. In newly infested groves the rate of mortality is much smaller as a rule.

The second important point established in the course of the investigations reported herein is that adult citrus white flies are so strongly attracted by growing leaves of umbrella trees that under certain conditions with umbrella and citrus trees growing side by side more adults collect on three or four umbrella leaves than are present on entire citrus trees of medium size.

It has been frequently observed that when the citrus white fly is first becoming established in a grove, if China trees or umbrella China trees are near, adults often can be found on these when none can be found on surrounding or intervening citrus trees. In order to obtain a more definite idea of the relative attractiveness of umbrella China trees and citrus trees, 4 records were made by the senior author on the laboratory grounds (fig. 2) at Orlando. In observation No. 1, the count on citrus was made on 4 trees, viz, 4 A, 4 B, 4 C, and 5 C, and the observations on umbrella China. trees were made on 2 small trees located about 6 and 20 feet, respectively, southwest of 4 A. These umbrella China trees were slender 2-year-old growths about 4 and 5 feet high and together bearing about 40 leaves.

Observation No. 2 was made on grapefruit and orange trees E 6, E 5, E 4, D 5, D 4, and F 3 and two stems of the umbrella China tree cluster in space F 7 nearest to tree E 6. Observation No. 3 was made on tree A 4 and the

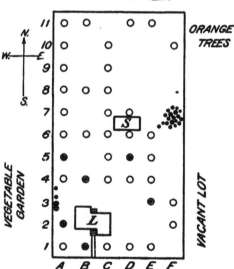

O = ORANGE,   ● = GRAPE FRUIT.
● = UMBRELLA TREES;
S = STORAGE SHED,   L = LABORATORY.

FIG. 2.—Diagram of the laboratory grounds at Orlando, Fla. (Original.)

nearest umbrella China tree sapling. This latter had been defoliated since observation No. 1 was made. Observation No. 4 was made on citrus trees D 7, E 6, E 5, E 4, E 3, and F 3 and on two stems of the umbrella cluster which had been defoliated since observation No. 2. The data obtained by the four observations are given in Table VII.

TABLE VII.—*Relative attractiveness to the citrus white fly of foliage of umbrella China trees and citrus trees.*

Observation No.	Date.	Citrus trees.			Umbrella China-trees.		
		Number of minutes.	Number of adults counted.	Number of trees examined.	Number of minutes.	Number of adults counted.	Approximate number of leaves examined.
1........	1909. May 18	5	257	4	5	508	25
2........	...do....	5	19	6	5	615	30
3........	June 11	3	34	1	3	477	6
4........	...do....	5	52	6	5	830	15
	Total .	18	362	17	18	2,427	76

In all, 2,789 specimens were counted, of which 88 per cent were on umbrella China tree leaves. It was estimated that in each record on a citrus tree approximately 2,000 leaves were examined, making 34,000 in all. The individual leaflets composing the 76 umbrella-tree leaves numbered approximately 6,000. For practical purposes these leaflets are more comparable to the citrus leaves although the latter have on the average fully twice as much surface. With this basis for comparison it can be figured from the above data that there was about one adult white fly per 100 leaves on the citrus trees while there were about 40 adults per 100 leaflets on the umbrella trees.

As has been indicated, the difference between the number of China-tree leaves in numbers 3 and 1 and between 4 and 2, respectively, represents the oldest spring growth, which was removed on May 24, leaving only a few growing leaves. No direct comparison was made between the attractiveness of the older growth of citrus and umbrella trees but apparently there is no striking difference between the two food plants in this respect. New watershoots were present on the citrus trees on both dates when observations were made but only in the case of one tree, 4 A, were many adults found on this growth. In observation No. 1 on the tree mentioned (4 A) 200 adults were counted on two watershoots. Except for watershoots there was no new growth on any of the citrus trees.

At Orlando the umbrella trees usually start to put on new foliage in the spring before new growth appears on citrus trees. As a consequence China and umbrella trees located near infested citrus trees receive large numbers of adults of the citrus white fly which migrate in search of attractive food. On February 22, 1909, the authors noted on the laboratory grounds that the shoots of the umbrella tree were beginning to put out new growth, the leaves not fully unfolded. The citrus white fly was found scatteringly on the umbrella leaves but on citrus trees specimens could be found only after careful search.

On March 27, 1907, near the laboratory, then located in the grove of Mr. J. M. Cheney, a striking example of the attractiveness of the umbrella tree was observed. The tree referred to was about 25 feet high and the leaves which were on the average only about half developed were estimated ·to number 5,000. Ten leaves were selected at random within 10 feet of the ground and the number of adults and eggs was counted, the former numbering 5.3 per leaf on the average and the latter 160 per leaf. The tree was cut down and an examination of the topmost leaves showed an average of 186 eggs per leaf; the adults, being disturbed, were not counted, but judging from the number of eggs present they evidently were more rather than less numerous than on leaves near the ground. Considering the average of 5.3 per leaf, however, the total number of adults on the tree would be estimated at 26,500, and at 160 eggs per leaf the number of eggs deposited would be estimated at 800,000. At the time of this observation about 50 per cent of the insects which overwintered on the citrus leaves had matured. The citrus white fly had been much reduced throughout the grove, in some sections by unexplained influences, in others by these influences and fumigation experiments combined, and on a few tangerine trees by a fungus parasite, red Aschersonia. The location of the umbrella tree did not seem to be a favorable one as regards opportunities for white-fly infestation, but examination showed the infestation to be at least 100 times greater, as regards the number of adults present, than on any citrus tree in the grove. There were, in fact, too few eggs deposited on the leaves of the citrus trees to allow of sufficient multiplication of the white fly during the season to cause any blackening of foliage or fruit.

CAPE JESSAMINE.

The cape jessamine has long been recognized not only as a favorite food plant of the citrus white fly, but as especially important economically on account of its retaining its foliage throughout the year. From a statement by Riley and Howard [1] concerning observations by Mr. H. G. Hubbard and statements by Dr. H. J. Webber, Dr. Montgomery, and others in the discussions on the citrus white fly at a meeting of the Florida State Horticultural Society, [2] it appears that the freezes of December, 1904, and February, 1905, which completely defoliated citrus trees when not especially protected, failed to defoliate cape jessamines. In many localities it is probable that this food plant was responsible for the survival of the white fly at the time referred to. According to Dr. Sellards, [3] temperatures as low as

---

[1] Insect Life, vol. 7, p. 282.
[2] Proceedings of the Florida State Horticultural Society, 1896, p. 78.
[3] Press Bulletin 56, Florida Agricultural Experiment Station, p. 2.

16° above zero at Lake City, between January 26 and January 29, 1905, failed to defoliate cape jessamine.

Except where grown for commercial purposes, as is the case at Alvin, Tex., where the blooms are shipped to northern markets, or where grown in nurseries, cape jessamines have not been observed growing in sufficient abundance to materially affect near-by citrus trees in sections where the white fly is already established. If overlooked in connection with the fumigation of citrus groves or defoliation of citrus trees by cold, cape jessamines might become a serious hindrance in the control of the white fly. The greatest economic importance of the cape jessamine as a food plant lies in the great danger it presents as a distributer of the white fly. This will be referred to again under the subject of methods of spread.

The subject of the adoption of the cape jessamine by the citrus white fly is not of sufficient importance to have been given more than incidental consideration. In general the degree of adoption seems to be less than is the case with the umbrella and China trees. On November 17, 1907, an examination made of 30 leaves picked at random from both old and new growth of a cape jessamine which appeared to be in an ordinary condition of infestation as observed when growing near infested citrus trees showed that there existed an average of 45.1 forms per leaf.

The extensive growth of cape jessamines, or gardenias, as the blooms are sometimes called, for commercial purposes is known to the authors and occasions a conflict of interests only in Alvin, Tex. From the orange grower's standpoint this, at the most, applies to a location adjoining an orange grove where the citrus white fly is uncontrolled. Fortunately, however, for the citrus growers, it is of great importance to the success of the florist's business that the white fly be kept in subjection in gardenias.

PRIVET HEDGES.

Privet hedges are not uncommon in citrus-growing sections, and heavy infestations by the citrus white fly occur in parts of Georgia and South Carolina, where no citrus trees are grown. As a food plant the privets are of economic interest in the same respects as is the cape jessamine, but they are more extensively grown and of proportionally greater importance. No studies have been made of the degree of adaptation and attractiveness, but the several species of privet observed in infested localities have shown the propriety of classing them with citrus, China trees, umbrella China trees, cape jessamine, and other preferred food plants. The senior author observed a migration of adults from privet hedges in Victoria, Tex., in the summer of 1904, which indicated that a hedge of this material might well be compared in its injurious influence on citrus-growing interests to one or more umbrella or China trees. The privet, like the cape jessamine, is hardy, and the disadvantages of the former in this connection are the same as those mentioned in discussing the latter food plant.

JAPANESE AND WILD PERSIMMONS.

Japanese and wild persimmons are attractive to the citrus white fly early in the season, but appear to be very little or not at all so late in the season. Being deciduous, their economic importance as white fly food plants is proportionally small. Under certain conditions the Japanese persimmons appear much more attractive to the citrus white fly than citrus trees. These conditions have not been investigated, but they are probably dependent upon the appearance of new growth in the spring a little earlier on persimmon than on citrus. On June 16, 1909, an examination of a large bearing persimmon tree surrounded by citrus nursery trees and bearing citrus trees of different kinds showed that the first spring growth of the persimmon was much more attractive to the first brood of adults than were the citrus trees. The second brood of adults, however, found the persimmon comparatively unattractive and showed a marked preference for the citrus trees. The earliest citrus growth of the spring had become fully matured, and no new growth appeared until after the second brood of adults had practically disappeared. The comparative condition of infestation is shown by counts made on leaves picked at random from the persimmon tree and from the surrounding citrus trees, including the sweet orange, sour orange, tangerine, and grapefruit. The average infestation with first-generation forms of the citrus white fly on 25 leaves each of persimmon and citrus was in the ratio of 10.9 to 1.3, while that of the same number of leaves by the second generation was in the ratio of no forms on the persimmon leaves as compared with 191 on the citrus, thus showing the great preference of the second generation of adults for citrus growth.

Neither the Japanese nor the wild persimmons are usually infested by the citrus white fly to the extent of causing noticeable blackening from sooty mold. The infestation, however, might be between from five to ten times as great as on the leaves from the trees referred to above without producing this result. Small wild persimmon bushes have been observed in a growing condition at the time the adults of the second brood are on the wing, and at such times they sometimes appear to be very attractive as food plants. Mr. W. W. Yothers has observed near Hawthorn, Fla., on April 29, 1909, the citrus white fly on wild persimmon bushes growing in pine woods at distances upward to one-fourth of a mile from any citrus grove, and the junior author has made similar observations along roadsides near Orlando, Fla., in June, 1909, the insects being in the adult stage only in this latter case. On the other hand, the senior author noted on June 18, 1909, that wild persimmon bushes growing in a vacant lot with China trees and abandoned citrus trees were only very slightly infested, although the citrus trees and the China trees were heavily infested. The wild persimmon had made vigorous growth, but its white-fly infestation consisted of less than 10 eggs per leaf and an occasional adult. The examination of leaves of the China tree showed hundreds of pupæ and pupa cases per leaf, with a few adults and newly deposited eggs. The old citrus leaves bore many larvæ, pupæ, and pupa cases, and the new leaves bore hundreds of unhatched eggs. The wild persimmon bush was as favorably located with respect to citrus trees as was the China tree. Notwithstanding the exceptions noted in degree of attractiveness, the Japanese and the wild persimmons very evidently rank well below citrus trees, China trees, and umbrella China trees.

In so far as observed the persimmons have little effect on the control of the citrus white fly, but in special cases they may rank as important food plants. The fact that the Japanese persimmon is a producer of fruit of some commercial value makes its ordinary lightness of infestation a matter of gratification. The wild persimmon, on the other hand, is of practically no value either for shade or fruit, and can easily be destroyed where advisable.

LILAC.

Lilac is not commonly grown in the citrus-growing regions of the Gulf States, and on this account, so far as observed, presents no element of menace to orange groves. In company with all of the ornamental plants listed as preferred food plants this one must be considered, however, as undesirable for introduction and growing in citrus-growing regions.

PRICKLY ASH.

Belonging to the family Rutaceæ, to which the genus Citrus also belongs, it is not strange that the prickly ash is a favorite food plant of the citrus white fly. This plant seems to be highly attractive to the adult flies, frequently being observed infested with more adults than many near-by citrus trees combined. The prickly ash is common in Florida and in some localities, where growing in abundance along roadsides, it constitutes a distinct menace to citrus groves through its connection with the spread of the white fly from city and town to country and from grove to grove.

COFFEE.

Dr. E. W. Berger has reported having observed a coffee tree thoroughly infested with as many eggs on its leaves as citrus leaves may have. This food plant is too rarely grown in the Gulf States in orange-growing regions to be of any importance economically as a white-fly food plant.

OCCASIONALLY INFESTED FOOD PLANTS.

The plants listed in Class II as a whole are of very little importance as regards their bearing on white-fly control. Banana shrub, cherry laurel, and cultivated pear might well be considered in a third class for rarely infested plants. Although not uncommon, their attraction for the citrus white fly is so slight as to make it safe to ignore them except in the matter of introducing the fly on them into noninfested districts. In unpublished notes Dr. Berger has recorded the wild olive as a food plant. He has observed the wild olive infested in comparatively isolated places. The junior author has observed wild olive heavily infested in Charleston, S. C., and in several places in Orange County, Fla. The wild olive, being an evergreen, if neglected may prove to be of considerable importance as a food plant when growing in abundance near a fumigated grove or when citrus trees have been defoliated by cold.

Dr. Berger has recorded pomegranate, allamanda, and smilax as food plants, and has verified Prof. Gossard's record of *Viburnum nudum* as a food plant of the citrus white fly. The positions of these plants as regards their attractiveness to the citrus white fly has not been fully determined, and further observations will perhaps show one or more of them to be of fully as high if not of higher rank in this respect than the persimmons. In general, however, like the coffee and lilac of Class I, they are not of sufficiently common occurrence in the Gulf coast citrus-growing regions to be of much economic importance as citrus white-fly food plants.

SPREAD IN THE UNITED STATES.

There is seldom positive evidence in regard to the means by which the citrus white fly has become established in a previously noninfested grove or locality. Such direct observations, however, as it is possible to make, aided by strong circumstantial evidence, give us a sufficient knowledge of the methods of spread to show the advisability of certain restrictive measures.

CHECKS ON SUCCESSFUL ESTABLISHMENTS.

Fortunately the chances are greatly against the successful establishment of the citrus white fly in a previously uninfested locality, which is outside the limits affected by large numbers of migrating adults. If this were not so the pest would have become established in every grove of the State long before the present time. Except for spread by direct flight and on nursery trees and ornamental plants, the chances are against more than a few insects being introduced into a particular grove by any of the other methods discussed hereafter.

In the case of a single adult there are two chances in three that it would be of the reproductive sex. If, as would be probable, the specimen were a female, there would be about one chance in three that it would not have been fertilized. In this case the second generation of adults would all be males, as shown by the observations recorded under the subject of Parthenogenesis. This would, of course, end the infestation directly due to the single specimen introduced, as the original female would have died several weeks before the first male matured. In case the originally introduced specimen were a female and fertile the chances of a male appearing among the second generation are not definitely known, but are with little doubt only small. The chances of such a male appearing at a favorable time to meet with and to fertilize a female of the same parentage are practically negligible, though possible as a result of the great variation in the length of the life cycle as recorded under life history. The third generation would, therefore, in all probability, be all males, and the infestation ended. The chances that a single adult specimen introduced into an isolated grove or into a previously noninfested community would successfully establish a permanent infestation are extremely small. The chances are only slightly increased by an increase to 5 or even 10 in the number of adults originally introduced into a single grove.

From the foregoing considerations it is evident that two or more distinct introductions of even a few individuals at proper intervals during a single season might greatly increase the chances for the successful establishment of the pest.

The flight of adults is the most important method of local distribution and is also an important element in its association with spread by means of winds and vehicles, railroad trains, and boats.

The distance to which the insect is capable of flying.—It would be almost impossible to obtain positive records on the distance the adult citrus white fly is capable of flying. Mr. W. W. Yothers, on April 29, 1908, found on wild persimmon first and second generations of this species of fly at a distance of one-fourth of a mile from the nearest orange grove, which was also the nearest point of the occurrence of a food plant upon which the insect could have passed the winter. The infested persimmons were in pine woods and the insects were in such numbers that it was evident that spread through pine woods might easily greatly exceed one-fourth of a mile. Mr. W. C. Temple, of Winter Park, Fla., states that he has observed adults migrating into one of his groves on Lake Maitland under circumstances plainly indicating that they had traveled over the water for 1¾ miles. Dr. Berger has recorded an instance which presents strong evidence that adult white flies have heavily infested citrus trees through flights of a mile or more. On the other hand, there are orange groves within three-fourths of a mile of the city limits of Orlando, Fla., and within 2 miles of the courthouse which have only so recently become infested that no blackening of the foliage has taken place, although the citrus white fly has occurred at Orlando for more than 10 years with migrating adults in summer about as abundant as in any town in the State. As regards the capability of flight of the citrus white fly, it may be said to be undoubtedly more than a mile and perhaps several miles when aided by a gentle breeze. Distances of even a mile, however, are not usually attained except under certain circumstances which are largely preventable and which are discussed in the following paragraph.

Cause of extensive migrations by flight.—Overpopulation of food plants, usually associated with the emergence of adults in large numbers at seasons when the new and attractive growth is scarce or entirely wanting, is the main cause for migrations from citrus, cape jessamine, and privet. Migrations from China trees and umbrella trees, probably the most potent factors in the spread of the pest, are not due directly to overpopulation, so far as observed, since leaves are never overcrowded in a manner comparable to the overcrowding on citrus leaves. An average of 25 live larvæ or pupæ and pupa cases per square inch of lower leaf surface would represent an unusually heavy infestation of a China or umbrella tree and is rarely exceeded, whereas an average of 50 or 60 per square inch is not unusual for citrus leaves. In the case of China and umbrella trees, migrations are evidently due to lack of attractiveness of the foliage to the adult white flies at the times when the migrations occur. There are comparatively few live larvæ and pupæ on the foliage after the middle of August at Orlando. The greater part of these represent delayed emergence from the second generation of white flies and not the result of eggs deposited by the third brood of adults. This supports direct observations to the effect that the third brood of adults, which is concerned in the most extensive migrations, deposits practically no eggs on the China and umbrella trees.

China and umbrella trees as a factor in dissemination.—Umbrella and China trees are extensively grown throughout the Gulf coast citrus-growing regions, and they are almost entirely responsible for the hundreds of millions of adults which in midsummer appear on the wing throughout most of the towns where the citrus white fly occurs. That these are principally those which have bred upon China and umbrella trees is shown clearly by the fact that at Gainesville, Lake City, Tallahassee, and other points in the northern part of Florida, where other food plants are too few to produce noticeable numbers of migrating adults, the numbers are apparently not less than where both citrus trees and China and umbrella trees are extensively grown, as at Orlando. On this point, Dr. Berger states:[1] "The principal food plants in Gainesville and north Florida are China and umbrella trees, there being only enough citrus, privet, and other evergreen food plants to bring about the restocking of the deciduous trees every spring." These considerations indicate very positively the main source of the enormous number of migrating adult flies on trees in midsummer, sometimes observed between the middle of May and the middle of June. These adults are the second brood of the season and the first to mature on the food plants mentioned. The newer growth of these trees is, as has been shown, very attractive to the adult flies, and if there is an abundance of it comparatively few migrate. The third brood, composed mainly of individuals of the second and third generations, matures over a more extended period, in general covering the months of July and August in different sections of Florida.

Estimates of the number of adult citrus white flies breeding on umbrella trees and on citrus trees as given under the subject of food plants have shown that a single umbrella tree of medium size may produce as many adult white flies by midsummer as could be produced on 7 acres of orange trees. The maturity of so many adults on single trees, and their migration therefrom in search of a more desirable food supply than China and umbrella trees afford in midsummer, cause the rapid spread of the pest throughout the towns, directly by flight of the adults and by mediums hereinafter discussed into the surrounding country and from town to town along railway lines and watercourses.

Dissemination by flight when citrus trees only are concerned.—It has been shown under the subject of food plants that the citrus white fly does not ordinarily increase to the point of overcrowding on grapefruit. Migrations of adults in noticeable numbers from solid blocks of these trees probably never occur under ordinary circumstances, and spread through such blocks or groves from the first point of infestation is very slow if no other food plants are concerned. The spread in groves of orange or tangerine trees or of both is more rapid, but not as much so as ordinarily considered. The white fly is rarely observed during its first year's appearance in a citrus grove. Attention is usually first attracted to its presence through the blackening of foliage on one or a few trees. This blackening of foliage in itself

---

[1] Press Bulletin 108, Florida Agriculture Experimental Station, February 13, 1909.

is almost positive evidence of the presence of the fly in the grove during at least the preceding two years unless the infestation is due to migrations from China or umbrella trees or from overstocked neighboring citrus groves. In such cases infestation may become quite general throughout several acres in one season and extensive blackening of the foliage may result early in the next season, or in about one year after the first introduction. In the case of new infestation in any locality, however, the beginning doubtless is usually the introduction of a few insects by some one of the means hereinafter discussed. In a mixed grove of tangerine and orange the pest, is discovered first as a rule on tangerine, and in a grove of seedling trees with a few budded trees intermixed usually the latter are first discovered to be infested. Many citrus growers who have groves, such as those mentioned, and who have watched carefully for the appearance of the pest in their groves, have finally found it well established on a single tangerine or budded orange tree before any evidence of the presence of the insect was observed elsewhere. Through the hindrances to successful establishment and the checks on multiplication, principally those discussed in connection with parthenogenesis and natural mortality, the white fly frequently develops so slowly after its first introduction that it may not increase to the point where it is usually first observed for three or four years. It is a common error to consider that the first discovery of the white fly in a grove is an indication of its very recent introduction. This may or may not be the case. Usually it is not the case. It should be borne in mind in this connection that in the most careful inspection, even by a competent entomologist, the failure to discover a single specimen of the white fly is not positive proof that it is not present. The foregoing generalizations are based upon many observations by the agents of this bureau who have been engaged in these investigations, more particularly the authors of this bulletin and Mr. W. W. Yothers.

The rapidity of spread into a citrus grove from neighboring infested groves is a subject which becomes temporarily important when a nonisolated grove becomes infested for the first time. It is a subject of more far-reaching importance in connection with fumigation, and it is in this connection that the most extensive studies in this line have been made. The result of these studies will be published in a final report on fumigation.

The slowness with which the citrus white fly increases in numbers and spreads from the first point of infestation has been noted by many citrus growers who have been observant enough to discover the white fly soon after its introduction into their groves. When the rate of spread of the white fly through the grove is affected by the presence of migrating adults from China or umbrella trees, the difficulties in effectively utilizing artificial checks, spraying and fumigation, are greatly increased.   As the infested area in a newly infested grove or locality becomes larger the rate of spread by flight increases, aided by secondary centers of infestation which become established by various means.

WINDS.

Light winds are an important adjunct to flight in the local distribution of adult white flies, but strong winds are ordinarily of slight consequence.   The effect of light winds is shown by the influence of almost imperceptible movements of the air on the direction of migrations.   This is especially noticeable in the vicinity of China and umbrella trees during a season when adults are emerging in abundance.   The principal effect of the movement of the air under such conditions is not in carrying the insects, but in causing the flight energy of the insect to be expended in one general direction rather than to be wasted in zigzag lines with comparatively little real progression.   Other conditions being equal, the adult white flies migrate in greatest abundance when the atmosphere is calmest, and conversely show the least tendency to migrate in strong winds. It is possible that isolated infestations may sometimes result from spread of adults by strong winds, but it is seldom that there is not a more plausible explanation obtainable.   With the white fly present in abundance for many years in Orlando, Fla., and other towns and cities in important orange-growing sections of Florida, the fact that there are still many noninfested citrus groves within a radius of 5 miles of nearly all such centers of infestation is in itself an indication of the minor influence of winds in this connection.   Strong breezes or winds exert some check on the spread of adults by causing them to cling tenaciously to their support, as pointed out by Prof. H. A. Gossard.[1]

VEHICLES, RAILROAD TRAINS, AND BOATS.

In towns in Florida where the citrus white fly occurs and China trees and umbrella trees are abundant it is a matter of common observation that during the periods of migration large numbers of adults alight upon automobiles, carriages, wagons, and railroad coaches. The authors have seen covered carriages with more than 100 adults resting on the inside of the top and sides.   In driving through a heavily infested citrus grove in late afternoon at certain seasons, hundreds of adults may be observed on the carriage (Pl. VI, fig. 2). Newly infested groves show the first infestation so frequently on trees close to a driveway or road that conveyance of the citrus white fly by means of carriages, wagons, and automobiles must be considered one of the most important methods of spread from town to surrounding

---

[1] Bulletin 67, Florida Agricultural Experiment Station, p. 13.

DISSEMINATION OF WHITE FLIES.

Fig. 1.—Nursery citrus trees infested with white flies set out in an isolated noninfested grove without having leaves removed. Fig 2.—Buggy in an orange grove; buggy top full of adult white flies ready to be carried to other groves. Fig. 3.—Train at station; adult citrus white flies swarming from near-by umbrella China trees into coaches ready to be carried for miles down the Florida east coast. (Original.)

country or from grove to grove. At Orlando, in July, 1906, adult citrus white flies were observed late in the afternoon alighting on the sides of coaches and flying into the windows and doors of coaches of a passenger train standing at a railroad station (Pl. VI, fig. 3). Hundreds of adults were carried west toward Wildwood through points which, so far as known, were not infested at the time. The presence of China, umbrella, or citrus trees near railroad stations increases the chances for successful introduction by railroad trains. In this connection the recent action of the Atlantic Coast Line Railroad and Seaboard Air Line Railway in destroying such trees along their right of way is to be commended. A map of Florida showing the distribution of the citrus white fly plainly indicates the relation between the railroads and the main lines of dissemination. This is shown in an incomplete way by figure 2, in which are given the points infested by the citrus white fly in Florida according to the records made in connection with the present investigation and such other records as are undoubtedly correct or which have been verified. The infestation at Arcadia, Fla., first discovered in January, 1907, but which probably resulted from an introduction of citrus white flies in 1905, was with little doubt due to the introduction of adult flies by means of railroad trains. An examination of the situation in February, 1907, by the senior author showed the center of infestation to be located near railroad stations, and careful inquiry concerning other possible sources showed that railroad trains were the most likely means of introduction. North of Arcadia no nearer infested point was known than Bartow and toward the south no nearer infested point than Fort Myers. The distance in each case was about 40 miles. So far as known there were at that time no intermediate points infested between Arcadia and the two points mentioned. Here again the factors unfavorable to the successful establishment of the pest in a previously uninfested locality play an important rôle, as shown by the fact that even at the present writing the citrus white fly is not generally distributed between Bartow and Fort Myers. As we have no record and have heard no report of the occurrence of the fly at any other point than Arcadia, it is unlikely that other infested points exist.

Steamboats are used quite extensively on the rivers and along the coast of Florida in transporting citrus fruits and have in a degree a similar status to railroad trains in transporting the citrus white fly.

CITRUS NURSERY STOCK AND ORNAMENTAL PLANTS.

The carriage of the citrus white fly in its egg, larval, and pupal stages by means of citrus nursery stock (Pl. VI, fig. 1) and ornamental plants has always been an important factor in the spread of the insect. The citrus white fly was without doubt introduced into the United States and distributed to the most important centers of infestation by this means. In Florida the white fly was probably introduced first on citrus nursery stock into some citrus grove on the St. Johns River in St. Johns County, and later by the same means into Manatee and Fort Myers. Gainesville, Ocala, Orlando, and Bartow were probably among the points to which the white fly was introduced on nursery stock. The distribution of the citrus white fly along the Gulf coast citrus-growing regions west of Florida has been largely due to shipments of infested citrus nursery stock, umbrella trees, privets, and cape jessamines. Of all methods of spread which are operative over greater distances than the flight of adults, introductions of live immature stages on trees or shrubs for transplanting purposes are by far the most certain to result in the successful establishment of the species. Fortunately it is practicable to prevent spread by this method by defoliating the trees as they leave the nursery. Much has been accomplished in the past by individual citrus growers, but more attention should be given to this matter in communities not now infested by both of the white flies treated in this bulletin.

ACCIDENTAL SPREAD BY MAN.

Carriage of the adult white flies on human beings.—Man is doubtless responsible to a limited extent for the spread of adult white flies. During migrating periods, when in heavily infested orange groves or in towns where there are infested China and umbrella trees, adults are frequently observed on the clothing. Prof. H. A. Gossard states that he has carried adult white flies for nearly half a mile on his clothing after standing beneath a heavily infested tree.

Introduction in pickers' outfits.—In some instances the citrus white fly is believed to have been introduced into previously uninfested localities by orange pickers. In this case the principal danger lies in introducing live pupæ on citrus leaves accidentally brought in with picking sacks and field boxes. The authors consider that there is practically no danger of the carriage of adults of the citrus white fly by pickers' outfits between December 1 and March 1. The few adults present in citrus groves during this period would rarely result in their transference to uninfested groves by such means, and the unfavorable factors heretofore discussed would almost certainly prevent the successful establishment of the pest. It would be almost impossible to conceive of any likely method by which a successful introduction of the citrus white fly into a noninfested grove could be accomplished by the carrying of leaves infested by eggs or larvæ. Leaves infested with live pupæ, however, particularly about the time of the beginning of emergence of the first spring brood, might readily produce a sufficient number of adults to successfully establish the pest. Such leaves, after introduction, would need to have a favored location, for exposure to much sunlight or to too much moisture would soon destroy the insects.

Introduction on leaves infected with parasitic fungi.—The matter of spread of the white fly in connection with the attempt to introduce parasitic fungi is a subject of considerable importance. The danger here is due to the failure to recognize the distinction between the citrus white fly (*Aleyrodes citri*) and the cloudy-winged white fly (*Aleyrodes nubifera*). The owner of a grove infested by the latter species only, would provide a very favorable opportunity for the introduction of the first and most destructive species if in introducing parasitic fungi he should obtain his supply of leaves from certain sections of Florida. The spread of the cloudy-winged white fly has been encouraged in a similar manner. The tree-planting method of introducing the fungi, especially the brown fungus, is the most dangerous practice in this connection. Of somewhat less danger in the individual cases, but of far greater danger on account of the more frequent opportunities presented, is the introduction of fungus-infected leaves for pinning or for spraying the spores. The pinning of leaves as a means of introducing the parasitic fungi has little more to recommend it than the tree-planting method, but it has without doubt been the means of introducing the citrus white fly on many occasions. Leaves introduced for the spore-spraying method of spreading the fungus parasites are an element of much danger under certain conditions. Some sections of Florida in which only the cloudy-winged white fly occurs are in more danger of having the citrus white fly introduced by some uninformed person in this way than they are of its introduction in any other manner. Specific examples might be cited where the introduction of either *A. citri* or *A. nubifera* was with little doubt due to introducing fungus-infected leaves or trees, but the danger is too obvious to require further discussion in this place.

LIFE HISTORY AND HABITS.

SUMMARY.

The eggs of the citrus white fly (fig. 3) are laid scatteringly, with few exceptions, on the underside of the leaves of the various food plants, and hatch in from 8 to 24 days, according to the season. During ordinary summer weather from 75 to 100 per cent hatch on the tenth to twelfth day. Infertile eggs hatch as readily as fertile eggs and produce adults of the male sex only. After hatching, the young larva (figs. 4–6) actively crawls about for several hours, when it ceases to crawl, settles upon the underside of the leaf, and begins to feed by sucking the plant juices. It molts three times before becoming a pupa. After the first molt (see fig. 7) the legs become vestigial; hence thereafter it is impossible for it to materially change its location upon the leaf. Larval life averages in length from 23 to 30 days. The pupa (fig. 9) closely resembles the grown larva (fig. 8) and requires from 13 to 304 days for development. The adult fly (fig. 10) has an average life of about 10 days, although several females have been known to live 27 days. Females may begin depositing eggs as soon as 6 hours after emergence and continue ovipositing throughout life. The maximum egg-laying capacity is about 250 eggs, although 150 more nearly represents the number laid under grove conditions. Unfertilized females deposit as many eggs as fertile females.

The entire life cycle from egg to adult requires from 41 to 333 days; the variation in the number of days required from eggs laid on the same leaf on the same day is very remarkable. During the course of the year the fly may pass through a minimum of two generations and a maximum of six generations. The generation started by the few adults that emerge during the winter is entirely dependent upon weather conditions and may or may not occur. Each generation except those started after the middle of August is more or less distinctly two-brooded.

METHODS OF STUDY.

As it is impossible to rear citrus white flies through their entire life cycle on detached leaves, a gauze-wire cage was devised by the senior author which has proved of great value and convenience in carrying on life-history studies under conditions as nearly normal as it is possible to get them. This cage (Pl. VII), which is cylindrical in shape and open at one end, may be made any size, but one 6½ inches long by 3½ inches in diameter has proved most convenient. It can easily be made by fashioning two rings of heavy wire to which is soldered the wire gauze, as shown in the illustration. To the open end is attached a piece of closely woven cheesecloth long enough to extend about 4 inches beyond the cage. After the leaf, or leaves, to be caged have been cleaned of all stages of the white fly by means of a hand lens and cloth, the cage is slipped over the foliage. The adult flies are then introduced, if desired, and the cloth attached to the cage wrapped around the stem of the shoot or petiole of the leaf, as the case may be, in such a manner that the flies can not escape nor the ants and other predaceous insects enter. To keep the entire weight of the cage from falling on the petiole of the leaf or its short stem, and to regulate the position of the leaf within the cage, a cord is tied around the outer end of the cage and attached by the loose end to a convenient branch.

CAGES FOR REARING WHITE FLIES.

Fig. 1.—Rearing cages in position on orange trees.   Fig. 2.—Enlarged rearing cage.   (Original.)

A very satisfactory method of definitely marking larvæ in order that no mistake may be made in identifying field notes with the individual larvæ to which they refer, is to scratch lightly on the epidermis of the leaf, with a thorn or pin, a bracket or other mark and outside this a number that shall correspond with that used in the note book. In marking larvæ care should be taken in scratching the leaf to allow for the future growth of the insect and not to injure the epidermis of the leaf too severely. In this manner a large number of larvæ were marked as soon as they settled and their growth noted by daily observations.

In determining the sums of effective temperature, 43° F. has been taken as a basis in accordance with Dr. Merriam's general law although this has led to certain inaccuracies of which the authors are aware. The determination of the effective temperature in the case of the white fly would require a special study which it has been impracticable to undertake.

The Egg.

DESCRIPTION.

Fig. 3.—The citrus white fly (*Aleyrodes citri*): Eggs. Greatly enlarged. (Original)

The eggs of the citrus white fly (fig. 3) are so small that they appear to the unaided eye as fine particles of whitish dust on the under surface of the leaves. Their minute size is emphasized by the fact that 118 placed end to end would measure but an inch, while about 35,164 could be placed side by side in one square inch. Under the magnifying lens they appear as smooth, polished, greenish-yellow objects shaped much like a kernel of wheat. Following is a more minute description:

Length, 0.2–0.23 mm.; width, 0.08–0.09 mm. Surface highly polished, without sculpturing, color pale yellow with faint greenish tinge when first deposited, paler than the under surface of the leaf. Egg elongate, subellipsoid, slightly wider beyond the middle or at about the point where the eyes of the embryo subsequently appear; borne at end of a comparatively slender brownish petiole or footstalk, slightly shorter than the width of the egg, and somewhat knobbed at base.

As the embroyo approaches maturity its purple eyes may be seen showing distinctly through the egg membranes at a point beyond the middle of the egg. At about this time, also, the hitherto uniformly colored egg contents become orange or golden at the proximal end and whitish translucent on the distal three-fourths. The egg surface sometimes assumes a white pruinose appearance, due to the presence of wax rubbed from the bodies of the adults while crawling over the eggs. Eggs deposited on leaves from which the adults have been excluded after egg deposition do not show this pruinose condition.

Eggs in which the embryonic development is normal do not turn dark in color, but those killed through attack by thrips or by other agency frequently become bronze colored, thus resembling the eggs of *A. nubifera* from which the waxy sculpturings have been rubbed.

DURATION OF EGG STAGE.

That no doubt might arise concerning the exact age of the eggs used in obtaining the data incorporated into Table VIII, suitable leaves were selected from which all eggs previously deposited were carefully removed by the aid of the hand lens and a cloth. Similar attention was given the leaf petiole and the stalk, and wads of cotton were tied about the latter both above and below the leaf to prevent crawling young from reaching the leaf along the petiole (Pl. VII, fig. 1). These preliminary steps completed, a rearing cage containing adult white flies was placed over the leaf and allowed to remain the length of time desired, usually from 1 to 24 hours, with preference shown the latter number. The cage was then removed and an empty one put in its place. By this method all doubt was removed as to the period of time over which deposition took place. As there is scarcely a leaf in a grove infested with the citrus white fly that does not bear from a few to many eggs, this becomes an important point and failure in its recognition has led in the past to statements greatly underestimating the minimum duration of the egg stage during the warmer months.

TABLE VIII.—*Duration of egg stage of the citrus white fly.*

Record number.	Date deposited.	Eggs under observation.	First egg hatched.	Last egg hatched.	Least number of days.	Largest number of days.	Daily rate of hatching in percentages with degrees of accumulative effective temperatures.																	
							8 days.	°F.	9 days.	°F.	10 days.	°F.	11 days.	°F.	12 days.	°F.	13 days.	°F.	14 days.	°F.				
	1909		1909	1909																				
1	Feb. 23	163	Mar. 13	Mar. 24	13	21	0		0		0		0		0		0		0					
2	Mar. 3	205	Mar. 17	May 6	14	16	0		0		0		0		0		0	423	.5	442				
3	Apr. 20	453	May 1	June 30	11	15	0		0		0		.5	401	90.6		96.2	434	14.8	462				
4	June 16	300	June 26				0		0		0		4.4	388			14.7							
	1907		1907	1907																				
5	July 17	927	July 26	Aug. 5	9	19	0		0.1	371	24.3	409	75.5		440		0		0					
6	July 19	1,225	July 30	July 31	11	13	0		0		8.		44.		454		0	497	0					
7	Aug. 1	448	Aug. 11		10		0		0		0	395	90.		438	51.0			0					
	1906		1906	1906																				
8	Sept. 18–21	250	Sept. 26	Oct. 2	8	11–16	1.3	283	1.9	318	1.9	335	15.		399	45.4	424	20.4	498					
9	Sept. 21	314	Oct. 2	Oct. 8	11	17	0		0		0		14.0	385	65.9	415	6.1	445	9.5	477				

Daily rate of hatching in percentages with degrees of accumulative effective temperature.

Record number.	15 days.	°F.	16 days.	°F.	17 days.	°F.	18 days.	°F.	19 days.	°F.	20 days.	°F.	21 days.	°F.	22 days.	°F.	23 days.	°F.	24 days.	°F.
1	0		0		0		3.0		1.6	563	1.6	626								
2	0	590	1.3	531	0		0		.3		.7	518	0	544	575		611		655	695
3	1.8	632	0		0		0		0		0		0		93.4				15.6	
4	4.5		0		0		0		.1		0	770	0							
5	0		0																	
6																				
7																				
8	2.8	597	1.0		1.0	538	0	570												
9																				

[1] The thrips *Aleurodothrips fasciapennis* Franklin is responsible for death of 93.5 per cent of the eggs.   [2] Records include eggs hatching on previous day.

The conclusions presented herewith are drawn from the data presented in Table VIII, based upon daily observations of over 5,000 eggs deposited at intervals from February to October. From these and other data not included, it can be stated that the eggs hatch during a period of from 8 to 24 days after deposition, according to the season of the year. While there are no data regarding the length of the incubation period for eggs deposited by the few female flies occasionally seen during the winter months, as noted under seasonal history, it is probable that hatching extends over even a greater number of days during the winter season. The deposition of such eggs is, however, a comparatively rare occurrence and will receive no further mention here. In no instance have eggs been seen to hatch before the eighth day from deposition, even during the months July and August, 1907, when the average mean temperatures were slightly above normal, while under the most favorable summer weather conditions from 75 to 100 per cent of the eggs hatch during the period from the tenth day to the twelfth day from deposition.

In general, the warmer the season the shorter and more nearly uniform is the period of egg development or incubation. During the months of July and August, when the normal monthly temperatures at Orlando range from about 72° F. as the mean of the minimums to about 93° F. as the mean of the maximum records, practically all the eggs hatch from the tenth day to the twelfth day. Even at this most favorable season, in one instance hatching was delayed for 19 days. During the somewhat cooler weather of late September and early October and the decidedly cooler months of February, March, and April, hatching is more or less delayed according to the prevailing temperature and is scattered over a larger number of days. This same result is brought about, only in a less degree, by a cool period occurring in an otherwise warm season, as shown under record 4 (Table VIII).

Reference to the daily rate of hatching in Table VIII, and to the accompanying degrees of accumulated effective temperatures, shows that regardless of the time of year deposited and the number of days required for incubation, over 90 per cent of the eggs, on an average, hatch between the accumulation of from 375° to 475° of effective temperature.

Exception to this statement must be taken in records 1 and 2 (Table VIII). The number of degrees of effective temperature required seems to be greater at this season, although this might not prove to be the case if, as is probable, an error has arisen from using 43° F. as the basis for calculating the effective temperature.

Reference to the two preceding tables shows that considerable variation exists in the length of the egg stage among eggs deposited on the same day, or even within the same hour, and subsequently subjected to identical conditions of heat and moisture.   Even when hatching was most concentrated during the heat of summer and 99.8 per cent of the eggs hatched on the tenth and eleventh days from date of deposition, hatching extended over a period of from 9 to 19 days.   Hatching over a period of from 6 to 7 days after the first crawling young appears is an ordinary occurrence during the cooler portions of the season of activity.   In this respect, white-fly eggs are markedly different from the eggs of most other insects deposited in batches which usually hatch within one or, at the most, a few hours of each other.

PARTHENOGENESIS.

The existence of parthenogenesis among aleyrodids was first recognized by the senior author[1] in connection with his investigations of the greenhouse white fly (A. vaporariorum).   His prediction at that time that this method of reproduction would ultimately be proved to occur among many if not all the species of Aleyrodes has been strengthened by the results of the present investigations.   While there are no definite data to the effect that parthenogenetic eggs are deposited under natural conditions, there is practically no doubt that such deposition does occur, especially by females not yet mated or by females appearing at unseasonable times or when males are decidedly in the minority.   Scattered females emerging during the winter, or resulting from the comparatively few pupæ surviving fumigation, either never have the opportunity to mate or deposit many of their eggs before such opportunity presents itself.

That virgin females of A. citri, emerging from pupæ kept separately in vials, and later confined in rearing cages under normal grove conditions, except for the exclusion of males, will readily deposit the normal number of eggs, and that these eggs will develop normally and will produce adults of the male sex, has been thoroughly demonstrated. Of the five separate cage experiments started with parthenogenetic eggs, all of 111 adults emerging in four of the cages were males, while of 208 more adults emerging from the fifth cage, all but 4 individuals were males; the 4 females emerging under such conditions as to lead to the supposition that they came from fertile eggs overlooked in preparing the leaf for the experiment.

HATCHING.

In hatching, the egg membranes rupture at the end opposite the pedicel, and then split down each side sufficiently to permit the young larva to crawl out.   The glistening eggshell, somewhat resembling in appearance a bivalve shell, eventually becomes shriveled and loses its original form.

---

[1] Notes on Some Aleyrodes from Massachusetts, with Descriptions of New Species. Psyche, April, 1903, p. 81.   Technical Bulletin No. 1, Mass. Agr. Exp. Sta., pp. 31–33.

Proportion of eggs that hatch.—Observations covering many thousands of eggs, both in the cage experiments and in the grove, have demonstrated that the number of eggs that fail to hatch is too insignificant and has too little practical bearing to warrant the collection of data on this point. It is safe to say that considerably less than 1 per cent do not hatch. In fact, it seems evident that no egg would fail to hatch except owing to the dropping of the leaf or unless subjected to attack from without. In many instances failure to hatch can be directly traced to attack by several species of insects and a fungous parasite.

Effect of drying of leaves on hatching.—In 10 instances leaves bearing many thousand eggs were so placed that the eggs were exposed to direct sunlight or to partial shade, and although frequent observations were made none of the eggs were known to hatch. In general the drying of leaves to which eggs are attached prevents hatching of all except those eggs containing nearly mature embryos. This feature is probably common to all aleyrodids, since the senior author has noted a similar occurrence in the case of the greenhouse white fly (*A. vaporariorum*).

THE LARVAL AND PUPAL STAGES.

DESCRIPTION OF STAGES.

THE LARVA.

The larvæ [1] are thin, translucent, elliptical, scalelike objects, found usually on the underside of the leaves, though more rarely upon the upper surface. When normally attached to the leaf they are so nearly transparent as to be seen with difficulty. They readily become visible, however, by either bending or rubbing the fingers along the opposite side of the leaf, thus loosening them and allowing the air to get beneath them. They then appear whitish (Pl. X, fig. 2). So very inconspicuous are the live larvæ and their attack so unaccompanied by any visible effects on the leaves, aside from the blackening of the foliage, that their presence is very frequently overlooked by the casual observer. A detailed description follows:

First instar larva (figs. 4–6). Length, 0.3 to 0.37 mm.; width, 0.182 to 0.22 mm. Body flat, scalelike, somewhat swollen ventrally, especially in the cephalothoracic region; margin entire, with 30 small tubercles, each bearing a horizontally directed spine of which 6 cephalic and 4 anal are proportionately longer. Spines of second pair, counting from anterior end of body, arising from tubercles not on, but slightly posterior to, margin on ventral surface. Relative lengths of the 15 pairs of spines as follows:

Pair	1	2	3	4	5	6	7	10	11	12	13	14	15
Spaces	$\frac{1}{11}$	$\frac{1}{9.8}$	$\frac{1}{11}$	$\frac{1}{6.5}$	$\frac{1}{5.4}$	$\frac{1}{5.5}$	$\frac{1}{6}$	$\frac{1}{4}$	$\frac{1}{4}$	$\frac{1}{4}$	$\frac{1}{18}$	$\frac{1}{5.4}$	$\frac{1}{18.5}$

[1] The larvæ and pupæ are frequently called by many growers the "egg" of the white fly. This misapplication of terms should be discouraged as it leads to undesirable confusion when referring to the various stages through which the white fly passes during its growth from egg to adult.

No marginal wax fringe appears before or after crawling young settles. Cephalo-thoracic and thoracic articulations invisible; 8 or possibly 9 abdominal segments are seen with little difficulty. Segments at posterior end of body modified by vasiform orifice. Latter nearly semicircular in outline, somewhat longer than wide, bordered laterally by chitinous thicken-ings which do not meet posteriorly; operculum semicircu-lar, nearly equaling in size the vasiform orifice itself, cover-ing the ligula and bearing on its median posterior margin what appear to be two pairs of small spines, the penulti-mate pair of which is about twice as long as the ultimate.

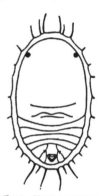

Ligula darker in color and broadly crescentic in shape. On either side of, and slightly anterior to, the vasi-form orifice is a short backwardly directed spine arising from a small tubercle. The two pairs of rounded, simple, reddish-brown eyes, less than 0.01 mm. in diameter and 0.096 mm. apart—a dorsal pair and a ven-tral pair—are situated mesad and slightly anterior to the fifth pair of marginal spines, the dorsal pair be-ing nearer the margin and slightly anterior to the ventral pair.

FIG. 4.—The citrus white fly: Crawling young; first in-star, dorsal view. Greatly enlarged. (Original.)

FIG. 5.—The citrus white fly: Crawling young; first instar, ventral view. Greatly enlarged. (Orig-inal.)

Antennæ, legs, and mouth-parts on the venter. Antennæ anterior and mesad to the anterior pair of legs, 0.1 mm. long, very slender; apparently 4-segmented, articulations between the segments seen with difficulty and frequently that be-tween the third and fourth entirely wanting, while in a few specimens the second segment appears to be divided into two parts: Segment 1 short, stout, fleshy; segment 2 one-half as wide and twice as long as segment 1; segment 3 narrower than segment 2 and about four times as long; segment 4 very slender, less than one-half as long as segment 3, and bearing on its proximal posterior side a minute spine, and distally a long spine. Legs

short, moderately stout, where ex-tended about one-third the width of the body; coxæ very short and stout, the two posterior pairs on the posterior inner side with a moderately stout spine about equal in length to the diameter of the coxæ and directed backward and inward; trochanters distin-guished with difficulty, about one-third as long as wide and collar-shaped; femora more elon-gate, slightly tapering distally, about four times as long as tro-chanters; tibiæ much narrower, somewhat longer than the femora,

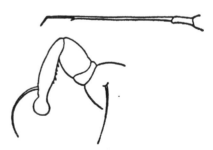

FIG. 6.—The citrus white fly: Antennæ and left hind leg, first instar. Highly magnified. (Original.)

with numerous short bristles, two on the outer proximal portion longer and more easily seen, on the outer distal portion with a long bristle forwardly directed and curving inward toward the tip of the tarsi; tarsi short, ending distally in an enlarged disk-like process.

Midway between the anterior pairs of legs in the middle of the body is the fleshy mouth papilla from which arise the mouth setæ, at first when bent backward reaching only to slightly beyond the posterior coxæ, but later becoming more elongate. Anterior to the mouth papilla is the semiovate prostomal plate, extending anteriorly as far as a line connecting the antennæ, and divided longitudinally by two curved sutures into one elongate median and two shorter lateral pieces. At the anterior end of the prostomal plate is a pair of small papillæ, each papilla bearing a small forwardly directed spine.

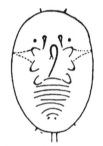

On the venter beneath and to the side of the vasiform orifice is a pair of spines arising from small tubercles, normally directed backward and outward, equal in length to the distal tibial spine.

Second instar larva (fig. 7).—Length, 0.37 to 0.43 mm ; width, 0.24 to 0.29 mm. Broadly ovate, dorsum densely rugose, all marginal tubercles and spines wanting except 2 cephalic and 4 anal, the three pairs, counting from the cephalic region, giving the relative lengths: $\frac{1}{9.5}$, $\frac{2}{4.5}$, $\frac{3}{10.5}$.

Eyes smaller and less regular in outline than in the first

instar, but distinctly evident. Antennæ greatly reduced, unsegmented, directed backward and slightly outward, tapering, reaching nearly to base of first pair of legs; on inside near base with a distinct spinelike projection, and on basal portion with numerous rougheuings; legs almost rudimentary, reduced to short, stout, fleshy processes without distinct segments, composed of a very stout, tapering basal portion, and a comparatively small, rounded, thick terminal disc; the second and third pairs of legs on the inner side at the base with a minute spine. Mouth parts as in previous stage; prostomal plate anteriorly indistinct and its pair of spines wanting. Spines on either side of vasiform orifice, both on dorsum and venter, as in first instar. A marginal pore, on either side of body opposite base of first pair of legs, and formed by an upward fold of the integument, becomes very evident in this instar.

Third instar larva (fig. 8).—Length, 0.62 to 0.78 mm.; width, 0.43 to 0.58 mm. Very similar to second instar but larger; the most striking difference presented by the antennæ, which have migrated backward so as to arise from a tubercle slightly anterior to base of first pair of legs. Antennæ immovable, directed mesad for about two-thirds of their length, and then suddenly doubled backward so that the distal third lies in the same plane as the basal portion. Legs smaller in proportion than in second instar and prostomal plate less developed, but the marginal pores and anal cleft more fully developed. A waxen rod is seen often protruding from the marginal pores. Relative lengths of the marginal spines: $\frac{1}{3-4}$, $\frac{2}{2.5}$, $\frac{3}{4.5}$

THE PUPA.

The introductory remarks regarding the general appearance of the larva apply with equal force to the young pupa (fig. 9, *a*, *b*, and *c*), with the exception that the pupa is larger, being nearly one-sixteenth of an inch long, is more easily seen, and on either side of the thoracic region 3 distinct curved lines representing the outlines of the legs are very distinct. As the pupa becomes older it becomes thicker, more rounded and opaque, and the outlines of the legs are obscured by the contents of the body. At the approach of maturity a bright red or orange spot develops on the back, and from three to eight days before emergence the eyes of the adult become visible. A detailed description is as follows:

Length, 1.10 mm to 1.40 mm; width, 0.60 mm to 1.0 mm. Body broadly elliptical, thin, not raised from leaf on vertical wax fringe, color pale yellowish-green, becoming more yellowish and thicker on approaching maturity; thoracic lobes, representing outlines of the three pairs of legs, and a line extending from between first two pairs of legs and from the vasiform orifice to edge of body distinctly more yellowish, as are also the lines representing the union of the body segments although these last are prominent. As body thickens thoracic lobes become less distinct due to body contents, a bright orange or red medio-dorsal spot develops at anterior end of abdomen, and later, a few days before emergence, the purple eyes of adult become very distinct, as also do the white developing wing pads; rim of vasiform orifice brown or yellowish. All marginal bristles lost except one anterior and one posterior pair of minute bristles. A low medio-dorsal ridge or carina and corresponding depressions on each side extend from the head to the anal ring, traversed by short transverse ridges on the thorax and abdomen, terminating in a low subdorsal ridge hardly perceptible; from these last numerous very fine granulated striæ radiate all around the body to the lateral margin. A short transverse ridge appears near posterior margin of head with a curved impressed line in front. A minute brown

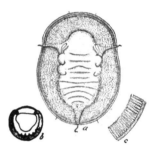

FIG. 9.—The citrus white fly: *a*, ventral aspect of pupa; *b*, vasiform orifice of same; *c*, margin of body of same. *a*, Greatly enlarged; *b*, *c*, highly magnified. (Original.)

tubercle at the anterior end of the subdorsal carina is sometimes to be seen. From a pore at the edge of the body, between head and thorax and top of anal slit, issues a very fine, glistening-white, curled thread of waxen secretion. These so-called "pores" in margin of the cephalo-thoracic region are formed by a slight upfolding of the body which extends from margin to cephalo-thoracic spiracle and forms an outlet for secretions from same. Location of spiracles· and respiratory system as already described for aleyrodids. Legs and antennæ easily seen with high-power lens. Antennæ located as shown in fig. 9, partially concealing front pair of legs, apparently 3-segmented but division into segments not distinct; last segment as long as other two combined, with quite a number of irregular annulations; tip provided with a stout spine. Legs short, very stout, especially the two posterior pairs; front legs projected forward; all without distinct segmentation; tarsus very short, stout, and rounded. Vasiform orifice nearly semicircular (for details and shape see fig. 9, *b*).

Pupa case.—White, firm, retaining definite shape, and remains firmly attached to leaf unless forcibly detached. (See Pl. VIII, fig. 1.)

DURATION OF STAGES

LARVAL INSTARS.

Data upon the duration of the larval instars nave been secured by daily observations of over 300 specimens marked as soon as the young larvæ had settled, supplemented by frequent counts of several thousand specimens in various life-history cages.

From these records those included in Table IX have been chosen as representative. A study of these will give a very accurate knowledge of this subject, and will impress upon one the considerable variation in the duration of the several instars of larvæ hatching at the same time, feeding upon the same leaf, and consequently subject to the same weather conditions. The data also emphasize the retarding effect of cool spring and fall weather upon the length of larval life, although this has not been found to be as great as many have thought. The period of larval growth ranges from an average of 23 days during the warmest months to an average of 30 days during the cooler months.

TABLE IX.—*Duration of larval instars of the citrus white fly.*

Specimen No.	Period of growth.	Instar 1.	Instar 2.	Instar 3.	Sum of effective temperatures.	Specimen No.	Period of growth.	Instar 1.	Instar 2.	Instar 3.	Sum of effective temperatures.
1	Mar. 22–Apr. 26	13	10	12	1,044	33	Oct. 3–Oct. 27	7	5	12	736
2	...do...	11				34	Oct. 3–Oct. 30	5	8	14	811
3	...do...	9				35	Oct. 3–Nov. 11	7	15	17	1,103
4	...do...	10				36	Oct. 3–Oct. 29	7	8	11	792
5	...do...	11				37	Oct. 3–Dec. 4	31	14	17	1,644
6	...do...	9				38	Oct. 3–Nov. 4	12	7	13	922
7	...do...	9				39	Oct. 3–Nov. 1	8	8	13	843
8	June 26–July 18	8	7	7	898	40	Oct. 3–Nov. 2	8	7	15	868
9	June 26–July 19	8	5	10	938	41	Oct. 3–Nov. 3	9	7	15	894
10	June 26–July 16	8	5	7	813	42	...do...	11	7	13	894
11	June 26–July 18	7	6	9	898	43	Oct. 3–Nov. 1	8	8	13	843
12	June 26–July 20	8	4	12	978	44	Oct. 3–Nov. 10	7	4	17	1,040
13	June 26–July 18	8	5	9	898	45	Oct. 3–Oct. 31	8	8	12	598
14	June 26–July 29	8	5	20	1,365	46	Oct. 3–Nov. 4	10	7	16	922
15	June 26–July 18	8	5	9	898	47	Oct. 3–Nov. 8	7	7	22	1,012
16	June 26–July 21	8	5	12	1,015	48	...do...	7	10	19	1,012
17	June 26–July 18	8	5	9	898	49	Oct. 3–Nov. 2	7	9	14	868
18	June 26–July 16	7	5	8	813	50	Oct. 5–Nov. 1	7	7	13	788
19	June 26–July 18	8	5	9	898	51	Oct. 5–Nov. 7	5	17	11	902
20	June 27–July 19	7	5	10	902	52	Oct. 5–Nov. 8	5	8	15	788
21	June 27–July 20	8	5	10	942	53	Oct. 5–Nov. 8	8	8	15	922
22	...do...	7	8	8	942	54	Oct. 5–Nov. 5	8	8	14	898
23	June 27–July 16	7	5	7	777	55	Oct. 5–Nov. 1	8	6	13	788
24	June 29–July 21	6	6	11	988	56	Oct. 5–Nov. 3	9	8	12	894
25	June 28–July 28	7	9	14	1,220	57	Oct. 5–Nov. 5	8	9	14	898
26	June 28–July 18	6	5	9	821	58	Oct. 5–Nov. 11	10	6	11	788
27	June 29–July 27	5	4	19	1,141	59	Oct. 5–Nov. 19	11	34	died.	...
28	July 28–Aug. 18	5	4	12	866	60	Oct. 5–Oct. 31	6	8	13	767
29	Sept. 28–Nov. 14	7	10	30	1,365	Average	{ June 26–Aug. 18	7.2	5.4	10.5	945.7
30	Sept. 30–Oct. 25	7	6	12	780		{ Sept. 28–Dec. 4	[1]7.8	[2]8.3	14.3	903.4
31	Sept. 30–Oct. 23	6	7	10	727						
32	Oct. 1–Oct. 21	7	9	7	635						

[1] Does not include No. 37.   [2] Does not include No. 59.

PUPA CASES OF THE CITRUS AND THE CLOUDY-WINGED WHITE FLIES.

Fig. 1.—Leaf showing pupa cases of *Aleurodes citri;* also a few pupæ and eggs   Fig 2.—Under surface of orange leaf, showing heavy infestation by citrus white fly.   Fig. 3 —Leaf showing pupa cases of *A. nubifera.*   Note delicate structure as compared with those of *A. citri.* (Original)

PUPAL STAGE.

One of the most interesting phases of life-history studies has been the wide range in the duration of the pupal stage; a range of from 13 to 304 days. Considering the relatively slight variation in the length of the larval life, this range among specimens passing into the pupal stage at practically the same time is remarkable. In view of the fact that the effect of this variation upon the duration of life and number of annual generations will be fully discussed under those headings and brought out in Tables XV and XVII and figure 12, only a few of the large number of records on file are given in Table X to illustrate this range in pupal life during different parts of the year.

TABLE X.—*Duration of pupal stage of the citrus white fly.*

Specimen No.	Period of growth.	Number of days.	Sum of effective temperatures.	Specimen No.	Period of growth.	Number of days.	Sum of effective temperatures.
1	Apr. 30–May 13.....	13	410	11	Aug. 15–Sept. 6....	22	885
2	Apr. 30–June 20.....	51	1,833	12	Aug. 16–Mar. 18....	214	5,414
3	Apr. 30–Aug. 3.....	64	2,564	13	Aug. 16–Mar. 20....	216	5,479
4	May 18–June 5......	18	664	14	Aug. 17–Mar. 25....	220	5,574
5	May 18–July 31......	74	2,866	15	Aug. 18–Sept. 10...	23	931
6	May 18–Mar. 18.....	304	....	16	Aug. 18–Apr. 1....	226	5,752
7	July 15–July 30.....	15	602	17	Sept. 30–Mar. 31...	182	4,473
8	July 15–Aug. 4.....	20	808	18	Oct. 28–Apr. 19....	173	4,167
9	Aug. 15–Aug. 27....	12	479	19	Nov. 1–Apr. 17....	167	4,009
10	Aug. 15–Aug. 28....	13	521	20	Nov. 8–Mar. 25....	137	3,256

It will be noticed that pupæ pass either a comparatively few or a comparatively large number of days in this stage and that ordinary temperatures and humidity do not have the power to determine which it shall be.

LOCOMOTION.

On hatching from the egg the young larva is provided with well-developed legs, as shown in figure 5, by the aid of which it crawls about the leaf for several hours and then settles and begins to feed. Because of the aimless way in which it crawls, frequently doubling on its own course and turning aside for the least obstacle, it travels over a very limited area. It is therefore improbable that the crawling larvæ ever leave the leaf upon which they were hatched, unless carried on the feet of birds or insects or blown or dropped from one leaf to another. After settling, the larva does not change its position on the leaf, while with the first molt its legs become vestigial (see fig. 7) and unfit for locomotion. Larvæ frequently move slightly, especially directly after or during molting when they merely describe an arc of 180°, using their mouth parts as a pivot. The larva passes into the pupal stage without materially changing its position on the leaf. The only time, then, during the life cycle when the white fly is capable of moving about from place to place is during the winged adult stage and the crawling larval stage.

Pronounced and striking growth in size occurs only at molting, when the soft flexible skin of the larva or pupa is able to stretch before assuming its normal rigid condition. With each successive molt the larva greatly increases its horizontal dimensions, until by the time it reaches the pupal stage these are about eighteen times as great as in the newly hatched larva. When first settled after molting the larva is very thin, papery, and transparent, being seen with difficulty except with the aid of a lens, but after feeding several days it slowly becomes thickened until, from two to five days, sometimes longer, before molting into the next instar, it is decidedly plump and whitish opaque in color. Oftentimes before molting the larva becomes very much swollen as though gorged with liquid. This appears to be an abnormal condition, since many that become thus unduly enlarged either fall or die without molting. During the increase in thickness following feeding, there is no increase in the horizontal dimensions. On the contrary, increase in the former is secured at a slight expense of the latter.

Daily observations on over 300 marked individuals from time of settling to emergence of adult have conclusively demonstrated that the larva passes through but three instars [1] before reaching the pupal stage, instead of four as has been previously supposed. Each larva, then, molts or casts its skin three times before becoming a pupa. The process of molting was first described by Riley and Howard [2] and as observed by the authors is as follows:

In preparing for a molt the insect curves the abdomen upwards at considerably more than a right angle, moving it also occasionally up and down. The margin of the abdomen has at the same time a slightly undulating motion. During these movements the insect is shrinking away from the lateral margin until it eventually occupies only about one-third of the original lateral space, causing a distinct dorsal and ventral median ridge. The skin then splits, not on the dorsum, as would be expected, but either at the anterior end or underneath the head. The head and prothorax are then pushed out and the skin is gradually worked backwards by means of the abdominal motions, the portion already out swelling as soon as it is free.

As the insect flattens after molting it appears milky white, the head, thoracic lobes, and abdominal segments being more greenish. At this time the legs, which resemble much the prolegs of a caterpillar, are very active, and there appears a pair of fleshy protuberances more or less movable, not as large as the legs, but apparently of the same

---

[1] This agrees with the senior author's observations on the greenhouse white fly (*A. vaporariorum*) and the strawberry white fly (*A. packardi*), which are the two species of the genus which have previously been studied in greatest detail. Tech. Bul. 1, Mass. Exp. Sta. and Can. Ent., vol. 35, pp. 25–35.

[2] Insect Life, vol. 5, p. 223, 1893.

structure, which act as sucking disks to aid the insect in reattaching itself. These protuberances are later withdrawn so that no trace of them remains. While becoming attached to the leaf the insect may be seen occasionally to rotate itself through an arc of 270°, in the meanwhile frequently raising and lowering the abdomen. The cast skins are usually blown away by the breeze or fall from the leaf as soon as molted, but not infrequently are found partially pinioned beneath the body of the insects. Molting occurs most actively during hours of high humidity. Newly molted larvæ are abundant during the early morning when the humidity ranges between 100° and 90°.

FEEDING HABITS OF LARVÆ AND PUPÆ.

As the white flies, or Aleyrodidæ, belong to the Hemiptera, or sucking insects, the larvæ and pupæ do not eat the tissue of the leaf, but insert their thread-like mouthparts and suck the plant juices by the aid of a suction apparatus located in the head. Their ravages are not accompanied by any visible effect upon the leaf itself, but may be detected by means of the sooty mold which develops after the fly becomes very abundant. Our only means of estimating the amount of sap taken up by the insect is by the amount of waste material, or honeydew, ejected by it. A first-instar larva, on being watched under the compound miscroscope for 20 consecutive minutes with the temperature at 90° F., was seen to eject honeydew 48 times, or an average of about 10 times every 5 minutes. A pupa with well-developed eyespots, in March, with the temperature at 85° F., ejected honeydew 4 times in 5 minutes. This difference in the amount of honeydew secreted is due in part to the different temperatures at which the observations were made as well as to the difference in the degree of development.

A very interesting observation on the amount of sap extracted by larvæ and pupæ of the white fly has been made by Dr. Berger [1] Leaves with live larvæ and pupæ were placed between glass plates so that the ejected honeydew was collected on the glass. By weighing it was found that each live insect had excreted about 0.0005 gram in 48 hours. At this rate a tree infested with 1,000,000 white-fly larvæ and pupæ would lose one-half pound of sap per day.

THE ADULT.

The adult citrus white fly is very small, measuring only about one-sixteenth of an inch in length, and with a wing expanse of less than one-eighth of an inch. The natural color of the body, antennæ, legs, and wings is entirely obscured by secretions of delicate white wax particles, so that the insect appears snowy white (Pl. IX; text fig. 10, a–i)

---

[1] Bulletin 97, Florida Agricultural Experiment Station, pp. 63–64, 1909.

without spots or traces of darker shades upon the wings. Only the purple eyes are free from the white wax, and are in sharp contrast to the color of the rest of the body. A detailed description of the adult, by Riley and Howard, follows:

DESCRIPTION.[1]

♀ .—Length, 1.4 mm.; expanse, 2.8 mm.; four-jointed rostrum about as stout as legs; joint 1 shortest, joint 2 longest, and about as long as 3 and 4 together; joint 3 somewhat longer than joint 1 and a little shorter than 4. Joint 1 of the 7-jointed antennæ very short, as broad as long, subcylindrical, slightly wider distally; joint 2 twice as long as 1, strongly clavate, and at tip somewhat broader than 1, bearing 3 or 4 short hairs arising from small tubercles; joint 3 longest, about twice as long as 2, slenderer than this and with a very narrow insertion, rather abruptly stouter at apical third, corrugated and terminating above in a small callosity resembling a similar organ in Phylloxera; joints 4

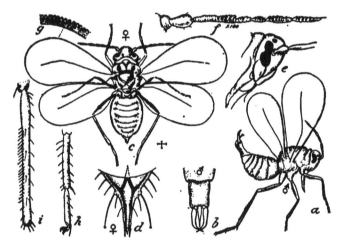

Fig. 10.—The citrus white fly. Adult. *a*, Male; *b*, claspers of male; *c*, female; *d*, ovipositor of female; *e*, side view of head of female; *f*, antenna; *g*, enlarged margin of wing; *h*, tarsus and claws; *i*, tibia. *a*, *c*, Greatly enlarged; *b*, *d–i*, more enlarged. (Adapted from Riley and Howard.)

and 5 subequal in length, each nearly as long as 2, joint 5 bearing a short spine anteriorly near apex; joints 6 and 7 subequal in length, each somewhat longer than 2, 7 with a stout spine at tip; joints 4 and 7 somewhat corrugate or annulate but less so than apical third of 3. The 2-jointed tarsi about half the length of the tibia, joint 1 of the hind tarsus bearing 6 rather stout spines on each side; joint 2 supporting at base 3 rather prominent claws, the middle one longest. Ovipositor short, acute, and retractile. Eyes divided into two by a curved pointed projection from middle of cheek, the upper portion being smaller than the lower portion. Wings clear, colorless; costa delicately serrate. General color, light orange yellow, tip of rostrum black, tarsi and part of tibia orange.

♂ .—The male resembles the female in all important respects except in being smaller. Claspers about as long as preceding abdominal joint, or one-fifth the length

[1] Riley and Howard, Insect Life, vol. 5, p. 222, 1893.

ADULTS OF THE CITRUS WHITE FLY ON FOLIAGE OF ORANGE.

Fig. 1.—Tender growth swarming with adults. Fig 2.—Leaf of same enlarged. (Original.)

of the abdomen, curved gently upward and inward, each bearing 4 or 5 equidistant minute cylindrical piliferous tubercles on upper and outer edge; style almost as long as claspers, rather stouter at base, more slender toward tip, terminating in a stout spine at upper end. Head and abdomen with heavy tufts of wax soon after issuing from pupa.

Examination of a large number of antennæ shows that the relative length of the antennal segments is subject to slight variations. The average relative lengths are about as follows:

$$\frac{\text{Segment}}{\text{spaces}^1} \quad \frac{1}{10}, \frac{2}{24}, \frac{3}{43}, \frac{4}{16}, \frac{5}{16}, \frac{6}{18}, \frac{7}{22}, \frac{\text{Spine.}}{3}$$

Although they have examined thousands of males both at and for some time after emergence and as they occur at all times throughout the grove, the authors have never been able to observe males with the tufts of wax on head and abdomen mentioned in the above description and illustrated in connection with its original publication.

EMERGENCE.

DESCRIPTION OF THE PROCESS.

The emergence of the adult occurs soon after its purple eyes and folded whitish wings can be seen distinctly through the pupal skin. About 20 minutes before the pupal skin is ruptured the body of the adult shrinks gradually away from it and assumes its natural shape. This gradual shrinking away from the edges of the pupa, and the accompanying thickening of the body, brings a pressure to bear on the pupal skin which causes it to split from margin to margin between the thorax and abdomen and along the median line from this rupture to the anterior margin. Through the T-shaped opening thus formed the insect first pushes its thorax, then its head, with little apparent exertion. The body now projects almost perpendicularly from the pupa case, as the pupal skin is called, with the antennæ, legs, and abdomen still in their pupal envelopes. By a series of backward and forward movements the antennæ and legs are freed from their membranes and are in constant motion. The abdomen is now so nearly out of the pupal case that the fly is practically free, holding on only by means of the end of the abdomen. With a sudden forward bend of the body the legs are brought in contact with the leaf, and with their aid the fly frees the rest of its abdomen and crawls away rapidly.

The period covered between the time the insect ruptures the pupal skin and the time it becomes entirely free from the case and is crawling is from 7 to 10 minutes. Not infrequently flies die during emergence.

---

[1] These represent the spaces read on eyepiece micrometer when 1-inch eyepiece and ½-inch objective are used, and the miscroscope tube is drawn to 160.

CHANGES AFTER EMERGENCE.

Immediately after emergence from the pupa case the adult differs from the more mature individuals in that the lemon-yellow color of the body is not obscured by the white waxy secretion that subsequently appears. Also the wings, which appeared as crumpled whitish pads when the thorax was first protruded from the pupa case, have had time only to partially expand. As the fly crawls away from the case the wings are held perpendicularly above the back, but as the wings gradually unfold and assume their normal shape they are lowered to their natural position. It requires about 7 minutes for the wings to become straightened after the fly leaves the case, and from about 14 to 17 minutes from the time they first begin to expand. When fully expanded, the wings are colorless and transparent, with the costa pale yellowish. The powdery whiteness so characteristic of the flies as seen in the grove gradually appears as the wax glands secrete their particles of wax. In about one and three-fourths hours the wings and body have become perfectly white.

CONDITIONS AFFECTING EMERGENCE.

Aside from that inherent influence affecting the development of the citrus white fly and determining whether the adult shall emerge during the first or second general emergence period, as hereinafter described under "Seasonal history," many field observations made at all seasons during the past three years, supplemented by laboratory experiments, have emphasized the great influence which temperature has on emergence. While a normal amount of humidity is necessary for emergence to occur, it is not so controlling a factor as temperature during ordinary Florida weather, as will be shown later. Light also seems to affect emergence under certain conditions.

EFFECT OF TEMPERATURE ON EMERGENCE.

Conclusions drawn from field notes, supplemented by laboratory experiments, show that emergence seldom occurs outside the range of 62° F. to 85° F., with preference to temperatures ranging from 70° to 85°. During the winter months of December, January, and February, when the average monthly mean is about 60° F., no emergence occurs except to a slight degree during warm spells of several days' duration. In January, 1906, when the average monthly mean temperature was 59.6°, or practically normal, no flies were noted on wing at Orlando, Fla., except in small trees beneath pinery sheds where the temperatures averaged several degrees higher than outside. During late December, 1908, and early January, 1909, the temperature had been sufficiently high to cause a limited amount of new growth to appear on some trees in Orlando, and on January 4 a comparatively large number of adult white flies were seen feeding and depositing eggs on new growth in a very sheltered place. The average mean temperature of the 6 days preceding this observation was 66° F., while, for the 6 days preceding these, when no white flies were on wing, the average mean was about 58.5° F. Adult white flies were not seen in the laboratory grove in February, 1909, until about the 20th, or until the temperature records for the grove showed an average daily mean of about 64.5°. It is from the above facts that the lowest temperature at which emergence occurs has been determined to be about 62° F. This conclusion, drawn from general field observations, is strengthened by emergence records kept in connection with cage life-history work during the period of active spring emergence of March, 1908, when the monthly mean was 71°. Reference to the data contained in Table XI brings out the fact that while

TABLE XI.—*Relation of temperature to emergence of the citrus white fly.*

Date.	Range in temperature.	Average mean temperature.	Emergence records.			
			No. 1.	No. 2.	No. 3.	No. 4.
1908.	°F.	°F.				
Mar. 19	62–89	75.5	4	....	6	2
Mar. 20	63–90	76.5	5	18	78	4
Mar. 21	54–65	59.5	0	0	1	1
Mar. 22	60–80	70.0	10	15	69	15
Mar. 23	66–90	78.0	17	35	71	4

emergence had been going on actively two days before and after March 21, when the average mean temperature was about 75° F., a drop in the mean temperature on the 21st to 59.5° F. practically prevented any white flies from emerging. The 1 white fly that is recorded under Nos. 3 and 4 may have emerged on the 20th after the daily record had been taken. Such emergences are not rare at this season of the year, as will be shown later. Of 2 lots of about 100 pupæ each, from which adults were nearly ready to emerge, 1 was placed in a refrigerator at about 56° F, and the other kept at room temperature which ranged between 70° F. and 80° F. while emergence was taking place. Of those kept on ice, but 1 white fly emerged during the first 12 hours, as compared with 17 from pupæ kept at room temperature. White flies continued to emerge on 3 consecutive days from pupæ kept at the latter temperature. No more emerged from the refrigerated pupæ. It is therefore evident that emergence may occur at as low a temperature as 56° F., though very rarely.

That white flies seldom emerge after the temperature reaches 85° F. may be concluded from the following facts: During the months of July and August, when the average daily mean is about 82° F., a newly emerged adult is rarely seen in the grove after 8 a. m. Practically all adults at this season emerge between 4 and 7 a. m. This is true both in the laboratory and in the grove. Of 233 white flies emerging separately in vials in the laboratory during August, 1907, 212 emerged between 3.30 and 8 a. m., and the remaining 21, with one exception, emerged between 8 and 9.30 a. m. In the grove over 95 per cent of the white flies emerge before 7 a. m. At this time of day the temperature ranges between 70° F. and 85° F. During the early spring, when the daily maximum temperature does not usually exceed 85°, emergence is not restricted to the early morning as during the heat of summer, but occurs at all times of the day. It may also be added that like conditions exist in October and November, but because of difference in seasonal history, they affect chiefly the spotted-wing white fly.

EFFECT OF HUMIDITY ON EMERGENCE.

Under normal Florida conditions at Orlando, at any season of the year, the relative humidity rises to nearly or quite 100 per cent by from 6 to 10 p. m., and there remains until about 6 a. m., when it normally drops rapidly, sometimes to as low as 19 per cent, though more often to from 35 to 60 per cent. It has already been stated that over 95 per cent of the white flies will have emerged before 7 a. m. or before the humidity has fallen far from the saturation point. That temperature and not humidity is the more important factor governing emergence in Florida, can be inferred by a comparison of the humidity and temperature records of Table XIII. It so happened that the cold wave of March 21, 1908, was accompanied by a higher average humidity, but the temperature and not the humidity prevented adults from emerging. Again, during the spring, when the daily maximum temperature is seldom above 85°—usually less—emergence goes on even at midday when the humidity has dropped to as low as 33°. In this connection attention should be called to the fact that the humidity in the corked vials mentioned under the preceding heading remained at about 100 per cent throughout the greater part of the experiment.

There are, however, times of abnormal weather conditions when lack of humidity seems to play an important part in preventing emergence. During the month of March there sometimes occur dry winds of several days' duration, accompanied by more or less heat, which seriously check emergence, and, as far as can be determined, cause many pupæ from which adults are about to emerge to die.

Two such periods occurred during March, 1909, from the 3d to the 6th, and from the 25th to the 27th, respectively. During these periods the relative humidity was extremely low, on one day dropping to 19 per cent. For 42 hours during the latter period the humidity ranged below 50 per cent and for 36 hours above 50 per cent. During these periods emergence was noted to be seriously checked and at the end of the latter upward of 30 per cent of the pupæ were dead, apparently from no other cause.

During the summer months light seems to have an influence on emergence. At this season emergence in the laboratory and grove begins at about daybreak. Observations made at hourly intervals on the emergence of 233 adults, from 3.30 and 4 a. m. show that white flies rarely emerge before this time. In one instance only about one-third as many white flies emerged from pupæ kept in the dark as from those kept in the open, and their emergence was noticeably delayed. During the cooler months the low morning temperatures prevent the white flies from responding to this apparent stimulation due to light, and they emerge at various times after the temperature has risen sufficiently high.

Without food.—In none of the experiments conducted to determine the length of adult life without food have white flies lived longer than 30 hours, and a very large percentage has died before the end of 24 hours. When confined on leaves of plants other than those recognized as food plants, life is usually longer than this, but never approaches the normal length. White flies confined on crape myrtle in July died as soon as those kept in empty cages, but flies caged on oak, in March, lived as long as 4 days; those on fig, in August, 3 days; and on banana shrub, in July, 2 to 3 days. In all these tests flies were placed only on the tenderest growth.

With food.—Adult life under normal outdoor conditions averages about 10 days, although individual white flies kept in cages have been known to live as long as 27 days. Adults are so fragile and so easily killed by winds and heavy showers and by numerous species of spiders and ants that their duration of life is at most very uncertain. Cage experiments during March, April, July, August, and September show that, in the cages at least, there is little difference in the length of life at various times of the year.

The courtship of the citrus white fly has been observed to begin within 2 hours after emergence, and in one instance even before the wings of either male or female had become whitened. There is no time in the day when the males can not be seen courting the females. The male appears unable to locate the female at a distance much greater than one-fourth of an inch, according to Prof. H. A. Gossard. Observations made during the present investigations show that when males and females are placed in separate receptacles and separated only by a very porous cheesecloth they show absolutely no attraction to each other. Mating, therefore, is not so likely to occur when the adults are scarce, as it seems to be the result of chance meeting upon the leaves rather than to such a definite attraction as exists between males and females of many moths.

Upon detecting the female, the male approaches her nervously, stopping at intervals, especially as the distance lessens, and swinging his body about excitedly in a semicircle, the head being used as a pivot, his wings in the meanwhile opening and closing spasmodically. While no movement is made by the female, she is repeatedly approached from many directions before coition occurs. More often the male lies alongside the female and courts her in this position, raising and lowering his wings as above described, and raising and swinging his abdomen from side to side. During these antics of the male the female remains quiet, only occasionally flittering her wings. While males may be seen courting females at all times of the day, it is seldom that one sees a pair in coitu except late in the afternoon and evening. Because of the uniformity of color and the ease with which adults are disturbed and made restless the duration of copulation can not be stated with certainty, but it probably lasts but a short time. Experiments to determine the duration of fertility have thus far proved unsuccessful.

OVIPOSITION.

AGE AT BEGINNING OVIPOSITION.

Virgin females in confinement have deposited eggs within 6 hours after emergence. In one instance 35 virgins deposited 58 eggs between 5 and 9½ hours after emergence during summer weather, with the temperature ranging from 80° to 92° F. However, even at this temperature single females occasionally did not deposit eggs for over 24 hours. Prof. H. A. Gossard[1] states that egg laying begins at from 18 to 30 hours after the emergence when the temperature ranges from 65° to 75° F. Laboratory tests have shown that lack of fertilization does not prevent a female from depositing eggs, but that she will readily deposit infertile eggs until opportunity for mating presents itself.

PORTION OF PLANT SELECTED.

If not numerous, the females deposit almost exclusively on the under surface of the leaves, laying over 75 per cent of their eggs on the half of the leaf bordering the midrib. It is only when very abundant and pressed for room that they deposit eggs thickly over the entire lower surface and more sparingly on the upper surface, the petioles, and the stems of twigs. Next to the portion bordering the midrib, the natural depressions and the curled margins of the leaf, especially

---

[1] Bul. 67, Fla. Agr. Exp. Sta., p. 609, 1903.

of the tender growth, are favorite places for oviposition, and not infrequently as high as 40 per cent of the eggs are there laid, even when the adults are not very abundant. Although eggs may be deposited along the leaf margin, it is seldom that they are laid on the margin itself, as is the case with the cloudy-winged white fly. Even when not crowded for leaf space, the adults sometimes settle upon the under side of young fruit, where they deposit eggs freely and apparently feed.

DAILY RATE OF OVIPOSITION.

Previous to these investigations no data have been published on the daily rate of oviposition. In obtaining the data given in Table XII, the females recorded were collected at random throughout the grove, without regard to age, and, together with males not mentioned, were caged over leaves cleaned of all eggs and larvæ of the white fly and allowed to remain the recorded time, when the adults were removed and the eggs counted.

TABLE XII.—*Daily rate of oviposition of the citrus white fly.*

Record No.	Date deposited.	Number of females.	Duration of egg laying.	Number of eggs deposited.	Average number of eggs per female per 24 hours.	Average mean temperature.
			Hours.			*° F.*
1	Feb. 23-24, 1909........	14	26	197	13	74.5
2	Apr. 11-15, 1907........	3	103	98	7.6	68.2
3	Apr. 20-21, 1909........	50	24	454	9.1	77.2
4	Apr. 21-22, 1909........	40	24	405	10.1	78.2
5	June 16-17, 1909........	30	24	360	12	82
6	July 17-18, 1907........	255	21	2,533	11.3	82
7	July 22-23, 1907........	105	24	1,216	11.6	85
8	July 24-26, 1907........	50	46	1,331	13.8	84
9	Aug. 17-18, 1907........	70	24	805	11.5	81
10	Sept. 21-22, 1908........	35	24	405	11.6	79

Number eggs per day per female, grand average, 11.2.

The generally uniform results obtained in the nine records when the average mean temperature was about 75° F. or above, together with the grand average daily rate of oviposition' for individual females whose age was definitely known, as shown in Table XIII, indicate that each female normally deposits on an average 10 or 11 eggs a day. Varying degrees of temperature above a daily mean of 75° F. do not correspondingly increase the number of eggs deposited. However, temperatures below an average mean of 72° F. (estimated) have a distinct checking effect upon oviposition, as shown by record No. 2.

Notwithstanding the general average number of eggs per day deposited by the females of all ages in Table XII, and the same for the females of known ages for the total number of days they lived, in Table XIII, reference to the daily oviposition records in the latter shows that as many as 14, 19, 27, or even 33 eggs may be deposited by a single female in one day. It will also be noted that much variation exists between the number of eggs deposited by several different females on the same day and by the same female on successive days without any apparent reason, and that there exists no appreciable difference between the rate of deposition by virgin and by fertilized females.

TABLE XIII.—*Number of eggs deposited by single females of the citrus white fly.*

Fly No.	Date of first daily record.	Condition of female.	Daily rate of oviposition by individual females.											
			1	2	3	4	5	6	7	8	9	10	11	12
1	Aug. 8, 1907..	♀+males 12 hours old.........	8	19	9	22	11	2	15	6	(¹)	.....	..v.	....
2	....do.......	....do....................	5	19	14	12	12	15	13	14	(¹)	.....	....	....
3	....do.......	....do....................	3	9	13	5	13	9	16	8	1	15	19	14
4	....do.......	Virgin 12 hours old...........	5	12	17	4	3	20	5	1	0	6	3	6
5	Aug. 15, 1907.	Virgin 8 hours old............	5	13	15	14	12	12	15	13	14	10	9	9
6	....do.......	....do....................	6	16	17	5	33	13	23	16	15	14	10	3
7	....do.......	♀+males 8 hours old..........	8	8	16	14	17	11	16	7	1	4	2	4
8	....do.......	....do....................	2	18	16	14	9	2	13	15	6	9	9	1

Fly No.	Daily rate of oviposition by individual females.												Total number of eggs laid.	Number of days lived.	Average number of eggs per day.
	13	14	15	16	17	18	19	20	21	22	23	24			
1	.....	....	.....	.....	.....	.....	.....	.....	.....	.....	.....	.....	92	8	11.5
2	.....	....	.....	.....	.....	.....	.....	.....	.....	.....	.....	.....	104	8	13
3	13	9	5	14	10	8	6	9'	6	1	(¹)	.....	206	22	9.4
4	5	10	2	2	3	0	8	(²)	.....	.....	.....	.....	112	19	6.0
5	5	20	11	2	(²)	.....	.....	.....	.....	.....	.....	.....	179	16	11.2
6	19	....	11	6	4	(⁴)	.....	.....	.....	.....	.....	.....	211	17	12.4
7	8	3	14	12	2	(¹)	.....	.....	.....	.....	.....	.....	146	17	8.6
8	1	27	3	(⁵)	.....	.....	.....	.....	.....	.....	.....	.....	144	14	10.3

¹ Dead.
² Dead; 13 eggs in abdomen.
³ Dead; 13 eggs in abdomen.
⁴ Dead; 11 eggs in abdomen.
⁵ Dead; 7 eggs in abdomen.

Number of eggs per day per female, grand average, 10.3.

The eight records in Table XIII are selected from about forty similar records on file and are considered as representing an average condition of oviposition. Although the general average of 10.3 eggs per day throughout life for the 8 females recorded in Table XIII agrees very closely with the similar average obtained in Table XII, there is sufficient evidence in the data in Table XIII to warrant the statement that the daily rate of oviposition for individual females is usually greater during the early part of the insect's life and decreases with each successive week of existence. Leaving out of consideration the first day, when the flies had not reached their normal egg-laying capacity, a little calculation shows that the average daily deposition for the three successive weeks is 12.8, 8.5, and 6.1, respectively. This same decrease in the number of eggs deposited with increase in age is perhaps better brought out by a study of the number of eggs deposited by the individual females over 5-day periods. Thus No. 6 averaged 16.2 eggs per day for the first 10 days, but for the next 5 dropped to an average of 7.8; Nos. 4 and 7 showed a sharp falling off during the second 5 days. No. 3 is an exception to the above statement, maintaining an average of from 9.8 to 11.8 eggs per day for the first three periods of 5 days each, and during the fourth period of 5 days deposited as many eggs as during the first 5 days. It will be noticed, however, that No. 3 deposited comparatively few eggs during the early part of her life. In view of the fact that the average adult life is only about 10 days, the higher rate of deposition during early life has an influence on multiplication.

NUMBER OF EGGS DEPOSITED BY SINGLE FEMALES.

First mention of the egg-laying capacity of the citrus white fly was made by Riley and Howard,[1] who based their conclusions on the number of eggs that could be counted in the abdomen of the females when mounted in balsam, and not upon daily counts of eggs deposited by the females throughout life. Their estimate of about 25 eggs as the probable total number of eggs deposited by a single female during life has been generally accepted by subsequent writers, none of whom has ever placed the maximum number deposited above this figure. The present investigations, however, have demonstrated that this estimate is far too low and that the number of well-developed eggs to be found in the abdomen of the female at any one time is not indicative of the number of eggs deposited throughout her life. Females have been known to deposit more than this number of eggs in a single day. As will be seen by reference to Table XIII, as many as 211 eggs have been actually deposited by one female, and should the 11 well-developed eggs found in her abdomen at death be added a total of 222 eggs would be obtained. As this female, No. 6, lived but 17 days and others have been known to live 28 days, it is even probable that as many as 250 eggs more nearly represent the maximum egg-laying capacity under most favorable conditions. However, it is seldom that a female lives sufficiently long to deposit her full quota of eggs. With the average length of adult life curtailed to about 10 days, the average of 149.2 eggs per female, as shown in Table XIII, is beyond doubt high. An average of 125 eggs per female is nearer the number of eggs deposited during life in the grove.

ACTIVITY IN OVIPOSITION DURING DIFFERENT PARTS OF THE DAY.

In order to determine that portion of the day when eggs are most freely deposited by females during summer weather, adults were inclosed in a rearing cage over leaves from which all previously deposited eggs had been removed, and allowed to remain for a period of two hours, when the cage with adults was removed to another leaf and the deposited eggs counted, with results shown in Table XIV.

---

[1] Insect Life, vol. 5, p. 222, 1893.

TABLE XIV.—*Activity of the citrus white fly in oviposition during different parts of the day.*

Time of day.	Mean temperature for period.	Number of eggs deposited.	Per cent of total eggs deposited.
	° F.		
6 a. m.–8 a. m...	82	20	2.6
8 a. m.–10 a. m..	89	55	7.3
10 a. m.–12 m....	91	35	4.7
12 m.–2 p. m.....	92	65	8.7
2 p. m.–4 p. m....	90	30	4
4 p. m.–6 p. m....	85	197	26.1
6 p. m.–8 p. m....	81	248	32.9
8 p. m.–6 a. m....	74	103	13.7
6 a. m.–6. a. m...	[1] 81.6	753	100

[1] Average temperature for entire day of 24 hours; not the average of the 8 periods.

From the data it will be seen that while oviposition occurs at all times of the day, nearly 60 per cent of the eggs are deposited between 4 p. m. and 8 p. m., and that oviposition does not cease on the approach of darkness. The variation in the number of eggs deposited during the periods from 6 a. m. to 4 p. m. has little significance. It was noted that the least number of eggs were deposited when the bright sun fell directly upon the cage.

In further evidence of the greater activity of oviposition during the latter part of the day, two other cages were started on August 1, 1909. One cage placed repeatedly over the tenderest growth resulted in 698 eggs being laid between 10.15 a. m. and 4.15 p. m., as compared with 895 eggs laid between 4.30 p. m. and 7.30 p. m. The second cage, covering spring growth, gave 115 as compared with 786 eggs deposited during the same periods.

Relation between oviposition and food supply.—As the egg-laying capacity of a single female is close to 250 eggs and but 25 well-developed eggs have ever been seen in her abdomen at any one time, it is necessary that she obtain nourishment sufficient to mature her numerous "potential" eggs. There remain many interesting observations and experiments to be made on the relation between oviposition and food supply. That females deposit fewer eggs when feeding upon many of the recognized food plants other than citrus than they do on the latter is a subject of considerable interest. While adults feed apparently as contentedly upon new growth of China trees and umbrella China trees, they do not appear to deposit as many eggs per female on these host plants as on citrus. The extremely small number of eggs laid by females swarming over new growth of wild persimmon in June at Orlando is even more astonishing considering the marked preference shown by the females for this growth over the spring growth of orange. Even on citrus itself oviposition is influenced by the ages and corresponding toughness of the leaves, though not as markedly as is that of the cloudy-winged white fly. In one instance equal numbers of adults were confined on a tender and an old leaf of orange for two hours, when the adults were removed and 576 eggs were found to have been laid on the tender leaf and but 25 on the old leaf. Again, under practically the same conditions, 364 eggs were deposited upon tender growth and but 2 on very old growth. The difference between oviposition on tender August growth and spring growth is not as great as this, though very marked, as about 90 per cent of the third-brood adults fly to the new growth put on by the trees late in July and early in August.

From the foregoing it is evident that the number of eggs deposited is strongly influenced by the nature of the insect's food. Females confined in empty cages never deposit eggs, neither do those resting upon thick bark, ladders, or picking boxes, and, as has been stated under "Food plants," oviposition is entirely checked[1] when females are confined with leaves of nonfood plants. This difference in the number of eggs deposited on various plants may prove of value from the standpoint of trap foods, and become a factor in the control of this pest.

PROPORTION OF SEXES.

Examination of thousands of adult citrus white flies at all seasons of the year has shown that after a grove has become well infested an equilibrium between the proportion of males and females is established from which there is under ordinary conditions little variation. In such groves it has been found that from 60 to 75 per cent of the adults are females. Of the records on file, about 66 per cent give percentages of 60 and over for females, while 66 to 76 per cent are more frequent percentages where adults are abundant.

In groves where the progress of the white fly has been very seriously and suddenly checked by natural or artificial causes, the proportion of sexes is subject to a much wider variation and there follow for a time fluctuations between a predominance of males and females. In one such grove where the white fly had been greatly reduced in numbers because of the scarcity of adults of the first brood, there was a very large percentage of males appearing with the second brood, which in turn resulted in the third brood of 90.5 per cent females. In a second grove, where over 99 per cent of the white fly were killed by fumigation, the few females of the first brood, because of their isolation due to scarcity in numbers, were forced to deposit mostly infertile eggs, which resulted, in the second brood, in a reduction of females to 18.6 per cent.

Dependence of sex upon parthenogenesis.—The proportion between the sexes is largely and evidently entirely dependent upon parthenogenesis. It has been shown that infertile females deposit eggs in as

---

[1] Three hundred adults of *A. citri* confined on the tenderest spring growth of oak for three days deposited 1 egg.

large numbers and as frequently when males are not given access to them as do fertile females, and that the adults developing from these eggs are all males. Whether the adults from fertile eggs are invariably females has not been proved, although the evidence leaves little doubt that they are. If otherwise, it would be difficult to account for the fluctuations in sexes mentioned under the preceding heading, or to explain the great predominance of females over males after the species has become well established.

INFLUENCE OF WEATHER CONDITIONS ON ACTIVITY OF ADULTS.

During the cooler portions of the year, when adults are present on the trees, very few are seen flying about from tree to tree unless abundant. The morning and evening temperatures easily chill them; hence their activities are confined to the warmer part of the day.

However, after summer weather has become established the white flies rest very quietly on the under surface of the leaves during the greater part of the day. They shun the bright sunshine and prefer leaves in shaded .places. When exposed to the sun without protection they soon die. As the temperature falls during the late afternoon, and especially after afternoon showers when the humidity has risen to 90° or even to 100°, they become very active, and about 4 o'clock begin to fly about from leaf to leaf and from tree to tree, and, when very abundant, swarm in such large numbers about the groves and town streets as to arrest the attention of pedestrians, to whom they become at times a source of much aggravation, becoming entangled in the hair, crushed upon the clothing, breathed in with the air and causing choking, and flying into the eyes.

FEEDING HABITS OF ADULTS.

The adult insects, having well-developed sucking mouth parts, feed upon the plant juices in the same manner as do the larvæ and pupæ, but with the advantages of not being confined to the same location. They do not leave any external evidence of the feeding except on very young growth, when the feeding of a large number of adults frequently produces a crinkling of the foliage.

It is difficult to determine positively whether or not an adult citrus white fly is feeding when it is resting on a leaf or stem. Adults rest contentedly during the warm portions of the day upon the underside of leaves of plants upon which they have never been known to deposit eggs. Under these circumstances they even appear to mate, and it seems probable that they feed to a limited extent. When on one of the principal food plants of the species, however, it is safe to consider that adults feed wherever eggs are deposited in noticeable numbers. It is because of this indiscriminate settling upon vegetation upon which they are not able to subsist, and upon which they never breed, that the belief has received such an unfortunately wide circulation among orange growers that the citrus white fly breeds on all kinds of hammock trees, shrubs, and grasses.   Regardless of the food plant, the adults feed almost exclusively upon the under surface of the leaves, more rarely upon the fruit, and never upon the woody portions of the tree.   When new growth is very young and the leaves have not expanded, adults often feed upon both surfaces of the leaf, the petiole, and even the tender shoots, but this lasts only for a short time.   At all seasons the newest growth is preferred, as indicated by the data under the caption of the relation of food supply to oviposition, and the portion of the plant selected coincides with that already discussed for oviposition.   It should be noted here that the decided preference of the adults for the new growth has a checking effect, as noted elsewhere, upon multiplication, as they are entirely lacking in instincts preventing over-oviposition   .

MULTIPLICATION.

The relation of multiplication to food supply and the restrictions upon multiplication due to overcrowding, natural mortality, dropping of leaves after freezes, parthenogenesis, and attacks by insects and other predaceous enemies and fungi will be found treated elsewhere.   It has been estimated that not more than 5 per cent, at the most, of the eggs deposited throughout the State result in the development of mature insects.   If each female deposited her full number of eggs and all the forms lived, it has been estimated, the progeny of a single pair of white flies emerging in January would amount to about 55,000,000,000 in one year.

LENGTH OF LIFE CYCLE.

Data concerning the duration of the egg, larval, and pupal instars of the citrus white fly have already been given, but not in a form readily showing the relation to the complete life cycle. From some of the more important and complete of the life-history studies the data in Table XV have been arranged to illustrate the important points in this connection:

TABLE XV.—*Length of life cycle of the citrus white fly at Orlando, Fla.*

Lot No.	Eggs deposited.	First fly emerged.	Last fly emerged in fall.	First fly emerged in spring.	Last fly emerged.
No. 1	Feb. 23	Apr. 30			
No. 2	Mar. 3	May 9		a	
No. 3	Apr. 3	May 30			
No. 4	Apr. 20	June 7			Mar. 15
No. 5	June 16	July 30			
No. 6	July 17	Aug. 27	Sept. 10	Mar. 16	Apr. 16
No. 7	...do.....	Sept. 4	Sept. 17	Mar. 17	May 4
No. 8	July 19	Sept. 2	Sept. 21	...do.....	May 1
No. 9	July 26	Sept. 6	Sept. 20	Mar. 18	May 10
No. 10	Aug. 1	Sept. 19	Sept. 26	Mar. 24	Apr. 6
No. 11	Aug. 3	Sept. 25	...do.....	Mar. 23	Apr. 15
No. 12	Aug. 8	Sept. 19	Sept. 27		
No. 13	Aug. 9	Mar. 30	0	Mar. 20	May 12
No. 14	Sept. 18	Mar. 16	0	Mar. 16	Apr. 26
No. 15	Sept. 21	Mar. 12	0	Mar. 12	May 10

Lot No.	Least number of days for development.	Largest number of days for development.	Per cent emerging before winter.	Per cent wintering over to emerge in spring.	Smallest number degrees effective temperature for development.	Degrees accumulating before spring emergence.	Degrees accumulating before last fly emerged.
No. 1	67		100.0	0	1,783		
No. 2	67		100.0	0	1,885		
No. 3	57		100.0	0	1,888		
No. 4	48	333			1,712		
No. 5	44				1,725		
No. 6	41	273	56.9	43.1	1,641	6,632	7,619
No. 7	49	291	30.8	69.2	1,972	6,665	8,253
No. 8	45	286	12.7	87.3	1,815	6,504	8,069
No. 9	42	288	29.8	70.2	1,703	6,322	7,931
No. 10	49	248	5.5	94.5	2,015	6,255	6,554
No. 11	53	255	3.6	96.4	2,153	6,107	6,858
No. 12	42		2.7	97.3	1,735		
No. 13	223	276	0	100.0	5,825	5,825	7,545
No. 14	179	222	0	100.0	4,552	4,552	5,835
No. 15	172	231	0	100.0	4,289	4,289	6,100

From this table it will be seen that the period of development for individuals hatching from eggs laid upon the same leaf within a few hours of each other is subject to an astonishing variation, ranging from 41 to 333 days. This variation is absolutely independent of both temperature and humidity influences. It will be noted that the sums of effective temperatures required for the minimum duration of immature stages for individuals developing from eggs deposited between February 23 and August 8 vary from 1,641° to 2,153°, with an average of 1,846°, which may be regarded as very nearly the normal for minimum development up to the time when all individuals winter over as pupæ. It should also be noted that the number of maximum degrees of effective temperature is more strongly influenced by the time of year the eggs are deposited—the nearer the winter months deposition takes place the fewer the degrees accumulating before the last fly emerges. This is due to the equalizing effect of the cooler winter temperatures.

This same equalizing effect of the winter temperatures upon the length of the life cycle for individuals developing from eggs laid on September 20 is brought out in Table XVI:

TABLE XVI.—*Duration of instars of the citrus white fly.*

Speci-men No.	Instar.								Total num-ber of days.
	First.		Second.		Third.		Pupal.		
	Duration of instar.	Num-ber of days.	Duration of instar.	Num-ber of days.	Duration of instar.	Num-ber of days.	Duration of instar.	Num-ber of days.	
	1908.		1908.		1908.		1908.		
1.....	Oct. 3–10.....	8	Oct. 10–15..	5	Oct. 15–27......	12	Oct. 27–Apr. 28	173	198
2.....	Oct. 3–10.....	8	Oct. 10–24..	14	Oct. 24–Nov. 10	17	Nov. 10–Apr. 17	158	197
3.....	Oct. 3–15.....	13	Oct. 15–22..	7	Oct. 22–Nov. 4..	13	Nov. 4–Apr. 8 ..	155	188
4.....	Oct. 3–Nov. 2.	31	Nov. 2–16...	14	Nov. 16–Dec. 3..	17	Dec. 3–Apr. 6...	128	190
5.....	Oct. 3–10.....	8	Oct. 10–18..	8	Oct. 18–29......	11	Oct. 29–Mar. 26.	148	175

From this table it will be seen that retardation in growth during any one instar does not affect materially or show a corresponding increase in the total number of days required for development when the individual passes the winter in the pupal stage. Also, that an unusually large number of days spent in one instar does not necessarily mean that the individual insect will be equally backward in the next instar. These records of daily observation on individual specimens from hatching to adult are only 5 of 85 similar observations for the same period. Nos. 2–5 were insects on the same leaf.

SEASONAL HISTORY.

GENERATIONS OF THE CITRUS WHITE FLY.

It has been generally understood in the past that there are three generations annually of the citrus white fly, although Prof. H. A. Gossard,[1] states that "four generations a year doubtless often occur, but not in sufficient numbers to obscure three well-defined broods as the rule." In the greenhouses at Washington, Riley and Howard[2] found that there were but two generations annually. The life-history work of the present investigations has shown that while the general

---

[1] Bul. 67, Fla. Agr. Exp. Sta., p. 612, 1903.
[2] Insect Life, vol. 5, p. 224, 1893.

observations of the past leading to the statement of three more or
less distinct periods of emergence are correct, the number of generations annually ranges from two to five, or, under unusually favorable
conditions, from three to six.   In figure 11 the maximum and minimum number of generations as actually known to occur in groves
at Orlando during 1907-9 has been plotted.   Figure 11 is based
upon the development of individuals in rearing experiments.   The
generation between January and March may or may not occur,
according to whether the winter weather is warm or cold, but when
present is numerically insignificant.   The other generations are
more confused than can be indicated diagrammatically.   As may

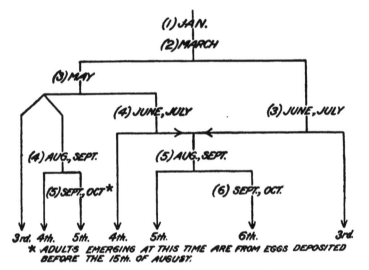

Fig 11.—Diagram showing generations of the citrus white fly.   (Original.)

have been inferred from data in Tables XV and XVII, the variation in
the number of generations is due almost entirely to the length of the
pupal stage, which, as has already been pointed out, is subject to
wide variation, the cause of which can not be traced to food, temperature, humidity, or location on the tree.   The most striking
variation in the length of life cycle, with its effect upon the possible
number of annual generations, is found among individuals developing
from eggs deposited in April and May.   In one instance eggs deposited on April 20 produced adults on June 5, July 31, and in the
following March.   In the main, each generation has two more or
less distinct periods of emergence, as reference to the data in Tables
XV and XVII will show.

TABLE XVII.—*Emergence of adult citrus white flies at Orlando, Fla.*[1]

Record No.	Eggs deposited.	April.		May.		June.		July.		August.	
		1-15	16-30	1-15	16-31	1-15	16-30	1-15	16-31	1-15	16-31
1	Feb. 22......	0	1	85	......						
2	Mar. 3.......	0	0	44	44						
3	Mar. 18-25...	0	0	1	51	30.3	2.9	0	4.9	9.9	0
4	Apr. 20......	0	0	0	0	50.1	28.6	0	8.2	0	0
5	June 16......	0	0	0	0	0	0	0	(²)		
6	July 17......	0	0	0	0	0	0	0	0	0	32.1
7	July 17......	0	0	0	0	0	0	0	0	0	0
8	July 19......	0	0	0	0	0	0	0	0	0	0
9	July 26......	0	0	0	0	0	0	0	0	0	0
10	Aug. 1.......	0	0	0	0	0	0	0	0	0	0
11	Aug. 8.......	0	0	0	0	0	0	0	0	0	0
12	Aug. 9.......	0	0	0	0	0	0	0	0	0	0
13	Sept. 18.....	0	0	0	0	0	0	0	0	0	0
14	Sept. 21.....	0	0	0	0	0	0	0	0	0	0

Record No.	Eggs deposited.	September.		October to March.	March.		April.		May.	
		1-15	16-30		1-15	16-31	1-15	16-30	1-15	16-30
1	Feb. 22......									
2	Mar. 3.......									
3	Mar. 18-25...	0	0	0	0	0	0	0	0	0
4	Apr. 20......	0	(³)	0	0	0	0	0	0	0
5	June 16......									
6	July 17......	24.4	0.4	0	0	43.1	0	0	0	0
7	July 17......	23.9	6.9	0	0	67.3	1.9	0	0	0
8	July 19......	8.7	1.7	0	0	89.2	.4	0	0	0
9	July 26......	23.8	6.5	0	0	64.3	5.4	0	0	0
10	Aug. 1.......	0	3.3	0	0	85.6	11.1	0	0	0
11	Aug. 8.......	0	2.7	0	0					
12	Aug. 9.......	0	0	0	0	98.7	1.3	0	0	0
13	Sept. 18.....	0	0	0	0	38.3	46.8	14.9	0	0
14	Sept. 21.....	0	0	0	0	38.6	47	14.7	0	0

[1] This table is introduced to demonstrate the two emergence periods for individuals developing from eggs deposited at the same time. It is not intended to represent the abundance of adults at different times of the year. Figures represent percentages.

[2] Only one fly survived to winter over.

[3] One fly emerged July 30. Leaf was broken off on same day. Development of flies indicated that at least 50 per cent would have emerged during August.

This makes it possible for adults emerging during the first period to deposit eggs for a second generation, a portion of the adults of which (first brood) will emerge at the time of the second emergence period (second brood) for the first generation, while a large proportion of the remaining individuals on the leaf to all appearances remain stationary in their development, though actively secreting honeydew, until the approach of the first emergence period of the second generation started by the second brood of the first generation, when they rapidly mature and emerge with this brood. However, this double-brooded character of each generation up to and including generations started in early August does not obscure the three well-defined "broods" of adults, to be discussed under "Seasonal fluctuations in the numbers of adults or so-called 'broods,'" but shows that the adults appearing during the three general emergence periods do not, strictly speaking, represent a single brood of one generation, but different broods of different generations. By far the greater number of wintering-over pupæ belong to the fourth, fifth, and sixth generations, with the last two most abundantly represented. The number of third-generation pupæ—or second generation should the first generation in figure 11 not occur—to winter over is insignificant.

SEASONAL FLUCTUATIONS IN THE NUMBERS OF ADULTS OR SO-CALLED "BROODS."

During winters of unusual mildness there is a tendency for continuous breeding, and adults in varying numbers can be found on the wing at different times, but these are as a rule too few in number to be of importance in effecting the general seasonal history of the citrus white fly. With the exception of the limited number of larvæ developing from eggs deposited by these unseasonal adults, the white fly passes the winter in the pupal stage. The first general spring emergence of adults begins after the daily mean temperatures have risen to about 65° F., which at Orlando in 1909 was about February 20.

There are three periods throughout the year when adult citrus white flies are so much more abundant than at other seasons that it is generally said there are three broods of white flies each year,

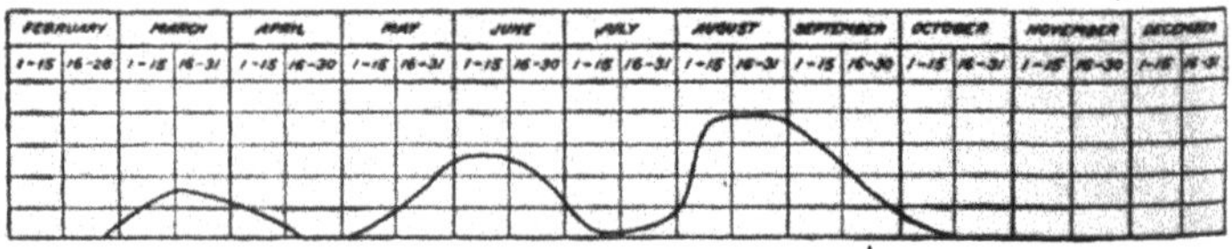

FIG. 12.—Diagram showing abundance of adults of the citrus white fly at Orlando, Fla., throughout 1909. (Original)

although, as already noted under the subject of generations, the term brood in this case is somewhat misleading. The dates at which these adults appear is subject to such variation in groves in the same county, town, or even on individual trees of the same grove, that no accurate statement of the dates between which the broods occur throughout the State as a whole can be made. The authors, therefore, have chosen to follow the history of the white fly in a single grove at Orlando during the season of 1909 as a specific example, as a basis, and present in figure 12 a curve representing the abundance of adults throughout that season. In all its essential features the curve is regarded as representing the appearance of adults in any grove, when it is remembered that variations of from one to three or four weeks may occur in the appearance of the broods.

While it is generally believed that adults appear earlier in the spring throughout southern Florida, it is a fact that there is very little difference in time of emergence between that and the central portion. Emergence throughout the northern portion of the State is, according to the season, from one to four weeks later than in the central and

southern portions. Prof. H. A. Gossard has stated that at Lake
City in 1902 white flies began to appear no earlier than April 14, and
continued to appear until late in May, although the majority of them
emerged during the latter half of April, while in 1903 the same trees
produced adults as early as March 12, or but one week later than
groves at Orlando and Palmetto. While the spring brood of adults
at Orlando in 1909 had begun to emerge as early as February 20 and
had reached their height and begun to decline by March 27, on the
latter date in St. Augustine only 5 per cent of the pupæ had developed the eyes of the adult and practically no adults had emerged.
Professor Gossard also is authority for the statement that at "Tampa,
30 to 40 miles north of the Manatee section, the spring brood of
white flies has in some seasons preceded their appearance about
Bradentown and Manatee by two weeks." By June 18, 1909, more
than twice as many adults of the second brood had emerged at
Manatee as at Island Grove about 125 miles north in Alachua County,
while by July 7 of the same year the white fly in a grove at Alva in
Lee County was no further advanced than at Orlando.

From the curve in figure 12 it will be noticed that there are two
periods of about three weeks during the summer between the broods
when adults are comparatively very scarce. While reference to
Table XVII shows that a few wintering-over individuals continue
to emerge as late as early May, the period between the first and second broods of adults is exceptionally free from adults of the citrus
white fly. This, however, is not true of the like period between the
second and third broods as before this time the generations of the
white fly have become somewhat confused, due to variation in life
cycle, and adults continue to emerge in appreciable numbers throughout the period.

In speaking of the entire citrus belt, including Florida and the
Gulf States, the greater part of the spring brood may be said to
emerge during March and April; the second brood to emerge during
late May, June, and July, and the third brood during August and
September. It should be noted here that the greater part of the
adult white flies appearing in October and November in the central
and southern part of Florida are the cloudy-winged white fly, *A.
nubifera*, although in the northern part adult specimens of *A. citri*
have been seen in small numbers on the wing in St. Augustine as
late as November 15.

THE CLOUDY-WINGED WHITE FLY.

(*Aleyrodes nubifera* Berger.)

HISTORY.

Specimens of the cloudy-winged white fly (eggs, larvæ, and pupæ) in the collection of the Bureau of Entomology show that this species occurred on oranges in the United States as early as 1889. The records in connection with the specimens show that it was collected in Mississippi and North Carolina in 1889, in Louisiana in 1890, and in Florida in 1895. Outside of the United States it is known to occur only in Cuba. Its introduction into the United States from Cuba does not seem as probable as its introduction into Cuba from the United States. At present there is no evidence concerning the probable origin of the insect except in the absence, so far as is known, of other food plants than citrus, which would seem to indicate the introduction of the insect with its only known food plant.[1]

Several writings on the citrus white fly (*Aleyrodes citri*) have in part included the cloudy-winged white fly (*A. nubifera*). Prof. H. A. Morgan,[2] in 1893, previous to the publication of the original description of *Aleyrodes citri*, briefly described the egg of *Aleyrodes nubifera* and figured it, the description of the pupa and adult given at the same time evidently being based on specimens of *A. citri*. The species to which Prof. Morgan referred the specimens was *Aleyrodes citrifolii* Riley MS. The original description of the citrus white fly,[3] while unquestionably defining the species generally recognized as *A. citri*, included in part reference to what is probably the cloudy-winged white fly. In the text the description of the first stage or instar of the larva was evidently based on a specimen of the cloudy-winged white fly and the illustration of the first instar[4] was also based on this species with little doubt. One figure of the pupa[5] and one of the pupa case[6] evidently were based upon specimens of the same species. In the writings of Prof. Gossard there are no references in the text which evidently refer to the cloudy-winged white fly, but what is probably this species is represented in an illustration of the first stage.[7]

[1] Its recent discovery on *Ficus nitida*, rubber tree, in greenhouses at Audubon Park, New Orleans, La., points to its possible introduction from India.

[2] The Orange and Other Citrus Fruits. Special Bulletin Louisiana Agricultural Experiment Station, p. 72, 1893.

[3] Insect Life, vol. 5, pp. 220–222, 1893.

[4] Id., vol. 5, p. 219, fig. 23, *d*.

[5] Id., vol. 5, p. 219, fig. 23, *h*.

[6] Id., vol. 5, p. 219, fig. 23, *i*.

[7] Bulletin 67, Florida Agricultural Experiment Station, pl. 2, fig. 1. See also Bulletin 88, pl. 2, fig. 1, and Bulletin 97, fig. 11, Florida Agricultural Experiment Station, and Circular 30, California Agricultural Experiment Station, pl. 2, fig. 1.

The white fly known as the cloudy-winged white fly was first determined as specifically distinct from the citrus white fly by Dr. E. W. Berger in 1908. Dr. Berger has recently given this species its scientific name in connection with a synopsis of the principal distinctive characters and illustrations of egg and larval and pupal stages.[1]

AMOUNT OF INJURY BY THE CLOUDY-WINGED WHITE FLY.

The injury caused by the cloudy-winged white fly is at present much restricted by several factors. In Florida the distribution of this species is limited as compared with that of the citrus white fly. Its food-plant differences and adaptations are such that orange trees[2] are not as a rule subject to as heavy infestations as by the citrus white fly, although with grapefruit trees this situation is usually reversed. Most important as a factor limiting the injury from the cloudy winged white fly is that when both occur in an orange grove the citrus white fly almost invariably predominates and the cloudy-winged white fly assumes a position of comparative insignificance. Owing to the difference in the seasonal history of the two species of white fly this latter point is not always apparent to the casual observer. An observation made between the broods of adults of the citrus white fly, or at any time after the middle of September up to December 1, may result in noting a great preponderance of the cloudy-winged white fly, leading one to conclude, perhaps, that it is this latter species which is causing the most injury. An examination of the leaves during the winter months, when there are practically no adults of either species, will probably show an entirely different situation. In many groves near Orlando and Winter Park in Orange County, Fla., both species of white fly are well established and practically have assumed their normal relative positions in point of numbers. Examinations of leaves varying in number from 85 to 400 picked at random in 11 of such groves furnish data which illustrate the general situation as regards the importance of the two species of white fly under the conditions mentioned. (See Table XVIII.) All the examinations were made during the winter months, using pupa cases and live pupæ as the basis of the comparison.

---

[1] Bulletin 97, Florida Agricultural Experiment Station. pp. 68–70, figs. 12, 14, 16, 18, 19.

[2] According to the latest statistics available (Ninth Biennial Report of Commissioner of Agriculture, State of Florida) there were more than five times as many orange trees as grapefruit in Florida, 1,786,944 orange trees being reported for 1905 as against 373,008 grapefruit trees.

TABLE XVIII.—*Comparative abundance of Aleyrodes citri and Aleyrodes nubifera in groves infested by both species.*

Grove Nos.	Tangerine.		Grapefruit.		Orange.		Grapefruit and orange.	
	Citri per leaf.	Nubifera per leaf.	Citri per leaf.	Nubifera per leaf.	Citri per leaf.	Nubifera per leaf.	Citri per leaf.	Nubifera per leaf.
1	44.2	7.5	1.6	1.4				
2					33.2	0.5		
3					4	1.2		
4					14.5	.4		
5					4.6	.2		
6			8.8	2.4				
7							11.2	1.1
8							21.	1.6
9					33	.66		
10					30	7.		
11	.9	.2	.8	3.7				
Average.	22.5	3.6	3.7	2.5	19.9	.59	16.1	1.3
Per cent......	86.7	13.3	59.6	40.3	97.7	2.3	92.6	7.4

Owing to the great attraction of new growth for the cloudy-winged white fly, which is discussed elsewhere, the scarcity of new citrus growth at certain seasons which causes concentration on water shoots, and other factors, this species, when it occurs by itself in a tangerine or orange grove, does not as frequently as the citrus white fly cause noticeable blackening of the foliage before the middle of June. At the end of the season the cloudy-winged white fly by itself may cause tangerine and orange trees to become as heavily blackened with sooty mold as the citrus white fly when the latter is at its greatest abundance. As has been stated, the cloudy-winged white fly is more likely to heavily infest grapefruit trees than is the citrus white fly. The cloudy-winged white fly seems to be subject to more extensive fluctuations from year to year, aside from the effects of fungus parasites, than is the citrus white fly, and frequently after infesting an orange grove for several years fails to cause enough injury to make washing of the fruit necessary or to make necessary the washing of more than one-fourth or one-third of the crop each year.

As a whole, the injury is not as extensive in groves where the cloudy-winged white fly occurs alone as in groves where the citrus white fly occurs alone. When the two species become well established, the former does comparatively little damage except to grapefruit. The authors would estimate that there are about 5 per cent of the orange and tangerine groves in the State infested by the cloudy-winged white fly that are not also infested by the citrus white fly, and that there are in addition 1 per cent of orange and tangerine groves infested by both species but in which the citrus white fly has not as yet attained injurious abundance. The average damage from the cloudy-winged white fly is estimated at about 10 to 15 per cent lower for oranges where that species alone infests the grove than where the citrus white fly is the species concerned. For injury to grapefruit the authors consider 25 per cent a fair estimate of the injury by the cloudy-winged white fly as compared with about 10 or 15 per cent by the citrus white fly. The total loss in Florida due to the cloudy-winged white fly is estimated by the authors at between $100,000 and $125,000 per annum at the present time.

DISTRIBUTION.

So far as known at the present writing the cloudy-winged white fly occurs in 12 counties in Florida. The locality list is given below:[1]

Brevard County: Mims, Sharpes, Titusville.
Dade County: Miami.
Hillsboro County: Riverview, Thonotosassa, Ybor City, Clearwater, Dunedin, Largo, Ozona, Safety Harbor, Saint Petersburg, Sutherland.
Manatee County: Bradentown, Oneco, Palmetto.
Monroe County: Key West.
Orange County: Geneva, Maitland, Ocoee, Orlando, Oviedo, Waco, Winter Park.
Palm Beach County: Palm Beach, West Palmbeach.
Polk County: Auburndale, Bartow, Lakeland, Winterhaven.
St. Lucie County: Fort Pierce.
Sumter County: Wildwood.
Volusia County: Haw Creek, Holly Hill, Port Orange, Pierson.

Outside of the State of Florida the only available records of the occurrence of the cloudy-winged white fly are those of the Bureau of Entomology in connection with specimens in the collection. Mr. A. L. Quaintance has identified as this species specimens from New Orleans (1890) and Baton Rouge, La. (1891), Pass Christian, Miss. (1889), and Raleigh, N. C. (1889). In a brief examination at Audubon Park, New Orleans, in August, 1909, the senior author was unable to find any evidence of the presence of this species, although the citrus white fly was prevalent on citrus trees, privets, and other food plants.

As stated in the footnote on page 27, the species occurring at Bakersfield, Cal., in 1907 was the cloudy-winged white-fly. Owing to the fact that the insect is, so far as known, confined to citrus as a food plant and only a limited number of these in an isolated location were infested, the thorough measures adopted by the agents of the State commissioner of horticulture met with complete success, and there is no record of this species occurring at present in this State.

Its occurrence in Cuba has already been noted, specimens having been received from Santiago de las Vegas in 1905.

The distribution of the cloudy-winged white fly in Florida, so far as now known, is shown in figure 13. The territory included in the

---

[1] The authors have determined as *Aleyrodes nubifera* specimens from all of the localities listed above except the following, which are listed upon the authority of Dr. E. W. Berger: Holly Hill, Ybor City, Bartow, Clearwater, and Safety Harbor.

infested area is not generally infested, and the same precautions should be observed within this area as outside of it to avoid unnecessary spread of the pest.

FOOD PLANTS.

The cloudy-winged white fly is not known in Florida to breed upon any other food plant than citrus. It has recently been discovered infesting the rubber trees (*Ficus nitida*) growing in the greenhouses in Audubon Park, New Orleans. Extensive examinations for possible food plants have been made by the authors and by Dr. Berger, and it is reasonably certain that no important food plant will be found in Florida citrus-growing sections which will interfere with the control of this species.

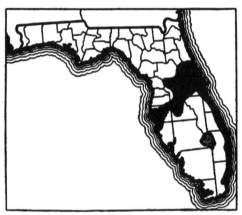

FIG. 13.—Map showing distribution of the cloudy-winged white fly (*Aleyrodes nubifera*) in Florida. (Original.)

Examinations of prickly ash (*Xanthoxylum clava-herculis*), the most common representative of the family Rutaceæ to which the citrus belongs, indicate that this species of white fly never breeds on this plant, regardless of the condition of infestation of neighboring citrus trees. Reports of blackening of the foliage of prickly ash by the white fly in sections where only the cloudy-winged white fly of the two herein treated occurs, are doubtless erroneous and probably based upon the blackening due to an aphis or to some other insect. In addition the following plants have been examined under favorable conditions to determine if subject to attack by the cloudy-winged white fly, but so far without results:

China trees and umbrella China trees, cape jessamine, privets, Japan and wild persimmons, oaks, wild cherry, guava, fig, grape, cherry laurel, blackberry, and magnolia.

SPREAD.

The dissemination of the cloudy-winged white fly is limited by the same factors which have been discussed as unfavorable to the successful establishment of the citrus white fly. Aside from these factors its chief limitation is in its lack of important food plants other than citrus. Migrations of adults are not an important factor in the spread of this white fly except between adjoining groves. Its spread throughout a newly infested grove and to adjoining groves is perhaps favored by its greater degree of attraction to new growth. It has been observed frequently in newly infested groves that it is found to be present in very small numbers over a considerable area, whereas the citrus white fly, when at a corresponding numerical status, would be expected to be more localized. Winds are doubtless concerned with the spread of the insect from local centers, but, as with the citrus white fly, they are evidently a factor of small consequence in spread to distant points.

Flying insects and birds as carriers of the crawling larvæ are necessarily of little or no consequence in the spread of the cloudy-winged white fly, as with the species previously discussed. Between groves vehicles of various kinds are of much importance in distributing this white fly, as is generally recognized. Without doubt infested citrus nursery stock has been the principal factor in the spread of the cloudy-winged white fly. Nursery stock from Mims, Fla., is quite definitely known to have been the source of this white fly at Sharpes, Fla. At the present time it seems to be the most probable source of this white fly at Fort Pierce, Palm Beach, and Miami.

Fig. 14.—The cloudy-winged white fly: Eggs. Greatly enlarged. (Original.)

The species here considered are apparently as likely to be carried on the person from an infested grove as is the citrus white fly, the degree of infestation being equal. The same may be said of pickers' outfits. The introduction of fungus-infected leaves into groves infested only by the citrus white fly in connection with the introduction of fungus parasites has doubtless assisted in the spread of the cloudy-winged white fly, but not to the same extent that this has assisted in the spread of the former species.

LIFE HISTORY AND HABITS.

The Egg.

The egg (fig. 14) of the cloudy-winged white fly differs from that of the citrus white fly in that it is not greenish-yellow and highly polished, but bluish or grayish black and roughened by a film of wax arranged in an hexagonal pattern. To the unaided eye the eggs appear as fine particles of blackish dust scattered over the leaves (Pl. X, fig. 1). Because of their dark color they are more readily seen on the tender citrus growth by the average observer than are the eggs of the citrus white fly. When first deposited they are not blackish, but are dull white or cream colored and under the microscope the waxy coating appears tinged with red. The eggs remain pale for three days, when about 96 per cent turn black, the rest taking sometimes as long as seven or eight days to darken. Meanwhile the waxy coating has turned gray. Because of the dark color and much tougher chorion, the eyes of the embryo are not as easily seen as are those of the citrus white fly. The eggs are attached to the leaf by a pedicel arising from the proximal end similar to that of *A. citri*.

A study of the contents of Tables VIII and XIX will prove that what has been said of the duration of the egg stage, variability under identical temperature conditions, and effect of temperature in general, regarding the citrus white fly, applies almost equally well to this species.

TABLE XIX.—*Duration of egg stage of the cloudy-winged white fly.*

Record No.	Date deposited.	Eggs under observation.	First egg hatched.	Least egg hatched.	Least number days.	Largest number days.	Daily rate of hatching of eggs of the cloudy-winged white fly in percentages, with accumulative effective temperatures.									
							8 days.	°F.	9 days.	°F.	10 days.	°F.	11 days.	°F.	12 days.	°F.
1	Apr. 20, 1909.	345	May 2	May 7	12	17	0		0		0		0.5	383	4.1	623
2	June 16, 1909.	349	June 26	July 1	11	16	0		0		.0		63.4	457	4.9	634
3	Aug. 23, 1909.	276	Sept. 3	Sept. 7	11	15	0		0		3.4	351	73.9	390	21.4	448
4	Sept. 4, 1907.	146	Sept. 14	Sept. 19	10	15	2.4	262	0	318	1.1	355	1.1	389	13.5	638
5	Sept. 18-21, 1908.	171	Sept. 26	Oct. 3	8	12-16	0		0		.3		1.5	396	40.7	634
6	Sept. 21, 1908.	225	Oct. 1	Oct. 9	10	18						349			0.0	415
7	Oct. 9-11, 1908.	888	Oct. 23	Oct. 20	14	18-20										

Daily rate of hatching of eggs of the cloudy-winged white fly in percentages, with accumulative effective temperatures.

Record No.	13 days.	°F.	14 days.	°F.	15 days.	°F.	16 days.	°F.	17 days.	°F.	18 days.	°F.	19 days.	°F.	20 days.	°F.
1	22.8	442	48.0		7.7	470	5.0	500								
2	45.5	462	44.9		4.0	466	.3	533	2.4	551	0	564	0			
3	4.3	534	7.3		3.6	571	.0	610	0	566	0		0			
4	19.8	466	2.1		1.8	501		538	0		0					
5	35.1	460	43.3		3.2	483	2.8	597	1.2	538	0.6	570	0	601	0.0	551
6	34.3	445	15.4		6.7	477	43.8	482	37.6	460	3.3	460	1.9	530	0.8	676
7	0		.8			407										

The duration of the egg stage, however, is in general slightly longer. As the bulk of the eggs hatch from 1 to 10 days later, a slightly higher number of degrees of accumulated effective temperature are necessary, and hatching is more evenly distributed though not always extending over a larger number of days.

The process of hatching, proportion of eggs that hatch, and the effect of drying of leaves in hatching do not appreciably differ from what has been stated of *A. citri;* in hatching, however, the egg membranes split only about one-third the length of the egg from the tip and on one side only, and on account of the tougher chorion do not shrivel, but retain their original form. Frequently, after the larva has escaped, the membranes spring back into their original position, thus causing the eggs to appear unhatched; as a rule, however, this does not occur, and the opening made by the escaping larva does not close. While no adults have yet been reared from larvæ hatching from infertile eggs, it has been proved that virgin females of the cloudy-winged white fly will deposit eggs and that these readily hatch and produce healthy larvæ, and the evidence in case of the citrus and greenhouse white flies leaves no doubt that adults resulting from infertile eggs will prove to be of the male sex.

Fig. 15.—The cloudy-winged white fly: Ventral view of crawling larva of the first instar. Greatly enlarged. (Original.)

THE LARVAL AND PUPAL STAGES.

THE LARVA.

The larva of the cloudy-winged white fly does not differ in general appearance from that of the citrus white fly except that it is a trifle larger. With the aid of the microscope the first instar may be separated from that of *A. citri* by the possession of 36 instead of 30 marginal bristles. No structural differences between the second and third instars of the two species have been discovered. Following is a more detailed comparative description:

First larval instar (fig. 15).—Length, 0.29–0.32 mm.; width, 0.19–0.22 mm. Similar to the corresponding instar of *A. citri*, but differing in being proportionately broader, in possessing 18 instead of 15 pairs of marginal bristles, and in developing soon after settling a marginal irregular wax fringe eventually equaling in width the length of the marginal spines. Relative length of marginal spines as follows:

Pair	1	2	3	4	5	6	7	8	9
Spaces	10.5	7.5	10.5	8.5	7.0	5.5	6.0	5.0	6.0

Pair	10	11	12	13	14	15	16	17	18
Spaces	5.0	5.0	5.0	5.0	5.0	6.5	15.5	6.5	15.5

The relative lengths and location of other spines of the body do not differ from similar spines on *A. citri*, neither do there appear differences in the structure of the antennæ, legs, vasiform orifice, or mouthparts when examined under a one-sixth inch objective. Particles of wax secretions are found in varying amounts on the ventral surface, sometimes in such abundance as to make microscopic examinations difficult.

Second larval instar.—Length, 0.42–0.51 mm.; width 0.28–0.37 mm. Except in point of size no differences have been discovered between this and the corresponding instar of *A. citri*.

Third larval instar (fig. 16).—Length, 0.66–0.9 mm.; width, 0.48–0.68 mm. Except in size no differences have been discovered between this and the corresponding instar of *A. citri*.

Fig. 16.—The cloudy-winged white fly: Ventral view of crawling larva of the third instar. Greatly enlarged. (Original.)

Fig. 17.—The cloudy-winged white fly. Pupa: *a*, Ventral view; *b*, enlarged vasiform orifice; *c*, enlarged margin. *a*, Greatly enlarged; *b*, *c*, highly magnified. (Original.)

THE PUPA.

In general appearance the pupa of the cloudy-winged white fly (fig. 17, *a*, *b*, *c*) resembles very closely that of the citrus white fly. No striking structural differences have been discovered between them. They are, however, very distinct, and one who has examined them carefully can readily separate them without the aid of a lens. The most important differences are in the larger size and thinner and flatter appearance of the pupa of the cloudy-winged white fly. The difference in outline is shown in figs. 9, *c*, and 17, *c*. Their skins are more membranous, making them more delicate and easily crumpled. Furthermore, after thickening before maturity they do not develop the bright red or orange spot on the middle of their backs, and the wing pads and body of the adult (fig. 18) are more easily seen. The pupa case (Pl. VIII, fig. 3) is much thinner, more membranous, and falls from the leaf more readily. Its walls do not remain rigid as do those of *A. citri*, but because of their more delicate structure collapse after the emergence of the adult and present the crinkled appearance shown in the illustration.

Fig. 18.—The cloudy-winged white-fly: Dorsal view of pupa, showing adult insect about to emerge. Greatly enlarged. (Original.)

DURATION OF INSTARS.

Larval instars.—By comparing the data in Tables IX and XX it will be found that the larvæ of *A. nubifera* are slower in maturing than those of *A. citri*. While this difference is not so pronounced during the warmer months of the year, the total average number of days being 25.9 and 23.1, respectively, during the cooler months it is very striking, the total average number of days then being 56.7 for *A. nubifera*, as compared with 30.4 for *A. citri*. In other respects the statements made on the duration of the larval instars for *A. citri* apply to *A. nubifera*.

TABLE XX.—*Duration of larval instars of cloudy-winged white fly.*

Specimen No.	Period of growth.	Number of days in—			Sum of effective temperature.	Specimen No.	Period of growth.	Number of days in—			Sum of effective temperature.
		Instar 1.	Instar 2.	Instar 3.				Instar 1.	Instar 2.	Instar 3.	
1	June 26–July 19...	9	6	8	938	33	Sept. 30–Nov. 20...	8	15	28	1,322
2	June 27–July 20...	6	6	11	942	34	Sept. 30–Nov. 23...	11	12	31	1,401
3	June 27–July 21...	8	6	10	979	35	Sept. 30–Dec. 20...	11	14	46	1,531
4	June 27–July 26...	9	7	13	1,179	36	Sept. 30–Dec. 3...	8	19	27	1,670
5	June 27–July 19...	9	7	6	902	37	Sept. 30–Nov. 23...	8	14	32	1,401
6	June 27–July 21...	8	8	8	979	38	Sept. 30–Dec. 3...	11	12	42	1,670
7	June 27–July 17...	7	6	7	820	39	Sept. 30–Nov. 27...	9	9	40	1,512
8	June 27–July 19...	5	7	10	902	40	.....do..........	8	14	36	1,512
9	June 27–July 20...	7	6	10	942	41	Sept. 30–Dec. 3...	8	14	42	1,670
10	.....do..........	7	6	10	942	42	Sept. 30–Nov. 23...	9	11	34	1,513
11	June 27–July 21...	7	6	11	979	43	Sept. 30–Dec. 10...	8	15	48	1,531
12	June 27–July 19...	7	6	9	902	44	Sept. 30–Dec. 3...	8	14	42	1,670
13	June 28–July 27...	14	8	7	1,179	45	Oct. 2–Dec. 7...	10	9	47	1,704
14	June 28–July 29...	7	6	18	1,260	46	Oct. 2–Nov. 10...	7	9	21	1,000
15	June 28–July 24...	11	8	7	1,056	47	Oct. 2–Nov. 30...	8	23	26	1,524
16	June 28–July —...	9	21	.....	.....	48	Oct. 2–Nov. 19...	9	10	27	1,235
17	June 28–July 19...	6	6	9	861	49	.....do..........	7	12	27	1,235
18	June 28–July 24...	7	9	10	1,056	50	Oct. 2–Nov. 23...	8	9	33	1,332
19	June 29–July 29...	6	15	9	1,222	51	Oct. 2–Nov. 19...	7	11	28	1,235
20	June 29–July 24...	7	7	11	1,018	52	Oct. 3–Nov. 27...	11	18	27	1,414
21	June 29–July 29...	8	8	14	1,222	53	Oct. 3–Dec. 3...	10	17	35	1,561
22	June 30–July 30...	8	8	14	1,229	54	Oct. 3–Dec. 7...	10	16	40	1,662
23	June 30–July 29...	12	7	10	1,188	55	Oct. 3–Dec. 2...	10	23	28	1,536
24	Sept. 30–Nov. 1...	5	10	18	912	56	Oct. 3–Dec. 1...	10	15	35	1,510
25	Sept. 30–Nov. 20...	8	16	28	1,322	57	Oct. 3–Dec. 3...	11	20	33	1,561
26	Sept. 30–Nov. 30...	12	12	38	1,593	58	.....do..........	11	20	31	1,561
27	.....do..........	14	15	33	1,593	59	Oct. 5–Nov. 11...	9	11	18	1,000
28	Sept. 30–Dec. 3...	12	23	32	1,670	60	Oct. 5–Dec. 10...	66	died.	.....	.....
29	Sept. 30–Nov. 3...	8	10	17	963						
30	Sept. 30–Dec. 3...	8	21	35	1,670	Average age.	June 26–July 30..	8	7.8	10.1	1,031.7
31	Sept. 30–Nov. 30...	8	13	40	1,593		Sept. 30–Dec. 10..	9.1	14.4	33.2	1,475.6
32	Sept. 30–Dec. 3...	8	14	42	1,670						

Pupal instar.—That little difference exists between the length of the pupal stages of the two species of white fly in question is shown by a comparison of the data in Tables X and XXI. The minimum length of the pupal stage (17 days) will average but a trifle above that of *A. citri*. But the maximum length is so dependent upon the seasonal history that a direct comparison is difficult; this subject, therefore, is more profitably discussed under the caption of seasonal history. What has been said in connection with the maturing of specimens of *A. citri* passing into the pupal stage at practically the same time is equally true of the cloudy-winged white fly, *A. nubifera*.

TABLE XXI.—*Duration of pupal stage of cloudy-winged white fly.*

Speci-men No.	Period of growth.	Num-ber of days.	Sum of effective tempera-ture.	Speci-men No.	Period of growth.	Num-ber of days.	Sum of effective tempera-ture.
1	May 24–June 11........	18	694	11	Oct. 12–Mar. 27.......	167	3,500
2	May 26–June 14........	19	741	12	Oct. 31–Mar. 23.......	23	2,893
3	May 26–July 19.........	24	2,107	13	Nov. 1–Mar. 25........	145	2,920
4	May 26...............	....	....	14	Nov. 14–May 29.......	196	5,102
5	July 16 [1]–Aug. 2	17	689	15	Nov. 20–May 4........	165	4,318
6	July 18–Aug. 5.........	18	714	16	Nov. 23–Mar. 28......	125	2,981
7	Oct. 9–Oct. 26.........	17	466	17	Dec. 3–Mar. 27.......	114	2,702
8	Oct. 9–Oct. 28.........	18	519	18	Dec. 3–Apr. 29........	147	3,662
9	Oct. 9–Oct. 30.........	21	562	19	Dec. 7–Mar. 28.......	111	2,681
10	Oct. 9–Oct. 26.........	17	466	20	Dec. 11–Mar. 24.......	104	2,476

[1] It is to be regretted that the falling of the leaf upon which Nos. 5 and 6 matured prevented gathering data on the maximum length of stage at this season of year.

GROWTH, MOLTS, LOCOMOTION, AND FEEDING HABITS.

Concerning growth, molts, locomotion, and feeding habits, there is little to add to that already stated in connection with the larvæ and pupæ of the citrus white fly. The two species are alike as regards the number of larval instars and in their crawling and sedentary habits. Their manner of feeding is similar also, with the exception that when crowded the larvæ of the cloudy-winged white fly settle freely upon the upper surfaces of shaded leaves, where they frequently reach maturity.

MORTALITY AMONG LARVÆ AND PUPÆ.

Remarks relating to mortality among the larvæ and pupæ of the citrus white fly apply with greater force to the cloudy-winged white fly. This mortality appears to result from the same causes in the latter as in the former species. Life-history work has shown that mortality due to spring droughts and dropping from leaves is prac-tically the same for the two species, but that general mortality including "unexplained" mortality is about 3 per cent higher for *A. nubifera*. In this last respect, however, observations throughout groves where infestation is much heavier than on leaves used in the life-history work, and counts of forms on leaves infested with both species of fly, show that the comparative susceptibility to the influ-ences producing mortality of all kinds is often at least twenty times greater for *A. nubifera*. This greater susceptibility appears to be due not only to the more delicate structure of the larvæ and pupæ and their need of more room for development because of their larger size, but also to the adults' habit of crowding the new growth with eggs far beyond·its capacity for maturing the larvæ hatching therefrom.

As may have been inferred already from statements upon the sub-ject of oviposition, it is this insatiable desire of the adults for feeding and ovipositing on new growth that is a most powerful factor lead-ing to the insect's control. While a large amount of data might here be presented illustrating the disastrous effect on the species resulting from overcrowding, the data itself would differ in no respect from that already presented under the general consideration of *A. citri*. Nevertheless, there is a great difference in the extent and practical bearing of this mortality among the immature forms of the two species.

THE ADULT.

The adult of the cloudy-winged white fly is similar to that of the citrus white fly, but is at once separated from it by the dark spot or shading on the outer portion of the upper wings (Pl. X, fig. 1). Except for the further fact that the female is appreciably more robust the adults of both species are structurally much alike. The antennæ of *A. nubifera* are not as highly corrugated as those of *A. citri*, but possess a terminal spine over three times as long as that of *A. citri*. The eyes of *A. nubifera* are more nearly divided in many instances than those of *A. citri*, although this is a character subject to variation in both species.

On nearly all features of life history and habits this species closely resembles the citrus white fly, and these subjects are therefore dealt with in a comparatively brief manner. The principal points wherein the cloudy-winged white fly differs from the citrus white fly may be stated summarily as follows: It is more closely restricted to citrus for its food supply as well as in oviposition; it shows a more strongly developed tendency to feed and deposit eggs on new growth; its arrangement of eggs and preferences for certain sections of leaves for ovipositing are characteristic, and it is slightly less prolific. Its apparent restriction to citrus as a food plant has been discussed under the subject of "Food plants." Its strong preference for new growth results in a situation which can be taken advantage of in the control of the pest by the pruning of water shoots.

The age at which oviposition begins and the activity in oviposition during different parts of the day are the same as for *A. citri*. The females, however, when not abundant deposit more readily along the outer margin of the under surface of the leaf and along the edge and upper surface, and not so freely along the midrib as is the case with *A. citri*. Not infrequently 90 per cent of the eggs will be deposited on the outer portion of the leaf while many are laid on the edge of the leaf itself, from which they often project perpendicularly. The depositing of eggs on the leaf margin and on the upper surface is peculiar to *A. nubifera* and is not the result of overcrowding. A count of 4,000 eggs on nine moderately infested leaves showed that 8.1 per cent of the eggs were laid on the edge of the leaf, 86.8 per cent on the lower surface, and 5.1 per cent on the upper surface. When adults are very numerous both surfaces of the leaves of tender growth and the petioles and shoot stems are thickly covered with

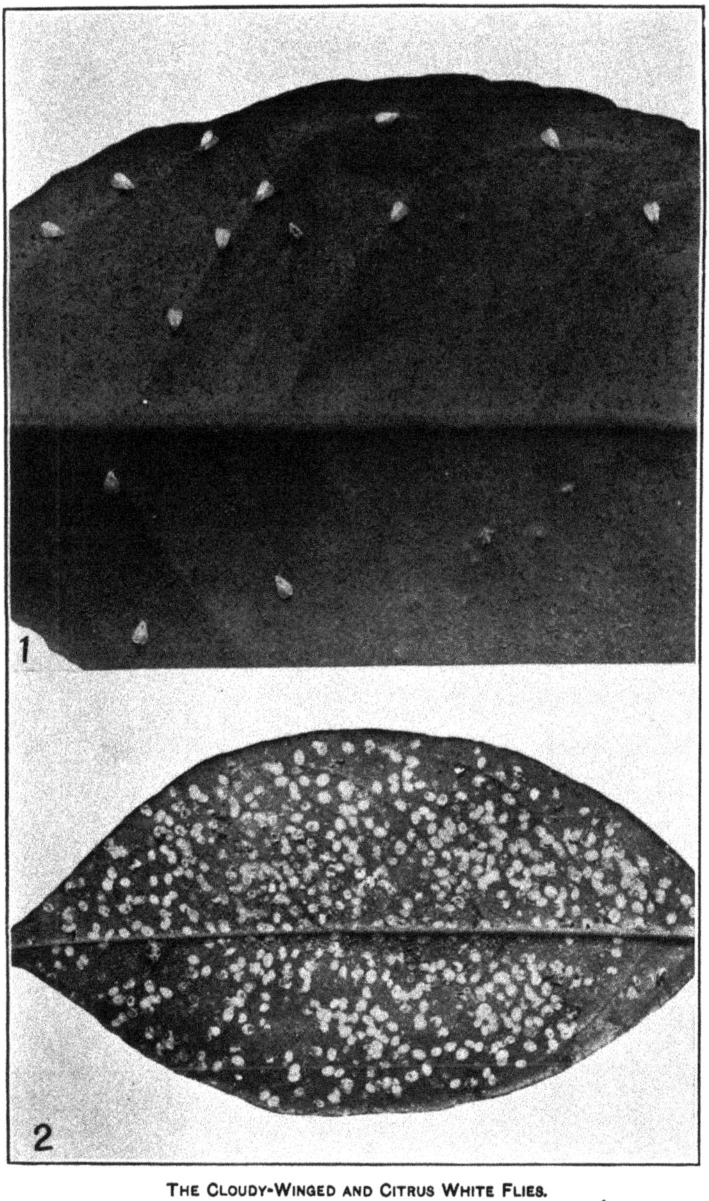

THE CLOUDY-WINGED AND CITRUS WHITE FLIES.

Fig. 1.—Adults of the cloudy-winged white fly, *A nubifera*, showing cloud or spot at tip of wings, and many eggs scattered about. Fig. 2—Larvæ and pupæ of both the citrus white fly and the cloudy-winged white fly killed by fumigation. During life they are nearly transparent and seen only with difficulty. Note eggs of *A. citri* along midrib. (Original.)

eggs. While the citrus white fly deposits her eggs without any definite arrangement, the cloudy-winged white fly, like many other species of Aleyrodes, very frequently lays hers in arcs of various sizes, and, as she is less restless while feeding, has a tendency to deposit her eggs in groups. This arrangement, together with the difference in color, makes easy the separation of the two species.

Reference to the data in Table XXII, especially when compared with that in Table XII, shows that the daily rate of oviposition for the cloudy-winged white fly is slightly less than for the citrus white fly. As much of the data in Table XXII was obtained before typical summer weather had set in, it is of more value as demonstrating the relative rate of oviposition between the two species.

TABLE XXII.—*Daily rate of oviposition of A. nubifera and A. citri compared.*

Record No.	Date eggs deposited.	Duration of egg laying.	Number of females of—		Number of eggs deposited by—		Average number eggs per female per 24 hours laid by—		Average mean temperature.
			Citri.	Nubifera.	Citri.	Nubifera.	Citri.	Nubifera.	
		Hours.							*°F.*
1......	Apr. 20-21, 1909	23	50	45	454	345	9.4	8	.76
2......	Apr. 21-22, 1909	24	40	44	405	432	10.1	9.8	80
3......	....do.........	24	....	16	....	200	....	12.5	80
4......	Apr. 23-24, 1909	46	....	36	....	662	....	9.2	78
5......	Apr. 24-26, 1909	48	....	26	....	516	....	9.9	80
6......	June 16-17, 1909	24	30	79	360	849	12	10.8	82
7......	July 16-17, 1907	24	....	150	....	1,558	....	10.4	82

The number of eggs deposited by single females has not been definitely determined. However, as experiments have shown that adults of *A. nubifera* are capable of living as long as those of *A. citri* and have been known to maintain unimpaired an average of about 1 egg per day less than *A. citri* for at least seven days, it is safe to say that the maximum egg laying capacity is not far from 200.

When all food plants other than citrus are eliminated, the remarks covering the relation between oviposition and food supply for *A. citri* hold for *A. nubifera*, with the exception that oviposition with the latter species is far more dependent upon new growth. This last fact, as discussed under mortality of larvæ and pupæ due to overcrowding, has a most important bearing on the control of this species.

After a grove has been well infested with the cloudy-winged white fly there exists the same high percentage of females as recorded under the same topic for *A. citri*. In fact, the same proportions of sexes, and the same fluctuations and dependence of sex on parthenogenesis, are found to occur with *A. nubifera*. A typical example is the condition found in one grove infested entirely by this species. During the summer preceding winter fumigation the ratio between females and males was 71.4 : 28.6 per cent. After fumigation, when over 99 per cent of the cloudy-winged white fly were killed, the females of the spring brood were so very few in number and so scattered that they deposited a very large percentage of infertile eggs, resulting in a second brood in September, 62.8 per cent of which were males. In other words, after the natural equilibrium between sexes had been disturbed by fumigation, there followed as the result of parthenogenesis a decided fluctuation between a predominance of females in one and of males in the following generations. Gradually this fluctuation diminishes until normal conditions obtain.

While less attention has been given the problems connected with the emergence of this species, observations have shown that the process and time required for emergence and the changes in color occurring thereafter are the same as for the citrus white fly, with the exception of the cloud at the tip of the wing already mentioned. Statements made concerning the conditions favorable and unfavorable for the emergence of the citrus white fly hold for this species. An examination of the extensive daily emergence records on file and summarized in Table XXIII show that even during October and early November emergence did not occur below an average daily mean temperature of 62° F. The emergence occurring later in the fall, and consequently during cooler weather, does not appear to be due to more resistance to cold, but to a difference in seasonal history.

LENGTH OF LIFE CYCLE.

From a study of the length of the egg, larval, and pupal stages already given one can obtain an accurate knowledge of the length of the life cycle. A general summary of the data already presented in connection with these various stages is presented in Table XXIII.

TABLE XXIII.—*Length of life cycle of cloudy-winged white fly at Orlando, Fla.*

Lot.	Eggs deposited.	First fly emerged.	Last fly emerged in fall.	First fly emerged in spring.	Last fly emerged.	Least number of days for development.	Largest number of days for development.	Per cent emerging before winter.	Per cent wintering over to emerge in spring.	Least number degrees effective temperature for development.	Degrees accumulating before spring emergence.	Degrees accumulating before last fly emerged.
No 1..	Apr. 20	June 10	.....	.....	.....	51	334	.....	.....	1,800	.....	.....
No.2..	June 16	Aug. 2	.....	.....	.....	47		.....	.....	1,849	.....	.....
No.3..	Aug. 23	Oct. 14	Oct. 31	Mar. 20	May 16	52	266	64.7	35.3	1,899	5,205	7,062
No.4..	Sept. 4	Oct. 23	Nov. 1	..do....	Apr. 1	49	209	26.8	73.2	1,693	4,706	5,055
No.5..	Sept. 18	Mar. 22	0	Mar. 22	May 5	185	229	0	100	4,650	4,650	5,915
No.6..	Sept. 21	Mar. 25	0	Mar. 25	May 20	185	241	0	100	4,638	4,638	6,318
No.7..	Oct. 2	Mar. 18	0	Mar. 18	.....	167		0	100	4,006	4,006	.....
No.8..	..do.....	Mar. 20	0	Mar. 20	.....	169		0	100	4,073	4,073	.....
No.9..	Oct. 22	Mar. 21	0	Mar. 21	Apr. 9	150	169	0	100	3,549	3,549	4,056

It will be noted that 47, the least number of days required for development, is but slightly higher than the minimum for *A. citri*, even during most favorable weather conditions. The greater average number of days required for development, shown especially by comparisons of lots 1, 3, 6, and 7 of Table XXIII and lots 4, 5, 14, and 15 of Table XV, is not the result of chance circumstances, but actually the result of slower general development under identical conditions. This fact is perhaps more forcibly brought out by the data in Table XXIV:

TABLE XXIV.—*Rate of development of A. citri and A. nubifera compared.*

Date.	Citri.				Nubifera.			
	Instar 1.	Instar 2.	Instar 3.	Instar 4.	Instar 1.	Instar 2.	Instar 3.	Instar 4.
	P. ct.	P. ct.	P. ct.	P. ct.	P. ct.	P. ct.	P. ct.	P. ct.
July 6	11.7	88.3	0	0	77.9	22.1	0	0
July 8	2.1	82.7	15.2	0	6.9	92.1	1.0	0
July 12	.7	10.5	89.8	0	1.4	30.8	67.8	0
July 16	0	4.8	86.1	9.0	0	8.1	91.1	0.8
July 21	0	2.9	20.6	76.7	0	5.9	62.2	31.9
July 25	0	0	6.1	93.9	0	.8	22.5	76.7
July 29	0	0	0	100.0	0	0	14.9	85.1
October 5	98.4	1.6	0	0	99.6	.4	0	0
October 11	8.0	92.0	0	0	35.5	64.2	0	0
October 15	.6	59.4	40.0	0	2.9	96.0	1.1	0
October 23	.6	4.3	90.7	0	2.8	48.8	48.3	0
October 31	0	2.4	17.2	4.4	1.8	20.9	75.4	1.8
November 11	0	.9	1.9	80.3	0	3.3	90.0	6.6
December 3	0	0	1.9	97.1	0	0	28.2	71.8
December 7	0	0	0	98.1	0	0	12.7	87.3
December 17	0	0	0	100.0	0	0	4.2	95.8
December 26	0	0	0	100.0	0	0	0	100.0

In this table is shown the corresponding progress of growth of both species on various dates after egg deposition. The data concerning development during July refer to larvæ hatching from eggs laid on June 16, 1909, and that during October, November, and December to larvæ hatching from eggs deposited on September 18, 1909. For these records, leaves on the same shoot were chosen for deposition of the eggs of each species; hence both species were subject to identical climatic and nutritive conditions.

A study of the data in Tables X and XXI will also prove that the same statements made for *A. citri* concerning the equalizing effect of winter on the length of the pupal stage for wintering-over pupæ are equally true for *A. nubifera*. The data show at a glance that eggs deposited in late October are capable of producing adults the following spring as early or even earlier than eggs deposited a month or. as sometimes occurs, five months earlier.

SEASONAL HISTORY.

GENERATIONS OF THE CLOUDY-WINGED WHITE FLY.

Aside from the fact that the adults of this species have never been seen by the authors on wing during January and early February, as have those of *A. citri*, there being therefore no winter generation

corresponding to that of *A. citri*, the statements made regarding the
number of annual generations of *A. citri* is true of *A. nubifera* when
the additional statement is made that the height of the various
emergence periods occurs usually about two or four weeks later than
the corresponding periods for *A. citri*. The emergence of adults
brings about the same complications in broods and generations
described for *A. citri*, resulting from variation in length of life cycle,
and the double-brooded character of each generation is also to be
found in the life history of *A. nubifera*. Of eggs laid August 23,
1907, 1.5 per cent produced adults between October 1 and 15 and 63.2
per cent between October 16 and 31; of the remaining pupæ wintering
over, 34.6 per cent emerged between March 16 and 30 and 0.7 per cent
between April 1 and 15. From eggs laid September 4, 1907, 24 per cent
of the adults emerged between October 16 and 31, 2.8 per cent between
November 1 and 15, 71.8 per cent between March 16 and 30, and 1.4 per
cent between April 1 and 15. From eggs laid September 18, 1908, 81.6
per cent of the adults emerged between March 16 and 30, 4.1 per cent
between April 1 and 15, 10.2 per cent between April 16 and 30, and
4.1 per cent between May 1 and 15. From eggs deposited March 29,
1909, 44.5 per cent emerged between June 1 and 15, no further records
being kept.

It might be inferred from the slower development of *A. nubifera*
that it would pass through a less number of annual generations than
A. citri. This, however, is not true, inasmuch as its slower development is offset by its seasonal history—it remaining active later in
fall and early winter.

SEASONAL FLUCTUATIONS IN THE NUMBER OF ADULTS OR SO-CALLED "BROODS."

Because the generations of the cloudy-winged white fly are of the
same general double-brooded character as those of the citrus white
fly, and are subject to the same unexplainable variation in the
length of the life cycle, the seasonal history of *A. nubifera* is not
unlike that of *A. citri* in nearly all essential features. In fact, the
same three periods of general emergence of adults occur as with *A.
citri*, but with the difference that the adults of each so-called "brood"
reach their numerical maximum usually from two to four weeks
later than the corresponding broods of *A. citri*. In figure 19 are
given curves representing the abundance of adults of *A. citri* and *A.
nubifera* at Orlando during 1909. As a result of this striking difference in the seasonal history of these two species, previous observations on this subject are considerably confused and should be disregarded unless one is positive of the species under consideration at the
time. As with *A. citri*, no one definite statement can be made to
cover the exact time when the emergence of various broods will begin.
Emergence is strongly influenced by local weather conditions. While
the curve in figure 19 represents the condition in one Orlando grove

in 1909, it is not meant to represent the abundance of adults in any other grove in that city, much less in groves in various parts of the State. The same variation in neighboring groves in the same county and in a lesser degree in different trees in the same grove occurs with *A. nubifera*, and this statement apparently holds for infested groves in any part of the State. For example, groves at Dunedin and Sutherland, in 1909, showed a difference of at least 10 days in the beginning of the active spring emergence of adults.

The most striking difference in the seasonal history between *A. citri* and *A. nubifera* which perhaps attracts most general attention and leads to more confusion between the two species in the minds of many is the much later appearance of adults of *A. nubifera* in the fall of the year. The last large "hatching" of *A. citri* is on a rapid decline at Orlando by the middle of September at the latest, while that of *A. nubifera* at that time is only approaching its maximum and lasts well toward the 1st of November, when its decline is rapid, although adults can be found during moderately warm falls as late as the middle of December. Thus at Orlando in October 18, 1907, when adult

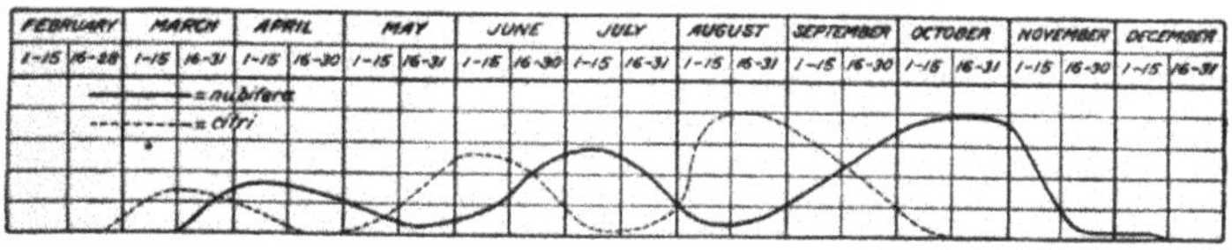

Fig. 19.—Diagram showing relative abundance of the adults of *Aleyrodes nubifera* and *A. citri*, throughout the year 1909, at Orlando, Fla. (Original.)

A. citri were practically off the wing and a large portion of the immature stages of *A. citri* had already reached the pupal stage, note was made that adults of *A. nubifera* were appearing in numbers and that pupæ of *A. nubifera* were rapidly developing eyespots on certain growths, and that new growth in places was crowded with ovipositing adults.

In consequence of the difference between the time of appearance of these fall broods, the immature stages of *A. citri* have largely reached the pupal stage and are prepared to winter over by the last of October. At this time females of *A. nubifera* are crowding the limited new growth with large numbers of eggs, and by far the larger proportion of this species will be found in the egg and larval stages up to the middle of December, and in a few instances third-instar larvæ may be found as late as the middle of February. It will be seen, therefore, from this and the foregoing data that there is no time during the season, except for about two months before spring emergence first sets in, that all stages can not be found in the grove in varying degrees of abundance.

INDEX.

PLATE I.

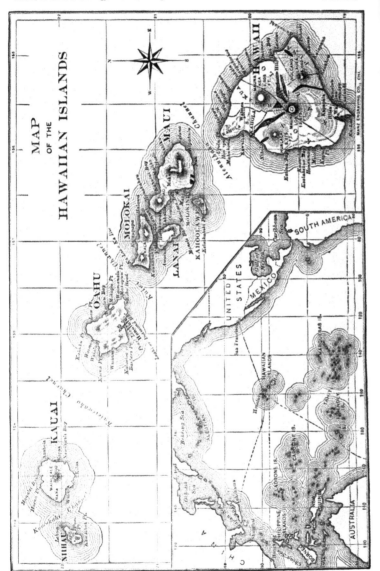

U. S. DEPARTMENT OF AGRICULTURE,

BUREAU OF ENTOMOLOGY— BULLETIN No. 93.

L. O. HOWARD, Entomologist and Chief of Bureau.

THE SUGAR-CANE INSECTS OF HAWAII.

BY

D. L. VAN DINE,

Special Field Agent.

ISSUED JUNE 15, 1911.

WASHINGTON:
GOVERNMENT PRINTING OFFICE.
1911.

LETTER OF TRANSMITTAL.

U. S. Department of Agriculture,
Bureau of Entomology,
Washington, D. C., December 22, 1910.

Sir: I have the honor to transmit herewith for publication a manuscript entitled "The Sugar-Cane Insects of Hawaii," by Mr. D. L. Van Dine, recently a special agent of this Bureau, and for several years entomologist of the Hawaii Agricultural Experiment Station. The manuscript includes a discussion of the present status of the sugar industry of the Hawaiian Islands and treats of the principal insect enemies to this important industry, which is rapidly assuming large proportions in our Southern States owing to the increased acreage which is being planted to cane. I would recommend its publication as Bulletin No. 93 of the Bureau of Entomology.

Respectfully,   .

L. O. Howard,
Chief of Bureau.

Hon. James Wilson,
Secretary of Agriculture.

3

PREFACE.

The acreage devoted to sugar-cane culture in the southern United States has increased rapidly in recent years. Some of the cotton lands, abandoned because of the depredations of the cotton boll weevil, are being planted to cane. New lands are being planted to the crop in the Rio Grande valley and in the reclaimed areas in the lower Mississippi valley. It is stated that quite an area of land in process of reclamation in the State of Florida will be planted to sugar cane. It is desirable that the experience obtained through investigations of insects injurious to sugar cane in the Hawaiian Islands be placed at the disposal of the planters in our Southern States in order that the sugar industry in those States may receive practical benefit therefrom.

The Hawaiian planters are well provided with expert advice and have at hand numerous reports dealing with the subject, which latter, unfortunately, are not available for general distribution. This report is written primarily, therefore, for the information of our mainland planters.

Acknowledgment should be made of the courtesies extended to the writer by the members of the entomological staff of the Hawaiian Sugar Planters' Association Experiment Station during his return visit to the Hawaiian Islands in March and April, 1909.

D. L. Van Dine.

5

CONTENTS.

ILLUSTRATIONS.

PLATES.

TEXT FIGURES.

THE SUGAR-CANE INSECTS OF HAWAII.

LOCATION AND CLIMATE OF THE HAWAIIAN ISLANDS.

The mid-Pacific Territory of Hawaii (see Plate I) is situated 2,100 miles to the southwest from San Francisco, the California coast being the nearest continental area. The islands are separated by channels varying from 20 to 58 miles in width. The 8 inhabitable islands, Hawaii, Maui, Oahu, Kauai, Molokai, Lanai, Kahoolawe, and Niihau, lie between 18° 54' and 22° 15' north latitude; that is, the northern limit of the islands is just within the Tropics. The climate of the entire group is, however, only subtropical, due largely to the prevailing northeasterly trade winds, the cool ocean currents from the north, and the relatively low humidity. The temperature varies according to the altitude and the location of the land as regards the higher mountains. The formation of the islands is of recent volcanic nature, with the exception of the low-lying coastal plains, which are of coral origin.

The annual maximum temperature ranges from 88° to 90° F., while the annual minimum temperature recorded ranges from 52° to 58° F. A temperature of 29° F. has been recorded at an altitude of 6,685 feet, and freezing temperatures are of frequent occurrence at these high altitudes. The rainfall varies in amount with the locality. Places within a few miles of each other are known to differ more than 100 inches in average annual rainfall. The sides of the islands exposed to the northeast trade winds have abundant rains, while the opposite sides have little and some localities hardly any.

The soils of the islands are exceedingly fertile and when properly cultivated yield abundant crops.

THE SUGAR INDUSTRY IN HAWAII.

The production of sugar is the leading industry in the Hawaiian Islands. Sugar cane is grown on four of the islands. The island of Hawaii has the largest acreage devoted to cane, Oahu, Maui, and Kauai coming next in importance in the order named. There are more than 200,000 acres planted to cane in the islands. In 1908 521,000 tons of raw sugar were produced, having a value of more than $40,000,000. The average yield of sugar per acre is 4½ tons.

The plant crop is taken off 20 to 22 months from the time of planting, and the first ratoon or stubble is harvested 18 to 20 months later. The second ratoon usually goes 18 months again before it is ground. Sometimes a "short ratoon" crop is made, in which case the cane runs about 14 months. The time given for growth depends on the maturity of the cane, which in turn is governed by the location and altitude of the land. To some extent also the time of harvest is governed by the labor supply, factory conveniences for taking off and grinding the crop, and the need of land for planting.

The sugar industry in Hawaii was placed on a basis to insure its becoming the leading industry by the reciprocity treaty of 1876 between the United States and the Hawaiian Government, the latter at that time being an independent monarchy. The effect of this treaty in removing the duty on raw sugar exported to the United States was to increase American influence in the islands and to strengthen the commercial relations between the two countries. A second great factor in the development of the sugar industry was the annexation of the islands as a Territory of the United States by an act of Congress passed July 7, 1898, by mutual agreement between the two countries, Hawaii at that time having overthrown the monarchy and become a republic. Annexation insured a free and protected market to the sugar output of the islands and gave confidence for the investment of capital. This is of prime importance, as the production of sugar in the islands is on a corporation basis and any disturbance in the market is felt at once by every plantation in the Territory.

Fundamental factors that have attended the development of the sugar industry are the equable climate of the islands, the natural productiveness of the soil, the resources of water for irrigation purposes, and the immunity from the more serious depredations by insects and diseases that retard the development of agricultural resources in less fortunate parts of the world. Further, there is to be found in Hawaii a class of progressive business men who have developed immense irrigation schemes, made use of the most modern agricultural and factory machinery, inaugurated advanced methods of cultivation, fertilization, and irrigation, and united their interests in a cooperative association.

This organization, the Hawaiian Sugar Planters' Association, has, since April, 1895, maintained a private experiment station, where important researches have been made and valuable results obtained. The work has applied to varieties and seedlings, propagation, cultivation, irrigation, the use of fertilizers, and the manufacture of sugar. These investigations, together with the perfection of factory methods and field machinery, have brought the sugar industry of the islands to the high standard it holds among the sugar-producing countries of the world.

SUGAR-CANE INSECTS.

The advent of a serious pest into the Hawaiian sugar-cane fields, the sugar-cane leafhopper (*Perkinsiella saccharicida* Kirk.), between 1900 and 1902 and the widespread injury of this insect throughout the sugar-cane districts in 1903 led to the establishment of an entomological division in the Sugar Planters' Experiment Station in September, 1904. In this division detailed studies have been made of the species of insects occurring in the Hawaiian cane fields, the investigations relating particularly to the leafhopper and its natural enemies.

Koebele[a] has earlier discussed the sugar-cane insects. Up to the time of the leafhopper invasion the sugar-cane borer ([*Sphenophorus*] *Rhabdocnemis obscurus* Boisd.) was the most injurious species. The sugar-cane aphis (*Aphis sacchari* Zehntner), the sugar-cane mealy-bug (*Pseudococcus calceolariae* Maskell), the leaf-roller (*Omiodes accepta* Butler), cutworms, and certain other pests occurred locally, but up to this time no detailed study of their injury had been made.

An insect enemy of sugar cane has exceptional advantages for development in the Hawaiian Islands. Approximately only one-half the total area is harvested at any one time. Thus the great extent of the plant gives an abundant supply of food, while the system of cropping provides a continuous supply. These conditions, together with the even climate, favor the uninterrupted breeding of any enemy of the plant. A further factor in the undue increase of the cane-feeding insects is the impetus to development arising from the absence of the special parasitic and predaceous enemies of the plant-feeding species. The absence of natural enemies is understood when it is known that the islands are isolated from all continental areas and that the economic plants are introduced forms for which the native flora has made way, carrying with it the endemic species of insects, while the insect enemies of a cultivated plant are of foreign origin, introduced into the islands with their host plant but without their natural enemies. These very facts, together with the almost total absence of secondary parasites as a group and the opportunity of eliminating them when introductions are made, furnish ideal conditions for the introduction and establishment of special parasitic insects. The greatest factor in the successful establishment of a special parasite is the absence of the secondary parasites of which it is the host. One can understand why emphasis has been placed on the use of natural enemies in the control of injurious species in Hawaii and why also greater success has been

---

[a] Hawaiian Planters' Monthly, vol. 15, no. 12, pp. 590–598, December, 1896; vol. 17, nos. 5 and 6, pp. 208–219 and 258–269, May and June, 1898; vol. 18, no. 12, pp. 576–578, December, 1899; vol. 19, no. 11, pp. 519–524, November, 1900.

attained in Hawaii than in continental regions where investigations of this character are under way. From the above remarks it is apparent that the entomologists of the Hawaiian Sugar Planters' Experiment Station are justified in placing emphasis on this phase of insect control. Indeed, their work has been almost entirely along this line.

THE SUGAR-CANE LEAFHOPPER.

(*Perkinsiella saccharicida* Kirk.)

DISTRIBUTION.

The Hawaiian sugar-cane leafhopper (*Perkinsiella saccharicida* Kirkaldy) was introduced into the islands some time prior to 1900 from Queensland, Australia. The species occurs throughout the sugar-cane areas both in Australia and in Hawaii and has been recorded from Java.[a]

APPEARANCE OF THE LEAFHOPPER IN HAWAIIAN CANE FIELDS.

The first appearance of the leafhopper in Hawaii is recorded by Mr. Albert Koebele in January, 1902.[b] Koebele notes the species under the heading "Leafhopper (Fulgoridæ)," the species at that time not having been described. Regarding its appearance Mr. Koebele says:

According to Mr. Clark a small homopterous insect appeared upon the sugar cane at the experimental station some twelve months since, affecting the Demerara and Rose Bamboo plants. Its presence is easily seen by the black and dirty appearance of the leaves and more or less red midribs.

The insect lives in company with its larva in large numbers behind leaf sheaths, which it punctures to imbibe the sap of the plant. When mature it is exceedingly active in its habits, springing with suddenness from its resting place at the least disturbance. The eggs are oviposited into the midrib over a large extent, most numerous near the base, in groups of about from four to seven, and large quantities are often present in a single leaf. The surroundings of the sting become red and in advanced stages the whole of the midrib becomes more or less of this color and brownish red

That the species caused little alarm at this time is indicated by Mr. Koebele's further statement in this same article. He says:

Should this insect become numerous on any plantation, they could be kept in check by careful and repeated stripping and burning, immediately after, of the leaves containing the eggs. I do not anticipate any serious results from the above insect, which may have been present upon the island for many years.

In May, 1902, Dr. R. C. L. Perkins under the title "Leafhoppers (Fulgoridæ)," in a report to Mr. C. F. Eckart, director of the Hawaiian

---

a KIRKALDY, G W.—A note on certain widely distributed leafhoppers. <Science, vol. 26, no. 659, p. 216, 1907.

b KOEBELE, A.—Report of the committee on diseases of cane. <Hawaiian Planters' Monthly, vol. 21, no. 1, pp. 20–26, January, 1902.

Sugar Planters' Experiment Station, mentions the doubtful origin and identity of the species.[a]   Doctor Perkins again records the insect under the heading "The leaf-hopper of the cane" in December of the same year and says: "This small insect is highly injurious to cane and its destructiveness threatens to exceed that of the cane borer beetle."[b]

In response to repeated requests made to the department the writer was detailed early in May, 1903, to make a report on the pest.   On May 11, 1903, specimens were forwarded by the writer to Dr. L. O. Howard, Chief of the Bureau of Entomology, Washington, D. C.   Under date of June 1, 1903, Doctor Howard replied that the species was new to science and that there was in press a description of the insect under the name *Perkinsiella saccharicida* by Mr. G. W. Kirkaldy of the British Museum.

DESCRIPTION OF THE LEAFHOPPER.

The species was described by Mr. G. W. Kirkaldy in 1903 and represents a new genus which was named after Dr. R. C. L. Perkins. The description of the genus and species is taken from Mr. Kirkaldy's article in The Entomologist, London, for July, 1903, pages 179–180, and is as follows:

Perkinsiella, gen. nov.

Closely allied to *Aræopus* Spinola, but distinguished by the first segment of the antennæ being distinctly shorter than the second; distinguished from *Dicranotropis* Fieber, to which it bears some resemblance, by the form of the frons, and by the flattened apically dilated first segment of the antennæ.   Type, *P. saccharicida* Kirkaldy.

Second segment of antennal peduncle about one-half longer than the first; flagellum about one-third longer than the entire peduncle, first peduncular segment much wider at apex than basally, flattened and explanate; second segment nearly as wide at base as the apex of the first segment [in *Aræopus* it is much narrower, while the first segment is more parallel-sided].   Exterior longitudinal nervure of corium forked near the base, and its exterior branch forked near its middle; interior longitudinal nervure forked near the apex.   Membrane with six nervures, the fourth (commencing inwardly) forked; the first area has an incomplete nervure reaching only to the middle.   Other characters as in *Aræopus*.

P. saccharicida, sp. nov.

Long-winged form ♂ ♀.—Tegmina elongate, narrow, extending far beyond apex of abdomen, interior half of clavus and corium more or less faintly smoky, a long dark smoky stripe on middle of membrane, three or four of nervures of the latter smoky at apex.

Short-winged form, ♀.—Tegmina reaching only to base of fifth segment, costa more arched, apex more rounded, neuration similar but shortened.   Tegmina hyaline, colourless; nervures pale testaceous brownish, with blackish brown non-piligerous dots (in both forms).

---

a ECKART, C. F.—Precautions to be observed with regard to cane importations. <Report to Hawaiian Sugar Planters' Association, May 9, 1902, p. 5.

b PERKINS, R. C. L.—Notes on the insects injurious to cane in the Hawaiian Islands. <Hawaiian Planters' Monthly, vol. 21, no. 12, pp. 593–596, December, 1902.

♂. Pallid yellowish testaceous. Abdomen above and beneath black, apical margins and laterally more or less widely pallid. Apical half of first segment and carinate edges of second segment of antennæ, flagellum, basal half of frons (except the pustules) and a cloudy transverse band near the apical margin of the same, longitudinal stripes on femora, coxæ spotted or banded near the base, a large spot on each pleuron, anterior and intermediate tibiæ with two or three annulations, apical segment of tarsi, etc., blackish or brownish. First genital segment large, deeply acute-angularly emarginate above

♀. Like the male, but abdomen above and beneath stramineous, irregularly speckled with brownish. Ovipositor, etc , blackish. Sheath not extending apically so far as the "scheidenpolster." Long   ♂ ♀ 4½ mill.; to apex of elytra in long-winged form, 6¼ mill.

DISPERSION OF THE LEAFHOPPER.

The spread of the insect over the cane districts of the Hawaiian Islands was apparently very rapid, although it had undoubtedly occurred in the fields unnoticed by the planters for several years. By February, 1903, the species became generally abundant throughout the cane fields of the entire Hawaiian Territory.

The main factor in the distribution of the pest is the habit of the female of depositing her eggs beneath the epidermis of the internodes of the cane stalk. It seems probable that the pest was introduced into the islands and to a great extent distributed over the cane districts in seed cane. In local distribution other factors present themselves. The leafhopper is an insect readily attracted by light at night, as its presence about lamps in the factories and homes on the plantations testifies. Passengers and steamship officers of the interisland steamers have frequently stated to the writer on inquiry that in many instances, especially at night, great numbers of the insects have come aboard in certain ports or when offshore from certain plantation districts. These adults have undoubtedly traveled in this manner from one locality to another so that an uninfested district might easily have become infested by adults flying ashore from a passing steamer previously infested while stopping at or passing by an infested locality. Railway trains have been equally active in the spread of the insect on land.

Another mode of distribution, during the general outbreak of 1903, under conditions of heavy infestation, was the migration of the pest from one locality to another during the daytime. These migrations were observed by many of the planters. The manager of one plantation in the Hamakua district of the island of Hawaii stated to the writer that in the early evening of April 26, 1903, the atmosphere was "thick with hoppers" for a distance of 2 miles and that the "hoppers" were traveling with the prevailing wind, about southwest. Similar migrations, described by the observers as "clouds," were mentioned by other managers.

LIFE HISTORY AND HABITS.

The writer spent two months in the cane fields during the outbreak and in the early part of July, 1903, presented a report to the Hawaiian Sugar Planters' Association on the occurrence and injury of the species. Later an account of these investigations was published, from which a part of the information on the leafhopper presented herewith is taken.[a]

"Leafhopper" is a popular term applied to a certain group of plant-feeding insects of the order Hemiptera. The family Fulgoridæ, to which the Hawaiian sugar-cane leafhopper belongs, is included in this group. Common characteristics of these insects are their peculiar habit of springing or jumping when disturbed; their feeding upon plants by sucking from the tissue the plant juice or sap through a beak or proboscis, a piercing organ by means of which they puncture the epidermal layer of the plant; their incomplete development (that is, the young upon hatching from the eggs resembles the adult, except that it is smaller in size, wingless, and sexually immature and by a gradual process of development acquires the characteristics of the adult); and the fact that their eggs are deposited in the same plant upon which the young and adult appear and feed.

The eggs of the sugar-cane leafhopper (Plate II, figs. 1, 2) are deposited beneath the epidermis of the cane plant in situations along the midrib of the leaves, in the internodes of the stalk, or, in the case of young unstripped cane, in the leaf sheath of the lower leaves. When deposited in the leaves, the eggs are inserted from either side, but usually from the inside, the greater number being in the larger portion of the midrib down toward the leaf sheath. The place of incision is indicated at first by a whitish spot, this being a waxy covering over the opening. The female accomplishes the process of oviposition by puncturing the leaf or stem with her ovipositor, which organ (fig. 1, b) is plainly visible on the lower side of the abdomen, attached to the body at the center behind the last pair of legs and extending backward along the median line of the abdomen, reaching nearly to the end. By the aid of this structure the female pierces the epidermis of the cane stalk and through the one opening forms a cavity or chamber to receive the eggs. The number of eggs deposited in each cavity varies, the writer finding the average to be between four and six. That a single female is responsible for many of these clusters has been verified by the writer by observation. As the growth of the cane continues and the new leaves unfold toward the top of the plant, the infested leaves naturally occupy

[a] VAN DINE, D. L.—A sugar-cane leaf-hopper in Hawaii, *Perkinsiella saccharicida.* <Hawaii Agr. Exp. Sta., Honolulu, Bul. 5, pp. 29, figs. 8, 1904.

the lower position on the stalk. The leafhopper, during a heavy infestation, will continue to puncture the midribs of the leaves as rapidly as the leaves unfold. The older egg chambers of the lower leaves are distinguished from the newly formed chambers of the upper leaves by a reddish discoloration.

Under laboratory conditions the writer found that the eggs deposited in cane growing in rearing cages hatched two weeks thereafter. The period of development of the young to the adult required 34 additional days, making the life cycle 48 days in length.

The length of the egg stage, under certain conditions, is much longer than the time given above. Mr. C. F. Eckart, director of the Hawaiian Sugar Planters' Experiment Station, records that hatching continued for 38 days from cane cuttings infested with eggs of the leafhopper.[a]

The fact that the eggs will hatch from cane cuttings during a period of at least 38 days is a very important point to bear in mind in the shipping of infested cane from one locality or country to another. Since practically the only means by which the Hawaiian leafhopper could be introduced into the cane fields of the Southern States is by the shipment of seed cane from New South Wales, Queensland, Java, or Hawaii to this country, the writer would emphasize the necessity of having all introductions made through officials engaged in sugar-cane investigations.

On issuing from the cavity, or chamber, the young, newly hatched leafhoppers appear at first small, slim, wingless nymphs, almost transparent. During the process of hatching or emerging from the egg chamber the insects slowly work their way head first to the surface of the leaf or stalk. The writer found, by timing the operation, that from 8 to 15 minutes were required, during which time the nymphs rest occasionally to unfold and dry their legs. When they become detached from their egg-cases and have emerged to the surface, they are at once active and scatter over the plant to feed, congregating at first down within the sheaths of the upper leaves. In a few hours the body becomes shortened and the outer covering, on exposure to the air, becomes darker in color. The habit of the very young in secluding themselves within the lower sheaths of the leaves renders them quite inconspicuous unless especially sought for. They may become very abundant and still remain undetected by an ordinary observer until the result of their feeding becomes apparent. (See nymphs, Plate II, figs. 3–6.)

Ordinarily when disturbed the adult leafhopper does not fly but moves off in an odd, sidewise fashion to another part of the leaf, or springs suddenly to another portion of the plant. (See adults, Plate II, fig. 7, and text fig. 1.)

---

[a] ECKART, C. F.—Report of the Hawaiian Sugar Planters' Association Experiment Station for 1903, Honolulu, 1904, pp. 78–79.

PLATE II.

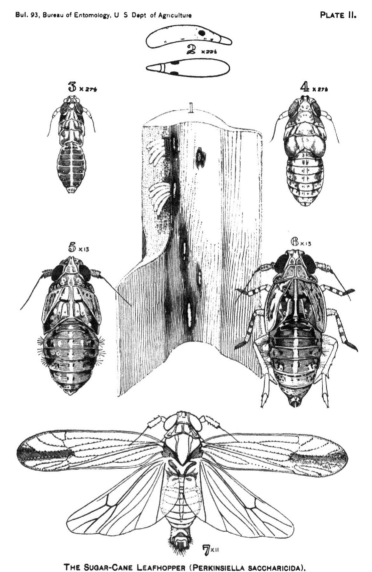

THE SUGAR-CANE LEAFHOPPER (PERKINSIELLA SACCHARICIDA).

Fig. 1.—Egg chambers in midrib of cane leaf, slightly enlarged. Fig 2.—Eggs, greatly enlarged.
Fig. 3.—First-stage nymph. Fig. 4 —Second-stage nymph. Fig. 5.—Third-stage nymph Fig. 6 —
Fourth-stage nymph. Fig. 7.—Adult male. (After Kirkaldy)

SYMPTOMS OF LEAFHOPPER INJURY.

The presence of the pest on the plantations was noticed first by the appearance of a sooty black covering on the lower leaves of the cane plant. This black covering became known as smut. It is a fungous growth and finds a medium for development in the transparent, sticky fluid secreted by the leafhoppers during their feeding on the plant. This secretion is commonly known as honeydew.

The black smut or fungous growth in the honeydew secretion of the leafhopper and the red discoloration about the openings to the egg chambers in the midribs of the leaves are the most pronounced symptoms of the work of the leafhopper on cane.

In the case of heavy infestation a further result is the appearance of the plant as a whole. The leaves on which the insects have been feeding develop a yellowish appearance, and as the work of the insects progresses they become dried and resemble the fully matured lower leaves of the plant. This premature death of the leaves is due to the excessive amount of juice extracted for food. As long as the cane plant is able to produce new leaves its life is not actually in danger, the injury being a check to the growth and indicated by the small, shortened joints in the stalk. Leaves thus prematurely ripened do not drop away from the stalk at the junction of the sheath, as is the case under normal conditions, but break and hang down at the junction of the leaf to the sheath, leaving the sheath still wrapped about the stalk. Leaves in such a condition remain green for some time, attached to the sheath by the midrib, and an attempt to strip the cane results in leaving the sheaths still adhering to the stalk and wrapped about it.

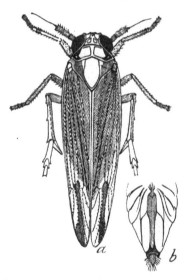

Fig. 1.—The sugar-cane leafhopper (*Perkinsiella saccharicida*): *a*, Adult female, much enlarged; *b*, ovipositor, greatly enlarged. (After Kirkaldy.)

In the last stages of an attack, when the plant is actually overcome by the pest, the young unfolded leaves at the top do not appear to have the vitality to unfold and the "bud" gradually dies out. At this stage the normal growth of the plant ceases. Many plants in

such a condition will then throw out sprouts from the eyes.   This is
a serious circumstance, since the growth of the sprouts is supported
by the stalk, and unless the cane is soon cut and ground the stalk is
rendered worthless.

CHARACTER OF INJURY TO THE CANE.

The first injury to the cane plant by the leafhopper occurs through
the piercing of the epidermal layer by the ovipositor (fig. 1, *b*) of the
female and the later rupturing of the tissue of the plant on the hatching of the young.   This injury to the tissue in itself is not serious,
but the many openings in the leaves and stalks allow excessive
evaporation to occur.   Through these wounds various diseases may
also gain entrance to the tissues of the plant, carried thereto by the
leafhoppers themselves in flying from infested to noninfested plants,
or by other insects, particularly certain flies, which frequent the
cane plant.

The most serious injury to the plant is the drain upon its vitality
caused by the young leafhoppers in feeding.   The structure of the
mouthparts of the leafhopper has been mentioned; that is, a piercing
organ, which is inserted through the outer covering of the tissue, by
means of which the insect sucks the juice or sap from within.   The
amount extracted in this manner by any particular individual is small
and of little consequence, but the result of a myriad of individuals working constantly in this manner upon a plant is readily conceived to be
serious in its consequences.   The leafhopper in feeding upon the
cane plant extracts therefrom an amount of juice greatly in excess
of its own needs for development.   This excess is excreted from the
body of the insect upon the cane plant in the form of a sweet, sticky
substance, known as honeydew.   It is in this substance that the
black smut develops.

The sooty covering or smut of the leaves referred to is a superficial fungus which bears a close resemblance to the fungi of the genus
Sphæronema.   The writer was informed by Dr. A. F. Woods, at that
time Pathologist of the United States Department of Agriculture,
that this fungus may be responsible for the dying back of canes
which followed heavy leafhopper infestation.   It is believed, however, that in the cane the smut affects the plant only by preventing
the assimilation of the elements taken up by the plant from the soil
as food, in cutting off the rays of direct sunlight, and also in closing
the stomata of the leaves, preventing the entrance and escape of
carbon dioxid and oxygen, respectively.   In damp localities another
fungus was taken in company with the smut, and was determined
by Dr. Woods as a species of the genus Hypochnus.   The resulting
injury to the plant from the leafhopper attack is also complicated by

the presence of the pineapple disease of sugar cane (*Thielaviopsis ethaceticus*) and the rind disease (*Melanconium sacchari*). The latter species, it is believed, gains entrance to the tissue of the plant through the wounds made by the leafhopper.

EXTENT OF INJURY.

It was estimated that the leafhopper caused a loss of $3,000,000 to the planters of Hawaii during 1903 and 1904.[a] In the writer's opinion this loss can not be attributed entirely to the leafhopper injury. Other species of insects and certain diseases were implicated. The leafhopper was directly responsible for the larger percentage of loss and indirectly responsible for the unusual development of certain diseases.

In speaking of the rind disease of sugar cane in Hawaii in 1907 Mr. L. Lewton-Brain says:[b]

> To bring before you the actual extent of the loss that the rind disease is now causing in your cane fields, I take the following fact obtained by Doctor Cobb from actual counts in the field. In one case the cane left on the ground represented about one ton of sugar to the acre. That is to say, that if the cane left on the field had been sound cane that portion of it left on an acre would produce about a ton of sugar. The area counted over, in this particular case, was representative of 200 acres.
>
> A few years ago, when the leaf-hopper was at the height of its glory in reducing the vigour and vitality of your canes, these figures would have been much higher. I have been assured that, at that time, there were acres and acres of cane to be seen on which the majority of the sticks had been ruined by rind disease.

Apart from the direct and indirect injury of the leafhopper (*Perkinsiella saccharicida* Kirk.), the sugar-cane borer (*Sphenophorus obscurus* Boisd.), the sugar-cane leaf-roller (*Omiodes accepta* Butler), and other minor pests contributed to the loss sustained.

The explanation of the undue increase on the part of the leafhopper is made clear when it is known that up to the time of the leafhopper invasion the sugar plantations had been particularly free from serious attacks of insect and disease pests. The planters were, therefore, unacquainted with the insect life to be found in their cane fields. They did not know the source or nature of the leafhopper attack and had at hand no general knowledge of insect warfare. The injury of the leafhopper, combined with that of the other species mentioned, and the complications arising through the development of certain diseases gave the leafhopper a favorable opportunity to develop great numbers in those localities where climatic influences or soil conditions were unfavorable to the sugar cane or where a deteriorated condition of the cane varieties prevailed.

---

[a] Report Governor of Hawaii for fiscal year 1907, p. 22.

[b] LEWTON-BRAIN, L.—Rind Disease of the Sugar-Cane. <Hawaiian Sugar Planters' Exp. Sta., Div. Path., Bul. 7, p. 15, 1907.

FACTORS RESPONSIBLE FOR THE OUTBREAK OF 1903.

On those plantations where the outbreak of the leafhoppers became epidemic the writer made careful observations to determine, if possible, the conditions of season, soil, varieties, or methods of cultivation which might have contributed to the leafhopper development. Some of these conditions noted will be mentioned.

(1) The season during which the attack was most serious was not the growing season, and in some localities the weather was most unfavorable for the growth of the cane. In one instance, for example, there were during one month 24 rainy days out of the 30; and since the temperature on a rainy day is some ten degrees lower than on a bright day, and because of the absence of sunshine to carry on the work of assimilation, a less vigorous growth of cane resulted.

(2) The long duration of prevailing high winds.

(3) An impoverished condition of the soil. Certain fields in which the leafhopper was epidemic had been planted continuously to cane for over 20 years. The soil in certain parts of some fields, also, where the leafhopper infestation was greatest was found to be in poor condition because of lack of drainage.

(4) As the rainy season was followed by a long period of dry weather, without the means of irrigation, the cane lacked sufficient moisture to enable it to put forth a vigorous growth. This point was demonstrated on an unirrigated plantation in the district of Kohala, Hawaii. A portion of a field was seriously attacked by the leafhopper during the month of September, 1903, after several months of dry weather. The manager of the plantation, Mr. E. E. Olding, was able to run water into this portion of the field and irrigated the cane four times at intervals of about a week, with the result that the cane, although showing the attack in the smallness of the joints grown during that time, recovered, and when the writer visited the field during the month of November of the same year was, in appearance, not unlike healthy portions of the same field.

(5) The presence of other pests, principally the cane borer (*Sphenophorus obscurus*) and the leaf-roller (*Omiodes accepta*).

(6) The lack of thorough cultivation.

(7) The injury to cane on the makai (seaward) fields by the salt spray or the check to the cane by the cold on the mauka (mountainward) fields.

(8) The deterioration of varieties.

(9) The complications due to the presence of certain diseases.

THE LEAFHOPPER AND BEEKEEPING.

An interesting condition of affairs arising from the leafhopper attack on sugar cane is the collection of the honeydew by honey bees. The increase in the production of Hawaiian honey of recent years corresponds with the advent of the sugar-cane leafhopper into the cane fields, and the recent extensive proportions which the bee-keeping business in the islands is assuming is in the vicinity of the immense areas of land given to cane culture.[a] (See fig. 2.)

The principal source of floral honey in the islands is the flowers of the algeroba. (*Prosopis juliflora*). The total production of this floral honey does not exceed 600 tons. The output of honey for 1910 in the islands exceeds 1,000 tons, and the remaining 400 tons consists almost entirely of the product gathered from the honeydew of the sugar-cane leafhopper. Some 100 tons of this forms a typical

FIG. 2.—An apiary near a sugar-cane field. (From Phillips.)

honeydew honey, the remaining amount consisting of natural blends of these two types.

Honeydew honey from the sugar-cane leafhopper is noncrystallizable and usually of a very dark color. The aroma is very similar to that of molasses and the taste insipid. The product is abnormally high in ash, the amount ranging from 1 to 2 per cent, and it has a decided right-handed polarization, while the floral or algeroba honey is low in ash and has a left-handed rotation, which is

---

[a] Van Dine, D. L.—The Source and Characteristics of Hawaiian Honeys. <Hawaii Agr. Exp. Sta., Bul. 17. Pt. I, pp. 1-12, 1908.

Phillips, E. F.—A brief survey of Hawaiian Bee Keeping. <U. S. Dept. Agr.. Bur. Ent., Bul. 75, Pt. V, Jan. 19, 1909.

characteristic of all floral honeys. The larger amount of honey-dew is obtained from the insects on the young plant cane, for there the leafhoppers are more abundant. The amount of honeydew gathered depends on the maturity of the cane and the amount of rain which washes the secretion from the leaves.

CONTROL OF THE LEAFHOPPER.

DIRECT MEASURES.

Insecticides.—Those familiar with the culture of sugar cane will readily understand the difficulty of getting in and through the fields after the cane obtains any height. This difficulty renders the use of insecticides as a remedy unpractical. In Hawaii such a method becomes still more difficult because of the prevailing slope of the cane lands and the manner in which the fields in many districts are laid out for purposes of irrigation. The feeding habits of the leafhopper are such that a contact poison or irritant would be necessary for its destruction, and the activity of the leafhoppers—that is, the suddenness with which they disperse at the least disturbance—still further prevents the successful application of a contact insecticide. Then, too, the cane fields of Hawaii are subject to prevailing winds, which greatly interfere with the use of any substance in the form of a spray. In the face of the above difficulties the writer attempted the destruction of the leafhopper by direct measures and found that an application of kerosene emulsion applied in the shape of a finely divided stream with considerable force was capable of killing only a small percentage. A mixture of lime and caustic soda was also applied, with negative results. Lime, prepared by reducing fresh stone lime to a powder by the use of solutions of copper sulphate and caustic soda, was applied as a dust on cloudy days, or just after showers, and while in comparison to spraying a much larger area was covered, and the dust came in contact with a large percentage of the leafhoppers, no appreciable beneficial results were observed.

Collection by nets.—Ordinary sweeping nets supplied with short handles were placed in the hands of the laborers, and the leafhoppers were collected by having the laborers go in a body through adjoining rows and sweep the nets over the cane leaves. The insects collected were dumped from the nets into buckets of water and kerosene at the ends of the rows. While immense numbers were captured in this way, the number collected and the area covered were so small in comparison to the abundance of the leafhoppers and to the extent of the infested area that this measure was also discarded.

Cutting and burning in the infested centers.—The direct measures of control advised by the writer were confined to the cutting down and burning over of those centers in the fields where the species had become numerous. In this practice it was observed that many of the adults were able to take flight from the burning cane and escape to adjoining fields. However, many adults and all of the unhatched eggs in the leaves and the immature wingless forms were destroyed. The center of infestation was destroyed, and this gave the ratoon crop over these areas a chance under more favorable conditions.

Stripping the leaves.—For agricultural reasons it was a common practice in Hawaii to strip the lower mature leaves from the cane stalk. It was believed at first that this operation would greatly lessen the numbers of the leafhopper by the exposure of the unhatched forms in the leaves of the cane and by removing a place of shelter for the active forms. Observations made during the summer months indicated that stripping was beneficial from the standpoint of the control of the leafhopper. Later observations made during the winter months, however, when growth of the cane practically ceases, showed a very serious condition of affairs, namely, that in heavy infestation the internodes of the stalk of stripped cane contained hundreds of punctures from egg laying, while the internodes of unstripped cane were protected from such injury by the leaf-sheaths.

Burning of trash after harvesting.—The thorough burning of the trash after the cane is harvested is the most effective method practiced for the control of the insects of sugar cane. In the case of the leafhopper many of the adults no doubt take flight, but the destruction to the eggs and immature forms in the trash is enormous. The place where the greatest numbers of the leafhopper were noted in 1903 was on a plantation where the practice of "burning off" had been discontinued for several years, and the manager attributed the unusual increase of the pest to the fact that the trash had not been burned. Both for the leafhopper and the cane borer, burning off has become general once more.

INDIRECT MEASURES.

PREVENTIVE METHODS

Selection of varieties of cane for planting.—There was noticeable in general throughout the plantations a marked difference in the power of the different varieties to resist the attack of the leafhopper. While the same variety would vary in different localities as regards growth and resistance, still the difference between any two varieties remained constant. For example, Yellow Caledonia was invariably the more resistant as compared to Rose Bamboo and Lahaina, and while the former was more seriously attacked in some localities than in others, wherever the opportunity offered itself for comparison with the latter, the Yellow Caledonia made the best showing. It is for the planter to decide whether or not the advantages of one variety over another are offset by the ravages of the leafhopper. If the loss from the leafhopper is greater than the gain in the yield between any two varieties in the absence of the leafhopper, then it is policy to select the more resistant cane.

The Yellow Caledonia (fig. 3) is a hardy cane and the plant makes a vigorous growth. These qualities, together with the showing which the variety made during the leafhopper epidemic, have made the cane a popular variety in the Hawaiian Islands. Mr. C. F. Eckart,

Fig. 3.—Yellow Caledonia sugar cane, a variety which is replacing Lahaina and Rose Bamboo in the Hawaiian Islands. Photograph taken during the leafhopper epidemic of 1903 (Original)

Director of the Hawaiian Sugar Planters' Experiment Station, reports as follows on this cane:[a]

Probably no subject pertaining to the cultivation of cane in the Hawaiian Islands during recent years has held more interest for the planters, in various localities, than that relating to the introduction and trial of new varieties

In the Hilo and Hamakua districts, the Lahaina first made way for the Rose Bamboo, and the latter, after a strong stand for many years, *is now being rapidly succeeded by the more vigorous Yellow Caledonia*. This cane with its upright growth and deep rooting propensities has proved a most valuable acquisition in wet and dry localities alike. Growing erect, with a natural tendency to shed its dried leaves, it becomes an admirable cane for rainy districts, where varieties that are prone to fall to the ground and remain in contact with a frequently saturated soil have shown extreme

[a] ECKART, C. F.—Varieties of cane. <Report of the Experiment Station Committee, Hawaiian Sugar Planters' Association, for the year ending September 30, 1904, Appendix IV, p. 31.

sensitiveness. The frequent stripping, required for Lahaina and Rose Bamboo in these wet places, has necessarily added to the cost of cultivation, and the ready manner in which Yellow Caledonia tends to strip itself is no small item in favor of economy. Again the manner in which it keeps down weeds, which were such a menace to its predecessors on the unirrigated plantations, is another strong point in its favor. In dry districts subject to occasional drought, it has amply demonstrated its hardihood over Rose Bamboo, which in turn is more resistant to such unfavorable climatic features than Lahaina. By sending its roots down deep into the soil it draws from a larger reserve supply of water than the older varieties, which are more shallow feeders and which soon feel the effects of a rainless period.

Dr. R. C. L. Perkins reports as follows on the relative immunity of different varieties of cane from leafhopper attack:[a]

It seems certain that some varieties of cane will stand the attack of leaf-hopper better than others. Mr. Eckart, Director of the Hawaiian Sugar Planters' Experiment Station, has furnished me with a list of new varieties of cane (see Appendix, Note II below), grown there, arranged in order, according to the relative injury that each sustained from leaf-hopper.

There may come, however, so severe an attack that no cane can resist it. Thus we have seen plants of "Yellow Caledonia" (at the extreme end of the list) which were of the strongest and most thrifty nature previous to the attack, some entirely destroyed and others very badly injured after a bad outbreak. It is, however, probable that from an attack of hopper which would entirely destroy a field of "Rose Bamboo," for instance, a field of "Yellow Caledonia" might recover.

The following is the note to which Doctor Perkins refers above:

The following list of new varieties (i. e., varieties other than the old standard ones of these islands) of cane at the Hawaiian Planters' Experiment Station has been drawn up for me by Mr. C. F. Eckart, the Director. They are arranged in order, according to the amount of damage sustained from leaf-hopper attack, Queensland 4 suffering most and Yellow Caledonia least: .

(1) Queensland 4	(10) Tiboo Merd
(2) Queensland 1	(11) Louisiana Striped
(3) Queensland 8A.	(12) Striped Singapore
(4) Louisiana Purple	(13) Big Ribbon
(5) Demerara 95	(14) Queensland 7
(6) Gee Gow	(15) Demerara 117
(7) Cavengerie	(16) White Bamboo
(8) Demerara 74	(17) Yellow Caledonia.
(9) Yellow Bamboo	

Cultural methods on the plantation.—The writer has already mentioned the fact that the epidemic of 1903 began during the winter months, in a wet season, and at a time when the cane was making practically no growth. The centers from which the infestation spread over the cane fields were invariably unfavorable locations for growth. It has been noted in this report that all varieties suffered in these unfavorable locations but that certain varieties made a better showing. The extension of the acreage of one variety in particular,

---

[a] PERKINS, R. C. L.—The leaf-hopper of the sugar-cane. <Bd. of Agr. and Forestry, Hawaii, Div. Ent., Bul. 1, p. 13, 1903.

Yellow Caledonia, will be a leading factor in preventing another epidemic. One other point was brought home to the Hawaiian planters as a result of the leafhopper epidemic, and that was the importance of intensive cultivation. The grass and weeds must be kept down by cultivation, the low places drained, and the impoverished lands fertilized. Those plantations which were in a high state of cultivation suffered less from the leafhopper attack, and the estates provided with the means of irrigation, in addition, suffered the minimum loss. There is a direct relation between intensive cultivation, fertilization, and irrigation and the amount of insect injury to any crop, showing that these operations are of great value in lessening insect damage.

Diversification of crops.—Sugar cane has been the leading crop in Hawaii since the days when the islands turned from the sandal-wood trade and the whaling fleet as a source of revenue. Some of the lands have been under cultivation to cane continuously for over twenty-five years. The time is at hand when the sugar-cane planters will find it both necessary and more profitable to diversify their crop. Some lands at present require a change from sugar cane, and the lands which are still highly productive will also require such a change as the years go by. When the general practice of inter-cropping cane with other plants does come, it will have a direct bearing on the control of the sugar-cane insects, the leafhopper included. The intermediate crop may be one of value in itself or one to be plowed under for green manure. Since it is not wise to cease the practice of burning off the trash after harvesting the cane, the planters can find no cheaper source of plant food, or no way in which the requisite texture and water-holding capacity of the soil can be more easily obtained than by removing their lands from cane cultivation in regular rotation and planting some nitrogen-gathering plant to be turned under when the land is put back into cane.

Control of the rind disease of sugar cane.—As has been mentioned, leafhopper injury is aggravated by the presence of the rind disease. In a discussion of the rind disease (*Melanconium sacchari*) Dr. N. A. Cobb says: [a]

> According to my observations on thousands of cuttings dug up on some twenty-five plantations a considerable part of the cuttings in some fields fail to grow on account of this disease, which, being present in the cuttings when they are planted, develops sufficiently to prevent germination. This is a difficult thing wholly to avoid by means of inspection of the seed, as the disease is sometimes present in cane that looks sound. It may be suspected to be present in any cane that has been attacked on the stalks by leaf-hopper or by borers. Other wounds that give admission to the rind disease fungus are those made by injudicious stripping, cracks at the bottom of the cane due to the effects of storms, and what are sometimes called "growth cracks."

[a] Cobb, N. A.—Fungus maladies of the sugar cane. <Hawaiian Sugar Planters' Exp. Sta., Div. Path., Bul. 5, p. 107, 1906.

Cane raised specially for seed and not stripped until wanted for planting is more likely to be free from insect punctures, and will therefore be less likely to develop rind disease after planting.

Mr. L. Lewton-Brain in a report on the rind disease thus describes the relation between the leafhopper and the disease: [a]

Under field conditions, of course, the spores gain access to the interior of the plant through natural wounds. Perhaps the most abundant wounds offered for this purpose are leafhopper punctures; even more favorable for the fungus are the tunnels of borers, leading as they do right into the heart of the sugar-containing tissue; other wounds may be made in stripping; in fact, it is a difficult matter to find a stalk of cane without a wound of some sort. The spores are produced in immense numbers on every stick of rotten cane. They are doubtless distributed partly by the wind, though the mucilaginous substance by which they are joined does not favor this; insects are certainly also important distributers of the spores, leafhoppers will get infected and deposit the spores in their punctures, ants will carry them into borer and other wounds in their search for food, flies may also serve the fungus in the same way.

The control of the rind disease of cane on the plantation will be another factor in reducing leafhopper injury. Since the leafhopper can not be exterminated and the punctures from this insect will always occur on a plantation to a greater or less degree, it becomes particularly essential for the planter to eradicate the disease.

On the control of the rind disease, Doctor Cobb has the following on pages 109 and 110 of his report referred to above:

The number of spores of this disease that exist on every plantation is past calculation, and almost inconceivable. This abundance of the spores of the disease tends of course to increase the losses. If there were no spores there could be no-rind disease. Anything that can be done to reduce the number of spores will tend to reduce the amount of the disease. Something can certainly be done in this direction. Stalks dead of the disease can be destroyed, and there can be no doubt that in some cases expenditure in this direction will be well repaid. There can be no doubt that the collecting and complete destruction of the stalks on the field would be a paying operation. How to destroy them is the question. The ordinary burning off destroys only a part of these rind disease stalks, leaving the rest untouched or only partially roasted, to go on producing their millions upon millions of spores.

It is the custom on all the Hawaiian plantations to leave on the ground after harvest the sticks of cane that have been attacked by borers or are worthless for other reasons. The reason for this is easy to understand. Such material is unsuitable to the highest efficiency of the mill as an extractor of cane juice. It is also of such a nature that it may interfere with the clarification, evaporation, or crystallization.

Notwithstanding this I think it would be advisable to consider whether this material, which is really a menace to the health of future crops, cannot in some way be run through the mill and burned. This is a practice adopted in some other parts of the world. On Saturday afternoons a special run of the mill is devoted to the milling of such refuse as I have mentioned, the "bagasse" being burned. The juice is allowed to run to waste, being first sterilized by heat.

In Hawaii it is usual to attempt to burn this diseased material, but from careful observation I am certain that this attempt often ends in failure, that is to say the disease that exists in the waste-cane is only partially destroyed.

---

[a] LEWTON-BRAIN, L.—Rind disease of the sugar cane. <Hawaiian Sugar Planters' Exp. Sta., Div. Path., Bul. 7, p. 21, 1907.

It may be that it would be better, at least from the disease point of view, if the harvesting of the fields were more in the nature of a clean sweep. If the diseased sticks are not too numerous they would not seriously interfere with the working of the mill. The advantage would be that whatever diseased material was thus dealt with would be dealt with in the very best manner, that is, it would be utterly destroyed.

NATURAL ENEMIES.

Species Already Present in the Islands.

Many beneficial species of insects, already present in the islands at the time of the leafhopper invasion, adapted themselves to the leafhopper as a source of food. The following species were noted during 1903:

A ladybird beetle, *Coccinella repanda* Thunb., one of Mr. Koebele's Australian introductions, was particularly abundant in the cane fields and the larva did good work against the young leafhoppers. An enemy of this species, the hymenopterous parasite *Centistes americana* Riley, has found its way to the islands and will no doubt reduce the effectiveness of the ladybird. The writer observed also the ladybird *Platyomus lividigaster* Muls. in the cane fields. A predaceous bug, *Œchalia griseus* Burm., was found in large numbers in the infested cane fields on the Island of Hawaii. The larvæ of two lacewing flies, *Chrysopa microphya* McLachl., and *Anomalochrysa* sp., were observed feeding on the young leafhoppers, the first species being particularly abundant in some localities.

Several species of spiders were abundant in the cane fields and were active enemies of the leafhopper. The writer collected two species, *Tetragnatha mandibulata* Walck. and *Adrastidia nebulosa* Simon. On the writer's advice large numbers of the egg-nests of spiders were collected in the localities where they were abundant and placed in sections where they had not as yet become established in the cane fields.

In the forest above the Kohala district, on the island of Hawaii, the writer found a fungous disease infecting to a great extent the common leafhopper *Siphanta acuta* Walk., a species belonging to the same family as the cane leafhopper. Quantities of this fungus were distributed in the cane fields in the hope that it would infest the cane leafhopper. No striking results were obtained, though diseased cane leafhoppers were found in some of the rainy districts.

Several species of ants were very active about the leafhoppers in the cane fields, the honeydew being an attraction to them.

Doctor Perkins mentions further in his early report a predaceous bug, *Zelus peregrinus* Kirk., and describes as new a hymenopterous parasite of the leafhopper under the name *Ecthrodelphax fairchildii* Perk.[a]

---

[a] PERKINS, R. C. L.—Bd. Comrs. Agr. and Forestry, Hawaii, Div. Ent., Bul. 1, pp. 20–22.

More recently the species of beneficial insects which were already present in the islands when the leafhopper was introduced and which have sought the leafhopper in the cane fields have been reported upon in detail by the entomologists of the Hawaiian Sugar Planters' Experiment Station.[a]

SPECIAL INTRODUCTIONS.

In 1903 Mr. Albert Koebele, after consulting with Dr. L. O. Howard, undertook extensive observations on the American parasites of leafhoppers. In Ohio Mr. Koebele had the assistance of Mr. Otto H. Swezey. A large quantity of living material was collected both in Ohio and in California and shipped to Doctor Perkins at Honolulu. The American material consisted in the main of insects belonging to the hymenopterous family Dryinidæ. The Hawaiian parasite *Ecthrodelphax fairchildii* Perkins is also a member of this family and, at the time of Mr. Koebele's American introductions, was being reared and distributed over the islands by Doctor Perkins. These introductions are discussed by Doctor Perkins in Part I of Bulletin 1, Division of Entomology, Hawaiian Sugar Planters' Experiment Station, 1905.[b]

Mr. Koebele also collected during his American investigations representatives of the order Strepsiptera (Stylopidæ) and a single species of an egg-parasite, *Anagrus columbi* Perk., belonging to the family Mymaridæ. [c]

In the spring of 1904 Messrs. Koebele and Perkins sailed for Australia to continue the search for parasites of the leafhopper. They reached Sydney in May and because of the cold weather which prevailed they proceeded to Brisbane. The results of the work in Australia are thus summarized by Doctor Perkins:[d]

Early in June we arrived at Brisbane, and on the first cane that we saw, a few plants in the public gardens, we at once observed the presence of the cane leaf-hopper. A

---

[a] Leafhoppers and their natural enemies. <Hawaiian Sugar Planters' Exp. Sta., Div. Ent., Bul. 1.

PERKINS, R. C. L.—Part I, pp. 1–60, May, 1905. (*Ecthrodelphax fairchildii.*)

PERKINS, R. C. L.—Part IV, pp. 113–157, pls. 5–7, September, 1905. (Pipunculidæ.)

TERRY, F. W.—Part V, pp. 159–181, pls. 8–10, November, 1905. (Forficulidæ, Syrphidæ and Hemerobiidæ.)

SWEZEY, O. H.—Part VII, pp. 207–238, pls. 14–16, December, 1905. (Orthoptera, Coleoptera, and Hemiptera.)

[b] PERKINS, R. C. L.—Leafhoppers and their natural enemies. <Hawaiian Sugar Planters' Exp. Sta., Div. Ent., Bul. 1, Part I, pp. 1–60, May, 1905. (Dryinidæ.)

[c] PERKINS, R. C. L.—Leafhoppers and their natural enemies. <Hawaiian Sugar Planters' Exp. Sta., Div. Ent., Bul. 1, Pt. III, pp. 86–111, pls. 1–4, August, 1905. (Stylopidæ.)

PERKINS, R. C. L.—Leafhoppers and their natural enemies <Hawaiian Sugar Planters' Exp. Sta., Div. Ent., Bul. 1, Pt. VI, p. 198, November, 1905. (*Anagrus columbi.*)

[d] PERKINS, R. C. L.—Leafhoppers and their natural enemies. <Hawaiian Sugar Planters' Exp. Sta., Div. Ent., Bul. 1, introduction, pp. III, IV, May, 1906.

short stay of about ten days gave ample proof of the existence in Australia of a considerable variety of Hymenopterous parasites of leaf-hoppers, of Dipterous parasites of the genus *Pipunculus*, and of Stylopid parasites of the genus *Elenchus*.

At Bundaberg, about twelve hours by rail north of Brisbane, we spent another ten days in June. Here is an extensive cane district with our leaf-hopper everywhere present, but never in numbers such as we are accustomed to in these islands. In fact we never saw the hoppers nearly as numerous as they are on our least affected plantations. From eggs collected here Mr. Koebele soon bred out specimens of the Mymarid parasites he had felt so confident of finding.

From our observations on the habits of the cane leaf-hopper in these islands, it seemed probable that in tropical Australia this species would be in its greatest numbers in the colder months, so after a brief stay in Bundaberg, we proceeded north to Cairns, which place we reached at the beginning of July. This plan seemed very expedient, for by retreating gradually towards the south, as the hot season advanced, we hoped to prolong the season during which natural enemies for the cane leaf-hopper could be obtained. It appeared likely that effective work could only be done at Cairns for a month or two, since without a reasonably large supply of hoppers, it was evident that the parasites could not be found in sufficient numbers for shipment. This indeed proved to be the case, and by the end of August, leaf-hoppers and their eggs had become so scarce in the cane fields, that we came south again to Bundaberg. At Bundaberg we made a long stay on this occasion, regularly sending off consignments of parasites, until here too, owing partly to the season and partly to the harvesting of the crop, the locality became unprofitable. After a short stay in Brisbane, at the end of the year, I returned to Honolulu, while Mr. Koebele proceeded to Sydney, where his attention was largely given to collecting beneficial insects for pests other than leaf-hopper. On the return journey Mr. Koebele spent one month in Fiji, the enemies of the cane-hopper in those islands being mostly similar to those already found in Australia. A fine consignment of the Chalcid egg-parasite (*Ootetrastichus*) of leaf-hopper was most important, as it enabled us to establish that important species without any doubt.

During January and February, 1906, Mr. F. Muir continued the work in the Fiji Islands begun by Mr. Koebele in the latter part of 1904. He reported as follows concerning the Fijian sugar-cane leaf-hopper and its parasites:[a]

The Fijian sugar-cane leaf-hopper (*Perkinsiella vitiensis*) I found all over the island, but it does no damage, being kept in check by several natural enemies.

The most important of these are the egg-parasites, *Ootetrastichus. Anagrus* and *Paranagrus* The first of these was introduced from Fiji into Hawaii by Mr. Koebele, and the other two appear to me the same as the Queensland species. In some fields as many as 90 % of the hopper eggs were parasitized, but in other fields it was lower. Observations extending over my six months' stay, and made at the various parts of the island visited, show that an average of 85 % of hopper eggs were destroyed by these parasites These figures are only approximate, as I have to estimate that one Chalcid (*Ootetrastichus*) destroys four hopper eggs, which is a low estimate. This Chalcid is more numerous, and on account of destroying the whole batch of hopper eggs, is of very much higher economic value than the Mymarids

---

[a] Muir, F —Notes on some Fijian insects <Hawaiian Sugar Planters' Exp. Sta., Div. Ent., Bul. 2, p. 3, November, 1906.

The Australian and Fijian material has been described in detailed reports with elaborate illustrations by Messrs. Perkins, Terry, and Kirkaldy.[a]

Regarding the effectiveness of the various parasites and enemies of the leafhopper, Dr. Perkins says:[b]

If we consider the effectiveness of the four egg-parasites, *Paranagrus optabilis*, *P. perforator*, *Anagrus frequens*, and *Ootetrastichus beatus*, in areas where all are well established, we must rate the first-named as *at present* by far the most effective. As I have previously pointed out, this species is capable by itself of destroying about 50 per cent of the cane-hopper's eggs and *Anagrus frequens* and *P. perforator*, extraordinarily numerous as they appear, where seen alone, are but as isolated examples in the crowd, where all are well established in one spot. The *Ootetrastichus* slowly but steadily increases in numbers, and on many plantations I expect that it will ultimately be the most efficient of all parasites. I do not think that it can show its full value till 1908, for each harvesting of the cane crop is necessarily a very great setback to its natural increase. *Anagrus frequens*, under which name are probably more than one species, or at least one or two distinct races of a single species, although it appears at a disadvantage, when in company with *Paranagrus optabilis*, is nevertheless a most abundant parasite. In Part VI of this Bulletin I have compared the habits of the two and need not refer to the matter here, but I may say that as many as eighty or a hundred exit holes of the *Anagrus* have been counted in a single cane-leaf, so that its great utility is unquestionable. *P. perforator*, common in Fiji, attacking eggs of hopper laid in thick stems of grass, more rarely those in cane, will probably gradually wander away from the cane-fields to attack the eggs of native hoppers, that are laid in stems and twigs, as it now chiefly attacks the cane-hopper eggs when these are laid in the stems.

Nor must it be forgotten, what valuable aid these egg-parasites receive in the control of leaf-hopper from other insects parasitic and predaceous, native or introduced. In fact, had there existed previously no restraint to the multiplication of the pest, no

---

[a] Hawaiian Sugar Planters' Exp Sta , Div. Ent .

PERKINS, R. C. L.—Bul. 1, Pt I, pp 1–69, May, 1905 (Dryinidæ).

PERKINS, R. C. L.—Bul. 1, Pt. II, pp 71–85, figs 1–3, June, 1905 (Lepidoptera).

PERKINS, R. C. L.— Bul. 1, Pt. III, pp. 86–111, pls. 1–4, August, 1905 (Stylopidæ).

PERKINS, R. C. L.—Bul. 1, Pt. IV, pp. 113–157, pls. 5–7, September, 1905 (Pipunculidæ).

TERRY, F. W.—Bul. 1, Pt. V, pp. 177–179, November, 1905 (Syrphidæ).

PERKINS, R. C. L.—Bul. 1, Pt. VI, pp. 183–205, pls 11–13, November, 1905 (Mymaridæ, Platygasteridæ).

PFRKINS, R. C. L.– Bul. 1. Pt. VIII, pp. 239–267, pls 18–20, January, 1906 (Hymenoptera).

KIRKALDY, G. W.—Bul. 1, Pt. IX, pp. 269–479, pls. 21–32, February, 1906 (Leafhopper).

PERKINS, R. C. L.—Bul. 1, Pt. X, pp. 481–499, pls 33–38, March, 1906 (Hymeuoptera, Diptera).

KIRKALDY, G. W.—Bul. 3, pp. 1–186, pls. 1–20, September, 1907 (Leafhoppers, Supplement).

PERKINS, R. C. L —Bul. 4, pp. 1–59, May, 1907 (Parasites of Leafhoppers).

[b] PERKINS, R. C. L.—Leaf-hoppers and their natural enemies. <Hawaiian Sugar Planters' Exp. Sta., Div. Ent., Bul. 1, introduction, pp. xv–xvii, May, 1906.

one who has paid the least attention to such matters can doubt that it would some time since have become impossible to raise any crop of sugar cane in the islands. The reason why these natural enemies have not alone got the upper hand of the hopper is due to various causes. In the first place, a number of the parasites such as the Dryinid *Ecthrodelphax fairchildii* and the parasitic flies of the genus *Pipunculus* are of local occurrence, and in many places cannot (for climatic or other unknown reasons) maintain their existence. This was well shown by the behavior of the first-named, which was distributed in thousands by the entomologists and the Plantation managers themselves to all the districts in the islands, but in many places did not thrive. Such, too, is the case with the predaceous black earwig (*Chelisoches morio*) which, a natural immigrant to the islands and no doubt acclimatised centuries ago, is found on comparatively few plantations   Other natural enemies are themselves periodically decimated by parasites, as is the case with the introduced green cricket (*Xiphidium varipenne*), which has its own egg-parasite (*Paraphelinus*). Other enemies like the common lady-bird (*Coccinella repanda*) introduced by Koebele years ago for other purposes, prey on young leaf-hoppers, in default of more favorite food, and this valuable predator too is itself subject to parasitic attack by the common Braconid (*Centistes*)   At present the whole number of parasites and predaceous insects that attack cane leaf-hopper to such an extent as to render their services worth noting is considerable, as the following summary shows.

The most valuable are the four egg-parasites, which there is every reason to hope will become still more effective with reasonable time, one (*Ootetrastichus*) having as yet had no chance to show its full effectiveness.

The two Pipunculus flies (*Pipunculus jusator* and *terryi*) are restricted to certain localities and are native species, which have transferred their attacks from native Delphacids to the cane leaf-hopper.

The ubiquitous lady-bird (*Coccinella repanda*) is valuable as a destroyer of leaf-hopper, though originally imported by Koebele to destroy Aphis. It is hoped that other lady-birds, especially *Verania strigula*, may become established and do good work, as in Australia and Fiji, whence they were imported.

The earwig *Chelisoches morio* is a local species, but no doubt useful where it exists in numbers.

The green cricket (*Xiphidium varipenne*) is very valuable, but is most unfortunately heavily attacked at certain seasons by an egg-parasite.

The Dryinid *Ecthrodelphax fairchildii* is locally valuable. At certain seasons in suitable, but limited, localities, it destroys a considerable percentage of hoppers. Its services are underestimated because for a large part of the year it lies as a dormant larva in the cocoon, and parasitized hoppers at such a time are naturally hardly to be found.

There are many other natural enemies of more or less importance, e. g. the various predaceous Hemiptera, and the several lace-wing flies (*Chrysopinæ*).

In addition to these insect enemies, we must mention the two fungous diseases of hoppers (amounting locally and at certain seasons to epidemics) which, long previously known to kill the native leaf-hoppers, have become transferred to the introduced pest. We also found one or more fungous diseases attacking leaf-hopper eggs in Fiji and Australia in all localities. With material imported from these countries, I easily infected eggs of the cane leaf-hopper under cover, and subsequently established the fungus at large in the field. As it was most probable that parasitized and healthy hopper eggs would be affected alike by the disease, and consequently many of the egg-parasites would be destroyed, it became a subject of discussion whether we should attempt to establish the fungus or not. As, however, throughout Australia, the fungus and parasite both attacked the eggs, Mr. Koebele was of opinion that we should try and establish the same conditions here. Consequently with the first cages sent to the plantations the cane cuttings and the cane itself were well sprayed with water containing spores of the fungous disease, so that these would be certainly carried abroad by the emerging hoppers and parasites. I imagine there is no doubt as to this disease becoming established in all suitable localities.

In speaking of the necessity for the continued propagation and distribution of the introduced parasites of the leafhopper, Doctor Perkins reports as follows:[a]

Owing to the manner in which cane is cultivated in these islands, the entomologist working along the lines that have been adopted to control the leaf-hopper pest, meets with a serious obstacle such as is not encountered in dealing with insects injurious to our other vegetation. I refer here to the universal custom of burning off the trash over great acreages, after the crop has been harvested. I have been told that on the Colonial Sugar Refining Company's estates in Australia no such burning off is allowed. If this is correct, it may help to account for the insignificant numbers of our cane-leaf hopper there, as well as of several other insects of the same group, which are fortunately not known in our cane fields. As, however, burning of trash is an established fact here, it becomes necessary to see what steps can be taken to provide against this serious disadvantage. I will first show whereof this disadvantage consists. The parasitic enemies of the leaf-hopper are mostly delicate and minute creatures, not accustomed to take prolonged flights. Their wings serve well to bear them from plant to plant, but for further distribution they are dependent on air-currents. If when a field of cane is cut the wind blows towards another cane field, no doubt some or many parasites will reach it, but if otherwise, probably none will do so. In burning over a field it is quite certain that almost every parasite yet present will be destroyed, but the adult leaf-hoppers on the other hand are well able to take care of themselves. When, as an experiment, a patch of about nine acres of cane, so heavily attacked by leaf-hopper as to be useless, was set on fire all around to destroy these, it was noticed that the adult hoppers rose from the cane in a cloud and spread to other fields; so this plan for destroying them was of no value. I have in an earlier publication shown how quickly the leaf-hoppers spread to new fields of very young cane, and with what regularity they distribute themselves over the young plants. It cannot be hoped that the parasites will (except under rare and fortuitous circumstances, such as constant favorable winds) spread themselves in like manner, and in the same time. Yet it is essential that the parasites should be on the spot when the leaf-hopper *begins to lay* in order to secure proper control. If the supply of laying hoppers at the beginning of the great breeding season is very small, it means that there is not time for the attack to become serious before that season is over. It is when the hopper is least abundant, that one wants to be assured that it is being attacked by all possible enemies. When a field is already seriously injured and swarming with hoppers, not much immediate help can be given for obvious reasons. It will be easier to prevent such a condition than to find a remedy. If one could provide that in each large area of cleared land, ready for planting, there should be in the middle a small patch of some variety of cane most susceptible to the attack of leaf-hoppers, that this cane should be kept well stocked with these, and with a variety of parasites and predaceous insects, and itself be of sufficient growth to afford good shelter to all these, the condition from an entomological standpoint would be ideal. This patch of cane, being already of suitable age and growth and stocked as aforesaid, at the time the much younger cane of the rest of the field began to be infested with hoppers, would

<hr>

[a] Perkins, R. C. L.—Leaf-hoppers and their natural enemies. <Hawaiian Sugar Planters' Exp. Sta., Div. Ent., Bul. 1, introduction, pp. xviii-xxi, May, 1906.

daily be distributing thousands of natural enemies, that should control these. Although such a plan or modification of it might be adopted on some plantations, on others (at least such as are under irrigation) it would either be difficult, or altogether impracticable. Only in the case of some fields of long ratoons would the matter be very simple, when a small area of the original ratoon growth in each field could be left uncut, and if well supplied with hoppers and their natural enemies would serve later on to stock the rest of the field. Unfortunately, owing to the fact that ratoons are (except in unusual cases) not severely attacked as compared with plant-cane, this matter becomes one of minor importance. Otherwise, in the majority of cases, owing to the clearing of large areas and the burning of trash, it is probable that new fields will have to be supplied by cages similar to those already used. Two things will be absolutely necessary: (1) that the new fields be well supplied with parasites; (2) that they be stocked immediately the hoppers enter them and commence laying. This plan, though less satisfactory than would be the other method, is nevertheless simple, and does not call for much expenditure of time, nor for skilled labor. The one thing necessary to be positively ascertained is that the spot whence the cuttings for distribution are taken is well supplied with *all* the kinds of parasites that it is desired to establish in new fields. It is now well known to us that *all* these destroyers are not yet established *in all parts* of all plantations, and therefore at present unless an entomologist previously test samples from the spot, whence distribution is to be made, it is quite likely that some of the most valuable parasites will not be taken to the new fields. If a sample be submitted to the entomologists, it can be passed as fit to supply all necessary parasites to new fields, or if not, cages of the deficient species can always be supplied from the cane in the grounds of the Experiment Station in Honolulu. As the parasites are continually spreading and increasing, such expert examination will at the most be necessary for a year or two; for it is perfectly certain that by that time all the species will be so general that it will be quite impossible to take any extensive sample of cane-leaves that bear eggs of leaf-hopper, which will not contain all. Such in fact is now the case in the cane at the Experiment Station. To sum up, the clearing of all cane from large acreages is a decided obstacle to the complete success of natural enemies of leaf-hopper, and the burning of trash aggravates the difficulty. As an offset to these conditions new fields should be supplied artificially with natural enemies, and they should be supplied as soon as any leaf-hoppers enter them. Of course future observation may prove this distribution unnecessary, but for the present it should be adopted.

RELATED SPECIES.

The Hawaiian sugar-cane leafhopper *does not occur on the mainland of the United States.* The insect is closely related to the corn leafhopper (*Dicranotropis maidis* Ashm.), common on corn in the Southern States.[a] A West Indian species of leafhopper is recorded as injurious to sugar-cane, by Westwood, in 1841, under the name *Delphax saccharivora* and is a member of the same family of insects as the Hawaiian sugar-cane and the corn leafhoppers.[b] Three further species of this same family, the Fulgoridæ, are recorded as sugar-cane pests in Java by W. van Deventer.[c]

---

a QUAINTANCE, A. L.—Fla. Agr. Exp. Sta., Bul. 45, 1898.

b WESTWOOD, J. O.—Mag. Nat. Hist , vol. 6, p. 407, 1841.

c *Phenice maculosa, Dicranotropis vastatrix*, and *Eumetopina krügeri*. Van Deventer, Handboek ten dienste van de Suikerriet-cultuur en de Rietsuiker-Fabricage op Java. II. De Dierlijke vijanden van het Suikerriet en hunne Parasieten. Amsterdam, pp. 167–169, 1906.

THE HAWAIIAN SUGAR-CANE BORER.

([*Sphenophorus*] *Rhabdocnemis obscurus* Boisd)

GENERAL CHARACTERISTICS.

The sugar-cane "borer" ([*Sphenophorus*] *Rhabdocnemis obscurus* Boisd.) (fig. 4), infesting the cane stalk in Hawaii is the grub of a beetle belonging to the weevil family Calandridæ. The sugar-cane stalk-borer of the southern United States is the caterpillar of a moth,

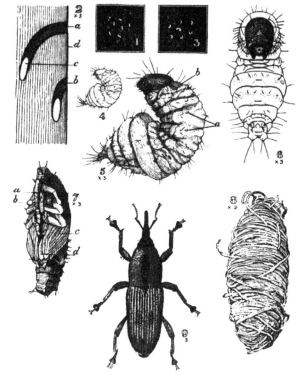

FIG. 4.—The Hawaiian sugar-cane borer ([*Sphenophorus*] *Rhabdocnemis obscurus*)· *1*, Eggs, natural size *2*, Eggs in situ, much enlarged *a*, Section of egg passage with egg, *c*, *b*, egg placed unusually near the rind, *d*. *3*, Larva, just hatched and older, natural size *4*, Full-grown larva, natural size *5*, Larva, side view, enlarged *a*, Spiracles, *b*, cervical shield *6*, Larva, front view, enlarged *7*, Pupa, enlarged, *a*, Rostrum or beak, *b*, antenna, *c*, elytron or wing cover, *d*, folded wing *8*, Pupal case or cocoon, enlarged. *9*, Adult, enlarged. (After Terry)

Diatræa saccharalis Fab. Entomologically the two species are widely separated, belonging to entirely different orders of insects, but in the character of their injury to the cane stalk these two insects are quite similar—that is, they both develop within the cane stalk, and by feeding on the interior cause great destruction to the plant.   Comparatively, the Hawaiian borer is more destructive and, because of the habits of the adult, a more persistent species to combat.   The adult beetle of the Hawaiian borer is a stronger flyer than the adult moth of the mainland borer and therefore has a wider range over any infested territory.   As the adult of the Hawaiian borer, too, can emerge from any reasonable depth when buried in the soil, this renders the question of infested seed cane a serious one in Hawaii, while on the mainland the careful covering of infested seed cane is effective in preventing the emergence of the adult moth.   These points are mentioned to bring out the fact that we are discussing here a species in no way related to the cane borer of the Southern States and in many ways not subject to the same means of control.

[Sphenophorus] *Metamasius sericeus* Oliv. is a species injurious to cane in the West Indies, being recorded from Jamaica, Barbadoes, St. Kitts, Antigua, St. Lucia, and British Guiana.

In Porto Rico *Sphenophorus sexguttatus* Drury is recorded by Busck[a] as boring in the stalks of sugar cane.

DISTRIBUTION.

The sugar-cane borer of Hawaii is recorded also from Fiji, New Guinea, New Ireland, Tahiti, Queensland, and the Malay Archipelago and probably occurs pretty generally throughout the islands of the southern Pacific.

OCCURRENCE IN HAWAII.

This species is a pest of long standing in the islands.   The insect is recorded from the Island of Oahu in 1885 by the Rev. Thomas Blackburn,[b] who found the species breeding in the stems of bananas in the mountains, and the files of the Bureau (then Division) of Entomology, record the receipt of the borer from the Hawaiian Islands, as early as 1888.[c]   It is believed that the sugar-cane borer was introduced into the islands from Tahiti in the stems of the banana plant during the early communications between the Hawaiian Islands and those of the South Seas.   Hon. H. P. Baldwin, of Puunene, Maui, informed the writer that to his personal knowledge the beetle was injurious to sugar cane in the vicinity of Lahaina, the ancient capital of the islands, as early as 1865.

Aside from the banana plant and sugar cane, the beetle infests the coconut palm, the sago palm, the royal palm, the wine palm, (*Caryota urens*), and the papaia (*Carica papaya*).

---

[a] U. S. Dept. Agr., Bur. Ent., Bul. 22, p. 89, 1900.

[b] BLACKBURN, REV. T., AND SHARP, D.—Memoirs on the Coleoptera of the Hawaiian Islands.   <Sci. Trans. Roy. Dublin Soc., 2 ser., vol. 3, pp. 119–290, pl. 1, 1885.

[c] General Notes, Bureau of Entomology, No. 4332b.

Until the recent injury by the leafhopper (*Perkinsiella saccharicida*) the sugar-cane borer was the principal insect affecting cane in the islands.

The species was determined by the Bureau of Entomology at Washington, D. C., in 1888 from specimens forwarded by the late King Kalakaua and was discussed under the title "The Sandwich-Island Sugar-cane Borer," in Insect Life, vol. 1, No. 6, pages 185–189, December, 1888.

In 1896 Mr. Koebele gave the following record on the work of the borer in Hawaii: [a]

This may be classed as the most injurious enemy of the sugar cane present on these islands. Its ravages will exceed those of all other insects combined. Its attacks on the sugar cane, however, seem confined to the more damp localities, whilst in drier places, such as Lahaina, the borer is hardly noticed. I have been informed that the Lihue Plantation has recently suffered severely from the attacks of the borer. Not only sugar cane is damaged by this insect, but many other plants are damaged by it, chiefly the bananas and cocoanuts. A grove of the latter was shown me in Hilo, in 1894, that was badly infested by the beetles. Setting fire to the dry leaves was recommended; this was done and the plants have since entirely recovered. Dying cocoanut palms were examined and in the tender heart of the palm were found great numbers of the insects, in all stages.

More recently (1907) Mr. F. W. Terry has discussed the sugar-cane borer in the Hawaiian Islands in a circular of the Hawaiian Sugar Planters' Experiment Station. [b]

LIFE HISTORY AND HABITS.

The eggs are found beneath the epidermis of the cane stalk, or more rarely in the tissue of the leaf sheath, having been placed singly in small cavities. The cavity is made by the female with her proboscis before depositing the egg.

The young grub or larva, on hatching from the egg, bores on into the stalk of the cane, completely honeycombing the interior with tunnels running lengthwise with the stalk (see fig. 5). The evidence of its work is not indicated by the outward appearance of the stalk. Many times a stalk, seemingly in a normal condition, is found on examination to be utterly destroyed. The life of the borer within the stalk of the cane is estimated to be about seven weeks by Mr. Koebele, [c] who points out the fact that the length of the larval life

---

[a] KOEBELE, ALBERT.—Report on insect pests. <Hawaiian Planters' Monthly, vol. 15, no. 12, p. 590, December, 1896.

[b] TERRY, F. W.—Hawaiian Sugar Planters' Exp. Sta., Div. Ent., Cir. 3, pp. 22, plates 2, fig. 1, December, 1907.

[c] KOEBELE, ALBERT.—Hawaiian Planters' Monthly, vol. 19, no. 11, p. 520, November, 1900.

FIG. 5.—Work of the Hawaiian sugar-cane borer in sugar cane: *a, a, a,* Emergence holes made by the larva before pupation; *b, b,* "rupture" holes, apparently accidental and made by the larva while feeding; *c,* holes made by the female borer for the reception of her eggs; *d,* cocoon; *e,* larva; *f, f,* "frass" or undigested cane fiber, passed by the larva. One-half natural size. (After Terry.)

depends to a great extent upon the condition of the food plant and climatic conditions; that is, the development will be more rapid in softer cane and during the warm summer months than during the low temperatures of winter.

When ready to pupate—that is, to transform to the inactive stage preparatory to emerging from the stalk as an adult beetle—the larva (fig. 5, *a*) forms about itself a cocoon (fig. 5, *b*) from the fiber of the stalk within the tunnels it has made in feeding. The adult beetle on issuing from this cocoon bores its way through the side of the stalk to the exterior, and this opening in the lower joints of the cane is the first distinct symptom of the presence of the borer. The length of the pupal period is as variable as that of the larval, the average time for transformation and emergence being from two to three weeks.

The beetles are night flying and hide during the day down within the sheaths of the lower leaves. The softer varieties of cane are more subject to attack than the hardier varieties, and the borer is more abundant in the wet districts than in the dry. Cane which has received an abundant supply of water by irrigation suffers more from the work of the borer than unirrigated cane. The borers occur in the largest numbers in young cane and the suckers are infested to a much greater degree than the stalks. The borers always occur in the largest numbers in the vicinity of the track used to haul cane to the factory, issuing from infested stalks that have dropped from the cars and have not been collected and destroyed afterwards.

The borer is a strong flyer and spreads from field to field in this manner. It is distributed in infested seed cane and also develops from the stalks left in the field after harvest or dropped from the wagons or cars in hauling to the factory.

CONTROL MEASURES.

SELECTION OF VARIETIES FOR PLANTING.

As has been mentioned, the softer varieties are more subject to attack than the hardier ones. The Yellow Caledonia, a variety which is replacing to a great extent the common Lahaina and Rose Bamboo in Hawaii, is injured to a much less extent than other varieties. The infestation is not necessarily less in Yellow Caledonia, but the borer meets with greater resistance in its feeding and consequent development because of the firmness of the fiber.

IRRIGATION.

Excessive irrigation favors the development of the pest, since cane in a succulent condition is more easily infested by the borer and its development within the stalk is more rapid. It is plain that in fields heavily infested by the borer the minimum amount of water should be used in irrigation.

BURNING OF TRASH.

The burning of trash after harvesting the cane is the most effectual method of keeping the borer in check. In this practice not only should the fields be burned over, but all the unburned stalks left in the fields and all stalks dropped from carts and cars along the roads and tracks used in hauling the cane to the factory should be collected and burned. One plantation found it necessary to collect such stalks in piles and use crude oil on them in order to destroy them completely, and by a careful estimate of the labor and cost of material found that the money had been well invested, as was shown by the reduction in the numbers of borers in the fields the following season.

SELECTION OF NONINFESTED SEED CANE.

The Hawaiian sugar-cane borer is able to emerge to the surface from any reasonable depth when planted with seed cane. For this reason great care should be exercised in the selection of cane for planting purposes, since new areas can in this way be readily stocked with the pest. It is not practical to treat successfully cane infested with the borer; since the borer is fully protected within the stalk. Therefore, next in importance to the thorough burning of all trash after harvest is the selection of noninfested seed cane.

PICKING AND BAITING.

The most effective direct measure employed against the cane borer is the collecting of the adults during the daytime from their hiding place within the lower leaf sheaths. The supply of labor will

influence the ability to use this method.   The method is more feasible
where the plantation is so situated that women and children can be
employed for the work.   Care should be exercised in this work in
order that the growing leaves may not be broken down.   It is
obvious that a larger number of beetles will be collected when the
wages are based on the numbers collected, but the results are more
satisfactory, as regards breaking down the cane, when the wages of
the laborers are fixed at a certain amount per day.

In the Fiji Islands a method of baiting the beetles is employed,
which consists of splitting cane stalks and placing pieces about the
edges of the field and within the rows at certain intervals.   The
method as practiced in Fiji is thus described by Mr. Koebele.[a]

> At the request of the Colonial Sugar Company we looked into the matter with a
> view of getting rid of the beetles the best way possible; all sorts of devices were em-
> ployed and none worked better than pieces of split cane about 12 inches long, placed
> along the edges of the field and through the same at intervals of 12-18 feet; thus with
> seven little Indian girls, I collected over 16,000 beetles in some four hours, and the
> same little girls alone brought in the following noon over 26,000 beetles.
>
> This method was kept up, and followed on all the plantations for the next three
> years, or until no more of the borers could be found   Tons of the same were brought
> in at the Nausori mill alone, and the expenses of collecting were practically nothing
> compared to the cost at Lihue, where such work has to be done by the day laborers.
> About four cents per pint of the insects was paid to the children.   The result has been
> highly satisfactory, for, ever since the last five years, the cane borer has not been a
> pest in those islands.

An important point regarding this split cane is that the females
usually infest these pieces heavily with eggs and the young resulting
grubs bore into the split stalks and perish as the pieces of cane become
dry.   In dry localities the pieces of split cane should be placed in the
irrigation ditches during the day and placed out as bait in the evening, otherwise they dry out rapidly and cease to attract the beetles.

RELATED SPECIES.

The Hawaiian sugar-cane borer is represented in the United States
by the "corn bill-bugs," of the genus Sphenophorus, several species
of which in the adult stage attack the leaves of corn, but rarely breed
in the stalk of corn as does the Hawaiian Sphenophorus in the stalk
of cane.   *The Hawaiian cane borer does not occur on the mainland of the
United States.*

---

[a] KOEBELE, ALBERT.—Hawaiian Planters' Monthly, vol. 19, no. 11, p. 522, November, 1900.

THE HAWAIIAN SUGAR-CANE LEAF-ROLLER.

(*Omiodes accepta* Butl.) (Plate III⁴)

EARLY HISTORY IN THE HAWAIIAN ISLANDS.

During the investigations relating to the leafhopper in 1903 the writer found the Hawaiian sugar-cane leaf-roller, the caterpillar of a native moth, doing serious damage to cane in the upper fields of plantations in the Kohala district, Island of Hawaii. The larvæ were collected also from Hilo grass (*Paspalum conjugatum*) growing wild above the cane areas. The species, primarily a grass feeder, occurs in the higher altitudes and invades the bordering fields from these locations. It is recorded by Meyrick [a] in 1899 from the islands of Hawaii, Maui, Molokai, and Kauai at elevations ranging from 1,500 to 5,000 feet. The caterpillar was described for the first time by Dr. H. G. Dyar, of the United States National Museum, from specimens collected by the writer on cane in the Kohala district. [b]

Swezey states that the leaf-roller occurs on practically all of the plantations of the islands, but is less abundant in the dry districts. Regarding its injury he says: [c]

It is present in some fields of cane sometimes in such large numbers as to do considerable damage; in fact, cases have been reported where the young cane has been entirely stripped of leaves. Such instances are not numerous, however, and even in the worst cases would not result in entire destruction of the crop of cane as it would grow again after the caterpillars had obtained their growth, or their parasites had got them checked. It is not usually to be considered a serious pest. Possibly it is not so abundant now as it was a few years ago when reports were made of cane fields having been entirely stripped by them.

At present there are a number of parasites preying upon this species and this keeps them well in check.

In this same report, page 10, the author describes the habits of the caterpillar as follows:

On sugar cane the very young larvæ feed in the crown of the plant where the young leaves have not yet unrolled. They are thus protected between the natural rolls of the leaf; later on they roll over the margin of a leaf forming a tube for their "retreat." When nearly full grown, they are usually found in tubes towards the tip of the upper leaves. These tubes are easily observed if the ragged leaves where the larvæ have fed, are examined. The work of the smaller larvæ shows as oval or elongate dead spots on leaves which have unrolled in the growing of the cane after the young larvæ have fed upon them.

When disturbed in its retreat, as by its being torn open, or violently shaken, or jarred, the larva wriggles very lively and drops to the ground for escape. This habit is

---

a MEYRICK, E.—Fauna Hawaiiensis, vol. 1, Pt. II, p. 204, 1899.

b DYAR, H. G.—Note on the larva of an Hawaiian pyralid (*Omiodes accepta* Butler). <Proc. Ent. Soc. Wash., vol. 6, no. 2, p. 65, 1904.

c SWEZEY, OTTO II.—The sugar-cane leaf-roller, *Omiodes accepta*. <Hawaiian Sugar Planters' Exp. Sta., Div. Ent., Bul. 5, p. 7, August, 1907.

probably to escape from parasites, many of which prey upon them. The retreat which it constructs is undoubtedly for the same purpose, as well as for protection from wasps and birds which prey upon it.

The caterpillars are full grown in about three weeks from hatching. They molt five times at intervals of about three to five days, and five to seven days between the fifth molt and the spinning of the cocoon and pupation. Pupation takes place within a slight cocoon of white silk in the "retreat" where the caterpillar has lived; however, the cocoon is sometimes made beneath the leaf-sheaths of cane, and in other favorable places.

CONTROL MEASURES.

No special remedies are employed in cane fields against this pest. Swezey suggests that in fields of young cane a spray of Paris green or arsenate of lead might be used with effect, and mentions that at times laborers have been sent over the field to pinch the caterpillars in their retreat between the folded cane leaves.

PARASITES.

The species is attacked, fortunately, by several introduced parasites. Regarding the natural enemies of the species of moths belonging to the genus Omiodes, Mr. Swezey reports as follows on pages 36 and 37 in his article above referred to:

Omiodes caterpillars are attacked by a large number of species of parasites, some of which are native, and several which are the most valuable have been introduced. The most of the species are kept in check by their natural enemies, so that they do not become very numerous; in fact, several of them are very rare. Two species feed so numerously on cultivated plants that they become serious pests; *accepta* on sugar cane, and *blackburni* on palms. These two species are preyed upon very extensively by the parasites and checked considerably, but not sufficiently to keep them from doing considerable injury in certain localities and at certain seasons. Apparently the moths are more prolific in the winter months (about December to March) and the parasites are scarcer owing to their having had fewer caterpillars for them to keep breeding on during the preceding summer. Hence, when the winter broods of caterpillars appear, there may be two or three generations of them before the parasites breed up to sufficient numbers so that they produce any noticeable check on the number of the caterpillars; then in another generation or two the caterpillars may be much reduced in numbers and a large percentage of them found to be parasitized; for example, on one occasion 75 % of the cane leaf-rollers in a field at Hutchinson plantation, Hawaii, were found to be destroyed by one species of parasite; at Olaa plantation, Hawaii, in a certain field, on one occasion a much higher percentage of them than that were killed; in Honolulu, of a large number of the palm leaf-roller caterpillars collected, 90 % were parasitized.

Since there are so many species of parasites preying on the leaf-rollers which are pests, it might be asked "Why do they not become exterminated, or at least cease to be pests?" Apparently, with all of the parasites, they are still not numerous enough to overbalance the prolificness of the pest, even though they do kill such high percentages of them at times. Since so many are killed by parasites, and yet there are enough left to do considerable injury at times, one cannot help but wonder to what extent these pests might increase were there no parasites preying on them, and how many times more serious would be the damage done by them. The extreme difficulty and impracticability of treating sugar cane fields, or large palm trees, artificially, for

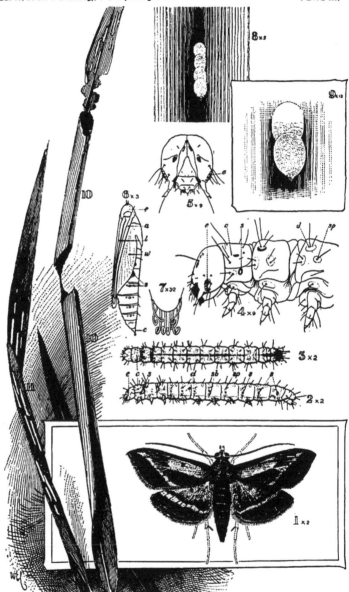

THE HAWAIIAN SUGAR-CANE LEAF-ROLLER (OMIODES ACCEPTA).

Fig. 1.—Adult moth. Figs. 2, 3, 4, 5.—Larvæ and details. Fig 6.—Pupa. Fig. 7.—Apex of cremaster, showing the curled spines by which the pupa is fastened to the cocoon. Fig. 8.—Cluster of 4 eggs in groove on surface of leaf. Fig. 9.—Eggs more highly enlarged. Fig. 10.—Leaf spun together for "retreat" or hiding place of caterpillar, showing where caterpillar has eaten. Fig. 11.—Leaf, showing spots where very young caterpillar has eaten, leaving one epidermis intact, instead of eating holes through the leaf. (After Swezey.)

the destruction of these pests, makes it all the more important that there are so many valuable parasites preying upon them; and shows the value of introducing natural enemies to control a pest, for the four best parasites of these leaf-rollers are introduced species, viz., *Macrodyctium omiodivorum*, *Chalcis obscurata*, *Frontina archippivora* and *Trichogramma pretiosa*.

THE SUGAR-CANE MEALY-BUG.

(*Pseudococcus calceolariæ* Mask.) (Plate IV.)

IDENTITY.

This insect (see Pl. IV, from photographs by Mr. T. C. Barber) is identical with the sugar-cane mealy-bug common on cane in the southern parishes of Louisiana. The species is recorded by Mrs. Maria E. Fernald from Australia, Hawaii, Fiji, Jamaica, and Florida.[a] Koebele earlier records this mealy-bug on cane in Hawaii.[b]

RELATED SPECIES.

The mealy-bug of the cane belongs to a very large family of insects, Coccidæ, which are world-wide in their distribution. Two other species of this family, *Pseudococcus sacchari* Ckll. and *Aspidiotus cyanophylli* Sign., have recently been recorded from Hawaii by Mr. J. Kotinsky.[c]

Three species, namely, *Pseudococcus calceolariæ*, *P. sacchari*, and *Aspidiotus sacchari* Ckll., are known to attack sugar cane in the West Indies.[d]

Van Deventer records several scale insects, among them *Lecanium krugeri* Zehntn., *Aspidiotus saccharicaulis* Zehntn., *Chionaspis* spp., and a species of Pseudococcus very similar to *P. calceolariæ*, on cane in Java.[e]

In Mauritius two species of related insects, *Icerya seychellarum* Westw. and *Pulvinaria iceryi* Guér., are reported as pests of sugar cane.[f]

FOOD PLANTS.

Mrs. Fernald gives the food plants of the sugar-cane mealy-bug as *Calceolaria, Danthonia, Phormium tenax, Cordyline australis*, and

---

[a] FERNALD, MRS. MARIA E.—A Catalogue of the Coccidæ of the World. <Bul. 88, Hatch Exp. Sta., Mass. Agr. Coll., p. 98, 1903.

[b] KOEBELE, ALBERT.—Hawaiian Planters' Monthly, vol. 15, no. 12, p. 596, December, 1896; vol. 17, no. 5, p. 209, May, 1898.

[c] KOTINSKY, JACOB.—Coccidæ not hitherto recorded from these islands. <Proc. Hawaiian Ent. Soc., vol. 2, no. 3, pp. 127–131, 1910.

[d] BALLOU, H. A.—Review of the insect pests affecting the sugar cane. <West Indian Bul., vol. 6, no. 1, p. 41, 1905.

[e] DEVENTER, W. VAN.—Handboek ten dienste van de Suikerriet-cultuur en de Rietsuiker-Fabricage op Java. II. De Dierlijke vijanden van het Suikerriet en hunne Parasieten, Amsterdam, pp. 227–266, 1906.

[f] FERNALD, MRS. MARIA E.—A Catalogue of the Coccidæ of the World. <Hatch Exp. Sta. Mass. Agr. Coll , Bul. 88, pp. 27, 133, 1903.

sugar cane. In Louisiana the mealy-bug infests, aside from sugar cane, the Johnson grass (*Sorghum halepense*) and the saccharine sorghums.

LIFE HISTORY AND HABITS.

plant.

and beneath the surface of the ground about the roots of 'the plant. In this latter location they hibernate during the cold months of winter on both cane and Johnson grass.

partly grown about the lower nodes of the stalk. The females are practically inactive, remaining in a mass about one of the nodes or beneath the leaves throughout their development and secreting about themselves in these locations the characteristic white covering (Plate IV, fig. 3). The young males do not remain stationary on the plant, but, after completing their development, spin a narrow white cocoon (Plate IV, fig. 4) within which they transform to a delicate winged adult.

CONTROL.

Selection of seed cane.—Since the common method of distribution is by the transportation of infested seed cane from plantation to plantation or from one part to another of the same plantation, care should be exercised to select clean stalks and not those which are infested, for seed cane.

Burning of the trash.— The practice of burning the trash after harvest is very effective in destroying this insect, since those remain-

PLATE IV.

THE SUGAR-CANE MEALY-BUG (PSEUDOCOCCUS CALCEOLARIÆ).

Fig. 1 —Adult mealy-bugs clustered about the base of young cane   Fig. 2.—Adult female,
twice natural size.   Fig. 3 —A single adult female, with white mealy-like covering
Fig. 4.—Cocoons of male mealy-bug   (Original.)

ing on the stalks are killed in the process of milling, and the remaining forms on the discarded stalks and leaves in the field are destroyed by the fire.

There is present in Hawaii a ladybird beetle, *Cryptolæmus montrouzieri* Muls., which is a special mealy-bug feeder. This ladybird is one of Mr. Koebele's introductions from Australia. It has proved particularly beneficial in feeding upon the sugar-cane mealy-bug in the Hawaiian cane fields, and through its work the numbers of the mealy-bug have been greatly reduced in recent years. This important predator has been established in California, and the Bureau of Entomology at Washington, D. C., has under way at present negotiations to import this beetle into the cane fields of southern Louisiana which are infested by the mealy-bug.

The ladybird is thus described by Prof. W. W. Froggatt, government entomologist of New South Wales.[a]

This beetle is very variable in size, measuring from under 2 to 3 lines in length, with the head, thorax, extreme tip of both wing covers light orange-yellow; the whole of the under surface reddish-brown, and both the upper and under surface clothed with fine hairs. In a number of specimens the under surface is variable in coloration, the middle and hind pairs of legs with the thorax dark reddish-brown to black.

The larva is of the usual smoky-brown tint, but so thickly clothed on the upper surface with white filaments that it appears to be of a uniform white, the pupa hidden beneath the larval skin and the immature beetle are pale yellow.

MISCELLANEOUS INSECTS AFFECTING SUGAR CANE IN HAWAII.

An aphis, *Aphis sacchari* Zehntn., is occasionally injurious to sugar cane. Köebele records an outbreak of the species on the Island of Kauai in 1896 under the name *Aphis* sp.[b] The species was determined by Kirkaldy in 1907.[c] This insect is known to occur on cane in Java. In Hawaii, the species is fed upon by the ladybird *Coccinella repanda* Thunb., though the benefit from this beetle is offset by the work of its braconid parasite, *Centistes americana* Riley.

In some districts where the cane fields are situated in moist locations, a mole cricket, *Gryllotalpa africana* Beauv., is sometimes abundant enough to be injurious. Another species of mole cricket, *Scapteriscus didactylus* Latr., is a most important pest of sugar cane

a FROGGATT, W. W.—Australian ladybird beetles. <Agr. Gazette of New South Wales, vol. 13, pt. 9, pp. 907, 908, September, 1902.

b KOEBELE, ALBERT.—Hawaiian Planters' Monthly, vol. 15, no 12, pp 596–598, December, 1896.

c KIRKALDY, G. W.—On some peregrine Aphidæ in Oahu. <Proc. Hawaiian Ent. Soc., vol. 1, pt. 3, pp. 99, 100, July, 1907.

in the island of Porto Rico.[a]   Regarding the work of the Hawaiian
mole cricket, Prof. Koebele reports as follows:[b]

A species of mole cricket has appeared in very large numbers in some of the moist
valleys on Oahu, it is likely another Asiatic introduction, as a rule these crickets
are found around the muddy borders of shallow ponds and watercourses where they live
in burrows resembling those of moles, and like that animal their food consists chiefly
of earth worms and the larva of various insects. The opinions as to its habits are as
yet divided; whilst some authorities claim that it is beneficial, others place it amongst
the injurious insects.

Specimens kept in confinement here with pieces of sugar cane would hardly touch
them, yet they readily devoured a large number of the larva of the Adoretus or Japanese
beetle, as well as those aphodius and a number of earth worms, all within 24 hours.

The ground infested by these crickets was examined and found to be very wet and
completely riddled with the burrows down to a depth of three and even four feet, as
many as three and four specimens were brought to light in a single shovel full of the
soil. In such localities there is no question as to the injurious effects of the crickets
on young cane plants, wherever they were numerous almost all of the seed cane was
destroyed; they would burrow into the seed from all sides, destroying all the eyes,
where the plants had made a growth of a couple of feet the cricket would burrow in
below the ground and eat to the center, killing the plant. This is the only instance
so far observed of the depredations of these crickets here. In rice and taro fields no
damage has been observed as yet, and the only damage that is likely to occur to cane
is when it is planted in wet swampy land, as the cricket can only live and thrive in
such places, and is not found in ordinary arable land; even in the swamp where the
cricket was very numerous, it did not attack the old cane but paid its attention solely
to the newly planted seed and very young plants.

This cricket, although living in marshy land, cannot live under water, yet it is a
good swimmer, the only remedy that can be recommended at present is to flood the
land with water and collect the crickets as they come to the surface, destroying them
by placing them in a vessel containing kerosene and water.

The fungoid so contagious to many insects and larva here, does not seem to have any
effect on this lively cricket, nor will he have anything to do with poison given in the
style of bran, sugar and arsenic.

Certain army worms and cutworms, among them *Heliophila uni-
puncta* Haw., *Agrotis ypsilon* Rott., and *Spodoptera mauritia* Boisd.,
are occasionally known to strip fields of young cane. These species
and related forms, together with their natural enemies, are discussed
in a recent report by Mr. O. H. Swezey.[c]

A bud moth, *Ereunetis flavistriata* Wlsm., is found generally
throughout the Hawaiian cane fields and at times is quite numerous.
Regarding its injury Swezey says:[d]

---

[a] BARRETT, O. W.—The changa or mole cricket in Porto Rico. <Porto Rico Agr.
Exp. Sta., Bul. 2, pp 19, fig. 1, 1902.

[b] KOEBELE, ALBERT.—Hawaiian Planters' Monthly, vol. 15, no. 12, pp. 594–596,
December, 1896.

[c] SWEZEY, O. H.—Army worms and cutworms on sugar cane in the Hawaiian
Islands. <Hawaiian Sugar Planters' Exp. Sta., Div. Ent., Bul. 7, pp. 32, pls. 3,
November, 1909.

[d] SWEZEY, O. H.—The Hawaiian sugar cane bud moth (*Ereunetis flavistriata*)
with an account of some allied species and natural enemies. <Hawaiian Sugar Planters'
Exp. Sta., Div. Ent., Bul. 6, pp. 40, pls. 4, October, 1909.

It is usually not particularly injurious as it customarily feeds on the dead and drying tissues of the leaf-sheaths of sugar cane; but when very numerous and on particularly soft varieties of cane the caterpillars do considerable eating of the epidermis, and also eat into the buds and destroy them, occasioning a good deal of loss where the cane is desired for cuttings to plant.

The grasshoppers *Xiphidium varipenne* Swezey and *Oxya velox* Fab. feed to some extent on the leaves of cane. The former species is also predatory in habit, attacking the young leafhoppers and the larvæ of the sugar-cane leaf-roller.

Two species of beetles which occasionally invade the cane fields from their common food plants and attack the leaves of the sugar cane are Fuller's rose beetle, *Aramigus fulleri* Horn,[a] and the Japanese beetle, *Adoretus tenuimaculatus* Waterh.[b]

RATS INJURING GROWING SUGAR CANE IN HAWAII.

The so-called roof-rat (*Mus alexandrinus*) in former years was very common in the cane fields of Hawaii and did considerable damage by eating the stalks. This is also the cane-field rat of the island of Jamaica. The species in Hawaii lives now for the most part in trees and the upper stories of dwellings, since it has been driven to a great degree from the cane fields by the introduced mongoose. The introduction of the mongoose was a benefit as regards its destruction to the rats in the cane fields, but the animal is an undesirable acquisition to the fauna of the islands for the reason that in recent years it has included in its dietary the eggs and young of ground-nesting birds and domestic fowls. The destruction of the ground-nesting birds is most regrettable.

---

[a] Van Dine, D. L.—Hawaii Exp. Sta., Press Bul. 14, p. 5, October, 1905.

[b] Koebele, Albert.—Hawaiian Planters' Monthly, vol. 17, no. 6, pp. 260–264, June, 1898.

INDEX.

O

OCT 2 1916

U. S. DEPARTMENT OF AGRICULTURE,

BUREAU OF ENTOMOLOGY—BULLETIN No. 94.

L. O. HOWARD, Entomologist and Chief of Bureau.

INSECTS INJURIOUS TO FORESTS AND FOREST PRODUCTS.

CONTENTS AND INDEX.

ISSUED SEPTEMBER 9, 1916.

WASHINGTON:
GOVERNMENT PRINTING OFFICE.
1916.

U. S. DEPARTMENT OF AGRICULTURE,
BUREAU OF ENTOMOLOGY—BULLETIN No. 94.

L. O. HOWARD, Entomologist and Chief of Bureau.

INSECTS INJURIOUS TO FORESTS AND FOREST PRODUCTS.

I. DAMAGE TO CHESTNUT TELEPHONE AND TELEGRAPH POLES BY WOOD-BORING INSECTS.

By THOMAS E. SNYDER, M. F., *Agent and Expert.*

II. BIOLOGY OF THE TERMITES OF THE EASTERN UNITED STATES, WITH PREVENTIVE AND REMEDIAL MEASURES.

By THOMAS E. SNYDER, M. F., *Entomological Assistant Forest Insect Investigations.*

WASHINGTON:
GOVERNMENT PRINTING OFFICE.
1916.

BUREAU OF ENTOMOLOGY.

L. O. HOWARD, *Entomologist and Chief of Bureau.*
C. L. MARLATT, *Entomologist and Assistant Chief of Bureau.*
E. B. O'LEARY, *Chief Clerk and Executive Assistant.*

F. H. CHITTENDEN, *in charge of truck crop and stored product insect investigations.*
A. D. HOPKINS, *in charge of forest insect investigations.*
W. D. HUNTER, *in charge of southern field crop insect investigations.*
————, *in charge of cereal and forage insect investigations.*
A. L. QUAINTANCE, *in charge of deciduous fruit insect investigations.*
E. F. PHILLIPS, *in charge of bee culture.*
A. F. BURGESS, *in charge of gipsy moth and brown-tail moth investigations.*
ROLLA P. CURRIE, *in charge of editorial work.*
MABEL COLCORD, *in charge of library.*

FOREST INSECT INVESTIGATIONS.

A. D. HOPKINS, *Forest Entomologist in charge.*

H. E. BURKE (in charge of Pacific Slope Station at Placerville, Cal.), JOSEF BRUNNER (in charge of Northern Rocky Mountain Station at Missoula, Mont.), T. E. SNYDER, W. D. EDMONSTON (in charge of Southern Rocky Mountain Station at Colorado Springs, Colo.), F. C. CRAIGHEAD, J. M. MILLER (in charge of seed insect station at Ashland, Oreg.), and A. B. CHAMPLAIN, *assistants in forest entomology.*
L. C. GRIFFITH, *assistant in shade tree insects.*
S. A. ROHWER, *specialist on forest Hymenoptera* (in charge of Eastern Station at East Falls Church, Va.).
A. G. BÖVING, *specialist.*
C. T. GREENE, *specialist on forest Diptera.*
W. S. FISHER, *specialist on forest Coleoptera.*
CARL HEINRICH, *specialist on forest Lepidoptera.*
JACOB KOTINSKY, *entomological assistant.*
WILLIAM MIDDLETON, *scientific assistant.*

II

PREFACE.

Bulletin 94, entitled "Insects Injurious to Forests and Forest Products," consists of two parts and an index.

Part I, "Damage to Chestnut Telephone and Telegraph Poles by Wood-Boring Insects," by Thomas E. Snyder, comprises the results of a special study of a serious damage to the base of standing chestnut telephone and telegraph poles by the wood-boring larva of a beetle designated by the author as the chestnut telephone-pole borer (*Parandra brunnea* Fab.).

Part II, "Biology of the Termites of the Eastern United States, with Preventive and Remedial Measures," by Thomas E. Snyder, is based mainly on investigations and experiments conducted during the past three years by Mr. Snyder in connection with his work in the Branch of Forest Insect Investigations. It also includes unpublished notes by Messrs. H. G. Hubbard and F. L. Odenbach. Termites are among the most destructive insects to both crude and finished forest products in North America, among which may be listed construction timbers in bridges and wharves, telephone and telegraph poles, hop poles, mine props, fence posts, lumber piled on the ground, railroad ties, and the woodwork of buildings. The sudden crumbling of bridges and wharves, the caving in of mines, and the settling of floors in buildings, are sometimes directly due to the concealed work of these insects. The use of untreated wood-pulp products, such as the various composition-board substitutes for lath, etc., is restricted in the Tropics and southern United States because of the ravages of termites. In the cities of Washington, Baltimore, St. Louis, Cleveland, New York, and Boston, and throughout the eastern and southern United States, damage by termites to the woodwork of buildings is occasionally serious.

Methods of prevention and control against injuries to finished and utilized forest products, etc., are based on the results of experiments conducted by this branch of the bureau.

A. D. HOPKINS,
Forest Entomologist.

III

CONTENTS.

III

ILLUSTRATIONS.

IV

U. S. D. A., B. E. Bul. 94, Part I.   F. I. I., December 31, 1910.

INSECTS INJURIOUS TO FORESTS AND FOREST PRODUCTS.

DAMAGE TO CHESTNUT TELEPHONE AND TELEGRAPH POLES BY WOOD-BORING INSECTS.

By Thomas E. Snyder, M. F.,
Agent and Expert.

OBJECT OF PAPER.

It has recently been determined through special investigations conducted principally by the writer that serious damage is being done to the bases of standing chestnut telephone and telegraph poles in certain localities by the grub or larva of a wood-boring beetle, here called the chestnut telephone-pole borer.[a] The character and extent of the damage under different conditions of site in several localities have been determined, and poles treated with various preservative substances have been inspected to compare the efficiency of both chemicals and methods of treatment. These investigations have resulted in the determination of practical methods of preventing injury to poles by wood-boring insects.

HISTORICAL DATA.

The first information of serious damage to standing chestnut poles by wood-boring insects was conveyed in a letter, dated December 15, 1906, from E. O. Leighley, a correspondent of this Bureau, reporting damage to telephone poles in Baltimore, Md., by borers. Mr. A. B. Gahan, assistant entomologist of the Maryland Agricultural Experiment Station, College Park, Md., who investigated the injury to the poles, stated that it was the work of a borer and was located just beneath the surface of the ground. Mr. Gahan brought specimens of the work and the insect to this office. The borers were identified as cerambycid larvæ, and later were determined to be the chestnut telephone-pole borer (*Parandra brunnea* Fab.).

On December 16, 1906, Mr. H. E. Hopkins, division superintendent of a telephone company, stated that the poles in West Virginia were

---

[a] *Parandra brunnea* Fab.; Order Coleoptera, Family Spondylidæ.

badly injured by borers and that these borers were abundant. On March 8, 1907, he collected larvæ from chestnut telephone poles at Pennsboro, W. Va. These were determined to be the larvæ of the chestnut telephone-pole borer.

The writer on October 3, 1909, inspected some chestnut telegraph poles which had been standing for about twelve years on New York avenue, in Washington, D. C. The poles had been taken down under orders from the city authorities, which necessitated the placing of wires in conduits under ground, and they had been lying in piles for about a month before they were inspected. The chestnut telephone-pole borer had been working in the base of the poles, and white ants, or termites, were associated with them. Twelve out of the 103 poles examined had been damaged, some more seriously than others.

On October 15, 1909, Mr. H. E. Hopkins sent a reply to a request by Dr. A. D. Hopkins for further information regarding insect damage to poles in West Virginia. He stated that in one line built twelve years ago (40 miles long, 36 chestnut poles to the mile, poles 20 to 40 feet long and 5 to 12 inches in diameter at the top) approximately 600 poles had been rotted off at the top of the ground, and inspection showed that 95 per cent of the damage was directly or indirectly due to insects. Other lines in this division were reported to be in about the same condition. It was later determined that most of the insect damage was the work of the chestnut telephone-pole borer.

Dr. A. D. Hopkins states in a recent comprehensive bulletin [a] that "construction timbers in bridges and like structures, railroad ties, telephone and telegraph poles, mine props, fence posts, etc., are sometimes seriously injured by wood-boring larvæ, termites, black ants, carpenter bees, and powder-post beetles, and sometimes reduced in efficiency from 10 to 100 per cent." Thus, while it has been known that almost all classes of forest products that are set in the ground are seriously injured by wood-boring insects, the problem of insect damage to standing poles, posts, and other timbers has never been made the subject of a special investigation.

In May, 1910, this study was assigned to the writer, and, in addition to a study of the insects involved, investigations in cooperation with telephone and telegraph companies have been conducted in the District of Columbia, Maryland, Virginia, Pennsylvania, New Jersey, and New York. Through the courtesy of the Western Union Telegraph Company several telegraph lines were inspected in July and August, 1910, in Virginia, where the poles were being reset or replaced. Here the butts of over 200 poles set under different conditions of site were thoroughly examined for insect damage, and sometimes the

---

[a] Insect Depredations in North American Forests. <Bul. 58, Part V, Bureau of Entomology, U. S. Department of Agriculture, p. 67, 1909.

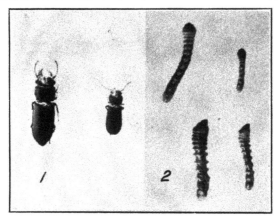

FIG. 1.—THE CHESTNUT TELEPHONE-POLE BORER (PARANDRA BRUN-
NEA): MALE AND FEMALE BEETLES. FIG. 2.—THE CHESTNUT
TELEPHONE-POLE BORER: YOUNG LARVÆ, DORSAL AND LATERAL
VIEWS. FIG. 1, SLIGHTLY ENLARGED; FIG. 2, TWICE NATURAL
SIZE. (ORIGINAL.)

FIG. 3.—DAMAGE TO AN UNTREATED CHESTNUT TELEGRAPH POLE
NEAR SURFACE OF GROUND BY THE CHESTNUT TELEPHONE-POLE
BORER. (ORIGINAL.)

entire pole was split open. In one line 10 to 12 years old (approximately 30 chestnut poles per mile, 25 feet long, about 6 inches diameter at the top, 10 inches at the base, and apparently of second quality), between Petersburg and Crewe, Va.—the poles had already been reset once, east of Wilson, Va.—serious damage by the chestnut telephone-pole borer rendered from 15 to 20 per cent of the poles unserviceable.

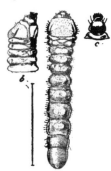

After the present second resetting it was estimated that the poles can not last more than four or five years longer. West of Wilson the poles were naturally in much worse condition, and many were broken off and only held up by the wires on the sounder poles. In another line examined, between Portsmouth and Boykins, Va. (poles 30 feet long and apparently of second quality), serious damage by this borer averaged about 10 or 15 per cent, and between Boykins, Va., and Weldon, N. C., according to a linesman, 50 per cent of the poles are badly decayed near the surface of the ground. Much of this damage, however, is due to fungous heart rot. According to a statement by the foreman of a resetting crew, between

Fig. 1.—The chestnut telephone-pole borer (*Parandra brunnea*): Full-grown larva. (About twice natural size. (Original.)

Asheville, N. C., and Spartanburg, S. C., hundreds of chestnut poles were badly decayed in the 67 miles of line reset, and were only held up by the wires. The line was 15 years old. There was serious damage by "wood lice" (termites) and also by "white wood worms."

THE CHESTNUT TELEPHONE-POLE BORER.

(*Parandra brunnea* Fab.)

CHARACTER OF THE INSECT.

The chestnut telephone-pole borer is a creamy white, elongate, stout, cylindrical, so-called "round-headed" grub or "wood worm" (fig. 1), which hatches from an egg deposited by an elongate, flattened, shiny, mahogany brown, winged beetle from two-fifths to four-fifths of an inch in length. (Plate I, fig. 1; text fig. 2.) The eggs are probably deposited from August to October in shallow natural depressions or crevices on the exterior of the pole near the surface of the ground; often the young larvæ enter the heartwood through knots. The young borers (Plate I, fig. 2) hatching therefrom eat out broad shallow galleries running longitudinally in the sapwood, then enter the heartwood, the mines being gradually enlarged as the larvæ develop. As they proceed, the larvæ closely pack the fine excreted boring dust behind them. This débris, which is characteristic of their work, is reddish to dunnish yellow in color and has a claylike consistency. The mines eventually end in a broad chamber, the entrance to which is plugged up by the excelsior-like fibers of wood chiseled out by the strong mandibles of the larva. Here the resting stage (fig. 3), or pupa, is formed, and in this chamber the perfect adult spends considerable time before emerging. Often all stages from very young larvæ only about one-fourth inch long to full-grown larvæ over 1 inch long, pupæ, and adults in all stages to maturity are present in the same pole. Adults have been found flying from July to September. As yet the seasonal history of this borer has not been completely worked out.

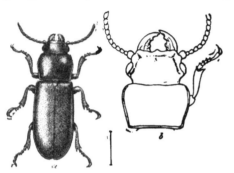

Fig. 2.—The chestnut telephone-pole borer: Female beetle, three and one-half times natural size. Head and pronotum of male beetle (Original).

DISTRIBUTION.

This insect is very widely distributed, ranging from Ontario, Canada, to Texas, eastward to the Atlantic coast, and westward to Arizona and southern California. It is common throughout the natural range of the chestnut—and in this connection it should be observed that most of the chestnut poles are purchased from local timber-land owners.

CHARACTER OF THE INJURY.

The injury to the poles consists in large mines in the wood near the line of contact of the pole with the ground, necessitating the frequent resetting or even the replacement of the damaged poles. These irregular galleries of the grub (Plate II, fig. 1) run both horizontally and longitudinally throughout the heartwood, and are sometimes 7 inches long, but vary with the individuals, which show great differences in size. The borers usually work in the outer layers of the wood at the base of the pole for a distance of from 2 to 3 feet below, and sometimes from 1 to 2 feet above the line of contact of the pole with the surface of the ground. The greatest damage is to that area just below and just above the surface of the ground (Plate I, fig. 3); here the conditions of air and moisture are most favorable. Often the entire butt up to a distance of from 4 to 6 feet and higher, according to the depth of setting, is mined. The numerous galleries, often very close together, completely honeycomb the wood in a zone

FIG. 1.   FIG 2.

WORK OF THE CHESTNUT TELEPHONE-POLE BORER.

Fig. 1.—Gallery of the chestnut telephone-pole borer, showing pupal chamber with the entrance plugged with excelsior-like wood fibers, work near base of pole, below ground Fig. 2 —Mines of the chestnut telephone-pole borer near surface of ground.   Natural size. (Original.)

3 to 4 inches in from the exterior of the poles; this so weakens the poles that they break off close to the surface of the ground. The basal 2 feet is usually sound. Even if the damage is not serious enough to cause the poles to break off under strain, they are likely to go down during any storm, and thus put the wire service out of commission; such damaged poles are a serious menace along the right of way of railroads. The beetle will attack poles that are perfectly sound, but evidently prefers to work where the wood shows signs of incipient decay; it will not work in wood that is "sobby" (wet rot), or in very "doty" (punky) wood. It has not yet been determined just how soon the borers usually enter the poles after they have been set in the ground. However, poles that had been standing only four or five years contained larvæ and adults of this borer in the heartwood, and poles that had been set in the ground for only two years contained young larvæ in the outer layers of the wood.

Poles that appear sound on the exterior may have the entire basal interior riddled, and the work of the borers is not noticed until the poles break off. If merely isolated poles are injured, the poles that are broken off are held up by the wires and can be detected by the fact that they lean over, but if several adjacent poles are affected, especially where there is any unusual strain, that portion of the line is very likely to go down. The presence of the borers in injurious numbers can be determined only by removing the earth from about the base of the pole; the exit holes of the borer are found near the line of contact with the soil. Often large, coarse borings of wood fiber project from the exit holes. Sometimes old dead parent adults are found on the exterior of the poles underground. During August the young adults may be found in shallow depressions on the exterior of poles below the ground surface.

FIG. 3.—The chestnut telephone-pole borer: Pupa. Slightly more than twice natural size. (Original.)

IMPORTANCE OF THE PROBLEM.

The subject of the relation of insects to the rapid decay of chestnut poles has not been thoroughly investigated in the past, but now that the supply is becoming scarcer it is especially important to know what are the various primary causes of the deterioration of these poles, hitherto described under the vague term "decay." Although the chestnut telephone-pole borer has not hitherto been considered an insect of any economic importance, and has been described in entomological literature as only living under bark, principally of pine, or in the decomposing wood of various species of deciduous and coniferous trees, the evidence is abundant that breeding in the bases of chestnut poles is not a newly acquired habit. It has also been determined that this beetle damages many species of living forest, fruit, and shade trees that have been previously injured by fire or other causes, and often leads to the destruction of trees that would otherwise recover from such wounds, and while not normally a primary enemy to trees, may thus become of more than secondary importance.

The damage by the chestnut telephone-pole borer is especially serious in consideration of the fact that in many parts of its range the chestnut is threatened with extinction as a tree species on account of the very severe ravages of the combined attack of an insect [a] and a fungous disease. Further unnecessary drain upon the supply of chestnut timber should be avoided by protecting that already in use and thus prolonging its length of service.

EXTENT OF DAMAGE AND LOSS.

As more than one-fourth of the 3,500,000 round poles exceeding 20 feet in length used annually by telephone, telegraph, and other electric companies are chestnut (Kellogg, 1909),[b] and as this borer has seriously damaged as high as 10 to 40 per cent, varying with conditions of site, of the chestnut poles which have been set in the ground for from ten to twelve years in lines in North Carolina, Virginia, West Virginia, Maryland, and the District of Columbia, it is evident that this insect is an important factor in decreasing the normal length of service of the poles.[c] In lines from twelve to fifteen years old the damage is much greater, and at the end of this number of years of service any line in which poles of this species are set has to be practically renewed. According to a statement in Forest Service Bulletin 78 (Sherfesee, 1909), "approximately 4 per cent, or 5,908 feet board measure of the 147,720 feet board measure of standing poles annually requiring replacement in the United States, is destroyed by insects." If only chestnut poles be considered, at least 10 per cent of the poles reset or replaced are injured by insects.

FAVORABLE AND UNFAVORABLE CONDITIONS FOR DESTRUCTIVE WORK.

The damage is apparently greatest and the borers are most abundant where the poles are set in high or level dry ground under good conditions of drainage. Such sites are the crests of railroad cuts through low hills, slopes of "fills," and in cultivated or other fields. Where the poles are in wet sites there is usually but little injury by

---

a *Agrilus bilineatus* Web.

b See list of publications, page 11.

c The average life of a chestnut pole is eight to ten years (Sherfesee and Weiss, 1909).

wood-boring insects except to that portion near the surface of the ground. Conditions of drainage are more important than different soil combinations, and the condition of the soil is more important than its composition; *i. e.*, where the soil is hard packed there is apparently less damage than where it is loose. The quality and condition of the poles before setting is a very important factor to consider before arriving at any conclusions as to the relative longevity of poles under various conditions of site. Green (unseasoned) or imperfectly seasoned poles are less durable than those thoroughly seasoned. Poles that are defective [a] before setting, as they very often are (i. e., showing evidence of incipient decay), and poles that have the heartwood mined by the chestnut timber worm,[b] the work of which is very abundant, will, of course, decay much more rapidly than poles that are in an absolutely sound condition. The galleries of the chestnut timber worm afford an entrance to the spores of wood-destroying fungi, and thus greatly accelerate decay. White mycelium compactly filled these galleries throughout many standing poles, thus clearly proving that these mines aid greatly in enabling the fungous heart rot more rapidly and completely to penetrate the entire heartwood of the poles. If the injury by both wood-boring beetles and wood-destroying fungi (between which there is a varying interrelation) be considered, then in several lines from ten to twelve years old in North Carolina, Virginia, and West Virginia at least 50 per cent of the poles are either rendered unserviceable or their length of service is much shortened.

ASSOCIATED WOOD-BORING INSECTS.

It is not to be concluded that this wood-boring beetle is the only insect that injures standing chestnut poles. Indeed, the most common injury is by the "wood lice" or white ants.[c] In lines from ten to twelve years old these insects have seriously damaged as high as 15 per cent of the poles, and their work is often present, at least superficially, in as high as 75 per cent of the poles under all conditions of site. However, the damage is usually to the outer layers of the wood, where it is moist or there is incipient decay, and is more superficial and localized than that of the chestnut telephone-pole borer. Nevertheless, white ants often completely honeycomb the sound heartwood of poles, especially at the base. They work both in sound wood, "doty" (dry rot) wood, and "sobby" (wet rot) wood. Sometimes a large channel runs up through the core of the heart

---

[a] Often this evidence is the old galleries of the destructive two-lined chestnut borer (*Agrilus bilineatus* Web.), showing that the tree must have been dead before it was cut for a pole, and hence is more likely to be defective throughout the interior; in other instances heart rot is clearly present.

[b] *Lymexylon sericeum* Harr.

[c] Identified by Mr. Theodore Pergande of this Bureau as *Termes flavipes* Kollar.

and the sides are plastered with clay, forming a hollow tube with several longitudinal galleries. Their work often extends from 2 to 4 feet above the surface of the ground. They leave the outer shell of the wood intact and work up through the longitudinal weathering checks, covering the exterior of the pole with earth to exclude the light. White ants will damage poles that have been set in the ground only two years. Evidently they enter the pole from below the surface of the ground. The habits and characteristics of these peculiar and interesting insects have been thoroughly discussed in Circular No. 50 of this Bureau by Mr. C. L. Marlatt.

A giant round-headed borer [a] is sometimes found in the poles, usually in association with the chestnut telephone-pole borer. In poles where the wood is sound this borer apparently works as a rule only in the outer layers of the wood, the galleries running longitudinally through the heart below the surface of the ground. In poles where there is decay it will completely honeycomb the heartwood near the surface of the ground.

In several poles where the wood was "doty" a large Scarabæid [b] which has before been found in decayed oak railroad ties was present and caused the poles to break off sooner than they otherwise would. The irregular galleries of the grub completely honeycomb the decayed heartwood near or just below the surface.

A flat-headed borer [c] and wireworms [d] were found in galleries locally in the more or less decayed heartwood of several poles. A large black carpenter ant [e] does some damage to sound poles set in dry ground through woodland. This ant often widens the longitudinal weathering checks and thus accelerates decay. A small black ant [f] was very numerous in many poles, but its work is usually confined to the outer layers of the wood. The work is often throughout "doty" poles. Injury by this ant is not primary, but it also widens weathering checks, enlarges other defects, and induces more rapid decay.

PREVENTION OF THE INJURY.

Doctor Hopkins makes the following statement in a recent bulletin:[g]

Insect damage to poles, posts, and similar products can be prevented to a greater or less extent by the preservative treatments which have been tested and recommended by the Forest Service for the prevention of decay. These should be applied

---

[a] *Prionus* sp.

[b] Identified by Mr. E. A. Schwarz, of this Bureau, as *Polymœchus brevipes* Lec.

[c] Identified by Mr. H. E. Burke, of this Bureau, as *Buprestis rufipes* Oliv.

[d] Species of the family Elateridæ The large larvæ of *Alaus* sp. were especially injurious.

[e] Identified by Mr. Theodore Pergande as *Camponotus pennsylvanicus* Mayr.

[f] Identified by Mr. Theodore Pergande as *Cremastogaster lineolata* Say.

[g] Insect Depredations in North American Forests. <Bul. 58, Part V, Bur. Ent., U. S. Dept. Agr., p. 84, 1909.

before the material is utilized for the purposes intended, or, if it be attacked after it has been utilized, further damage can be checked to a certain extent by the use of the same substances.

It is often of prime importance to prevent injury from wood-boring insects, for the reason that such injuries contribute to more rapid decay. Therefore anything that will prevent insect injury, either before or after the utilization of such products, will contribute to the prevention of premature deterioration and decay.

Through the courtesy of the American Telephone and Telegraph Company and the Forest Service, about 40 chestnut poles set in a test line near Dover, N. J., were inspected by the writer on July 15, 1910, in company with engineers of the telephone company and Mr. H. F. Weiss, Assistant Director, Forest Products Laboratory, Forest Service, to determine the relative merits of various methods of preventing damage by wood-boring insects to the bases of poles. In this line, which is eight years old, variously treated poles alternated with untreated poles in order that each chemical preservative and method of treatment might be given an absolutely fair test under the same conditions of site. The poles were 30 feet long, 7 inches in diameter at the top, and 33 inches in circumference 6 feet from the base. In this inspection the earth was removed (to a depth of about 1 foot) from the base of the pole, and then the pole was chopped into to determine the rate of decay. This method of inspection for insect damage is not very satisfactory. The various methods experimented with in this test line were brush treatments with a patented carbolineum preservative and spirittine, charring the butt, setting the pole in sand, and setting it in small broken stone. It was found that, although these methods may temporarily check the inroads of wood-boring insects, they will not keep the insects out of the poles. The most serious damage to the poles in this line was by white ants. Other insect damage was by a large black carpenter ant[a] and the larvæ of a round-headed borer.[b]

An inspection was made, between September 6 and 14, 1910, of the bases of over 400 chestnut poles set in a similar test line near Warren, Pa., and Falconer, N. Y. These poles were treated by the creosote "open-tank" method of impregnation, and brush treatments of creosote, wood creosote, creolin, two different carbolineum preservatives, and tar; they had been set in the ground for a period of five years. All these treatments, except the brush treatments with creolin and tar, were efficient in preventing the attacks of wood-boring insects, at least for a five-year period, in this northern climate. There was but little damage by insects to the poles in this test line. The most common injury to the untreated poles was by the large black carpenter ants which widen the longitudinal weathering checks, and hence induce more rapid decay. The work of the chestnut tele-

---

[a] *Camponotus pennsylvanicus* Mayr.
[b] *Prionus* sp.

phone-pole borer was found in several poles, and this beetle was evidently just beginning to attack these poles. There was some damage by a round-headed borer.[a] No white ants or termites were present, and this is evidently too far north for these destructive borers. A report by inspectors of the American Telephone and Telegraph Company and the Forest Service on the remainder of the poles in this test line (between Jamestown and Buffalo, N. Y.) not personally inspected by the writer, showed that these conclusions can be applied to all the poles in the line with the exception that there was superficial injury by small black ants to two poles treated by brush treatments of carbolineum avenarius and to two treated with wood creosote; also, as the inspection progressed, injury by the chestnut telephone-pole borer became more abundant and serious, and the borers seemed to be established in the poles. The poles treated by the creosote "open-tank" method of impregnation and by brush treatments with creosote and with "S. P. F." carbolineum remained uninjured.

Methods of treating poles superficially by brushing with various preservatives have proved to be temporarily efficient in keeping wood-boring insects out, if the work is thoroughly done and not only the butt, but also the base, is treated. If the pole is not thoroughly brushed, insects enter through the untreated or imperfectly treated portions, especially through weathering checks and knots. Where the base is left untreated, insects, especially white ants or termites, enter the pole from below ground and, avoiding the treated portions, come right up through the pole.

The few poles of southern yellow pine in a line near Bartley, N. J., inspected on July 15, 1910, which had been impregnated with creosote by the Bethell cylinder-pressure process, 12 pounds of oil to the cubic foot, and had been set in the ground since February, 1903, were apparently absolutely free from signs of decay or damage by wood-boring insects. In another line, running between Norfolk, Va., and Washington, D. C., the few poles (12 years old, of squared—with the sapwood cut away—southern yellow pine) inspected on August 10, 1910, near Portsmouth, Va., which had been impregnated with creosote by the Bethell cylinder-pressure process, were also apparently absolutely sound.

Thus, it is evident that impregnating the poles with creosote by some standard process (either the open-tank or the cylinder-pressure processes) will keep wood-boring insects out and preserve the poles for a much longer period than they would last untreated. In the open-tank method only the area most subject to the attacks of wood-boring insects and deterioration in general (i. e., the basal 8 feet) is treated, while by the cylinder-pressure processes the entire pole is impregnated. Alternating less susceptible juniper (red cedar)[b] poles

---

a Priorus sp.   b Juniperus virginiana.

or pine poles thoroughly impregnated by some standard process in the line with the chestnut poles would be a safeguard in holding up an old line where the damage is found to be serious on resetting.

A list of some available publications on wood preservation is appended.

PUBLICATIONS ON WOOD PRESERVATION AND STATISTICS ON POLES.

1903. Von Schrenk, H.—Seasoning of timber. <Bul. 41, Forest Service, U. S. Dept. Agr.

1906. Grinnell, H.—Prolonging the life of telephone poles. <Yearbook U. S. Dept. Agr. for 1905, Extract No. 395.

1907. Crawford, C. G.—The open-tank method for the treatment of timber. <Cir. 101, Forest Service, U. S. Dept. Agr.

1907. Crawford, C G.—Brush and tank pole treatments. <Cir. 104, Forest Service, U. S. Dept. Agr.

1907. Grinnell, H.—Seasoning of telephone and telegraph poles. <Cir. 103, Forest Service, U. S. Dept. Agr.

1908. Sherfesee, W. F.—A primer of wood preservation. <Cir. 139, Forest Service, U S Dept Agr.

1908. Weiss, H. F.—Progress in chestnut pole preservation. <Cir. 147, Forest Service, U. S Dept. Agr.

1909. Sherfesee, W. F.—Wood preservation in the United States. <Bul. 78, Forest Service, U S Dept Agr., pp. 24, 25, Table I.

1909. Sherfesee and Weiss.—Wood preservation. <Rep. Natl. Conserv. Com, vol. 2, p. 663.

1909. Kellogg, R. S.—The timber supply of the United States. <Cir. 166, Forest Service, U S Dept. Agr., pp. 20–21.

1910. Willis, C. P.—The preservative treatment of farm timbers. <Farmers' Bul. 387, U S. Dept. Agr.

O

U. S. DEPARTMENT OF AGRICULTURE,

BUREAU OF ENTOMOLOGY—BULLETIN No. 94, Part II.

L. O. HOWARD, Entomologist and Chief of Bureau.

INSECTS INJURIOUS TO FORESTS AND FOREST PRODUCTS.

BIOLOGY OF THE TERMITES OF THE EASTERN UNITED STATES, WITH PREVENTIVE AND REMEDIAL MEASURES.

BY

THOMAS E. SNYDER, M. F.,

*Entomological Assistant,
Forest Insect Investigations.*

ISSUED FEBRUARY 17, 1915.

WASHINGTON:
GOVERNMENT PRINTING OFFICE.
1915

CONTENTS.

ILLUSTRATIONS.

TEXT FIGURES.

INSECTS INJURIOUS TO FORESTS AND FOREST PRODUCTS.

BIOLOGY OF THE TERMITES OF THE EASTERN UNITED STATES, WITH PREVENTIVE AND REMEDIAL MEASURES.

By Thomas E. Snyder, M. F.,

Entomological Assistant.

INTRODUCTION.

The following notes on the biology of the common termites,[a] or "white ants," of the eastern United States were, for the most part, made while conducting investigations to determine the character and extent of damage by termites and other forest insects to various classes of crude and finished forest products and to devise methods of preventing the injury. During 1910 and 1911 special investigations were conducted by the writer as to the character and extent of damage to telephone and telegraph poles and mine props by wood-boring insects. This contribution is thus based, largely, on these investigations, as well as on additional experiments conducted during the past three years by the writer in the branch of Forest Insect Investigations and on some of the unpublished notes of the late H. G. Hubbard and those of Rev. F. L. Odenbach.

"White ants" are among the most destructive insects of North America to both crude and finished forest products, among which may be listed construction timbers in bridges, wells, and silos, timbers of wharves, telephone and telegraph poles, bean and hop poles, mine props, wooden cable conduits, fence posts, lumber piled on the ground, railroad ties, and the foundations and woodwork of buildings, etc. The sudden crumbling of bridges and wharves, caving in of mines, and settling of floors in buildings are sometimes directly due to the hidden borings of termites. The use of untreated wood-pulp products such as the various "composition boards" used as substitutes for lath, etc., is restricted in the Tropics and portions of the southern United States because of the ravages of termites. In the cities of Washington, Baltimore, St. Louis, Cleveland, and New York, and throughout the Southern States damage by termites to

---

[a] Order *Platyptera*, Packard (1886), suborder *Isoptera*, Brullé (1832), family *Mesotermitidæ*, Holmgren, genus *Leucotermes* Silvestri.

the woodwork of buildings is occasionally serious. As far north as Boston, Mass., damage of this sort occurs, and in Michigan cases are reported in which furniture in buildings has sunk through the floors, mined by "white ants."

The preventives and remedies against injury to forest products and nursery stock herein given are based on the results of experiments conducted by the Bureau of Entomology. The species to be considered in this paper are *Leucotermes flavipes* Kollar and *Leucotermes virginicus* Banks. The former is widely distributed [a] over the United States, but the recorded distribution of the latter is more limited.[b]

CLASSIFICATION.

Termites are naturally to be classed among that most interesting group of social insects comprising the ants, bees, wasps, etc., since they live in colonies which are made up of various highly specialized forms or castes. Each of these forms has a distinct rôle in the processes of the social organization, as there is a well-defined division of labor. In the systematic classification of insects, however, termites are widely separated from the other social insects. These latter represent the highest and most specialized development, whereas termites represent the lowest and are among the oldest of insects. Furthermore, in many points in their life habits termites are widely different from the other social insects.

The tropical genus Termes, from which the family and generic names of termites are derived, is a Linnæan creation [c] and appears for the first time in 1758, in the tenth edition of Systema Naturæ, where it was placed among the Aptera, between the genera Podura and Pediculus. Since then termites have been classed [d] among the orders Neuroptera, Corrodentia, and Pseudoneuroptera, although Brullé, in 1832 (Expéd. Sc. Morée, t. 3 (Zool.), p. 66, Paris, 1832), had founded the distinct new order Isoptera.[e] Packard,[f] in 1863, stated that "seven out of the eight well-established families of the Neuroptera sustain a synthetic relation with each of the six other suborders." Hagen,[g] in 1868, stated in regard to the error that he made in describ-

---

[a] Marlatt, C. L. The White Ant (*Termes flavipes* Koll.). U. S. Dept. Agr., Bur. Ent., Circ. 50, pp. 8, figs. 4, June 30, 1902. See p. 4.

[b] Banks, N. A new species of Termes. Ent. News., v. 18, no. 9, p. 392–393, November, 1907.

[c] Desneux, J. Isoptera, Fam. Termitidæ, pp. 52, figs. 10, pls. 2. (Wytsman, P., Genera Insectorum, fasc. 25, Bruxelles, 1904, p. 1–3.)

[d] Feytaud, J. Contribution à l'étude du termite lucifuge. Arch. Anat. Micros., t. 13, fasc. 4, p. 481–606, figs. 34, 30 juin, 1912.

[e] Desneux, J. Loc. cit.

[f] Packard, A. S. On synthetic types in insects. Boston Jour. Nat. Hist., v. 7, no. 4, p. 590–603, fig. 4, June, 1863.

[g] Hagen, H. A. Proc. Boston Soc. Nat. Hist., v. 12, p. 139, October 21, 1868, Boston, 1869.

ing [a] a damaged earwig (Forficula) from Japan as a wingless termite, "that the three families, Termitina, Blattina, and Forficulina, are coordinated and very nearly allied." Packard,[b] in 1883, placed termites in a new order, the Platyptera, including the Termitidæ, Embiidæ, Psocidæ, and Perlidæ, and three years later he dismembered [c] the Pseudoneuroptera into the Platyptera, Odonata, and Plecoptera. Knower,[d] in 1896, stated that in the development of the embryo of termites there is a resemblance to that of the Orthoptera. Desneux, in 1904,[e] stated that termites are derived phylogenetically from the "Blattides," an idea accredited to Handlirsch in 1903, although Hagen, in 1855, had already formulated this theory based on purely biological considerations. Comstock,[f] in 1895, also placed termites in the order Isoptera. Enderlein,[g] in 1903, placed termites with the Embiidæ in the order Corrodentia, suborder Isoptera. In this order he also placed the Psocidæ and Mallophaga. Handlirsch [h] contests the affinity of the Termitidæ with the Embiidæ; he further states [i] that termites are derived from all deposits from the lower Tertiary on, but that all older fossils formerly mistaken for termites by Hagen, Scudder, and Heer do not belong to this order. Banks [j] (1909) considered termites to be in the order Platyptera, suborder Isoptera, with two other suborders, the Mallophaga and Corrodentia. Holmgren states[k] that he believes both groups, the "Termiten" and "Blattiden," are offshoots of a more primitive group, the "Protoblattoiden." The oldest "Blattoiden" occur in the first part of the middle, upper Carboniferous (Pottsville, North America), the oldest "Protoblattoiden," in the last

---

[a] Hagen, H. A. On a wingless white ant from Japan. Proc. Boston Soc. Nat. Hist., v. 11, p. 399–400, illus., February 26, 1868.

[b] Packard, A. S. Order 3, Pseudoneuroptera, U. S. Dept. Agr., U. S. Ent. Com., 3d Rept., p. 290–293, 1883.

[c] Packard, A. S. A new arrangement of the orders of insects. Amer. Nat., v. 20, no. 9, p. 808, September, 1886.

[d] Knower, H. McE The development of a termite—*Eutermes (Rippertii?)* A preliminary abstract. Johns Hopkins Univ. Circ., v. 15, no. 126, p. 86–87, June, 1896.

[e] Desneux, J. Loc. cit.

[f] Comstock, J. H. Manual for the Study of Insects. Ithaca, N. Y., 1895, p. 95–97, fig. 104–106.

[g] Enderlein, G. Über die Morphologie, Gruppierung und Systematische Stellung der Corrodentien. Zool. Anzeig., vol. 26, p. 423–437, figs. 4.

[h] Handlirsch, A. Zur Systematik der Hexapoden. Zool. Anzeig., vol. 27 (1904), p. 733–769.

[i] Handlirsch, A. Die fossilen Insekten und die Phylogenie der rezenten Formen, pt. 8, p. 1240, Leipzig, 1908.

[j] Banks, N. Directions for Collecting and Preserving Insects, pp. 135, figs. 188, Washington, 1909. U. S. Nat. Mus. Bul. 67. Platyptera, p. 6–7.

[k] Holmgren, N. Termitenstudien 1. Anatomische Untersuchungen. K. Svenska Vetensk. Akad. Handl., Bd. 44, No. 3, pp. 215, Taf. 1–3, Uppsala & Stockholm, 1909. Die Verwandtschaftsbeziehungen der Termiten, p. 208–213.

part of the coal measures (Allegheny, North America), but the youngest "Protoblattoiden" occur in the lower Permian formation of Europe. Holmgren considers the termites to be in a distinct order, the Isoptera.

The order Isoptera, according to Holmgren (1911),[a] is divided into three families, the Protermitidæ, the Mesotermitidæ, and the Metatermitidæ. All of these families are represented in North America, and while this paper is restricted to a discussion of species of the genus Leucotermes Silvestri, subfamily Leucotermitinæ Holmgren, family Mesotermitidæ Holmgren, a species in the genus Termopsis Heer, subfamily Termopsinæ Holmgren, family Protermitidæ Holmgren, is briefly mentioned. Thus it will be seen that the species under observation occupy a middle position between the highest and the lowest genera in the systematic classification of termites.

HISTORICAL.

According to Desneux, Smeathman's marvelous descriptions in Some Account of the Termites Which Are Found iń Africa and Other Hot Climates (London, 1781) are the real basis of scientific researches on the biology of termites.[b] Hagen [c] gives a résumé of the researches on termites up to 1860. Drummond's Tropical Africa (London, 1889), in chapter 6, gives an interesting account of "white ants." Froggatt, in Australian Termitidæ,[d] gives a résumé of the studies of various workers on termites. Saville-Kent, in Chapter III of the Naturalist in Australia (London, 1897), describes the habits of termites in the Tropics; Smeathman, Drummond, and Saville-Kent in their popular accounts of termites stimulated interest in the habits of these insects and led to scientific researches. Grassi and Sandias,[e] in their classical work, have outlined the results of the more important biological researches. Sharp[f] gives an excellent

---

a Holmgren, N. Termitenstudien 2. Systematik der Termiten. K. Svenska Vetensk. Akad. Handl., Bd. 46, No.6, pp. 86, Taf. 1-6, Uppsala & Stockholm, 1911. Ordnung Isoptera, p. 10-11.

b Smeathman, H. Some account of the termites, which are found in Africa and other hot climates. In Philos. Trans. London, v. 71, p. 139-192, 3 pls., 1781. (Smeathman's observations were afterwards confirmed by Savage in 1850 and the late G. D. Haviland.)

c Hagen, H. A. Monographie der termiten. Linnæa Entomologica, v. 10, 1855, p. 1-144, 270-325; v. 12, 1858, p. 1-342, pl. 3; v. 14, 1860, p. 73-128.

d Froggatt, W. W. Australian Termitidæ, Part I. Proc. Linn. Soc. N. S. Wales, v. 10, ser. 2, p. 415-438, July 31, 1895. Distribution, p. 416-426.

e Grassi, B., and Sandias, A. The constitution and development of the society of termites; observations on their habits; with appendices on the parasitic protozoa of Termitidæ and on the Embiidæ translated by F. H. Blandford. Quart. Jour. Micros. Sci. [London], v. 39, pt. 3, n. s., p. 245-322, fold. pl. 16-20, November, 1896, and v. 40, pt. 1, p. 1-75, April, 1897.

f Sharp, D. Cambridge Nat. Hist., vol. 5, Insects, Pt. 1, chap. 16, p. 356-390, London, 1901.

summary of the various writings on termites up to 1901. Escherich [a] gives a résumé of the more important work up to 1909, and also an extensive bibliography. Holmgren's treatise [b] deals with anatomy, systematic classification, development, and biology, and gives a summary of the effect of the social life, as shown in the progressive and retrogressive development of termites. Bugnion [c, d] has recently made observations that lead him to state that (in *Eutermes lacustris* Bugnion and *Termes* spp.) the differentiation of the three castes occurs in the embryo. Feytaud [e] has published an account of the life history of *Leucotermes lucifugus* Rossi, a species closely related to our common species of the eastern United States. His observations confirm those of Heath on the same species in this country. He also made studies of the internal anatomy. The more important contributions to our knowledge of termites are the results of researches by Bobe-Moreau, Bugnion, Czervinski, Desneux, Doflein, Drummond, Escherich, Feytaud, Froggatt, Grassi and Sandias, Hagen, Haldemann, Haviland, Holmgren, Jehring, Latreille, Leidy, Lespès, Fritz Müller, Pérez, Perris, Petch, Rosen, Savage, Silvestri, Sjöstedt, Smeathman, Wasmann, and many others. In the United States, Banks, Buckley, Hagen, Heath, Howard, Hubbard, Joutel, Knower, Marlatt, Porter, Schaeffer, Schwarz, Scudder, Stokes, Strickland, and Wheeler have contributed to the knowledge of the habits of our native species. The following paragraph gives a brief résumé of the more important researches on the biology of the common species of Leucotermes in the United States.

H. A. Hagen contributed several articles on the habits of *flavipes*, and has described forms [f, g] found by the late Mr. H. G. Hubbard, of the Division of Entomology. Hubbard, many of whose unpublished notes are herein included, was the first to find royal individuals, both

---

[a] Escherich, K. Die Termiten . . . eine biologische Studie, p. 2–7, Leipzig, 1909.

[b] Holmgren, N. Termitenstudien 3. Systematik der Termiten. Die Familie Metatermitidæ. K. Svenska Vetensk. Akad, Handl., Bd. 48, No. 4, 166 p., Taf. 1–4, Uppsala & Stockholm, 1912.

Blick auf dem mutmaßlichen, stammesgeschichtlichen Entwicklungsverlauf der Termiten, p. 129–153.

Holmgren, N. Termitenstudien 2. Systematik der Termiten. K. Svenska Vetensk. Akad. Handl., Bd. 46, No. 6, pp. 86, Taf. 1–6, Uppsala & Stockholm, 1911. Ordnung Isoptera, p. 10–11.

[c] Bugnion, E. La differenciation des castes chez les termites [Nevr.]. Bul. Soc. Ent. France, 1913, no. 8, p. 213–218, April 23, 1913.

[d] Bugnion, E. Les termites de Ceylan. Le Globe: Memoires, Soc. Geog. Geneva, t. 52, p. 24–58, 1913.

[e] Loc. cit.

[f] Hagen, H. A. The probable danger from white ants. Amer. Nat., v. 10, No. 7, p. 401–410, July, 1876.

[g] Riley, C. V. Social insects from psychical and evolutional points of view. Proc. Ent. Soc. Wash., v. 9, p. 1–74, figs. 12, April, 1894.

true and neoteinic, in the United States, and his material forms an important auxiliary to later investigations.   Much of Hubbard's collecting in Florida and Arizona was done in company with Mr. E. A. Schwarz, of the Bureau of Entomology, who has since published[a] some of his observations on the habits of termites in southwestern Texas. The Rev. F. L. Odenbach, S. J., of Cleveland, Ohio, has studied the habits of our native species of termites since 1893, and has contributed many manuscript notes.   Mr. L. H. Joutel, of New York, has studied the habits of our common species[b] and has contributed some unpublished notes.   Dr. H. McE. Knower,[c] late of Johns Hopkins University, in 1894 published new and important contributions on the embryology of termites (Eutermes).   His observations are not in accord with those of Bugnion, since he determined that the nasutus develops from a worker-like larva nearly as large as a young nasutus, having 13 joints to the antenna and worker-like head and jaws.   This worker-like larva had a small head gland with no "corne frontale" on the outside of the head, although sections show essentially the same structure in the gland as that of the nasutus.   Mr. C. Schaeffer, of Brooklyn, N. Y., was the first to record[d] the presence of a fertilized true queen of *flavipes* in a colony.   The researches[e] of Dr. Harold Heath, of Stanford University, Cal., on the habits of California termites confirm the statements of Pérez and Perris that some colonizing individuals of *Leucotermes lucifugus* succeed in establishing new colonies.   Mr. Nathan Banks, of the Bureau of Entomology, has contributed several articles on our native termites, and has described a new species of Leucotermes (*virginicus*)[f] from the eastern United States.   Mr. C. L. Marlatt, assistant chief of the Bureau of Entomology, has described the distribution, life history, and destructiveness of *flavipes* in the United States and figured the castes.[g]   Mr. E. H. Strickland,[h] Carnegie scholar in economic entomology at Bos-

[a] Schwarz, E A.   Termitidæ observed in southwestern Texas in 1895.   Proc. Ent. Soc. Wash., v. 4, No. 1, p. 38–42, Nov. 5, 1896.

[b] Joutel, L. H.   Some notes on the ravages of the white ant (*Termes flavipes*).   Jour. N. Y. Ent. Soc., v. 1, No. 2, p. 89–90, June, 1893.

[c] Knower, H. McE.   Origin of the "Nasutus" (soldier) of Eutermes.   Johns Hopkins Univ. Circ., v. 13, No. 111, p. 58–59, April, 1894.

[d] Schaeffer, C.   Jour. N. Y. Ent. Soc., v. 10, No. 4, p. 251, December, 1902.

[e] Heath, Harold.   The habits of California termites.   Biol. Bul., v. 4, No. 2, p. 47–63, January, 1903.

[f] Banks, N.   A new species of termes.   Ent. News, v. 18, No. 9, p. 392–393, November, 1907.

[g] Marlatt, C. L.   The White Ant (*Termes flavipes*, Koll.).   U. S. Dept. Agr , Bur. Ent., Circ., No. 50, rev. ed., pp. 8, figs. 4, June 27, 1908.

[h] Strickland, E. H.   A quiescent stage in the development of *Termes flavipes* Kollar.   Jour. N. Y. Ent. Soc., v. 19, No. 4, p. 256–259, December, 1911.

ton, Mass., has described and illustrated a quiescent stage in the development of the nymph of the first form of *Leucotermes flavipes*, which stage was discovered by Prof. W. M. Wheeler, of Bussey Institute.

While investigating damage to the bases of telegraph poles by wood-boring insects a large fertilized true queen of *Leucotermes flavipes* was found by the writer on August 12, 1910.[a] The finding of this interesting form, usually considered to be rare in colonies of our native termites, served as an incentive to further biological study on the habits of our common species. On August 11, 1911, molting

FIG. 4.—View at Falls Church, Va., showing a portion of the treated experimental stakes under test, as to the relative effectiveness of preventives against termite attack. (Original.)

larvæ in the quiescent stage were observed by the writer for the first time, in a colony in Illinois. Special investigations were begun at Falls Church, Va., in March, 1912, to determine (1) the habits of our common termites, (2) the effectiveness of various methods and chemical wood preservatives in preventing attack by our native termites (fig. 4), and (3) the "immunity" (?)[b] or relative resistance of native and tropical species of wood.[c]

a Snyder, T. E. Record of the finding of a true queen of *Termes flavipes* Kol. Proc. Ent. Soc. Wash., v. 14, No. 2, p. 107–108, June 19, 1912.

b It is doubtful if any species of native wood of economic importance is absolutely immune to termite attack.

c Impregnation of wood to resist insect attack. Amer. Lumberman, Nov. 15, 1913

THE TERMITARIUM.

Early in March, 1912, work was begun on a large outdoor termitarium, at Falls Church, Va. The projects outlined for study in connection with the termitarium were (1) to observe the progressive development of the termite from the larva to the adult, especially in the case of the colonizing form; (2, to watch the insects when they swarmed, to determine whether this swarming is a nuptial flight, i. e., whether or not the sexes leave the nests in separate swarms and at what time copulation occurs; (3) to observe how the new colony is established, (4) in what proportion the pairs survive, and (5) the conditions in the parent colony after the swarm is over.

The termitarium was merely a large cage in which the termites could swarm under conditions as nearly similar to those in nature as possible. The cage consisted of a bottom of galvanized iron in the form of a rectilinear box sunk in the ground to the depth of a foot, a framework of wood to support the wire netting, and a wooden roof covered with tarred paper. The bottom of the cage consisted of a galvanized-iron box 10 by 6 feet and 1 foot deep, of 27-gauge galvanized sheet iron with an inch flange turned inwards. (Fig. 5, a.) This box was raised off the ground on a wooden floor of seven-eighths inch material, 4 inches wide, which was laid on 2 by 4 inch studding. There was an air space of 1 inch between the sides of the box and the earth, as well as a similar space between the sheet-iron bottom and the earth. The top of the sheet-iron box was nailed to a framework of 2 by 4 inch studding, which was supported on cedar posts, the top of the box being several inches above the surface of the ground. Loose earth was placed in the box to about half its depth, over which was spread a shallow layer of black earth (leaf mold) from the forest floor, for the purpose of retaining moisture. The wooden framework consisted of 2 by 4 inch studding, which was covered on the inside with galvanized-wire netting (14 squares mesh per inch) and reinforced at the side and ends by boards 4 inches wide by seven-eighths inch thick. The cage is about 7 feet in height, and the roof, slanting away from a shed against which it is built, acquires sufficient pitch to shed rainfall. (Fig. 5, b.)

Two chestnut logs which had been cut and allowed to season about one month before the experiment was begun were placed in the cage. One was placed on end in the dirt, the top of the log approximating the height of a stump, while the other, a tangential section, was partially buried in the ground. The bark was loosened, but was left intact on both logs. The logs were both sound, but were kept moist. A number of small decaying branches and strips of decaying chestnut boards were also laid flat upon the earth in the

Fig. 5.—a, Unfinished termitarium showing iron bottom; b, finished termitarium; c, interior view of same termitarium showing infested log in foreground, infested stump, and trap logs. (Original.)

termitarium, after a careful examination to determine that no termites were already present. The earth in the cage was kept sufficiently moist in the endeavor, in so far as possible, to approximate natural conditions. (Fig. 5, c.)

The termitarium was ready for occupancy on April 8, and a chestnut log infested with a colony of *Leucotermes flavipes* was introduced. This log had the bark on and was partially buried. On April 9 a decaying oak stump containing a colony of the same species was also placed in the cage and partially buried in the earth. Several termite colonies in logs and stumps in the forest were kept under observation at the same time, and seasoned logs and slabs with loosened bark were placed near by under conditions similar to those in the cage. The following notes are based on observations of colonies of termites in the termitarium, colonies in small tin boxes, and colonies in nature.

COMMUNAL ORGANIZATION.

SITUATION OF THE NESTS.

Termites in the eastern United States usually make their nests in decaying stumps or in logs and even small pieces of wood on the forest floor, although they also inhabit dead standing trees as well as injured living trees. They never form mounds as in the Tropics. These soft-bodied insects always conceal themselves within wood or in earth as means of protection against sunlight and their enemies, in consequence of which much of the damage they do is hidden. Termites of the genus Leucotermes are essentially wood destroyers and infest and seriously injure a great variety of crude and finished forest products which are in contact with the ground. The longitudinal excavations usually follow the grain of the wood,[a] and in the more sound wood their work is confined to the outer layers, where there is abundant moisture and incipient decay. A protective outer shell of wood is always left intact, since all except the winged, sexed, colonizing forms shun the light and are blind. Small, transverse, round tunnels which nearly pierce the outer shell are to be found when their tunnels closely approach the exterior. Sometimes the thickness of this protective shell is less than one-half millimeter. These may be soundings to see how nearly the surface is being approached, or merely unfinished excavations for the exit of the sexed adults, or possibly they may be feeding burrows. Termites often take advantage of the burrows of other wood-boring insects, enlarging and adapting them to their purposes; by these means they are

---

[a] Termite work can be readily distinguished from that of carpenter ants, whose excavations do not follow the grain. Sometimes in decayed wood, however, termites construct long, deep, but narrow, transverse galleries across the grain, forming ledges or shelves.

LEUCOTERMES VIRGINICUS.

All castes in heartwood of a maple tree infested by *Parandra brunnea.* (Original.)

Bul 94, Part II, Bureau of Entomology, U. S. Dept. of Agriculture.

PLATE IV.

DAMAGE BY TERMITES.

A book destroyed by termites, Georgetown, D. C.   (Photograph by W. R. Cline.)

DAMAGE BY TERMITES.

Pine barn sill cut into ribbons by *Leucotermes* sp. at Mayfield, Kans.   (Original.)

DAMAGE BY TERMITES.

Living stag-headed chestnut tree, 50 feet in length, Falls Church, Va. *a,* Complete length of tree showing stag top and lightning scar on side; *b,* view of interior showing heartwood completely honeycombed , *c,* view of north side of tree showing width of scar, and honeycombed interior   (Original.)

able to penetrate more rapidly to the heartwood and honeycomb the interior.

Termites quickly disintegrate the wood of dead trees and stumps, which soon becomes converted to humus, the rapidity with which this is done depending on the relative resistance of the species of wood. This beneficial rôle in nature, however, is offset by the enormous destruction they accomplish in rapidly rendering insect, fire, and disease killed timber unmerchantable and by the damage they inflict to the roots and trunks of injured living trees. Termites will infest the heartwood of living trees injured at the base by fire, disease, or other insects (Pl. III), and sometimes in such trees they excavate upward, throughout the dead heartwood, longitudinal tunnels, irregular in diameter, the sides of which are lined with earth mixed with excrement. These insects also infest the roots of living trees, finding ingress through abandoned burrows of the large, roundheaded (Prionid) borers. Sometimes they girdle young trees—forest-tree nursery stock, for example—eventually cutting the trees off near the ground, examination disclosing that the stems were honeycombed. This is not necessarily due to the presence of dead wood near by, since termites will tunnel for long distances underground. While usually confining their work to moist or decaying timber or to vegetable material of any sort, and to books (Pl. IV) and papers that are somewhat moist, termites will attack seasoned, dry wood, provided there is access to moisture elsewhere; i. e., they use moist frass and earth in extending the burrows, thus creating more favorable conditions. In the Southern States termites will infest the bark and outer layers of the wood of the base of yellow pines killed by barkbeetles before the foliage has all fallen; trees that have been killed in the spring and show reddish-brown needles and much fallen foliage being infested by the middle of August. Trees killed in the spring will also have the outer layers of wood of the base honeycombed by the following December. (Fig. 6.) The larger-celled, thin-walled spring wood is eaten away first, leaving the smaller-celled, harder summer wood uneaten. (Pl. V.)

Where the heartwood is decayed in a standing living tree termites will work for a considerable distance above the ground, completely honeycombing the heartwood. In a living chestnut tree at Falls Church, Va., with the decayed heartwood exposed in a long scar, termites had infested the heartwood to a height of from 45 to 50 feet above the ground. The outer shell of living sapwood was intact. (Pl. VI.)

Termites are quite effective in clearing fields of old snags and stumps, but this benefit is offset by the damage they do to posts and buildings.

Termites (*Leucotermes* spp.) also inhabit subterranean passages. Drummond has compared tropical species of termites to earthworms and declared that they are equally as beneficial to man in aerating the soil.   After swarming, many of the sexed adults excavate shallow cells in the earth under.small pieces of decaying wood, and later enter the wood.   The royal cell is constructed in decaying wood or in the earth slightly below the surface of the ground.   Termites usually infest wood by entering from the ground underneath, rather than directly on the exposed surface, the latter being usually the habit at the time of the swarm.   (*Termopsis angusticollis* Walk. usually infests wood by gaining ingress through wounds and abrasions.)   None of the sound traplogs with loose bark in the termitarium or in the forest was infested except at the point of contact with the ground, but termites in pairs have been found under loose split bark on decaying logs where more moisture was present.   Workers and soldiers are frequently to be found in the spring in small pieces of decaying wood lying on the ground, and termites probably extend old colonies or establish new colonies by means of subterranean tunnels.   During the winter the members of the colony are to be found in a labyrinth of undergound passages.

FIG. 6 —Work of termites in insect-killed southern yellow pine. Tree killed in the spring; wood at base honeycombed by following December. Spartanburg, S. C. (Original.)

These excavations are of varying size and shape, and extend in all directions.   Some of the tunnels are partitioned off into separate chambers, while others are unpartitioned runways.   In the main runways the very young are absent.   The partitions consist either of uneaten portions of the wood or small conical piles of moist earth mixed with frass (excreted, finely digested wood) of clay-like consistency.   Sometimes in broad, shallow channels a small irregular mass nearly blocks the channel.   The sides of the channels are smooth, and the uneaten masses of wood which serve as barricades appear as little islands and are distinct because of the rough appearance due to the pores and cell walls of the wood. The walls of these channels are spotted with little piles of finely digested, excreted wood, giving the wood a characteristic mottled appearance.

Large cavities encountered by termites when working in the wood of logs, poles, or trees with decayed heartwood or hollow core are usually filled up with moist earth mixed with frass, the whole having a clay-like consistency and a conglomerate appearance due to the irregular deposits of excreted, finely digested wood. The "doty" hollow cores in the bases of infested poles are filled up in this manner.

In all their operations the termites carefully wall up and conceal themselves. Usually there is but little evidence on the exterior of infested wood to indicate the presence of termites, and they may not be detected until the interior is .completely honeycombed. Sexed adults swarming from infested buildings are always a warning of their presence. Again, the outer surface is covered over with longitudinal viaducts of small diameter constructed of earth (Pl. VII, figs. 1 and 2), and whereas the interior channel is smooth, the exterior has a rough granular appearance. These viaducts are resorted to in order to extend the colony or to reach some object, such as decayed wood, or when their pathway crosses some impenetrable substance, the object, of course, being always protection from the light—all except the winged sexed adults shun the light—and their numerous enemies. Viaducts, or "sheds," can be seen running up to a considerable height above the ground in the longitudinal weathering checks on poles and posts, as well as between the crevices in the bark of infested dead and living trees, and uncovered viaducts are found under the loose bark on dead trees. Viaducts in the interior of hollow-cored poles or trees may be of large diameter and may consist of several irregular longitudinal interior channels.

Termites sometimes resort to another type of viaduct, which for descriptive purposes may be called suspended tubes. On May 26, 1912, at Elkins, W. Va., Dr. A. D. Hopkins found termite tubes of earth, 5 to 6 inches long, hanging from the end of a Virginia scrub pine sapling which had fallen, leaving the broken base 2 feet from the ground. The termites had evidently infested the base of the sapling through the ground before it fell and were trying to make connection with the ground by means of these suspended tubes. (Figs. 7 and 8.)

These covered viaducts or sheds, the uncovered viaducts, and the tubes—all constructed by termites of earth and excreted wood—are fragile.

NUMBER OF INDIVIDUALS IN COLONIES.

Young colonies of *Leucotermes flavipes*—that is, colonies but recently established, in decaying wood or in the ground under decaying wood, by sexed couples that have swarmed—are small, and since the rate of egg-laying by the young queens is remarkably slow, the increase in numbers is also correspondingly slow. Observations of such incipient colonies in the spring of 1912 and 1913, after the swarm, the time of which varies with the season, from the middle of June to early July, indicate that from 6 to 12 eggs are normally to be found with pairs of *flavipes*. While the brood first hatched is relatively small, contrary to the habits in the other social insects coition is repeated and the young colony gradually increases in numbers.

On April 25, 1912, at Falls Church, Va., a small colony or sub-colony of *Leucotermes flavipes*, the branch of a larger colony, was found under a small chestnut slab sunken in the ground. The day was warm and bright and many members of the colony were congre-

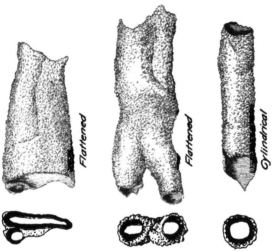

FIG. 7 —Suspended tubes, constructed by termites of earth and excreted wood.
(Original.)

gated under the slab above ground. An attempt was made to capture the entire colony in order that the relative proportions of the various castes might be ascertained. Although nymphs of the first form constituted the most abundant caste, some few transforming nymphs and a few sexed adults were present, as well as a few workers and soldiers. Workers constituted the second most abundant caste. Nymphs of the second form appeared more active than the nymphs of the first form. The following possibly is not a fair statement, either of the number of individuals in the colony or the relative proportions of the different castes, since (1) the nymphs were probably above ground to take advantage of the warm afternoon sun which would aid their development—a common procedure; (2) only a few workers and soldiers were necessary to attend them and no doubt many more were still in the subterranean passages; (3)

WORK OF LEUCOTERMES VIRGINICUS.

Maple tree on Plummers Island, Md., infested by *Parandra brunnea* and by termites; showing shells constructed by termites to cover up galleries exposed by sawing. (Original.)

many escaped and the workers would escape more easily than the nymphs; (4) probably, unconsciously, more effort was made to capture nymphs and soldiers than workers. The following figures, however, will probably approximate the relative abundance of castes. Nymphs of the first form, 279; nymphs of the second form, 86; nymphs of the first form in the quiescent stage, 31; individuals in the stage following, that is, with wings unfolded and held away from the body, 3; immature sexed adults without pigmentation, 17; immature sexed adults with gray pigmentation, 26; nearly mature adults, 5; nymphs of the second form molting, 4; workers, 93; soldiers, 24; the total being 568. Nymphs of the second form were only one-fourth as numerous as nymphs of the primary form, and, including all stages to sexed adults, gave a count of 90 to 361, respectively.

On March 22, 1913, at Black Mountain, N. C., nymphs of the second form (*flavipes*) taken in a small decaying oak branch on the ground greatly outnumbered nymphs of the first form, although usually the latter is by far the more numerous form.

Fig. 8 —Broken-off pine sapling from basal end of which tubes in figure 7 were suspended by termites toward stump. (Original)

Due to the wandering habits of species of this genus, it is difficult to estimate the size and extent of an old, well-established colony, which may branch out over several acres of ground.[a] However, the number of individuals in well-established, permanent colonies probably runs up into the ten thousands, since from 5,000 to 10,000 (estimated) eggs, scatteringly or in clusters the size of a pea, were found in a large colony of *Leucotermes virginicus* near Chain Bridge, Va., on June 19, 1913. This colony, which was in a large decaying black oak log, consisted of a large number of workers and soldiers, and numerous larvæ.

THE DIFFERENT CASTES—POLYMORPHISM.

In a termite colony there are several different forms, or castes, of mature individuals, as well as those of different castes in the various stages of development. The castes are the workers, the soldiers, the

---

[a] The spreading out of a colony is largely due to increase in numbers and consequent need of fresh supply of food; that is, decaying wood.

colonizing winged adults of both sexes, the supplementary or neo-teinic[a] reproductive forms (often in large numbers), both "ergatoids" and nymphal "neotenes," and the single true royal pair. Besides these mature forms there are freshly hatched, undifferentiated larvæ, differentiated larvæ, and nymphs of the first and second forms. Of course, all these forms are not present in a colony at the same time, as there is seasonal variation.

Workers are developed from larvæ that will not mature the sexual organs, but, unlike the bees, are of both sexes. They are dirty white in color, are large-headed and soft-bodied, with 10 segments to the abdomen, and at maturity are approximately 5 millimeters in length in *flavipes* and 3.5 millimeters in *virginicus*. The antennæ consist of 15 to 17 segments exclusive of the base in *flavipes* and 13 to 15 in *virginicus*. The workers constitute the most injurious wood-destroying form and have well-developed mandibles. The left mandible has five pointed teeth, the fifth tooth with a broad base, and the inner margin having parallel carinæ. The right mandible has two pointed teeth, the third and fourth teeth being broad, and the fourth with parallel carinæ. The mandibles evidently have both tearing and rasping functions. The labrum is rounded. The worker termites possess a large intestinal paunch and the contents enable the outline of this paunch to be seen through the tissue of the abdomen. Workers are blind.

Soldier termites (fig. 9, *a*) are more highly specialized workers, being also developed from large-headed workerlike larvæ that will not mature the sexual organs, and the caste is represented by both sexes. While they are soft-bodied, the head, which is pigmented yellowish-brown, is chitinized and is more oblong and elongate than in the worker, tapering slightly toward the apex, or being slightly broader at the base. The mandibles, which are enormously developed, are long, slender, saberlike, with no marginal teeth, chitinized, and of a yellowish-brown color. The body is of a dirty white or pale yellowish color. The labrum of the soldier is more narrow than that in the worker, elongate, and subelliptical. The "menton," which is chitinized, is convex, elongate, and more slender in comparison with the worker. Mature soldiers are from 6 to 7 millimeters in length in *flavipes* and 4.5 to 5 millimeters in length in *virginicus*. The antenna has from 14 to 17 segments in *flavipes* and 15 segments in *virginicus*. The soldiers, as well as the sexed adults, of these two

---

[a] Grassi, B., and Sandias, A. Op. cit., p. 249: "The term neoteinia has been introduced by Camerano (Bul. Soc. Ent. Ital. [v. 17], 1885, pp. 89–94) to denote the persistence during adult life of part or all of the characteristics normally peculiar to the immature, growing, or larval stages. * * * Neoteinia, or the persistence of larval characteristics, does not necessarily imply that anticipation of sexual maturity which is usually connoted with the use of the term pædogenesis, which, moreover, is strictly applied to agamic reproductions."

species can be differentiated. The soldiers, like the workers, are
wingless and blind. Individuals of both castes complete their development in less than a year.

The nymphs of the reproductive form develop from larvæ that will
mature the sexual organs. The term "nymph" is applicable,

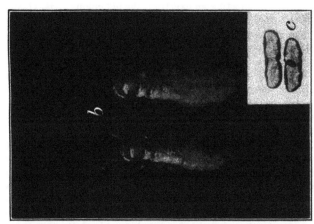
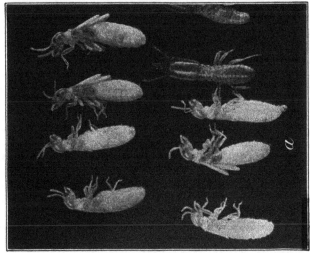

FIG. 9.—*Leucotermes virginicus*: a, Nymphs of the first form, and soldiers, Maryland, May 18, 1912; b, young nymphs, Maryland, May 18, 1912; c, view of brownish-black bands on dorsum of the abdomen of worker (enlarged 20 times); these bands occur on both *flavipes* and *virginicus*. (Original.)

according to Lespès and Hagen, to individuals "with the wing-
rudiments easily distinguishable to the naked eye."[a] Individuals
the wing-rudiments of which are present, but only distinguishable
when under magnification, are termed larvæ. Nymphs are white

---

[a] Grassi, B., and Sandias, A   Op. cit., p. 264.

and soft-bodied, and when fully developed and ready to molt for the last time are from 7 to 7.5 millimeters in length in *flavipes* and 4.5 to 5 millimeters in *virginicus*. In *flavipes* the antenna consists of from 16 to 17 and in *virginicus* from 14 to 15 segments. The mandibles are practically the same as in the worker, and are probably very necessary in effecting an exit from the old colony. R. D. Grant states [a] that in a Missouri Pacific Railroad Co. engine house the rafters were injured and the cement of the brick walls built 14 years previously was perforated. Mr. C. L. Marlatt has specimens of plaster which was laid on metal lathing in a building at Charlotte, N. C., which had been mined in order to allow the winged adults to escape from heavy wooden beams which had been honeycombed. Also, in the establishment of the new colony the young royal couple have the excavating to do.

There are two forms of nymphs (Pl. VIII, figs. 3, *a*, *b*), namely, the primary form, with elongate wing pads, that develops into the winged sexed adult, and the "second form" (Lespès), with short wing pads—mere buds—which represents an arrested early stage of the nymph of the primary form, or even a larva. Nymphs of the secondary form are slightly more elongate, and develop the sexual organs without progressing further, instead of completing their normal development to the winged, pigmented, sexed adults that swarm. These nymphs, after becoming sexually mature and attaining a straw-colored pigmentation—normally after the nymphs of the primary form have developed to sexed adults—become supplementary royal individuals, kings, and queens, but never (?) leave the parent colony. They do not possess functional eyes.

The sexed individuals, when ready to swarm, are castaneous-black or light brown in color, have two pairs of long, filmy wings, and are so chitinized that they can bear full sunlight. They possess both functional compound eyes and simple eyes or ocelli. The body—excluding the wings, which are slightly longer than the entire insect and project some distance beyond the abdomen when "in situ"—is slightly less in length than in the case of the nymphs. The entire body of the sexed individual is from 9 to 10 millimeters in length in *flavipes* and from 7.5 to 8 millimeters in length in *virginicus*. There are from 16 to 18 segments to the antenna in *flavipes* and from 15 to 16 antennal segments in *virginicus*. The sexes are as easily distinguishable as in the case of fully developed nymphs. "(1) The seventh [abdominal sternite] (the apparent sixth) is strongly developed and semicircular, with the rounded edge posterior in the female and very short in the male. (2) The eighth sternite, which is reduced to two lateral lobes in the female, is small and entire in the male.

---

[a] Grant, R. D. Jour. Proc. Acad. Sci St. Louis, v 3, p. cclxix, November 19, 1877.

FIG. 1.—*a*, Lateral view of fully developed nymph of primary form; *b*, lateral view of neoteinic king—compound eyes without pigmentation.

FIG. 2.—Dorsal view of same: *a*, Showing pubescence and tapering abdomen of neoteinic king; *b*, nymph of the primary form.

FIG. 3.—*a*, Fully developed nymph of the first form with elongate wing pads before final molt; *b*, nymph of the second form with short wing pads before final molt; *c*, neoteinic king after final nymphal molt. (Original.)

STAGES OF LEUCOTERMES FLAVIPES.

(3) The ninth sternite nearly resembles the eighth."[a] The genital "appendices" in the male are attached to "what is apparently the eighth, but is really the ninth sternite, the first being fused with the metasternum."[b] There are two segmented appendages, or cerci, attached to the abdomen. After swarming, the sexed adults become royal individuals.

Larvæ, or young, are undifferentiated individuals which, after further development, attaining chitinization and pigmentation, change to differentiated individuals. The larvæ, like all the other stages of termites, are active. When young or freshly molted the individuals are more transparent and white and the segmentation of the body at this stage is more sharply defined. "The younger the individual, or the more recent its ecdysis, the thinner is the chitin."[c] The term larva is applied (1) to any individual which has not yet attained full size, mature chitinization, and pigmentation; (2) in the case of the sexed individuals, to those with the wing-rudiments not easily distinguishable to the naked eye; and (3) in case of the soldier, to the large-headed, undifferentiated, workerlike forms.

The eggs are white, slightly reniform, and those laid by true queens (queens that have swarmed) are approximately 1 millimeter in length in *flavipes*. The eggs are usually found in clusters or scattered singly in the galleries.

THE SENSE ORGANS.

Termites are essentially subterranean in habit and in consequence all castes of Leucotermes are blind except the colonizing individuals. The soldiers have compound eyes, but without pigmentation; in some neoteinic royal individuals the pigmentation of the compound eye is not visible to the naked eye (Pl. VIII, fig. 2, *a*), but in most cases there is a slight pigmentation of variable intensity. All castes except the colonizing individuals shun the light.

Although blind, termites are known to possess other sense organs.[d] The antennæ are important tactile sense organs, and often individuals may be seen feeling their way by means of these appendages. The antennæ of the colonizing individuals are pitted, and from these pits, which appear as white depressions on the pigmented antennal segments, prominent hairs arise. A. C. Stokes, in an article entitled "The sense organs on the legs of *Termes flavipes* Koll.,"[e] describes

---

[a] Grassi, B., and Sandias, A. Op. cit., p. 306.

[b] Ibid , p. 271.

[c] Ibid., p. 256.

[d] Müller, Fritz. Beiträge zur kenntniss der Termiten. Jenaische Ztschr, Bd. 9 (n. F. Bd. 2), p. 241–264, pl. 10–13, Mai 8, 1875. See p. 254, pl. 12, figs. 32, 34.

[e] Stokes, Alfred C. The sense-organs on the legs of our white ants, *Termes flavipes* Koll. Science, v. 22, no. 563, p. 273–276, illus., November 17, 1893.

and figures "sensory hairs"; "sensory pits"; "tibial spurs with prominent basal apertures, across which extends a delicate membrane (auditory organs?)"; "pilose depressions"; "surface markings"; etc. Grassi [a] mentions tactile, "very long, fine, readily vibratile hairs" on the body, and states that the cerci also appear to be "essentially tactile." It is believed that there is a relation between the convulsive movements frequently observed, that is, the sudden jerking of the whole body, and these sense organs, and that individuals are thus enabled to communicate, or at least give danger or distress signals. The convulsive movements made by the workers and soldiers, when the royal pair are disturbed in the cell, are very violent and indicate great agitation.

There is a characteristic musty or acrid odor which can be easily detected in colonies of Leucotermes, and individuals frequently can be seen to follow directly in the path taken by others, but as termites usually travel in well-worn channels this may be due to tactile sense alone.

THE FUNCTIONS OF THE CASTES.

The social economy of termites is somewhat similar to that of other social insects, although in many respects totally different. Undifferentiated young hatch from the eggs, are active, and in turn transform to the differentiated individuals of the castes after a series of quiescent stages and molts.

As the name implies, the workers are those individuals that make the excavations, extend the colony, and care for and protect the royal couples and young.

The soldiers, more highly specialized workers, are of less importance functionally than the workers—just as the anther transformed to the petal in the common pond lily, *Castalia* sp., is less important functionally than the other anthers—yet both serve a purpose. Just before the time of swarming, the members of colonies become restless, and as the sexed adults emerge numerous workers and soldiers congregate on the outskirts of the colony near the exit holes with heads toward the exterior. The duty of the soldiers is apparently entirely protective, but they do not appear to be very effective, at least when the colony is opened and they are exposed to the attack of ants, etc.

The pigmented, winged, sexed adults, developed from nymphs of the first form, are the colonizing individuals which swarm in enormous numbers in the spring and found the new colonies, and when finally established at the head of a new colony, which they have reared, they become the royal couple. Unlike other social insects, the male continues to live, even after the fertilization of the female, and the king and queen inhabit the royal cell together, there

---

[a] Grassi, B., and Sandias, A. Op. cit., p. 267–269.

being repeated coition. Termites are sometimes polygamous, at least in incipient colonies. The young queens care for the young and carry the newly hatched larvæ in their mouths to safer places in the nest when the colonies are disturbed. (Nymphs of the first form also have been seen performing this duty in old colonies.) The workers attend the larvæ in old colonies.

In the life cycle of termites, however, there is so much variation in the development of the castes that it leads to a rather complex life history. Young are kept in a retarded, undifferentiated state and can speedily be turned into substitute reproductive forms of both sexes. These neoteinic royal individuals are used (1) as substitutes for the true royalty that have swarmed and (2) in splitting up the old colony into new and independent colonies. Fully developed nymphs of the second form are to be found in colonies in the spring. They never complete their normal development, which consists of the acquisition of wings and the mature pigmentation, but assume the characteristic pale straw color after the final molt. The sexual organs are developed after the acquisition of pigmentation. They become neoteinic reproductive forms, but always (?) remain in the parent colony.

Each caste has a distinct function, there being a well-defined division of labor, which, however, is not strictly adhered to. Termites properly represent the phenomenon of polymorphism, i. e., "species in which one or both of the sexes appear under two or more distinct forms,[a]" both sexes being equally polymorphic.

THE LIFE CYCLE.

THE METAMORPHOSIS—CASTE DIFFERENTIATION.

The metamorphosis of termites, while formerly considered to be a contrastingly simple type of insect development, is in reality very complex; indeed, there may be said to be different types of development for the castes. In the development of the worker, as in the order Thysanura, there is no external change or metamorphosis, the freshly hatched worker larva being active and of the same form as the adult worker. In the development of the soldier, however, marked changes in form occur, the mature soldier, with pigmented head and saberlike mandibles, being developed from a large-headed, white, workerlike larva (Pl. IX). The winged, pigmented, sexed adult is developed from a small-headed, white larva, there being a still more radical change. However, what may popularly be termed the "antlike" form can be distinguished in all stages in the development of all the castes, hence termites can hardly be classed with insects with "complete metamorphosis' as the true ants, where there

---

[a] Wheeler, W. M. Concerning the polymorphism of ants. Bul. Amer. Mus. Nat. Hist., v. 23, p. 50–93, 6 pls., January, 1907. See p. 50.

are the three successive stages of larva, pupa, and adult, although termites pass through pupalike "quiescent" stages in the molting of larvæ and nymphs, with temporary periods of inactivity. Furthermore, great variation is possible in the metamorphosis, as is seen in the development of "neoteinic" or "supplementary" reproductive forms, as the "neotenes" and "ergatoids." Therefore, in general, termites may be classed with insects having "incomplete metamorphosis," as the locusts and roaches. In this type of development the form of the body is always essentially the same as that of the adults. The development of the larva and nymph to the winged sexed adult termite greatly resembles the development of the locusts and roaches.

In the genus Leucotermes there is no apparent metamorphosis in the case of individuals of the worker caste. The freshly hatched worker larva is active, there being a very simple development. The

FIG. 10.—*Leucotermes flavipes:* Quiescent stage of molting larvæ. Enlarged 10 times.
(Original.)

change from larva to adult is a gradual growth, the white, large-headed larva passing through a series of quiescent stages and molts. The adult worker is of a dirty grayish-white color. In the soldier caste and the reproductive forms the metamorphosis is more complicated, and marked changes in form occur during the development. Young molting larvæ in the "quiescent stage"[a] were first observed by the writer on August 11, 1911, near Jerseyville, Ill., in a large, wide, longitudinal channel in the decaying heartwood of the butt of a white cedar telegraph pole.[b] Previous to the molt the larva falls over on its side and passes through a quiescent stage, the head being bent down to lie on the ventral side of the body along which the antennæ and legs, also, lie extended in a backward direction. This gives the larva the appearance of being doubled up (fig. 10). After the skin has been shed the larva resumes its normal activity.

[a] Strickland, E. H. Loc. cit.
[b] Snyder, T. E. Changes during quiescent stages in the metamorphosis of termites. Science, n. s., vol. 38, No. 979, pp. 487–488, October 3, 1913.

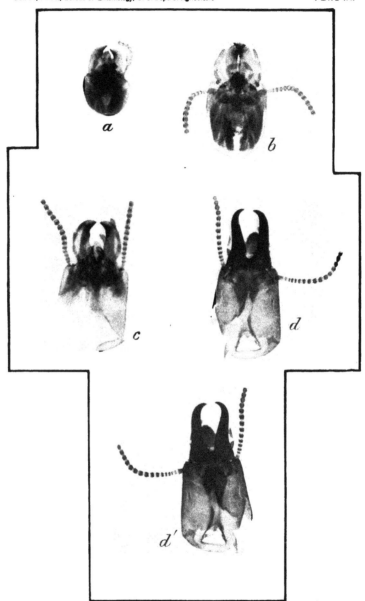

LEUCOTERMES VIRGINICUS AND L. FLAVIPES.

Molting soldier nymphs: *a, L. virginicus,* head showing mandibles and labrum of soldier nymph just molted from workerlike larval form; *b, L. flavipes,* same at later molt, *c, L. flavipes,* same at later molt, *d, L. flavipes,* mature soldier; *d',* ventral view showing convex, slender, chitinized, pigmented "menton." Photograph from specimens in balsam slides. Enlarged 16 times. (Original.)

The antennæ increase in the number of segments by the subequal division of the third segment [a] independent of the molts in L. lucifugus Rossi.

In the metamorphosis of the soldier, as has been shown by Grassi [a] (Calotermes flavicollis Fabricius and Leucotermes lucifugus Rossi), Knower [b] (Eutermes sp.), and Heath [b] (Termopsis angusticollis Walker), there is a radical change from a large-headed larva to the soldier caste. This change takes place during [c] a quiescent stage. The worker, like the soldier, also develops from a large-headed larva; that is, in the colony there are two types of larvæ, the large-headed and small-headed, the former normally developing to workers and soldiers, while the latter become the reproductive forms. According to Grassi there are four molts in the development of the asexual forms of Leucotermes lucifugus, whereas the sexed forms pass through five molts. Holmgren states [d] that in Leucotermes the soldiers are polymorphic and originate [e] from at least five different larval stages, while the polymorphic workers originate from three different larval forms. The undifferentiated larvæ present possibilities of development in all directions, but the development of Leucotermes shows that the breeding possibilities, while great, are by no means so great as in Calotermes. This would indicate progressive development toward stability, as is also indicated by the absence or rarity of neotenes and ergatoids in the more highly developed termites (Termes and Euter-

[a] Grassi, B., and Sandias, A. Op. cit., p. 270.

[b] Op. cit.

[c] Snyder, T. E. Loc. cit.

[d] Holmgren, N. Termitenstudien 3. Systematik der Termiten—Die Familie Metatermitidæ. K. Svenska Vetensk. Akad. Handl., Bd. 48, No. 4, pp. 166, 4 pls., Uppsala and Stockholm, 1912. Blick auf dem mutmasslichen, stammesgeschichtlichen Entwicklungsverlauf der Termiten, p. 129-153.

[e] The recent discovery by McClung, Stevens, and Wilson (Wilson, E. B., Heredity and miscroscopical research: Science n. s., v. 37, No. 961, p. 814-826, May 30, 1913) of the association of an odd number of chromosomes, in the divisions of the spermatocytes—that is, cells formed by the division of the "spermospore," the male germinal cell—with the determination of sex may also be applicable to caste differentiation in termites. Wheeler (Wheeler, W. M., The polymorphism of ants, with an account of some singular abnormalities due to parasitism: Bul. Amer. Mus. Nat. Hist., v. 23, p. 1-93, pls. 6, Jan., 1907), however, states, with reference to ants, that nourishment, temperature, and other environmental factors merely furnish the conditions for the attainment of characters predetermined by heredity, that is, with Weismann he believes that the characters that enable us to differentiate the castes must be represented in the egg, but with Emery he believes the adult characters to be represented in the germ as dynamic potencies or tensions rather than morphological or chemical determinants. Holmgren states (op. cit., p. 140) that in termites, as the result of the method of feeding, three potential germ plasms are released in at least three directions * * * and that there must be a germ plasm correlation which finds its expression in the caste correlation.

mes), wherein, however, there may be several true queens [a] in the one colony and where polygamy exists. In insects supposed to represent the most primitive, or lowest and least developed, types this is a rather complex metamorphosis. Here also there is much variation in the life cycle and no strict rule is followed. (See chart.) Apparently the individual development is entirely subservient to the needs of the colony. This ability of adaptation of individual to circumstance leads to a complex economy.

A representation [1] of some of the successive stages in the development of the various forms or castes in the life cycle of Leucotermes flavipes Kollar as found in colonies in the eastern United States. All these reproductive forms are not present in the same colony at the same time.

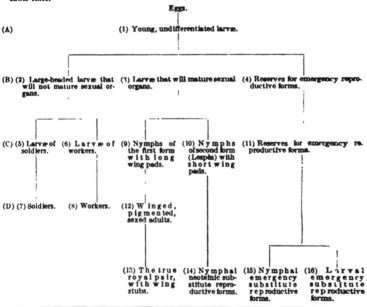

[1] See correlated forms of *Leucotermes lucifugus* Rossi. Grassi, B., Ein weiterer Beitrag zur Kenntnis des Termitenreiches. Zool. Anz, Bd 12, No. 311, pp 355-361, July 8, 1889. Übersicht der im Termitenstaate vorkommenden Formen- *Termes lucifugus*, p. 36[1]), id., Ein weiterer Beitrag zur kenntniss der Termiten-reiches. Ent Nachr., Jahrg. 15, No 14, pp. 213-219, July, 1889; Holmgren,N ., Termitenstudien 3. Systematik der Termiten- Die Familie Metatermitidæ. K. Svenska Vetensk. Akad. Handl., Bd. 48, No. 4, p. 148, Scheme B, Uppsala & Stockholm, 1912 (table showing parallel evolution and at what molts changes or development occur).

PROGRESSIVE DEVELOPMENT OF THE NYMPHS.

Colonies of *Leucotermes* spp. in the northeastern United States are dormant during the winter, the insects retiring to the more remote

[a] John, O. Notes on some termites from Ceylon. Spolia Zeylanica, v. 9, pt. 34, p. 102-116, 1913.

Riley, C. V. Termite economy. Proc. Biol. Soc., Wash., v. 9, p. 71-74, April, 1894.

Escherich, K. Termitenleben auf Ceylon, Jena, 1911, p. 45-46.

galleries in the wood or to the subterranean passages of the nests, At Falls Church, Va., in 1912 it was not until March 11, in 1913. February 20, and in 1914, March 23 that signs of activity were observed in colonies of *flavipes*. Nymphs of the first form with "short wings" (Fritz Müller) or "wing pads" (Hagen) were present on March 11, 1912. At this time the wing pads were well developed, being about two and one-half times the length of the segment from which they originated, and had a yellowish tinge. The antennæ consisted of from 16 to 17 segments, excluding the base, and the line of demarcation between the basal segments was less distinct. On March 29, 1912, the ocelli were visible in the nymphs; antennæ, head, and thorax were acquiring a tinge of yellowish-brown, and the compound eyes were becoming pigmented—a reddish-brown. On April 18, 1912, and April 8, 1913, at Falls Church, Va., nymphs of the first form when fully developed and ready for the final molt could be readily distinguished by the opaqueness of the elongated wing pads, the filmy, yellowish-brown, loosening skin, which becomes separated from the body, particularly posteriorly, and the reddish-brown pigmentation to the compound eyes. (Pl. VIII, fig. 1, *a*.) The nymphs gradually increase in size and pass through a series of molts and quiescent stages until the final molt, when the wings are unfolded. Packard, who figures [a] the stages in the growth of the wing in *flavipes*, states that the wings are simply expansions. During the latter part of April, 1912, nymphs with short wing pads, or those of the second form (Lespès) (Pl. VIII, fig. 3, *b*), were found in colonies. Their antennæ had from 16 to 17 segments. These nymphs appear to be more active than nymphs of the first form and have but slight pigmentation of the compound eye.

During the final molt—which occurred from April 18 to 27, 1912, April 8 to 17, 1913, and April 22 to May 2, 1914, at Falls Church, Va., in case of the nymphs of the first form (*flavipes*)—nymphs of both the first and second forms pass through a "quiescent stage" [b] (Pls. X and XI), which closely approaches the pupal stage of insects with complete metamorphosis. This quiescent stage apparently serves the same purpose as the pupal stage, since the most marked changes, both external and internal, take place during this molt; it is, however, of short duration. On April 18, 1912, the first nymphs (*flavipes*) in this stage were observed in the outer layers [c] of a decaying stump. All of the nymphs in this colony were fully developed, but very few had

---

[a] Packard, A. S. A textbook of entomology. New York, 1903, p. 140.

[b] Strickland, E. H. Loc. cit.

[c] Larvæ or nymphs in the quiescent state are usually to be found isolated in small but deep, transverse conical nitches or shelves in the nest where they are not liable to disturbance by the movements of the other members of the colony. Possibly they seek out such secluded places, usually near the outlying galleries, beforehand, or may be carried there while helpless, by the workers. Clusters of eggs are also found in similar nitches.

yet begun to molt.   A large number were placed in a covered tin box 3½ by 5 by 1½ inches, with disintegrated wood and moist earth in the bottom.   This was to be a check box, and the conditions here were more natural than in the small corked vials, in which over 100 other nymphs had been kept in the dark, one nymph to each vial.   The development of each nymph was watched and the time necessary for the various changes from nymph to adult was noted.   On June 8 fully developed nymphs of *virginicus*, with opaque wing pads, were observed to molt in a similar manner.   Nymphs in all stages described for *flavipes* were observed molting until June 11.

As E. H. Strickland has already figured and described these changes in the nymphs of the first form (*flavipes*), his accurate description is here quoted in detail, with a few comments.   Until this description appeared, the quiescent stage in the development from nymph to adult for this species was undescribed.   His observations were made from nymphs collected in the neighborhood of Boston, Mass., May 7, 1910.

The mature nymph becomes very sluggish and finally all movement ceases; it then falls over on its side and the head is bent down till it lies on the ventral side of the body, along which also the antennæ and legs are extended in a backward direction, * * * while the wing pads are bent downward till they lie laterally along the sides of the body. * * * It will be at once noticed that while in this position the nymph is to all appearances a quiescent "pupa libera." There does not appear to be an ecdysis immediately prior to this quiescent period, however, so I would hesitate to describe it as a true pupal state though it undoubtedly has the same physiological function.

This quiescent stage lasted in the few specimens observed for a period varying from four to about nine hours.[a] The duration in time seems to be controlled to a large extent by the amount of moisture in the earth surrounding the pupa for when specimens were placed in perfectly dry earth they were unable to pass beyond this stage of development,[b] while the greater the amount of moisture, the shorter the period. During this stage the last nymphal skin splits across the head and along the dorsum, and is slowly worked downward and backward till a large portion of it hangs freely from the apex of the abdomen on the ventral side. The legs are the last part [c] of the body to be freed from this skin, which then becomes detached as a much crumpled mass. As soon as the wings are liberated they begin to move away from the body at their base. This is apparently due to the tracheæ in the basal portion of the wing becoming inflated. The inflation, however, does not extend beyond the suture along which the wing is subsequently broken off, and the distal portion remains tightly folded. * * *

The ecdysis described above is the last in the development of the imago, for the insect now disclosed is the sexually complete[d] adult; it does not, however, become

----

[a] Varying from about 3½ to 12 hours at Falls Church, Va.—T. E. S.

[b] A considerable amount of moisture apparently is essential to normal development. Specimens placed in small, individual, corked vials molted normally, while others placed in vials the mouths of which were lightly plugged with cotton developed abnormally, with distortions, or not at all.—T. E. S.

[c] Sometimes the antennæ are the last part to be freed of the cast skin in case of nymphs of both the first and second forms.—T. E. S.

[d] At this stage the sexual organs are not yet fully functionally matured.—T. E. S.

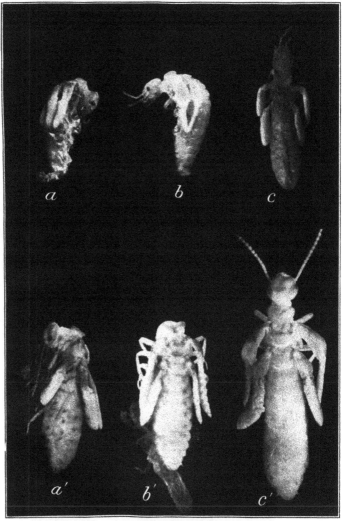

LEUCOTERMES FLAVIPES.

a and *b*, Lateral views of the quiescent stage of nymphs of first form; *c*, dorsal view of active
molted nymph showing how wings are held away from the body.　Photographs from etherized specimens.　*a'*, Lateral view of quiescent stage of nymphs of the first form (skin cast);
b', dorsal view of same; *c'*, active molted nymph of first form with wings unfolding.　From
alcoholic specimens.　(Original.)

LEUCOTERMES FLAVIPES.

a, Quiescent stage of molting female nymph of second form; b, molted male nymph of second form or neoteinic king—cast skin still attached to legs; c, mature neoteinic king not sexually mature and just molted. (Original.)

active as soon as it emerges but remains for about a quarter of an hour in the same position as that in which the ecdysis occurred. During this time, however, the head is slowly drawn upward to its normal position and the insect finally struggles to its feet. Its movements are at first very awkward and uncertain but after a few minutes it is actively running about. As before mentioned, the greater portion of the wings remains closely folded together so that at first sight they appear as abnormally placed wing-pads. A close examination with a hand lens shows them to consist of the very compactly folded wing. * * *

After these young adults have been running about for an hour or so the main portion of the wing begins to expand; the basal portion becomes expanded before the apical part begins to unfold, but the inflation gradually works toward the apex till the typical fully winged though pigmentless adult is produced. The wings continue to be held away from the body till this process is complete, after which they are folded from the base in an overlapping position over the abdomen. The ensuing pigmentation of the body is gradual[a] and does not appear to be affected by the presence or absence of light; the entire body turns black through shades of yellow and brown till in about twenty-four hours the sexually complete imago is ready for swarming.

It will be seen that the whole period intervening between the normal nymphal stage and the typical pigmentless adult stage[b] occupies only some nine to ten hours, and this apparently accounts for its not having been recorded before, even though it appears to be perfectly normal,[c] for it has occurred in different localities in two successive years and all nymphs taken passed through these stages before completing their development.

An illustration of the thorax of *Leucotermes flavipes* with unexpanded wings was given by Packard in his Text-book of Entomology, but he here described it as a late nymphal

Fig. 11.—*Termopsis angusticollis:* Quiescent stage of nymph of the first form. Enlarged 7 times. (Original.)

wing pad, otherwise there seem to be no references to either of the stages herein figured and described.

---

[a] The borders of the chitinized parts first take on pigmentation, passing through shades of gray to castaneous to the castaneous-black of the mature adult, the abdomen being the last to take on pigmentation; there are often abnormalities in development.— T. E. S.

[b] The whole period intervening between the fully developed nymph and the mature pigmented adult is about one day and one-half to two days for individuals, and about 1 week to 10 days for the colonies.—T. E. S.

[c] These stages are absolutely normal and necessary to the progressive development of nymphs of species of both the genera Leucotermes (*L. flavipes* and *virginicus*) and Termopsis (*angusticollis*) (fig. 11).—T. E. S.

During this final molt the females of nymphs of both the first and second forms normally lose the genital appendices;[a] that is, the genital appendices are present in nymphs of both sexes before this molt, but afterwards only in mature winged males and supplementary kings, developed from nymphs of the second form; these appendices can be readily seen on the cast skins. In egg laying complementary queens of *Termopsis angusticollis* (with no indication of wing pads), genital appendices were present, though absent in true queens.

On April 25, 1912, at Falls Church, Va., molting nymphs (*flavipes*) of the second form were found. The progressive development of over 100 nymphs was noted, and apparently the molting is similar to that of nymphs of the first form. Before the quiescent stage is reached, the nymphs assume a straw-colored hue. Various stages of molting nymphs were observed through April 28. After the molt, the head and prothoracic segments darken, especially on the borders. The abdominal segments also darken. Sometimes there are grayish-black, longitudinal, pigmented markings on the head. Some of the nymphs have very short wing pads, mere buds, while in others the wing pads are more elongate, the pair on the metathorax reaching down to and slightly overlapping the (apparent) second abdominal tergite. This difference can be seen in supplementary or neoteinic reproductive forms and illustrates the fact that the growth of nymphs of the first form is arrested in various stages of development, as does (?) the presence of genital appendices in complementary queens of *Termopsis*.

Sometimes, due to unfavorable moisture conditions, there is an abnormal development of nymphs of the first form. Individuals may be observed with partial pigmentation to the chitinized parts, acquired before the quiescent stage or before the wings are unfolded; that is, the wings may be in various stages of development, from the opaque-colored, elongate wing pads to partially unfolded, or unfolded but still crumpled, wings. There is great individual variation in the manner of molting. Adults with mature body pigmentation but with distorted, poorly developed wings, or even with opaque wing pads, have been observed emerging from the parent colony at swarming time. A swarm[b] of *flavipes* occurred at Falls Church, Va., on May 8, 1912, and another on April 25, 1913, while in the case of *virginicus* the swarm occurred on June 1, 1913.

This quiescent stage, or "Ruhestadium," has been described by Hagen.[c] He states (in regard to the final molt of nymphs of the first form) that the skin bursts on the prothorax; and in order to

---

[a] Grassi and Sandias. Op. cit., p. 306.

[b] It is from 7 to 10 days after the last sexed adults in colonies have acquired wings and mature pigmentation that the swarm occurs.

[c] Hagen, H. A. Monographie der Termiten. Linnaea Entomologica, v. 12, p. 337–338, 1858.

get out of the old skin, the insect doubles up. The insect lies on the ground during the molt. Grassi, while he did not describe this stage, knew that during certain ecdyses in the development of nymphs and soldiers important changes took place. Odenbach, January 13 to 24, 1896, observed in an artificial nest indoors (manuscript notes) molting larvæ of *flavipes* in the quiescent stage, as if dead. He states that the molting process lasts three and one-half hours, that workers assisted, and that the skin is eaten. His observations are practically the same as those of Strickland. Holmgren describes this stage in the larva of *Rhinotermes taurus* Desneux, Escherich[a] figures larvæ of *Termes obscuripes* Wasmann, and Bugnion[b] figures a soldier of *Termes horni* Wasmann in this quiescent stage. Holmgren was the first to state that a quiescent stage occurs in connection with each molt, and to note the internal as well as external changes that occur during these molts.

The writer has observed quiescent stages of undifferentiated (?) larvæ, larvæ of nymphs of the first form, nymphs of the first and second forms, and larvæ of workers and soldiers of *Leucotermes flavipes* and *L. virginicus* and soldiers and nymphs of the first form of *Termopsis angusticollis*.[c] Differentiated nymphs of the first form of *L. virginicus* only 2.5 mm. in length have been observed. Bugnion states[d] that since he has found nasuti larvæ of *Eutermes lacustris* Bugnion 1.3 mm. in length he believes that the differentiation is effected in the embryo for the three castes. The young nasutus with the distinct "corne frontale" is figured. This is not at all in accordance with Knower's statements and drawings of the development of the nasutus of Eutermes (rippertii?) which developed from a worker-like larva, and with Grassi's description in *Calotermes flavicollis* Fabricius and *Leucotermes lucifugus* and the writer's description in *L. virginicus* of the development of soldiers from worker-like larvæ. Bugnion further states[e] that in the higher termites the differentiation of caste reaches perfect expression.

Observations by the writer of molting soldier larvæ of *Leucotermes* spp. show that the differentiation takes place during a "quiescent stage" rather late in the life cycle.

From the first to the middle of August, 1913, freshly molted pigmentless soldier nymphs of *flavipes* in the stage preceding maturity were noticeable in colonies in Virginia. From the middle of June

a Escherich, K. Op. cit., p. 43.

b Bugnion, E. Le Termes Horni Wasm. de Ceylan. Rev. Suisse Zool., t. 21, no. 10, p. 299–330, pl. 11–13, juin, 1913. See p. 305–309.

c Snyder, T. E. Loc. cit.

d Bugnion, E. Les termites de Ceylan. Le Globe: Memoires Soc. Geog. Geneva, t. 52, p. 24–58, 1913.

e Bugnion, E. La differenciation des castes chez les Termites [Nevr.]. Bul. Soc. Ent. de France, 1913, no. 8, p. 213–218, April 23, 1913. See p. 217.

In incipient colonies nymphs are not produced during the first year that the colony is established, but are developed every year in old, well-established colonies. Nor are nymphs of the first form produced the first year in "orphaned" colonies. Nymphs of the second form are commonly to be found during the latter part of April to early May in Virginia (March in North Carolina), and they occur in varying numbers associated with nymphs of the first form. Nymphs of both forms are well developed by the middle of September of the year in which sexual maturity is attained, in *flavipes*.

Colonizing individuals of *flavipes* appear from the early part of April to May in Virginia and Maryland (earlier in infested buildings). In the case of *virginicus* they do not appear till a month later, or the end of May or early June in Virginia and Maryland. After having attained the mature pigmentation they soon swarm.

The royal individuals of *flavipes*—the pairs of winged sexed adults that have swarmed—are to be found together in the royal cell. In incipient colonies of termites, unlike the other social insects, the male assists in the establishment of the colony and continues to cohabit with the queen, there being repeated coition.

NEOTEINIC REPRODUCTIVE FORMS.

Neoteinic reproductive forms are normally(?)[a] developed from nymphs of the second form after the swarm; the males continue to cohabit with the females. Immature neoteinic reproductive forms are to be found at the time the winged sexed adults are attaining mature pigmentation—the end of April in Virginia. Maturity is probably attained shortly after the swarm, namely, May to June, although mature neoteinic reproductive forms have been found before the swarm.

In case of *virginicus*, neoteinic reproductive forms, produced from nymphs of the second form, are matured, fertilized, and egg laying by July to August. In case of *flavipes*, these reproductive forms are matured by May to June. There is a seasonal variation.

WORKERS.

Workers are always present in colonies except those just established by colonizing sexed adults; they are permanently present in the colony and constitute the most numerous caste.

---

[a] These forms can be produced at any time necessary. On April 9, 1912, a decaying stump infested with *flavipes* was removed from the forest and placed in the termitarium; this was at a time when the nymphs of the first form were nearly mature. On November 18 the termites had entered the ground and two neoteinic ergatoid(?) forms were found about 3 to 4 inches below the surface in chambers in the earth. No nymphs of the first form were present. They could be distinguished from the workers by the larger size, distended abdomen, straw-colored pigmentation, and sharper segmentation of the abdomen. Rudimentary wing pads were present. The antennæ had 15 to 16 segments. Workers solicited drops of liquid from the female by stroking the abdomen with their palpi.

Soldiers are also always present in colonies, except those just established by colonizing sexed adults, usually a few being present in incipient colonies; they are always relatively much less numerous than the workers. From the first to the middle of August, 1913, freshly molted, pigmentless soldiers of *flavipes* were common in colonies in Virginia, where they were found as late as the middle of October. On August 17, 1913, near Chain Bridge, Va., molting larvæ and nymphs of soldiers of *virginicus* were found.

LOCATION OF THE COLONY IN WINTER.

By the middle of November to December, depending upon the season, termites retreat to the subterranean passages of the colony, the earth under infested logs being riddled by a labyrinth of galleries. In case of very large logs, termites may remain in the more impenetrable inner galleries in the heartwood. In Virginia they remain in this retreat until the last of February or first or last part of March, depending on weather conditions.

Indeed, the center of activity in termite colonies changes with the seasons, due to the varying needs of the colonies as to conditions of warmth and moisture, which are essential to life and development. In spring when there is abundant moisture, open, wooded southern exposures are favorable, and the outlying galleries of colonies are teeming with developing nymphs, whereas during the heat of summer conditions would be too dry. Consequently in summer termites bury themselves more deeply in the wood or earth in less exposed galleries, in moist, shady sites, and in winter usually enter the ground to escape the cold.[a] In autumn, developing larvæ of the castes, and nymphs of soldiers and sexed adults, are to be found in the outlying galleries where the warmth will enable more rapid development. Therefore colonies apparently depleted at certain seasons, at others will be infested. Again, the Leucotermes colony readily migrates, and the site is liable to abandonment if conditions become unfavorable. A single colony may extend to and inhabit several adjacent stumps or trees, and it is often impossible to define the limits of a colony or the line of demarcation between different colonies in a region where termite colonies are abundant and there are many decaying stumps or logs; hence what is apparently a separate colony may be only a branch connected with the main colony by subterranean passages. If colonies are cut into and the reproductive forms removed, the colony quite frequently abandons the nest. The reproductive forms are capable of movement, and it may be that old colonies branch out by means of neo-

---

[a] As the higher altitudes are attained, termite colonies in the earth under stones are more common, that is, in the North Carolina mountains and in cañons in Arizona.

teinic reproductive forms which eventually establish new colonies. However, this is only a theory, but otherwise what becomes of the large number of nymphs of the second form in colonies in the spring? Surely they are not needed in the parent colony any more than the winged sexed adults, and it may be that they are impelled to leave the colony by the same irresistible force that induces the swarm. However, it is probable that workers and soldiers accompany them from the parent colony and that by means of subterranean passages they establish the new nest. Indeed, these forms may be the nucleus of the small bands of foraging workers and soldiers infesting decaying branches mentioned frequently in literature. The alternative is, of course, that such bands become isolated from the parent colony and rear the reproductive forms. It seems that both methods may be possible and necessary.

DURATION OF DEVELOPMENT AND LIFE.

The eggs of *flavipes* hatch in about two weeks after they are laid. Workers developed from eggs laid on July 15, 1912, were 4.5 mm. in length by the following December, with 13 segments to the antennæ. Both workers and soldiers complete their development within one year.

Definite data on the duration of life of any individuals of Leucotermes, not excluding even the royal pair, are lacking. However, the males or true kings of *flavipes* continue to live with the true queens after copulation, the royal individuals probably living at least five years;[a] neoteinic queens live at least one year and probably as long as true queens. Smeathman conjectures that a queen of *Termes bellicosus* Smeathman when 3 inches long is about 2 years old.

CANNIBALISM.

There are many instances to show that termites are cannibalistic in their habits; all dead or injured individuals are eaten; also, according to Odenbach, larvæ that have difficulty in molting.

It is not at all rare to find, especially in cases of workers, a narrow, grayish band, with black, scalloped, turned-up edges usually on the dorsum of the abdomen, but also sometimes present on the ventral surface and as a plate on the legs. This band is sometimes present on the abdomen of soldiers, and it also occurs on the thorax and head of workers. Possibly these black bands are healed-over wounds where the insects have bitten one another (fig. 9, c), or they may be due to a bacterial or fungous disease, or to both wound and disease.

These bands occur on workers and soldiers of both *flavipes* and *virginicus*.

---

a Heath, H. The longevity of members of the different castes of *Termopsis angusticollis*. Biol. Bul., v. 13, no. 3, p. 161–164, August, 1907.

LEUCOTERMES FLAVIPES.

FIG. 1.—*a*, Abandoned burrow of *Lymexylon sericeum* in solid wood of chestnut telegraph pole, in which a fertilized true queen was found; *b*, cell excavated in decaying wood by young royal couple.

FIG. 2.—Royal cell in solid chestnut in which 40 neoteinic royal individuals, for the most part queens, were found. (Original.)

On August 12, 1914, several workers with these bands on the body were taken from a colony of *flavipes* at Falls Church, Va., and placed in a small tin box with decayed wood and earth. Normal soldiers were also placed in the box. On October 2, 1914, the workers with the black bands were still alive and apparently in the same condition; the soldiers had no bands.

Chanvallon, according to Hagen,[a] recommends placing arsenic in termite nests, and since the insects are cannibals and the dead are eaten, a large number can be killed in this manner.

SITUATION OF THE DIFFERENT FORMS IN THE NEST.

The reproductive forms are not necessarily to be found in a "royal" cell situated in the more remote parts of the nests, as in tropical species, but are usually in the more sound or solid wood (Pl. XII, figs. 1, a and 2). In colonies recently established by colonizing individuals the eggs and young are present in a definite royal cell, where they receive the care of the queen. In well-established colonies no forms are permanently present with the reproductive forms, and there apparently is no well-defined royal cell.

The royal cell, excavated in decayed wood by the sexed adults that have swarmed, is a broad, oval chamber, the entrance to which is but slightly larger in diameter than the abdomen of the queen at a period 14 months after swarming (Pl. XII, fig. 1, b).

Most of the 40 neoteinic reproductive forms found at Falls Church, Va., May 27, 1912, were congregated in a single chamber in the solid wood of a chestnut slab. This chamber was a broad but shallow longitudinal cell in the solid, sound wood. The entrances to this chamber were but slightly larger in diameter than the abdomens of the fertilized queens. Other neoteinic individuals were found in shallow, broad, oval cells in the wood and in earth under the slab (Pl. XII, fig. 2).

The nymphs are usually present in the more remote passages of the nest, except during the spring, when they are in the outlying channels,[b] where the warmth of the sun will hasten their development.

---

[a] Hagen, H. A. Monographie der Termiten. Linnæa Entomologica, Bd. 10, 1855, p. 35.

[b] Developing larvæ, nymphs, or immature adults are normally to be found, temporarily at least, in that part of the nest where there are the most favorable conditions of heat and moisture for their rapid development—changing with the seasons. In the spring and autumn these forms occur under the bark on decaying stumps and under decaying wood or bark sunken in the ground in open sunny sites, always being in the outlying galleries where the warmth will enable more rapid development.

There are apparently no permanent sites used as "nurseries," as is the case in tropical species of termites. However, young larvæ are seldom found in the main unpartitioned runways, but rather in partitioned galleries, where they will not be disturbed by the activities of the other members of the colony. Often they are in broad but shallow unpartitioned galleries in the more sound wood of the interior heartwood of logs, etc.

Unpartitioned channels or runways usually contain only workers and soldiers; the partitioned channels contain larvæ and nymphs besides. The young, freshly hatched, developing larvæ are often found in broad but shallow moist galleries in the interior of the heartwood. Extensions of an old colony, consisting of subterranean passages leading to small pieces of decaying wood, contain only workers and soldiers.

As previously stated, during the winter all the castes seek the more remote or subterranean passages of the nest.

THE SWARM OR SO-CALLED NUPTIAL FLIGHT.

Under normal conditions in the Southern States,[a] in early April and May, in the case of the more common species, *flavipes*, and in June in the case of *virginicus*, the colonizing individuals emerge in enormous numbers from small holes in the wood of stumps, logs, poles, fence posts, foundation timbers in buildings, and the roots of trees and from the ground. It is 7 to 10 days after the last sexed adults have acquired wings and mature pigmentation that the swarm occurs. The winged insects usually crawl upon some elevation before taking flight; before swarming they teem over the tops of infested stumps (fig. 12) and festoon brush lying on the ground in order to get a start. If a sexed adult loses its wings while at a height above ground (as on the top of a stump) it jerks itself up in the air in endeavoring to get down. Numerous workers and soldiers are congregated in the outer layers of the wood near the exit holes at the time of emergence. These colonizing individuals differ from the other soft-bodied castes in that they are larger, of a castaneous color, and are highly specialized and developed for the purpose of swarming and starting new colonies. In addition, they are sexed and have eyes and wings. While it is true that they are weak fliers and are preyed upon by many insectivorous animals—birds, lizards, insects, etc.—yet some escape to found new colonies.

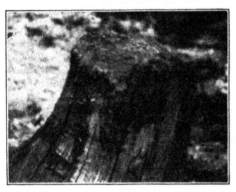

FIG. 12—*Leucotermes flavipes:* View of a swarm of sexed adults emerging from a stump at Falls Church, Va; a portion of the enormous numbers constituting a swarm. (Original)

[a] The first swarm which contains the greatest number of individuals does not occur outdoors until the ripening of the pollen of the flowers of dogwood (*Cornus florida*), which is also influenced by seasonal and geographical variations.

After the adults have flown a short distance in an irregular "wobbly" manner, they fall to the ground, and by catching the tips of the wings against some object and turning sideways they pry them off at a suture or line of weakness near the base, leaving stubs. This triangular, basal portion of the wing, or stub, is thickened and more chitinized than the wing and is also pubescent up to the suture, a possible aid in breaking off the wing after flight.

The male follows the female tirelessly and persistently, with head close to her abdomen, and touches her abdomen with the antennæ. Often the male and female run in a circle of small diameter, and sometimes the pursued turns pursuer, apparently attracted by some secretion at the posterior end of the body.[a] Sometimes as many as three individuals may be seen running off together. This is apparently due to sexual attraction, an amatory procedure preliminary to pairing, which accomplishes the purpose of bringing the sexes together. This continues for several days after the flight. The sexes are attracted to each other at a period several days before swarming, as is evidenced by the fact that when a colony is broken into there is a short flight, followed by loss of the wings, after which the male follows the female in the same manner as after normal swarming.

Neither of the terms "swarm" or "nuptial flight" is appropriate in referring to the emergence of the colonizing sexed adult termites, since the insects after a short flight separate into pairs, or the males and females may even "pair" (but do not "mate" sexually till later) with individuals of other colonies, and never congregate again in the same colony, but form many new colonies. In the case of bees, on the contrary, after the swarm subsides the insects all together form one new colony. Furthermore, copulation does not take place at the time of the swarm, which is not a "nuptial flight."

THE ESTABLISHMENT OF NEW COLONIES.

Many investigators have considered that the foundation of new colonies by winged sexed adults was impossible, and was not the purpose of the swarm, but Perris (1876), Pérez (1894), and Heath (1903) disproved this.

For several days after swarming the now wingless sexed adults can commonly be found together under small pieces of decaying wood, lying on the ground, ultimately disappearing, either to excavate shallow cells in the ground, in decaying wood (Pl. XII, fig. 1, b), or to take possession of an old abandoned insect burrow. The entrance to this now royal cell is but slightly wider than the abdomen. It

a Heath, H. The habits of California termites. Biol. Bul., v. 4, no. 2, p. 47–63, figs. 3, January, 1903. See p. 54.

is here that copulation probably takes place. Doubtless many colonizing pairs, even after escaping their numerous enemies at the time of the annual swarm, fail to become established in a new colony owing to unfavorable moisture conditions. Sometimes several pairs are found together in the same cell in a piece of wood, or perhaps one male and two females, or vice versa, but these reproductive forms, unlike the neoteinic forms, are not normally polygamous. The young royal couple, after finding suitable shelter, forage for themselves and the abdomens of both sexes increase slightly in size, becoming swollen. This is probably due to feeding [a] and development of the sexual organs.[b] Heath states that "everything is apparently sacrificed to lightness of body" at the time of the swarm. This results in wider dispersal.[a] Copulation probably does not take place till about a week after the swarm, when the couples are established together in the royal cell, as sexed adults of *flavipes* that swarmed on May 8, 1912, were in royal cells on May 15, the male no longer following the female about.

While it is not essential that the colonizing pairs, the young kings and queens, be adopted by foraging workers and soldiers, it is possible that this sometimes occurs. It is significant that small branch colonies of workers and soldiers are to be commonly found under decaying pieces of wood and in the ground after the swarm.

COPULATION AND THE RATE OF EGG LAYING.

Actual copulation was not observed during these investigations, but observations indicate that copulation does not take place till after the male and queen are established in the royal cell, and copulation at the time of swarming outside the nest is not very probable, as the genitals of the males are in a very imperfect stage of development.[c] Other observers have noted the process in the case of *flavipes*.

Haviland states,[d] "In *Termes malayanus* I have reason to think that the king fertilizes the eggs after they are laid; indeed, copulation in the case of kings and fully grown queens of most species of the genus *Termes* is apparently impossible." [?]

---

[a] Heath, H. Loc. cit.

[b] Müller, Fritz. Beiträge zur kenntniss der Termiten. Jenaische Ztschr. f. méd., Bd. 7, Heft 3, p. 333–358, fig. 11, pls. 19–20, März 7, 1873. See p. 337–351.

I. Die Geschlechtstheile der Soldaten von Calotermes.

II. Die Wohnungen unserer Termiten, p. 341–358.

[c] Hagen, H. A. Some remarks upon white ants. Proc. Boston Soc. Nat. Hist., v. 20, p. 121–124, December 4, 1878.

[d] Haviland, G. D. Observations on termites, with descriptions of new species. [Read 3d June, 1897.] Jour. Linn. Soc. (London), Zool., v. 26, p. 358–442, pl. 22–25, April 1, 1898.

Grassi states:[a] "On April 17, 1891, about 11 a. m., I detected the king and queen in coitu in a glass containing a small Calotermes nest * * *. They stood end to end in a straight line with the tips of their abdomen applied to each other * * *. It is therefore certain that the connection takes place in the nest, and is repeated at intervals." Sandias states[b] that he observed what appears to be a similar process between substitute forms still far from maturity, being only about a fortnight old.

The Rev. F. L. Odenbach, of St. Ignatius College, Cleveland, Ohio, has noted (MSS.) the copulation of neoteinic royalty (*Leucotermes flavipes*), namely, the mating of supplementary queens, time and again in artificial nests. On March 11 and 29, 1898, he observed many sexed adults pairing, coitu lasting about three minutes. He further states that the same queen has connection repeatedly with different males. On December 27, 1899, in describing the pairing of neoteinic royalty with short wing pads, he states: "The introduction is a lively play with feet and feelers, heads looking in opposite ways, the bodies curved together so as to make a circle, then the male slips along the body of the female until the organs meet; then they stand in one line, heads looking in opposite directions. The body is moved backward and forward, hinging on the legs and finally to both sides, as if they wished to telescope the abdomens. Time of connection, about one minute."

Heath states[c] * * * "almost a fortnight after swarming, I have on several occasions seen the royal pair of Termopsis in coitu. With their bodies closely appressed end to end in a straight line they remain from one to ten minutes in contact."

Egg laying in the case of *flavipes* begins about the middle of June or July, varying with the season, or about one and one-half months after the swarming. While the eggs hatch in about 10 days after they are laid, larvæ of varying sizes are often present, since they do not all hatch uniformly. Most of the first broods develop workers and a few soldiers, as the workers constitute the caste most necessary to the conduct of the young colony. At this time the queen and the male still occupy the royal cell together and the queen, with abdomen only slightly distended, cares for and carries about the eggs and later the young larvæ in her mouth, when the colony is disturbed. The royal cell is kept clean and the sides are now smoothed.

Recently hatched larvæ are fed on prepared food and do not eat wood until later in their development. On January 8 and 15, 1896, Odenbach observed workers to draw a white substance of the consistency of butter from the anus of neoteinic queens and devour it.

---

[a] Grassi, B., and Sandias, A. Op cit., p. 285–286.

[b] Ibid., p. 386.

[c] Heath, Harold. Op. cit., p. 52.

On November 18, 1912, workers were observed to solicit liquid from the anus of a larval or ergatoid (?) queen, at Falls Church, Va. On August 5, 1913, a fertilized, fully developed queen, collected at Veitch, Va., about 14 millimeters in length, ejected a clear white liquid from the anus, when disturbed.

The rate of egg laying of the young and active queen is not very rapid, as clusters of eggs in varying numbers from about 6 to 12 were observed in several cells with single pairs. The new colony at first is very small, and even after the rearing of the first brood of workers and soldiers the increase in numbers is not rapid. In July, 1912, at Falls Church, Va., about 12 small white eggs in a cluster were observed in a royal cell with young royal individuals of *flavipes*. At least three were observed, probably 2 males and 1 female. These had been captured after the swarm on May 8, in the earth under a small piece of decaying wood. On July 29 the newly hatched larvæ were observed, and on October 30 seven workers and one soldier surrounded a single royal pair. Fragments of the chitinized parts of another adult were found near the royal cell. The abdomen of an egg-laying female under observation, 13 months after swarming, was oblong and somewhat distended, the segments of the abdomen being slightly separated and showing white between. On October 30, when the female in the royal cell was disturbed, she continually moved the end of the abdomen, curving it ventrally under the body. This alternate rising and falling of the abdomen has been described by Smeathman as a constant "peristaltic" movement, in the case of tropical species. No eggs had been laid since the first were deposited in July, and it is believed that this so-called "peristaltic" movement in case of Leucotermes is merely the result of alarm, and has no direct bearing on egg laying. During this time the male still occupied the cell with the female and both were active. Eventually the abdomen of the female becomes immensely distended through the development of the ovaries, but in the case of certain species of Leucotermes the queen still retains the power of locomotion.

It will thus be seen that development under the foregoing conditions is at best a slow process and not at all comparable to that which takes place in tropical species, where growth of the queen and the rate of egg laying is correspondingly rapid.

On February 21, 1913, nine or more additional eggs were observed in a cluster near the royal cell of the above-mentioned pair. This cell was in a small decayed branch, placed on moist earth, and isolated in a tin box. On February 24 the first freshly hatched larva was observed. The abdomen of the queen at this time was not markedly distended. On May 16 freshly hatched larvæ were again present in this colony. On August 15 six eggs, as well as newly hatched young, were present in the royal cell. The male still cohabited with the queen, and the abdomen of the queen was not as yet markedly distended.

While the recently hatched young are active, they are dependent on the care of the parents or upon the workers for food.

Wheeler [a] states, "In incipient ant colonies, the queen mother takes no food, often for as long a period as eight or nine months, and during all this time is compelled to feed her first brood of larvæ exclusively on the excretions of her salivary glands. This diet, which is purely qualitative, though very limited in quantity, produces only workers and these of an extremely small size (micrergates)." In incipient termite colonies (Leucotermes and Termopsis [b]) the young royal couple share the royal cell, excavated in decaying wood, at which time the abdomens of both the sexed adults increase slightly in size, and they take food—that is, wood. The first larvæ develop to workers and a few soldiers, both forms being smaller than normal individuals or those in well-established colonies. No nymphs of sexed adults are produced during the first year. The rate of egg laying of a fully developed true queen is much more rapid. On August 5, 1913, at 5 p. m., a true queen, about 14 millimeters long, which had been taken in the root of a dead chestnut tree above ground, was isolated with the king in a cell in wood. By 9 a. m. on August 6, more than 12 eggs had been laid. When captured in the tree there were several hundred eggs as well as numerous recently hatched larvæ near by.

The antennæ of some true royal pairs that have swarmed are apparently entire at a period of seven months after swarming; however, the segments were not actually counted. In other pairs the antennæ of both sexes are mutilated.

THE ROYAL PAIR AND OTHER REPRODUCTIVE FORMS.

OCCURRENCE IN THE UNITED STATES.

Feytaud[c] gives a historical account of the frequency of occurrence of reproductive forms of *L. lucifugus* in Europe and figures the reproductive forms. Between April and September in the eastern United States the several types of reproductive forms of our common species of termites are to be found in colonies in decaying wood above ground; that is, the pigmented, true royal pair with wing stubs, developed from the sexually mature adults; the supplementary neoteinic forms, with pale straw-colored pigmentation and short wing pads, developed from nymphs of the second form; and the ergatoids and neoteinic larval forms, with straw-colored pigmentation and no wing pads or rudiments, developed from mouldable larvæ. It is believed that since these forms are mobile and that in

---

[a] Op. cit., p. 68.   [b] Heath, H. Op. cit., p. 57.   [c] Loc. cit.

old, well-established colonies there is apparently no definite permanent royal cell, these forms inhabit the subterranean passages in
wood or in the earth below the frost line during the winter. During
the warm months, probably to facilitate the processes of reproduction and development of the young, they inhabit the passages in
decaying wood above ground. The occurrence of true royal pairs
is not rare,[a] but supplementary or neoteinic reproductive forms are
apparently more common in colonies.

HISTORICAL.

The following is a historical record, in chronological sequence, of
the occurrence of reproductive forms in colonies of *Leucotermes
flavipes* and *L. virginicus* in the United States, together with notes on
the conditions under which they were found.

The first queens of *flavipes* taken [b] in the United States were found
by the late H. G. Hubbard in a colony in Florida and were of the
neoteinic type, or supplementary form, with short wing pads. Hubbard also found the first neoteinic kings, although he makes no mention of them and may not have recognized the fact that he had found
both sexes of neoteinic reproductive forms. Some of the neoteinic
queens that Hubbard collected are deposited in the Hagen collection
at Cambridge, Mass., but most of the specimens are in the collection
of the United States National Museum. The number of specimens now
present in the vials at the museum is probably not as great as when
originally collected, since Hubbard gave certain of the royal individuals
to Hagen. Hubbard's note, dated "Enterprise, Fla., May 19, 1875,"
recording the finding of the first reproductive forms in the United
States, reads: "*Termes flavipes* (determined by Hagen), females with
their eggs from small, rotten log in road near lake shore; females not
in separate cells, several together." [This vial also contained two
supplementary kings and nymphs of the first form.] Another note
dated April 4, 1882, Crescent City, Fla., reads: "*Termes flavipes*,
nymphs, queens, and eggs from galleries in pine log; Trichopsenius
and Anacyptus were found in this nest." [A neoteinic king was also
present in the vial.] A note dated "May 11, 1883, Crescent City,
Fla., in pitch pine," is in a vial containing 11 neoteinic queens, 5 neoteinic kings, and 631 eggs with the embryos in various stages of development. [Nymphs of the first form were also present in this vial.]

---

[a] It was formerly thought that true queens did not exist in colonies in the United
States and Europe (proper). According to E. A. Schwarz (Termitidæ observed in
southwestern Texas in 1895. Proc. Ent. Soc. Wash., v. 4, no. 1, p. 38–42, Nov.
5, 1896), there are but few permanent nests, headed by true royalty, of Leucotermes,
due to the wandering habit of the genus; that is, the frequent moving of colonies
would necessitate such rarity.

[b] Hagen, H. A. The probable danger from white ants. Amer. Nat., v. 10, p. 401–
410, July, 1876. See p. 405.

Mr. Louis H. Joutel,[a] of New York City, found a number of fertilized, egg-laying neoteinic queens occupying the same cell, 9 in one colony and 14 in another. The number varies with the colony, as has since been ascertained, and many neoteinic royal individuals may be present in a small colony, where they would be more needful. Mr. Joutel spent several years in the study of termites, and in correspondence with the writer states that on two separate occasions he has found true queens—the first shortly after the finding, in 1893, of the above-mentioned neoteinic queens. This would be the earliest record of the finding of a true queen of *flavipes* in this country. He further states that on a later occasion he found two true queens at Peekskill, N. Y., on the same day, July 14, under about the same conditions. The queens were located in a "* * * dead hickory stump, about 12 inches diameter, and were in the upper part of the tunnels among the workers. There was nothing to suggest a queen cell and no eggs to be found, although I looked over them carefully. They were found in stumps about 20 to 30 feet apart." In commenting on a statement made by the writer [b] that a true queen was inactive in a burrow when discovered, Mr. Joutel states: "The three [true queens] that I found were very active, and while they did not move quite as fast as the larvæ [workers], it was due only to their size." Mr. Joutel is quite right, as I have since found out, and the instance cited was probably due to the queen being caught in a burrow too narrow for her distended abdomen, in trying to escape, rather than being confined in a cell the entrance to which was narrower than the size of her abdomen. While Mr. Joutel was the first actually to find a true queen, Mr. C. Schaeffer,[a] of Brooklyn, N. Y., was the first to record the finding of a fertilized, true queen of *flavipes*, at Moshulu, N. Y., July 16, 1902. He has kindly loaned the specimen to me for study. This queen is approximately 8 millimeters in length, the abdominal tergites and sternites not being widely separated, is markedly pubescent, and the antennæ are mutilated.

In describing the true queens which he found, Mr. Joutel states that they resembled the queen that Mr. Schaeffer found. He further states, "* * * they were all apparently broader than the one you figure in relation to its length. The heavy chitinized parts looked like little dots on the surface and did not take up as much space in relation to the rest of the surface as those parts do in the one you figure. [True, in living specimens.] The first one had its antennæ complete; of the other two, one had two segments missing in one antenna and the other had three segments wanting—

[a] Loc. cit.

[b] Snyder, T. E. Record of the finding of a true queen of *Termes flavipes* Kol. Proc Ent. Soc., Wash., v. 14, no. 2, p. 107–108, June 19, 1912.

these happened to be on the same side." He also says, "I have repeatedly taken a pair at swarming time and bred the eggs from them, but have never got them to colony size," and that "* * * in stating that I found only three queens, I had reference to the fully developed ones. On Staten Island, in company with Mr. W. T. Davis, I came across a small log that had eight separate cells with a king and queen in each. A few had three and one four (individuals). They had eggs but no young. Some had perfect antennæ and others had segments missing." He states: "One small colony that I kept alive several years that had only about two or three dozen workers *swarmed* each year, a few individuals only, but I broke up and examined each fragment of wood and sifted the ground several times but never found a trace of either kind of queen; the whole colony was contained in a gallon jar. Other colonies that had thousands of workers, which I kept alive about 22 months from time of collecting, never swarmed or had young, but finally died of old age, I presume. They got smaller and seemed to have more fat in their makeup." Mr. Joutel also says, "Light, I found, was not objected to by termites, unless it was too strong, as long as they felt that they were *covered*, that is, they worked under cover, not necessarily in the dark." This is true, but even in metal termitariums with sliding, thick, red glass covers, termites, while at first they actively wandered about on the surface of the earth, and were apparently unaffected, soon sought cover under decaying pieces of wood or in the earth.

The Rev. F. L. Odenbach, S. J., has, since 1895, made observations on the habits of *Leucotermes flavipes* in artificial colonies. He has roughly sketched in his notes two types of neoteinic queens that he has found. The neoteinic kings he describes as being compressed laterally, and therefore seeming to have a ridged back. One neoteinic queen, with short wing pads, that he found pairing was 10 millimeters in length and had a markedly distended abdomen. Another gravid queen that he figures is apparently adapted from a nymph of the first form, since long, well-developed wing pads are present. The abdomen is distended and the abdominal tergites are separated. He states:

These reproductive forms were from a very large nest I took up in South Brooklyn, near Cleveland, Ohio, in the fall of 1895

I placed the termites in a large glass globe. In September, 1897, I discovered a large mass of eggs from which were developed all the different nymph forms known to me, *true winged males and females* also in large numbers. These latter chased about the nest in such wild disorder for a few days that the workers fell upon them and destroyed them to the very last one. It reminded me of the slaughter of drones in a beehive. At the time it seemed to me that the wild orgies, [?] which as a rule occur outside of the nest in midair, disconcerted the workers and soldiers who did the next best thing to restore order and quiet in their household.

From the above large nest I caused a colony to migrate into a Lubbock nest, and in this nest I found the different neoteinic reproductive forms. First, the nymph with the long wing pads. She laid eggs, and I repeatedly induced her to do so by the same means by which I first caused ant workers to lay eggs (1885).

If they seem numb with cold, I place my hand on the glass plate and this induces the activity.

This queen was quite different in shape and color from those I will mention below, being larger and of a lighter color. She was slow in her movements and did not change her location very often. She was tended by the workers, which could hardly be said of the others, since they were too restless.

The others, reproductive individuals, were nymphs of different kinds, with different shaped wing pads, but none with as long ones as those of the individual mentioned above. They also laid eggs, but during one of their wild rushes around the nest were given their quietus. I now had only the above nymph and the one with no visible wing pads left. This latter I thought to be a true queen (?), since the treatment of her was nothing like that accorded to the nymph with long wing pads. She laid eggs, but seemed to be disregarded by the workers.

On March 26, 1895, at Haw Creek, Fla., Hubbard found an imago (a male) with the wings gone, in a colony. The antennæ have segments missing. His note reads, "In old, rotten pine log in swamp hummock; soldiers, larvæ, and imago with wings gone."

King [a] also states that he has found a black form (male) with wing stumps in a colony and that it is a swift runner. "Again, there is associated with *Termes flavipes* a clear black form, variable in size, some with wing stumps, and others, so far as I can see now, without being cleared, appear to have none. [?] I have only met with five of these forms so far; one measured 6 mm. in length, another 5 mm., and two of these measure 4 mm." He states that the fifth form was sent, not long since, to Dr. L. O. Howard, with notes, who referred it to Mr. E. A. Schwarz, the latter stating that it was a fully developed male with wing stumps.

Mr. King further says:

" * * * When I first observed its appearance with *Termes flavipes*, and in the nest with it, I supposed it to be a species of a staphylinid beetle, so swift were its movements that they made them quite deceptive. They are very swift runners and hard to capture. Further observations will be necessary to determine whether these are new species or not. It is my impression, however, that they are of a different type."

He does not state at what period of the year he found this form.

These forms found by Hubbard and King were possibly royal individuals, and in the case of the specimens collected by Mr. King they were possibly of both species, *flavipes* and *virginicus*.

H. G. Hubbard found a true queen of Leucotermes, probably *lucifugus* (a species very similar to *flavipes* [b]). As in *virginicus* the ocelli are nearer to the compound eyes than in *flavipes*. *L. virginicus*

---

[a] King, G. B. *Termes flavipes* Kollar and its association with ants. Ent. News, v. 8, No. 8, p. 193–196, October, 1897. See p. 194.

[b] Hagen, H. A. Monographie der Termiten. Linnæa Entomologica, v. 12, 1858, p. 174–180 and 182–185.

Banks has as yet been found only in Virginia, West Virginia (Hopkins), Maryland, North Carolina, and Illinois (Snyder). *Leucotermes lucifugus* Rossi occurs in the United States and "is found in Texas, Kansas, Colorado, and southern California, and perhaps elsewhere." [a] The species *lucifugus* of Mediterranean Europe, according to E. A. Schwarz,[b] is probably native to America (Mexico). According to Dr. Knower, *flavipes* has been introduced into Japan and is firmly established. This species has also been introduced into Europe and has been destructive in the vicinity of Vienna.

This true queen, found by Hubbard, is slightly over 13 millimeters in length, has the abdominal tergites and sternites more projecting, and has not as greatly distended an abdomen as the true queen found in Virginia; that is, the chitinized parts are less fused than in the older queen. (Pl. XIII, *b*.)

The note in Hubbard's field diary recording the discovery of the first fertilized true queen reads as follows:

June 20th, 1898, Santa Rita Mountains, Madera Cañon (Southern Arizona). We ascended a ravine filled with majestic sycamore trees under which the ground was wet with numerous springs, but entirely tramped by cattle and devoid of smaller vegetation. * * * I found under a stone in a little dry mound in a wet springy spot on the mountain side, a colony of true Termes, among which, in a cell cavity just beneath the stone was a single matured gravid female, or queen, which certainly had been winged; took eggs, larvæ, workers, and soldiers with the queen. This is the first instance known of a true queen in the genus Termes. There were no supplementary or nymphal queens in this colony and no male was found. I explored the entire colony, which was not a large one.

Prof. Harold Heath, of Stanford University, Cal., has described the life cycle of *Leucotermes lucifugus* in California,[c] which is very similar to that of our common species of the East. He figures a true "primary" queen, as only 8 months old, however, that has a markedly distended abdomen and the abdominal tergites separated—probably an error, due to transposed descriptions under the figures, according to observations by Feytaud [d] and the writer.

A true queen, with wing stubs (*flavipes*), found by the writer was in an abandoned burrow of *Lymexylon sericeum* Harr. (Pl. XIII, *b*) in the decaying butt of a chestnut telegraph pole near Portsmouth, Va., on August 12, 1910. This queen was inactive, since the burrow was no wider than her abdomen, and apparently she was unattended. The abdomen was greatly distended and oblong in shape, the abdominal tergites and sternites being widely separated. In the bright sunlight the abdomen appeared to have a yellowish green tinge. The length of the queen was approximately 14 millimeters. The antennæ were mutilated.

---

[a] Howard, L. O. The Insect Book. p. 356. New York, 1901.
[b] Loc. cit. See p. 39.
[c] Heath, Harold. Loc. cit.
[d] Feytaud, J. Op. cit., p. 567, fig. 21.

FIG. 1.—a, Supplementary queen with straw-colored pigmentation to chitinized parts; b, true queen with castaneous pigmentation to chitinized parts. (Original.)

FIG. 2.—Supplementary and true queen, showing deeper pigmentation of chitinized parts of the true queen. (Original.)

LEUCOTERMES FLAVIPES.

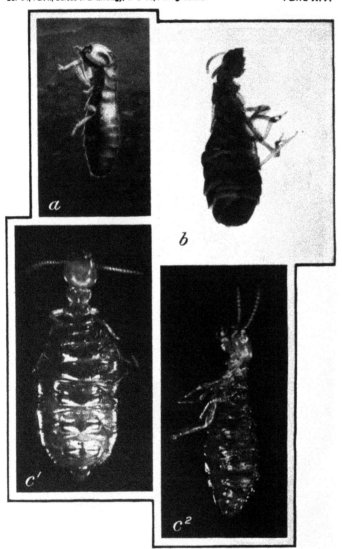

LEUCOTERMES VIRGINICUS AND L. FLAVIPES.

a, *L. virginicus*, neoteinic larval reproductive form; *b*, *L. flavipes*, neoteinic nymphal reproductive form (queen), *c¹*, *c²*, *L. flavipes*, neoteinic larval queens, dorsal and ventral views (note absence of wing pads).   From etherized specimens.   Enlarged 10 times.   (Original.)

Mr. W. B. Parker, of the Bureau of Entomology, has found two large fertilized true queens of *Leucotermes lucifugus* in redwood hoppoles in California. These queens were found in May, one being approximately 8 millimeters in length, with 11 segments to the antennæ, the other approximately 6 millimeters in length, with antennæ broken. Like the queen found by Mr. Schaeffer, the abdomens were not fully distended.

At Falls Church, Va., during the seasons of 1912, 1913, and 1914, numerous nymphs of the second form, in stages before, during, and after molting, were found by the writer; also young and mature neoteinic reproductive forms and young and mature true royal individuals.

On April 23, 1912, at Falls Church, the first colony of *flavipes* was found with a nymph of the second form present. The abdomen was noticeably broader, and this nymph was active and associated with nymphs of the first form. The nymphs of the first form were fully developed and in various stages to the sexed adults with immature pigmentation. Freshly hatched young larvæ were present. The day was sunny and warm.

On the same day a colony of *flavipes* was found in a decaying tulip tree stump, established well in the heartwood. Sexed adults with mature castaneous-black pigmentation were present and flew when the colony was disturbed. They were not present in large numbers, and all were in the outer layers of the wood. No fully developed nymphs in the stage just before the final molt were present, which is remarkable, as they were apparently present in all other near-by colonies. Very young nymphs and half-grown nymphs, as well as larvæ just hatched and young to medium-sized larvæ, workers, soldiers, and two neoteinic males (mature kings?) were present. They were active and in the outer layers of the wood. The abdomens of these males tapered markedly, being narrower at the end of the abdomen. They had a yellowish brown pigmentation, and in one king the compound eyes were without a trace of pigmentation. (Pl. VIII, fig. 2, *a*.)

A small colony of *flavipes* was found on April 25, 1912, under a small chestnut slab lying on the ground, exposed but sunken compactly into the moist earth. The day was sunny and warm. Fully developed nymphs of the first form were the most abundant stage, although some few transforming nymphs and a few sexed adults with dark pigmentation, as well as a few soldiers and workers, were present. The next most abundant caste to nymphs of the first form was nymphs of the second form, of which 90 were present and 4 were molting. They were active, much more so than nymphs of the first form. Most of these forms were either in shallow tunnels in the earth or on the earth, but a few were found in the decaying wood.

On the same day a single nymph of the second form was taken in a small colony under a strip of chestnut bark compactly pressed into the ground; this nymph was associated with nymphs of the first form. In two other colonies single nymphs of the second form were found associated with sexed adults under the bark on decaying chestnut stumps. Normally, but relatively few nymphs of the second form, as compared to nymphs of the first form, are present in colonies.

In another colony about one dozen nymphs of the second form, one of which was molting, were found associated with adults with pigmentation in various stages of development to maturity. A few of the latter with mature pigmentation had only one pair of wings unfolded, whereas others were present with the wings at the tips still compactly folded.

In a near-by colony, nymphs of the second form that had completed the final molt were present with sexed adults, but unlike the foregoing they are darker, some being pigmented. Very young larvæ were present in the outer layers of the wood. The acquisition of pigmentation in neoteinic royalty apparently is a sign of maturity and old age.

On May 27, 1912, 40 neoteinic forms, for the most part fertilized queens, together with females in which the abdomen was not greatly distended, were found in the more solid wood of a decaying chestnut slab on the ground. (Pl. XII, fig. 2.) Smaller forms, males, with the abdomen not oblong or distended were present in the colony. Most of these forms were congregated in the longitudinal royal chamber, in which young were also present. According to Grassi,[a] in case of *lucifugus*, "The colony must therefore rear fresh (complementary) kings every year [?], which become mature in August and September, fertilize the queens, and die." In colonies in Virginia neoteinic reproductive forms of *flavipes* are being developed from nymphs of the second form during April and May and are matured by May to June. In case of *virginicus* these reproductive forms are matured by July to August. There is a seasonal variation. There is no conclusive evidence that the kings do not live as long as the queens. They are not always present with the queens in colonies, but being so much more active are more likely to elude capture.

Two sexed adults, a royal couple, were captured by C. T. Greene at Falls Church, on November 12, in the frass in an old burrow of *Prionoxystus robiniæ* Peck, in the base of a living chestnut tree. Three royal individuals were present in the frass, and were located from 2 to 3 inches deep in the burrow. Four to six (at most) young were present, and, according to Mr. Greene, apparently all were workers.

---

a Grassi, B , and Sandias, A., op. cit., p. 298.

On November 18, 1912, two larval (ergatoid?) reproductive forms with a yellowish pigmentation and rudimentary wing pads were found in passages in the ground under a decayed stump at Falls Church.

Nymphs of the second form greatly outnumbered nymphs of the first form in a colony of *flavipes* taken in a decaying oak limb on the ground on March 22, 1913, at Black Mountain, N. C.

On April 17, 1913, at Falls Church, molting nymphs of the second form of *flavipes* were found together with molting nymphs of the first form, some of the former of which had already molted and attained the characteristic yellowish pigmentation.

Molting nymphs of the first and second forms of *virginicus* were found on April 30, 1913, at Falls Church, under decaying slabs of wood on the ground.

Mr. H. S. Barber, of the Bureau of Entomology, on May 30, 1913, on the Virginia shore of the Potomac River, opposite Plummers Island, Md., found a neoteinic reproductive form of *flavipes*. This was a queen exceeding 6 millimeters in length, with wing buds and 17 segments to the antennæ. This queen was associated with workers, soldiers, and young in decaying pine wood on the ground. (Pl. XIV, *b*.)

On May 26, 1913, nine neoteinic queens of *flavipes* with distended abdomens and one king were found near Charteroak, Pa., in the interior of a decaying maple stump. Freshly hatched young and young larvæ were numerous. The queens were not in separate cells, but were active and associated with workers and soldiers. (Pl. XV.)

Mr. H. G. Barber, under the title "Another Queen of the White Ant Found," states:[a]

> While on a collecting excursion last Fourth of July with Mr. Charles E. Sleight, at Lake Hopatcong, N. J., I found a small colony of white ants beneath a small log where they had made some tunnels along the ground beneath the log. Among the individuals was captured a fully developed queen, which was preserved and presented to the local collection of insects, at the American Museum of Natural History.

A large fertilized true queen and a king of *flavipes* were found on August 5, at Veitch, Va., in a large colony associated with workers, soldiers, and young and several hundred unhatched eggs in the sound outer layers of wood, about 1 inch in from the exterior, near the base of a dead standing chestnut tree. This tree had died at least two years previous, but this may not approximate the age of the colony, since termites often infest the bases of living trees by obtaining entrance through old abandoned insect burrows. The king was hidden beneath the queen, which was about 14 millimeters in length. The antennæ of the queen in this instance were mutilated

---

[a] Barber, H. G. Another queen of the white ant found. Jour. N. Y. Ent. Soc., v. 22, no. 1, p. 73, March, 1914.

and the workers and soldiers were very solicitous. It is evident that
the queen is fed on food prepared by the workers, since her abdomen
became markedly shrunken when she was isolated.

On August 7, 1913, another true royal couple of this species was
found near Chain Bridge, Va., in the interior of a decayed yellow
poplar (tulip tree) log lying on the ground. The colony was small
and the abdomen of the queen was not fully distended, being more
flat than oblong. (Pl. XVI, *b*, king.)

On August 15, 1913, at Falls Church, a young neoteinic reproductive form? (female) of *virginicus* was found in an experimental
stake of decaying yellow pine which had been set in the ground a
year previous. The queen (?) was of a dirty yellow color with four
grayish-black longitudinal markings on the epicranium, and grayish-
black markings between the coxæ and mesothoracic and metathoracic
tergites. The length was 4.5 millimeters. The segments to the
antennæ were mutilated, one antenna having 11 segments. (Pl.
XIV, *a*.) There was no pigmentation to the compound eyes.

On September 17, 1913, near Chain Bridge, Va., numerous nymphs
of the second form of *flavipes* of mature size were found associated
with nymphs of the first form in the sound heartwood of a decaying
black locust stump. The nymphs of the first form had well-developed wing pads.

On November 3, 1913, at Falls Church, nymphs of the second form
of *flavipes* were found associated with those of the first form in the
sound wood of a decayed oak stump near the ground. Young were
present in this colony.

In 1914, on April 17, the first fully developed nymphs of the second
form of *Leucotermes flavipes* were found in colonies at Falls Church,
Va.

A large true queen of *flavipes* was found in a decayed oak stump
about 5 feet high on July 17, 1914, at Falls Church. The tree had
been dead for at least three to four years, and the stump was about 14
inches in diameter and still had the bark on. This queen, which was
oblong but somewhat quadrate—4 millimeters in width—was 14½
millimeters in length (measured while alive [a]), and in color had a
slight tinge of greenish-yellow or opaqueness. The antennæ had segments missing. The colony was very large. Numerous eggs were
present in the galleries in the decayed wood, and the queen was
found in an elliptical cell about 2 inches in the wood from the exterior; the sides of the cell were cleanly eaten out. The royal cell was
situated about 1½ feet above the surface of the ground. The male
was not found.

---

[a] The abdomens of queens become further elongated during the killing and fixing
process, and hence in photographs of preserved specimens there is an apparent loss
in width.

LEUCOTERMES FLAVIPES.

a, Dorsal, b, ventral, and c, c′, lateral views of abdomen of fertilized nymphal neoteinic queens. a and c′, Enlarged 7 times; b and c, enlarged 10 times. (Original.)

PLATE XVI.

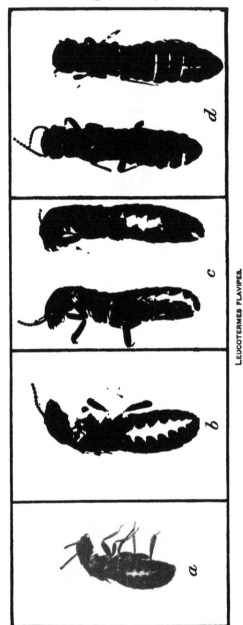

LEUCOTERMES FLAVIPES.

a, Mature king, nine months after the swarm and cohabitation with a true queen that had reared a brood of young—abdomen of queen not markedly distended; *b*, mature king, probably several years old, cohabiting with fertilized queen with abdomen markedly distended and tergites and sternites separated; *c*, fertilized true queen that swarmed April 25, 1913, and by September 25, 1913, was rearing a brood of young; *d*, Same, dorsal view. *a*, Enlarged 7 times; *b, c, d*, enlarged 10 times. (Original.)

In a near-by dead oak tree about 1 foot in diameter, which had been dead probably for two years, another true queen of this species was found on the same day. This queen was oblong and approximately 12½ millimeters in length (measured while living). The royal cell was in the decayed wood about one-half inch from the exterior at the base of the tree near the roots. The colony was large in numbers. Hundreds of eggs in clusters were in shallow galleries under the bark. The male was not found. Sometimes eggs in large clusters were found in deep, transverse, conical ledges or notches, attended by workers.

When the royal cell is cut into and the queen removed, the large number of attending soldiers and workers become very much excited, as evidenced by constant convulsive movements or sudden jerking of the whole body.

In this same woodland, on July 20, four neoteinic queens of *flavipes* were found in the more solid wood of a low (the wood above ground being nearly disintegrated) decayed oak stump, near the surface of the ground. One queen was of the normal neoteinic type, developed from a nymph of the second form, being 9 millimeters in length, with wing pads present. The compound eyes were pigmented, and the antennæ consisted of 16 segments. The other three neoteinic queens were of a type not previously found in colonies by the writer. These queens, of a straw-colored pigmentation, were 7, 6, and 5 millimeters in length (measured while alive), respectively. In shape the abdomens are oblong quadrate, like those of the true queens. The antennæ comprise 15 segments, which have a tinge of grayish-brown pigmentation to the dorsal surface. No wing-pads or rudimentary buds are present. The compound eyes are without pigmentation. The head, thoracic segments, and tergal and sternal "nota" (chitinized tergal and sternal areas) are not as broad as in neoteinic reproductive forms developed from nymphs of the second form. In these larval queens, as in true queens, the segments of the abdomen are less projecting than in normal neoteinic queens, and the "nota" are less semicircular. The mouth parts and legs are also less gross in structure. Indeed, these queens more nearly approach the true queens as to these points. (Pl. XIV, *c*.) They were probably developed from larvæ of the sexed forms. The outline of rows of numerous eggs in various stages of development could be seen through the body tissue under a high-power Zeiss binocular. The body tissue of normal neoteinic queens is coarser and thicker. This is the first time that reproductive forms of apparently different types have been found in the same colony of this species. The colony of termites in which these queens were found was small in numbers.

All the queens found by the writer in July, 1914, at Falls Church were captured within an area of ground less than an acre in extent.

It is believed that the spread or distribution of a colony may be largely dependent on the supply of decaying wood near by; that is, if there is a large amount near by (as colonies in large dead trees) the colony will not branch out over a large area.

On August 5, 1914, at Falls Church, a true royal pair of *flavipes* was found in a cell in the more solid wood of a decaying oak chopping block, that is, a section of a log that had been put to this use in the woods. The cell was in a knotty area of solid wood about 1½ feet from the ground but in the outer layers. The king was hiding beneath the queen, and is 6 millimeters in length. The abdomen is distended and the antennæ mutilated. The queen was of large size (probably 10 to 12 millimeters in length) but was crushed in cutting into the royal cell. The colony was large, and galleries extended from the ground up through this block and another similar block placed on top of it. The termites had filled in the crevices between the two blocks with clay, and larvæ, pupæ, and adults of *Homovalgus squamiger* Beauvois, a scarabæid beetle, were present in the termite galleries in the clay or in pupal cells.

In the same woods several young royal pairs of *flavipes* were found established in incipient colonies in the outer layers of a decaying chestnut slab partially sunken in the ground. Each pair was in a shallow cell excavated in the wood and was surrounded by a few young larvæ (a half dozen to a dozen). A few unhatched eggs were present in some of the cells. In one of the cells three adults were present instead of simply one pair. This is quite often the case in incipient colonies.

At Veitch, Va., on August 12, eight neoteinic reproductive forms of *flavipes* were found in a decaying yellow pine stump. They were in the more solid wood about 1 foot from the ground, but in the outer layers. Five were females and three were males or kings. The females were all about 7 millimeters in length, with abdomens distended with eggs, but with the segments not markedly separated except near the end of the abdomen, where the latter was lumpy and distorted, due to distension. The males, with narrow, slender abdomens, were all about 6½ millimeters in length, but had, instead of the straw-colored pigmentation of the females, a darker pigmentation streaked with grayish-black markings on the head and borders of the tergites and sternites and between the coxæ and the mesothoracic and metathoracic tergites. The antennæ of these reproductive forms consisted of 16 or 17 segments. These forms were developed from nymphs of the second form. The colony was not very large, but numerous unhatched eggs and young larvæ were present.

At Lake Toxaway, N. C., on August 27, 1914, an ergatoid queen of *Leucotermes virginicus* was found under a large flat rock in a shallow cell in the earth. Workers, soldiers, and young were aggregated under the stone, but the colony was not large. This queen had no trace of wing pads or pigmentation to the compound eyes, if present. The abdomen is oblong-ovate, being distended; its length is 5 millimeters. The antennæ are mutilated. Workers and soldiers surrounding the queen ran up and touched her with the antennæ and then evidenced excitement and alarm by the convulsive jerking of the body backward and forward.

This locality is a wooded, rocky hillside at an elevation of about 3,400 feet above sea level. There had been surface fires through the forest. It was noted that termite colonies were unusually abundant in the earth under stones in this locality, which is apparently the case as the higher elevations are reached.

On September 15, 1914, near Chain Bridge, Va., well-developed nymphs of both the first and second forms of *flavipes* were found in a colony in the more solid wood of a decaying chestnut stump. There is an annual seasonal variation in the degree of development the nymphs of the first form have attained by late fall; sometimes the wing pads are long and the nymphs apparently nearly mature; in other years in March these nymphs will still have comparatively short wing pads. There is sometimes also a variation in the different colonies.

DESCRIPTION OF THE REPRODUCTIVE FORMS.

As has been previously stated, the abdomens of both the young queen and male increase slightly in size and become distended after swarming. The abdomen of queens (*flavipes*) that had laid eggs during July, in January (9 months after swarming) were oblong and somewhat distended, the segments of the abdomen being slightly separated and showing white tissue between. The queens are dark castaneous in color and the males more blackish. The legs, tarsi, and tibiæ are markedly light yellowish in color in both sexes. After the swarming the abdomens of the males become only slightly distended.

The gradual distension of the abdomen of the queen, brought about by ovarian development, necessitates the separation of the abdominal tergites and sternites, and the connecting tissue between the abdominal tergites, pleurites, and sternites becomes remarkably distended, or more probably, according to Hagen,[a] as stated by Riley, there is actually further growth after the insect has reached the imaginal stage. It may be noticed from the illustration of the true queen found by the writer on August 12, 1911 (Pl. XIII, *b*), that there are two brownish scars located on the pleural tissue on the right side of the abdomen. These were not noticed until the queen was

---

[a] Riley, C. V. Termites, or white ants. Proc. Biol. Soc. Wash., v. 9, pp 31–36, Apr., 1894.

removed from the cell, and were probably due to injury received in shipment, as the queen was not taken from the royal cell, but a small block of wood was cut out of the pole, the whole having been placed in a vial of alcohol. The queen was partly out of the cell and the scars were probably due to abrasions by small fragments of jagged, projecting wood. This is stated in detail, because it might be thought that the wounds were "battle scars."

Attached to the mesothorax and metathorax in true queens are the stubs of wings, lost at the time of swarming. The head, thorax, and scutellar area ("nota") of the abdominal segments of true queens are more heavily chitinized and more deeply pigmented with castaneous brown than in the neoteinic queens, developed from nymphs of the second form, which are straw-colored (Pl. XIII, a); consequently, the nonfunctional eyes and ocelli are not so prominent in neoteinic queens, which never develop wings and which always[a](?) remain in the parent colony. The head, thoracic tergites, and abdominal tergites and sternites are both longer and broader than in the true queens, which same differences are apparent in the nymphs. (Pl. VIII.) In true queens the mesothoracic and metathoracic tergites have a distinctively irregular shape. The chitinized "nota" of the abdominal tergites and sternites more markedly approach the semicircular in shape and are much more projecting in neoteinic queens developed from nymphs of the second form. However, in younger (smaller) true queens (*flavipes* and *lucifugus*) the tergites and sternites are slightly more projecting than in older queens. Furthermore, the legs are more slender and the mouth parts slightly smaller (less gross) in true queens. The mesothorax and metathorax and pigmented, chitinized "nota" have a distinctive shape in the neoteinic larval queens. (Pl. XIV, c.)

In matured true queens (of both *flavipes* and *lucifugus*) ribbons of parallel ovules of various sizes and stages of development are visible (under high-power Zeiss binocular) through the tissue of the abdomen, where there are no deposits of fat. Sections through the body show an enormous ovary development, with ribbons of ovules in progressive stages of development. In a lateral or dorsal view of the abdomen of the true queen, small, round spiracles can be seen, set in at the base of the lateral slope of the tergites. (Fig. 13, a and c.) The spiracles are approximately similarly placed on queens of species in the genera Calotermes Hagen, Termopsis Heer, and Eutermes Fr. Müller. In some tropical species, as in *Termes bellicosus*, the spiracles are located in the pleural tissue of queens with enormously

---

[a] It may be possible that subcolonies or offshoots of large old colonies are established by these mobile queens and workers and soldiers by means of subterranean passages to decaying wood.

distended abdomens, an indication of displacement by actual post-adult growth (?).   In tropical species of Eutermes the enormous development of the ovaries in old queens crowds the digestive and excretory organs to the ventral surface of the abdomen.

Situated on the vertex of the epicranium is a small depression which appears as a rather prominent white spot, under the binocular microscope, in workers, nymphs, and neoteinic royal individuals of *flavipes*.   This depression, slightly larger in diameter than an ocellus, is also present in the soldiers and in colonizing individuals.   This is the "retrocerebral gland"[a] mentioned by Grassi,[b] as a "* * * [gland of unknown function," [existing in *lucifugus*] "(only ?) in the nymph of the first form, the perfect insect, and the soldier.   It eliminates a transparent secretion which can be squirted out for some distance."]

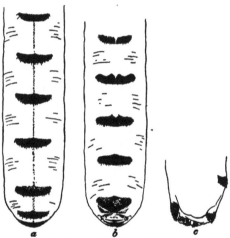

While the abdomens of neoteinic queens apparently never become as elongate as in true queens, they become as much distended, but have not the oblong or quadrate shape, being more oval, or wider near the end of the abdomen, which tapers markedly.   At the end of the abdomen of true queens the chitinized

Fig. 13.—*Leucotermes flavipes:* Fertilized, true queen; dorsal (a), ventral (b), and lateral (c) views of abdomen. Drawn from specimen preserved for three years in alcohol. Note position of spiracles. (Original.)

parts are more compressed or fused than in the neoteinic forms.

Fertilized, neoteinic queens, developed from nymphs of the second form, range in length from 9 to 12 millimeters.   The males are of the same length as the nymphs from which they develop; their abdomens are compressed laterally and taper toward the end, and "therefore seem to have a more narrow ridged back" (Odenbach).

Neoteinic or supplementary royal individuals are obtained from an arrested early stage in the development of the nymphs of the first

---

a Homologous to the small head gland of worker-like larvæ of Eutermes, which larvæ develop to nasuti, with a nose-like process (?).

b Grassi, B., and Sandias, A. Op. cit., p. 317.

form, or even larvæ, as young are kept in an undifferentiated state, which can be speedily turned into reproductive forms that serve as substitutes. Fritz Müller*a* compares the modes of diffusion and reproduction to those of plants which continue the species by means of cleistogamic as well as perfect flowers, the neoteinic forms corresponding to the cleistogamic flowers which are supplementary, emergency forms for use in case the perfect flowers (or winged colonizing forms) should fail. The winged forms would furnish a possible, but not probable, escape from interbreeding, since, usually, males and females from the same colony pair. Neoteinic supplementary forms are produced not only to counterbalance the loss of true royalty but also for the purpose of extending the colony.

DATES OF THE SWARMING OF LEUCOTERMES.

Colonizing individuals of *Leucotermes flavipes* usually swarm in the forenoon during the first part of April and May in the southern part of the eastern United States. Farther north the swarm occurs later, usually the last of May or early June. Hagen *b* mentions an exceptionally large swarm which occurred in Massachusetts. The sexed adults normally emerge earlier in infested buildings. According to E. A. Schwarz, sexed adults swarmed from infested beams in the floor in the basement of the old United States National Museum on March 15, 1883, the second year after the opening of the museum. The following year they swarmed during the latter part of March.

On April 16, 1910, sexed adults, possibly from the same colony, swarmed from crevices between the bricks in the sidewalk opposite the old National Museum.

On March 30, 1908, sexed adults came up through cracks in the floor in a building at Philadelphia, Pa.

Mr. Schwarz states, in an article entitled "Termitidæ Observed in Southwestern Texas in 1895": *c*

Termes flavipes Kol.—Common throughout southwestern Texas and very destructive to woodwork in houses. An immense swarm of winged individuals [*lucifugus?*] issued from several houses at San Diego [Tex.], on October 25. I was informed that in early spring another flight takes place in buildings infested by termites [*flavipes?*]. What appears to be the same species is also common in sticks and branches lying on the ground in the chaparral, but I failed to get the winged form from such situations.

---

a Müller, Fritz. Beitrage zur Kenntniss der Termiten. III. Die Nymphen mit kurzen Flügelscheiden (Hagen), "nymphes de la deuxième forme" (Lespès). Ein Sultan in seinem Harem. Jenaische Ztschr., Bd 7, Heft 4, p. 451–463, figs. 3, November 18, 1873.

b Hagen, H. Some remarks upon white ants, Proc. Boston Soc. Nat. Hist., vol. 20, p. 121–124, December 4, 1878.

c Proc. Ent. Soc. Wash., vol. 4, no. 1, p. 38–42, November 5, 1896. See p. 38.

W. D. Hunter states [a]:

Termes flavipes K. is not uncommon in Texas, where swarms occur ordinarily during the early part of the season. In 1911, however, the insect did not come into notice until about the middle of October [*lucifugus?*]. At that time much more than usual numbers were to be seen throughout the State.

Often several swarms emerge from the same colony in the same year. At McDonogh, Md., in April, 1913, according to a correspondent, sexed adults swarmed from the woodwork of a house, the beams being honeycombed. A very large swarm issued on April 6, another on April 13, a third on April 18, and a fourth swarm issued April 25, comprising in all four distinct swarms. In size, however, the first swarm, from my observations, is usually the largest. (Fig. 12.)

Leucotermes virginicus swarms during the forenoon in the vicinity of Washington, D. C., usually one month later than *flavipes*, or in early June. However, on August 11, 1913, at Falls Church, Mr. William Middleton observed a large swarm to emerge from a small chestnut corner stake which was set in the ground. The colony, which was not large, swarmed at 12.30 p. m.

Leucotermes lucifugus, according to Heath,[b] swarms "* * * at different times between the months of October and April * * *." "It usually takes place about 11 a. m."

During the early part of May, 1909, sexed adults of *Leucotermes* sp. (probably *flavipes*) were observed by the writer emerging in great numbers from the ground in the fenced yard of an old house located in a pine grove near Woodville, Tyler County, Tex. The swarm occurred in the forenoon of a warm, sunny day.

The following notes were made by members of the branch of Southern Field Crop Insect Investigations, in charge of Mr. W. D. Hunter:

On September 12, 1910, a light shower fell, and before the rain had entirely stopped the air was full of flying termites; this lasted one hour. On December 16, 1910, there was a swarm [*lucifugus*] at Dallas, Tex. After a rain at Dallas, October 13, 1911, there was a swarm; another. rain brought them out on October 16, 1911. On October 18, 1912, there were swarms of termites at Dallas, following a two-day's rain.

It will be seen from the foregoing that reliance, in the determination of species, can be placed on the dates of swarming, since the species *flavipes* never swarms in the fall in the eastern United States. Apparently rainfall has no influence on the time of swarming, as is the case in Texas.

---

[a] Hunter, W. D. Some notes on insect abundance in Texas in 1911. Proc. Ent. Soc. Wash., vol. 14, no. 2, p. 62–66, June 19, 1912. See p. 63.

[b] Op. cit., p. 52.

ASSOCIATION WITH ANTS.

Grassi [a] and Escherich [b] have given interesting accounts of the relations between termites and ants. S. A. Forbes [c] states that *flavipes* has been found associated with *Formica schaufussi* Mayr. H. C. McCook [d] states that *flavipes* occurring under stones in the neighborhood of the Alleghenies were seized and carried off by the mound building ants (*Formica exsectoides* Forel) when disturbed. J. C. Branner [e] refers to the common ants as being enemies of the termites. King [f] has noted the association of termites with ants. The following brief notes have been made while investigating damage by termites to trees and forest products:

Termites and ants are commonly to be found inhabiting the same log or stump, yet ants are the enemies most to be feared by termites, as they will capture and carry away the members of a disorganized colony. Ordinarily the relations between termites and ants seem to be neighborly and peaceful. If the termite colony is opened up and disorganized, the ants at once take advantage of the opportunity and carry away the termites, which offer but little resistance. Ants of several species may be attracted to such a helpless colony from a distance. The soft-bodied soldiers are apparently not very effective in such emergencies, although in the narrow channels of the colony, where the powerful head with open mandibles is the only front presented to the marauding ants, they afford some protection to the colony.

Two species of carpenter ants (*Camponotus pennsylvanicus* Mayr and *Cremastogaster lineolata* Say[g]) are the ants which more commonly have been found associated with termites in the eastern United States. The latter species, due to its small size and rapidity of movement, is a most formidable enemy.

Ants greatly diminish the number of the colonizing individuals at the time of the swarm, carrying them away as they are running about on the ground. Soldiers and workers guard the breaches from which the sexed adults have emerged, possibly to keep the greedy ants from following their prey to the parent nest.

[a] Grassi, B., and Sandias, A. Op. cit., p. 282–283.

[b] Escherich, K. Die Termiten oder weissen Ameisen. Eine biologische Studie, p. 122–126, Leipzig, 1909.

[c] Forbes, S. A. Nineteenth Report of the State Entomologist on the Noxious and Beneficial Insects of the State of Illinois, p. 198, Springfield, Ill., 1895. The white ant in Illinois (*Termes flavipes*, Kollar), p. 190–204.

[d] McCook, H. C. Note on mound-making ants. Proc. Acad. Nat. Sci. Phila. for 1879, p. 154–156, August 12, 1879.

[e] Branner, J. C. Geologic work of ants in tropical America. Bul. Geol. Soc. Amer., v. 21, p. 449–496, figs. 11, pl. 35, August 20, 1910. See p. 478–479.

[f] Loc. cit.

[g] Identi. ed by Mr. Theodore Pergande, of the Bureau of Entomology.

TERMITOPHILOUS INSECTS.

The presence of termitophilous insects or "guests" in colonies of *Leucotermes flavipes* in the United States has been recorded by several writers on termites. Mr. E. A. Schwarz[a] has published an extensive list of Coleoptera associated with *flavipes*. Inquilines, or guests, are found only in permanent colonies and, as brought out by Mr. Schwarz,[b] might be of importance in establishing the original habitat of a species; that is, if a termite species had peculiar inquilines (guests that do not occur among other species of termites) in one country and none in another, it would indicate that the termite species was native to the country where the inquilines occur in its colonies. Mr. H. G. Hubbard found the staphylinid beetles *Trichopsenius* and *Anacyptus* in a colony with supplementary royalty, April 20, 1882, near Crescent City, Fla. He also found peculiar wingless psocids which resemble young termites in a colony of the latter, near Haw Creek, Fla., March 26, 1895. The note in his field diary reads:

Termitophilous psocid found with termites in large log of pine, swampy hummock of Prairie Farm. Several specimens in alcohol, together with worker of termite. * * * The resemblance to a young termite is perfect, especially in mature specimens. * * * The psocid is, however, much more active than the termite and very difficult to capture. Immature specimens were not rare. * * * The immature specimens inhabit the galleries of the termites, but are not so apt to be found among the termites as in their immediate vicinity.

King[c] records three inquilinous staphylinid beetles as associated with *flavipes*, *Philotermes pilosus* Kraätz, *Homalota* sp., and *Tachyporus jocosus* Say.

All stages of small scarabæids, *Homovalgus squamiger* Beauvois and *Valgus canaliculatus* Fabricius, are commonly found associated with termites in decaying wood in Maryland, Virginia, West Virginia, and North Carolina, and are probably truly inquilinous. From the middle of July, 1914, prepupal larvæ and pupæ of *Homovalgus squamiger* were commonly found in decaying wood infested by termites and in the galleries of termites in Virginia. Adults of this beetle begin to mature about the middle of August. The larvæ construct oval pupal cells in the decayed wood or make them of earth; the interior is smooth and glossy. This beetle is probably a true inquiline.

---

a Schwarz, E. A. Termitophilous Coleoptera found in North America   Proc. Ent. Soc. Wash.. vol. 1, no. 3, p. 160–161, March 30, 1889.
Schwarz. E. A. Additions to the lists of North American termitophilous and myrmecophilous Coleoptera. Proc. Ent. Soc. Wash., vol. 3, no. 2, p. 73-78, January , 1895.
b Schwarz, E. A. Termitidæ observed in southwestern Texas in 1895. Proc. Ent. Soc. Wash., vol. 4, no. 1, p. 38–42, November 5, 1896.
c Op. cit., p. 196

Beetles of the family Pselaphidæ are frequently found in decaying wood near termite nests, and some are known to be their guests. Adults of *Tmesiphorus carinatus* Say [a] were found in decaying wood in which colonies of *flavipes* were present at Falls Church on March 18, 1912. Mr. E. A. Schwarz has included this species in his list of myrmecophilous beetles as "often found among ants of various species." [b] Adults of *Batrisus virginiæ* Casey [a] were found in decaying wood infested with *virginicus* on the same day.

Adults of the staphylinid, *Philotermes pennsylvanicus* Kraätz [a] were collected with *flavipes* near Kane, Ill., August 11, 1911, in the butt of a decaying white cedar telegraph pole. The species is a true inquiline, and the beetles are very active. On August 16, 1913, near Chain Bridge, Va., an adult of *Philotermes* sp. (possibly *fuchsii* Kraätz) (determined by Mr. H. S. Barber of the Bureau of Entomology) was found in a colony of *virginicus*. Blatchley [c] records *Philotermes pilosus* Kraätz and *P. fuchsii* Kraätz in the nests of *flavipes* in Indiana.

PARASITES.

Termites are infested externally with mites and internally with protozoan parasites, but no internal or external feeding insect parasites have been recorded by Leidy,[d] Grassi,[e] or Porter.[f] Grassi states that the presence of these protozoa in the intestine retards sexual development, as evidenced in the case of workers and soldiers. He further states that they are normally absent in the reproductive forms and newly hatched larvæ.

SUMMARY AND CONCLUSIONS BASED ON THE RESULTS OF THE EXPERIMENTS.

The following conclusions are based on observations of colonies in the termitarium, colonies in small tin boxes, and other colonies in the forest at Falls Church.

Colonizing individuals of both sexes swarm together from colonies of *Leucotermes flavipes* and *L. virginicus* from about 11 a. m. to 1 p. m., the length of time occupied by the adults in emerging being about one hour. No evidence of the separate swarming of the sexes

---

[a] Identified by Mr. E. A. Schwarz.

[b] Schwarz, E. A. Myrmecophilous Coleoptera, found in temperate North America. Proc. Ent. Soc. Wash., v. 1, No. 4, p. 237–247, May 15, 1890.

[c] Blatchley, W. S. The Coleoptera or beetles of Indiana, p. 343–344, Indianapolis, 1910.

[d] Leidy, J. On intestinal parasites of *Termes flavipes* Proc. Acad. Nat. Sci. Phila. [v. 29] for 1877, p. 146–149, June 26, 1877.

Leidy, J. The parasites of the termites. Jour. Acad. Nat. Sci. Phila., ser. 2, v. 8, p. 425–447, pls. 51–52, February, 1881.

[e] Grassi, B., and Sandias, A. Op. cit., p. 11–13.

[f] Porter, J. F. Trichonympha and other parasites of *Termes flavipes*. Bul. Mus. Comp. Zool., v. 31, no. 3, p. 45–68, pls. 6, October, 1897.

or seasonal dimorphism has been observed.   However, all the individuals do not necessarily emerge from the same colony at the same time, since there may be several swarms from the same nest.   On May 14, 1912, at Falls Church sexed adults of *flavipes* emerged, although not in great numbers, from colonies, whereas there was evidence, by the discarded wings on the ground, of an earlier swarm. Most of the colonizing individuals had swarmed from other colonies on May 8; yet some individuals, not in the enormous numbers of the first swarm, were observed swarming from the same colonies on May 14.   While there may be as many as four swarms, the first is usually the largest.   This may be explained by the fact that there is an uneven development of individuals.   Indeed, a few retarded, winged, sexed adults (*virginicus*) may remain in the colony till July 24 (near Kane, Ill.) and early August (District of Columbia) or be found, as individuals, flying.   A large swarm of *virginicus* has emerged from a colony as late as August 11, 1913, at Falls Church.

The winged insects usually crawl up to some elevated place before taking flight.   There is an enormous mortality of the colonizing individuals, and insectivorous animals destroy them in great numbers.   The swarm is not a "nuptial flight."

The so-called "amatory passages" possess a sexual significance, and there is a mutual attraction between the sexes several days before the normal period of swarming.   This can be observed if the colony is disturbed and the colonizing forms emerge prematurely, and is evidenced even before the loss of the wings.   This attraction, probably due to some secretion, continues till after the royal pair is established in the royal cell and copulation has taken place.   Copulation probably occurs a week after the swarm of *flavipes;* that is, on May 15 adults that had swarmed on May 8 were in the royal cell and apparently no longer following each other about, head close to abdomen, as previously.   There is evidence that individuals of neither sex are sexually mature at the time of swarming, and that there is further development before copulation, as can be noted in the increase in size of the abdomen of both sexes.

The colonizing individuals are not all irretrievably lost.[a]   The establishment of new colonies by these forms is a normal process. Although there is an enormous mortality at the time of the swarm, and a still further diminution in numbers due to inability to become established under favorable conditions, yet some pairs do become established.   These sexed adults are not necessarily monogamous.

---

[a] Perris, E.   Nouvelles promenades entomologiques, *Termes lucifugus.*   Ann. Soc. Ent. France, sér. 5, t. 6, p. 201–202, 1876.

Pérez, J.   Sur la formation de colonies nouvelles chez le termite lucifuge (*Termes lucifugus*).   Compt. Rend. Acad. Sci. (Paris), t. 119, No. 19, p. 804–806, Nov. 5, 1894.

Heath, Harold.   Loc. cit.

True royal pairs are independent of workers in the foundation of new colonies; that is, it is not necessary that they be found and established in royal cells by foraging workers and soldiers, although it is possible that this occurs, which would then constitute an independent colony. New colonies established by sexed adults that swarmed were found both in the termitarium (where conditions were similar to those in nature) and in the forest. The king and queen are equally important; they continue to cohabit and coition is repeated. The first brood reared by young true queens that have swarmed consists of workers and soldiers of smaller size than the normal form.

Neoteinic royal individuals are to be found commonly in colonies of *flavipes* in the eastern United States; they are "ergatoids" or "neotenes" and are developed by retarding the normal development of nymphs of the first form at an early stage, and from young larvæ; sometimes as many as 40 or more, consisting of many queens and a few males, may be present in the same colony. The males are polygamous. It is believed that these forms are provided (1) when overcrowded old colonies are split up and new, independent colonies are to be established, or (2) through the actual loss of the true royalty or by the accidental separation of some members of the colony from the royalty. This method of the formation of new colonies is sure, and probably much more rapid in the case of establishment by the true colonizing forms, as the royalty would receive the care of the workers, would not have to forage for themselves, and their only function would be reproduction. The number of eggs laid and the rate of increase would necessarily be much more rapid, due to the following facts: (1) The abdomens of mature, fertilized, supplementary queens are nearly as fully distended as in the case of true queens; (2) the large number of supplementary royalty possible to be present in a single colony and consequent increase in the number of eggs laid, and (3) the proper care and nourishment the royalty would receive.

It is often noted that the antennæ of the reproductive forms are mutilated, that is, have a greater or less number of segments missing. However, this is not always the case in young reproductive forms. The loss of the segments in individuals in long-established colonies might be due to the treatment the latter receive from the workers, or from being dragged about by them, but the antennæ are sometimes mutilated in individuals in incipient colonies.

There is a series of molts and "quiescent stages" in the development of the larvæ of the castes; caste differentiation occurs during such a stage, which corresponds somewhat to the pupal stage in insects with complete metamorphoses.

In conclusion, the more important facts may be summarized as follows: There is great variation in the life cycle of Leucotermes;

FIG. 1.—Chestnut telephone pole, with base charred but basal area untreated. This pole has been standing near Dover, N. J., for eight years. (Photograph loaned by United States Forest Service.)

FIG. 2.—Pine flooring honeycombed by termites at New Iberia, La. (Original.)

DAMAGE BY TERMITES.

these insects are adapted to meet emergencies successfully and overcome obstacles without the disorganization of the colony. New
colonies may be established (1) by the sexed colonizing adults that
invariably swarm and leave the parent colony; (2) by neoteinic royal
individuals, produced from nymphs of the second form which never
(?) leave the parent colony, or from young larvæ, as in colonies
orphaned after the nymphs of the first form have nearly completed
their development; (3) by neoteinic reproductive forms supplied to
orphaned colonies, which may be derived from nymphs of the first or
second forms, or larvæ. Nests headed by true royalty are not rare,
but many difficulties are to be surmounted in their establishment by
sexed adults; such recently established colonies are small in number.
Nests headed by neoteinic reproductive forms are more commonly
to be found, as this is a more sure and more rapid method of establishment. Colonies established by neoteinic reproductive forms
necessarily increase in size more rapidly due to the numerous egg-
laying queens and the care and food they receive from the workers.
Subcolonies or temporary colonies are frequently found with only
workers and soldiers present; these subcolonies, which furnish
increased facilities for habitation and food supplies, are possibly
offshoots from the parent colony or nest and are established by
means of subterranean passages, which are extended for long distances by foraging workers and soldiers.

THE DAMAGE TO FOREST PRODUCTS.

Termites seriously injure construction timbers in bridges,[a] wharves,
and like structures; telephone and telegraph poles [b] (Pl. XVII, fig. 1),
hop poles,[c] mine props,[d] fence posts and rails or boarding; lumber
piled on the ground, railroad ties set in the ground (not where there
is stone or slag ballast or heavy traffic), woodwork (Pl. V; Pl. XVII,
fig. 2) in new and old buildings,[e] and especially seriously damage the
wooden boxing or "conduits" of insulated cables placed in the ground
(to the detriment of the insulation); tent pins and ridge poles, wooden

a Hagen, H. A. The probable danger from white ants. Amer. Nat., v. 10, no. 7,
p. 401–410, July, 1876.

b Snyder, T. E. Insects injurious to forests and forest products. Damage to chestnut telephone and telegraph poles by wood-boring insects. U. S. Dept. Agr., Bur.
Ent., Bul. 94, pt. 1, pp. 12, figs. 3, pls. 2, December 31, 1910. See p. 9–10.

c Parker, W. B. California redwood attacked by *Termes lucifugus* Rossi. Jour.
Econ. Ent., v. 4, no. 5, p. 422–423, October, 1911.

d Snyder, T. E. Insect damage to mine props and methods of preventing the
injury. U. S. Dept. Agr., Bur. Ent., Circ. 156, pp. 4, July 13, 1912. See p. 2–3.

e Marlatt, C. L. The white ant. (*Termes flavipes* Koll.). U. S. Dept. Agr., Bur.
Ent., Circ. 50, rev. ed., pp. 8, figs. 4, January 27, 1908.

Hopkins, A. D. Insect injuries to forest products. U. S. Dept. Agr. Yearbook
1904, p. 381–398, 1905. White ants, or termites, p. 389–390.

beehives and tree boxes; wooden electrotype blocks, and books (Pl. IV) and documents stored in damp, dark places, etc.; timber in contact with the ground being especially liable to serious damage. Often the damaged material has to be removed and replaced, or rebuilt. The wood of no species of native tree of commercial importance is "immune" to attack, although some are relatively more resistant than others. Such damage has occurred as far north as Boston and the shores of the Great Lakes, but greatly increases as the Tropics are approached. In the Southern States termites are especially destructive to wooden underpinning, beams, and flooring (Pl. XVII, fig. 2), and all other material of wood accessible in buildings. They enter buildings by means of tunnels through the ground, by way of wooden joists, or by means of covered paths (minute "sheds" constructed of earth and excrement of the superficial consistency of sand), leading to the woodwork over the surface foundations of stone or other material which they can not penetrate. This enables them to avoid the light. Thus termites silently, secretly, and ceaselessly work their insidious damage, instinctively never perforating the exposed surface of timber, except to enable the sexed adults to swarm. Sometimes the emergence of these winged forms is the first indication of their presence, but at other times joists and floors collapse without warning.

PREVENTIVES, REMEDIES, AND "IMMUNE" WOODS.

Forest products in contact with the ground should be impregnated with coal-tar creosote, which is a permanent preventive against attack by our native termites. Coal-tar creosote has many properties which would recommend its use in this respect, for it is also a fungicide, and, being insoluble in water, will not leach out in wet locations. These requirements furnish objections to many chemicals that otherwise are very effective insecticides. The various methods of superficially treating timber, as by charring, by brushing, or by dipping with various chemical preservatives, among which are creosotes, carbolineums, etc., have proven to be temporarily effective in preventing attack,[a] if the work is thoroughly done. If not thoroughly done, termites enter through the untreated or imperfectly treated portions, especially through weathering checks and knots. Where the bases of poles, mine props, etc., are left untreated termites enter the timber from below, and, avoiding the treated portions, come up through the interior. Charred timber is effective against termite attack for a period less than a year, although it is not seriously damaged at the end of one year. It will readily be seen that neither brushing nor spraying the exterior after place-

---

[a] Snyder, T. E. Insects injurious to forests and forest products. Damage to chestnut telephone and telegraph poles by wood-boring insects. U. S. Dept. Agr., Bur. Ent., Bul. 94, pt. 1. pp. 12, figs. 3, pls. 2, Dec 31, 1910. See p. 9–10.

ment, as is sometimes practiced, is effective in keeping out termites, since the portion that sets in the ground could not be treated, and it is usually at this point that termite attack occurs.

Before treating timber with chemical preservatives, especially where the brush method is employed, it is essential that the timber be thoroughly seasoned, otherwise penetration by the preservative will be retarded.

A treatment with "blue oil" is recorded as apparently effective in protecting wood against the attacks of "white ants," or termites, besides acting as a preservative (fungicide) generally. "Blue oil"[a] is the residue left in the distillation of mineral oils after the isolation of kerosene (petroleum) and paraffin; (a) the oil to be a shale product; (b) its specific gravity (at 60° F.) to be 0.873 to 0.883; (c) its flashing temperature to be not lower than 275° F. (close test).

Many patented wood preservatives, advertised as effective against wood borers, often merely contain simple preservatives, as for instance, linseed oil, to which a slight odor of oil of citronella has been imparted, or contain simple poisons. For timber to be set in the ground, brush coatings with linseed oil are not effective against termites.

An English firm manufactures a saccharine solution which probably contains a salt as arsenic [b]; this patented treatment is supposed to be efficient against wood-boring insects, especially termites. The wood is seasoned by immersing in the saccharine solution at 120 to 140° F. This process is being tested.

Impregnation with chlorinated naphthalene may prove effective against termites, as a preservative for woodwork, in interior finish, where a requirement is that the preservative should not "sweat" out, or stain the wood. Treated wood blocks buried in the ground with termite-infested logs were not attacked after a test of nearly six months.[c] Impregnation with paraffin wax was not effective (fig. 14). If the wood is not in contact with the ground, impregnation treatments with bichlorid of mercury and zinc chlorid are effective. The mercury and zinc in this form are both soluble in water.

H. W. Bates, in a paper entitled "On the prevention of destruction of timbers by termites," Transactions of the Entomological Society of London, 1864, Vol. I, p. 185, cites preventive measures. M. J. Berkeley, in The Technologist, (London), 1865, Vol. V, p. 453, gives remedies based on the report by the committee of inquiry into

[a] The protection of timber against white ants. Trans. Roy Scot Arbor Soc., v. 23, pt. 2, p. 227–228, July, 1910.

Dixon, W. B. Protection from "white ants" and other pests. Nature, v. 85, no. 2148, p. 271, December 29, 1910.

[b] Chemical abstracts, v. 7, no. 2, p. 408, January 20, 1913.

[c] Impregnation of wood to resist insect attack. Amer. Lumberman, no. 2009, p. 32, November 15, 1913.

the ravages by white ants at St. Helena.   Froggatt gives[a] preventive
measures applicable in New South Wales.

FIG. 14.—Red oak block, impregnated with paraffin wax, honeycombed by termites after five months'
test   This was buried in the ground with termite-infested logs. (Original.)

[a] Froggatt, W. W.   White ants, with some account of their habits and depredations.
Misc. Pub. no. 155, Dept. Agr., Sydney, N. S. Wales, Sydney, 1897. See p. 6–8,
account of depredations and methods of prevention.

The wood-pulp products and various patented "composition boards" used as substitutes for lath, etc., might be made termite proof by adding during their manufacture such poisons as white arsenic, antimony, bichlorid of mercury, zinc chlorid, etc; tests are being conducted.

In general, serious damage by termites to the wood of fire or insect killed, standing, merchantable timber can be prevented if the timber is utilized within from one to two years from the time that it was killed, depending on the species of wood and the locality—one year for pine (less in the Southern States) and two years for chestnut.

Forest tree nursery stock should be planted in ground that has been plowed deep late in the fall of the year in a region where injury by termites is common.

Marlatt [a] states regarding white-ant infestation in buildings that setting foundation beams or joists in concrete is only a partial protection, since in the settling of the house the concrete will crack and afford entrance to the insects. Some protection is afforded by removing decaying stumps or posts, etc., adjacent to buildings, by drenching infested timbers with kerosene, and by removing infested joists in cellars and drenching the ground where they were set with kerosene (or carbon bisulphid). Where the injury is confined to exposed woodwork in buildings, hydrocyanic-acid gas fumigation [b] is to be recommended, the infested beams and joists beneath being exposed, if possible, by opening up the floors.

Certain species of wood appear to be naturally highly resistant to termite attack. Species of wood that, so far as determined by test,[c] have been resistant to attack by our native termites are all tropical species and woods too expensive for ordinary use, including teak (*Tectonia grandis*) from Siam and Burma, greenheart (*Nectandra rodiæi*) [d] from South America and the West Indies, "peroba" (several species of *Aspidosperma*) from South America, and mahogany (*Swietenia mahoghani*) from tropical America. Hagen [e] states that, according to Kirby, "Indian oak" or teak (*Tectonia grandis*) and "ironwood" (*Sideroxylon*) of India are immune to attack by termites. This immunity (?) or relative resistance of ironwood is not due to hardness, since Asiatic termites attack the hardest wood, *Lignum-vitæ*, but to the presence in the wood of substances (as oils or alka-

[a] Op. cit.

[b] Howard, L. O., and Popenoe, C. H. Hydrocyanic-acid gas against household insects. U. S. Dept. Agr., Bur. Ent., Circ. 163, p. 8, November 29, 1912.

[c] Impregnation of wood to resist insect attack. Amer. Lumberman, no. 2009, p. 32, November 15, 1913.

[d] This wood has been superficially attacked by termites after 12 months' test; the wood was eaten to the depth of one-eighth inch.

[e] Hagen, H. A. Monographie der Termiten. Linnæa Entomologica, Bd. 10, p. 44–45, 1855.

loids) repellent or distasteful to termites.[a] The presence of tyloses
or of gums may be factors in determining[b] the durability and
resistance of hardwood species. Hagen further states that, while teak
is not destroyed by termites, he believes that even teak will be
attacked when it has become old or been long exposed to the air.
Hagen states that it is useless to consider tannin as a preservative,
since, according to Williamson, termites will destroy leather. Some
species of woods native to the Philippine Islands are apparently
immune to termite attack.

Capt. Ahern, Chief of the Philippine Forestry Bureau, quotes
Foreman, p. 390, as stating[c] that the "anay," or native termite, "eats
through most woods (there are some rare exceptions, such as 'molave,'
'ipil,' 'yacal,' etc.)." Capt. Ahern states that—

The following woods are not subject to attack by anay: "Dinglas" [*Eugenia brac-
trata* Roxb., var. *roxburghii* Duthie, family Myrtaceæ], "ipil" [*Afzelia bijuga* Gray,
family Leguminosæ], "molave" [*Vitex littoralis* Dene, family Verbenaceæ], and
"yacal" [*Hopea plagata* Vidal, family Dipterocarpeæ].

"Tindalo" [*Afzelia rhomboidea* Vid.] is attacked by "anay" when there is no other
wood in the vicinity.

"Baticulin" [*Litsea obtusata* B. and H., F. Vill, fam. Laurineæ] is attacked by
"anay," but is not damaged or destroyed, except such parts as are buried underground.

TEST WITH THE WHITE ANT.

Mr. D. N. McChesney, master mechanic at the depot quartermaster shops in Manila,
found last February that his trunk (made of an American spruce) had been invaded
by white ants and was almost entirely destroyed; the clothes contained in the trunk
were also eaten. He placed the trunk on the ground and near it pieces of the following
woods:

Result of 30 days' contact with ants.

American woods:

Oregon pine	Entered and eaten; a mere matter of time for complete destruction.
Bull pine	Eaten more readily than Oregon pine.
Spruce	Do.
Western hemlock	Not touched.
California redwood [*Sequoia semper-virens*].	Ants tried, but discontinued after a slight effort.
California white cedar	Do.

Native woods:

Molave	Ate a little of it; deepest hole about one-fourth inch.
Narra [*Pterocarpus indicus* Wild., family Leguminosæ].	Ate a little of it; deepest hole about one-half inch.
Painted wood	Ants worked under paint and ate the wood readily.

---

[a] The quality of hardness in wood, while not rendering a species of wood immune
to termite attack, is an important factor in determining the relative resistance of
species of woods. Hardness is a factor in the grading of mahogany lumber.

[b] Gerry, M. "Tyloses: Their occurrence and practical significance in some American woods." Jour. Agric. Research, vol. 1, no. 6, p. 462–464, March 25, 1914.

[c] Ahern, G. P. Important Philippine woods. 112 pp., col. pls. Forestry Bur.,
Manila, P. I., January 2, 1901. See p. 91

Hemlock and redwood [a] are badly honeycombed by our native species of termites on the Pacific coast, and California white cedar is honeycombed by the termites of the Pacific coast.

European cypress (*Cupressus sempervirens* Linn.) is reported [b] damaged by termites at Rochefort, France.

Cedrus deodar, from India, and *Cedrus atlantica*, of the mountains of northern Africa, are reported to be immune (?) to termite attack.

Red "deal" is less liable to attack than white "deal." [c]

The following three species of wood remained untouched by termites for three years in the District of Pretoria, Transvaal: [d] "Leadwood" (*Combretum prophyrolepsis*), "black ironwood" (*Olea laurifolia*), "vaalbosch" (*Brachylæna discolor*).

In case of certain species of pines, with an extremely resinous heartwood, as the "fatwood" of the longleaf pine (*Pinus palustris*) of the Southern States, while termites honeycomb the sapwood, the heartwood apparently is resistant. Odenbach states that turpentine is very repellent to termites in artificial nests. In southern Rhodesia [e] the wood of the "mopani" tree (*Copaifera mopani*) withstands termite attack for years, and is therefore very suitable for straining posts for fences, though unfortunately not a timber that can be cut and squared.

Tests of the relative resistance of various native and exotic woods have been begun, but as yet no definite conclusions can be drawn. The heartwood of the following native species of wood is relatively more resistant to attack by our native termites: Black locust; black walnut [f] (cases on record where heavy beams supporting flooring in a building in Baltimore, Md., were completely honeycombed); eastern white cedar (*Chamæcyparis thyoides*); eastern red cedar or juniper; bald cypress of the Southern States; western red cedar (*Thuja plicata*) of Washington, Oregon, and California; incense cedar (*Libocedrus decurrens*) of Oregon and California; and Monterey cypress (*Cupressus macrocarpa*) of California. All these species of woods, however, are attacked by termites.

---

[a] Parker, W. B. Loc. cit.

[b] Hagen, H. A. Monographie der Termiten. Linnaea Entomologica, Bd. 10, p. 1–144, 1855. See p. 133.

[c] French, C. Handbook of the destructive insects of Victoria, pt. 2, p. 141, Melbourne, 1893.

[d] Howard, C. W., and Thomsen, F. Notes on termites. Transvaal Agr. Jour., v. 6, No. 21, p. 85–93, illus., October, 1907; v. 7, No. 27, p. 512–520, April, 1909; v. 8, No. 29, p. 86–87, October, 1909.

[e] Jack, R. W. Termites or "white ants." Rhodesia Agr. Jour., v. 10, No. 3, p. 393–407, pl., February, 1913.

[f] Impregnation of wood to resist insect attack. Amer. Lumberman, No. 2009, p. 32, November 15, 1913.

Hagen [a] briefs the records of various travelers as to the immune (?) or termite-resistant wood species of the countries of the world. Froggatt [b] states that in Australia red pine is more resistant than clear pine; that "jarrah" is said to be resistant, though not immune; and that desert cypress when sawn up appears to be resistant, but in the form of logs is not immune. The Rev. Joseph Assmuth, S. J.,[c] of Bombay, India, states that "deal" and "pukka" teak are injured by termites; he gives photographs of the damaged specimens of wood. In answer to a letter of inquiry he states:

The "pukka" teak is what is called here in India "Burmese teak," *Tectona grandis* L. "Pukka" means genuine; it is used here in opposition to the less reliable timber of "Malabar teak," though both come from the same species of wood. The difference of both lies, I am told, in the seasoning of the timber, or rather in the different mode of felling the trees. In Malabar they cut off a ring of bark from near the base of the tree, so that the tree dries up while standing still erect, and is then felled. This seems to cause a gradual withdrawal of the oils contained in the wood, which makes the wood more liable to the attack of white ants. In Burmah the tree is felled as it stands and allowed to dry up lying on the ground. Thus the peculiar oils remain in the wood and are preserved in it, and consequently this timber is less palatable to the white ants, and shunned by them until in course of time the oils evaporate also. Then the white ants go for it too. Such, at any rate, is my theory. I can't explain otherwise why the one sort is at once attached by white ants, whereas the other remains for a longer period, at least, immune. The case therefore is briefly this: Malabar teak (here also called "jungle teak") is attacked by the white ants from the beginning; Burmah or "pukka" teak remains safe for a certain period, sometimes longer, sometimes shorter, but it is not absolutely safe either.

"Dealwood" is the common European wood used for boxes and the like; it is usually timber of Abies, Picea, or Pinus. It is the wood most readily attacked by termites of different species.

Kanehira,[d] tabulates the results of experiments with a large number of species of woods as to their relative resistance to termite attack in Formosa, Japan. Reasons are advanced as to the probable causes which render the wood "termite proof" (?). The conclusions he draws can not be accepted without further details as to how the tests were conducted and after a longer period of experimentation. Certain of the woods he lists as "immune" are known to be attacked by termites.

METHODS OF OBTAINING PHOTOGRAPHS FOR THE ILLUSTRATIONS.

The photographs of the insects reproduced in this paper were made after a method devised by Mr. H. S. Barber, of the Bureau of Entomology. The specimens were placed either between horizontal

[a] Loc. cit.

[b] Froggatt, W. W. White ants. (Termitidæ) Misc. Pub. 874, Dept. Agr. N. S. Wales, Sydney, 1905, p. 43–44.

[c] Assmuth, J. Wood-destroying white ants of the Bombay presidency. Jour Bombay Nat. Hist. Soc., v. 22, no. 2, p. 372–384, 4 pls., September 30, 1913.

[d] Kanehira. On some timbers which resist the attack of termites. Indian Forester, v. 40, no. 1, p. 23–41, January, 1914.

glass slides submerged in alcohol or immersed in water in a chamber made by gluing glass slides to thin slips of rectangular cover glass with heated balsam. In this water-tight compartment the specimens can be placed in vertical or horizontal positions. Ether is used to clear out the cell, which is filled with boiled water, thus avoiding the formation of air bubbles. If the specimens are placed head downward, movements of the antennæ, due to vibration, can be avoided. Cells of various sizes are used, or one large cell can be utilized by placing thin slips of cover glass back of the specimens to hold them firmly in position. By clamping the cell to the lens holder of a dissecting microscope placed horizontally, focusing can be conveniently accomplished by adjustments of the screw.

The best results are obtained by etherizing recently killed specimens that have not been placed in alcohol. A 72 millimeter focal length lens brings out the characters to the best advantage, if the enlargement is 6 to 8 diameters.

Many of the photographs were taken by Mr. H. S. Barber and some by Mr. H. B. Kirk at the eastern field station. Most of them, however, were made by the photographic laboratory of the United States Department of Agriculture.

BIBLIOGRAPHY.[a]

This bibliography consists mainly of publications on our native termites or closely allied species; some little-known publications are included. Hagen, up to 1860, Froggatt, Holmgren, Escherich, and Feytaud give extensive bibliographies.

ANDREWS, E. A., and MIDDLETON, A. R. Rhythmic activity in termite communities. Johns Hopkins Univ. Circ., n. s., 1911, no. 2 (whole no. 232), p. 26–34, illus., February, 1911.

ATKINSON, G. F. Some Carolina insects. 1st Rpt. So. Car. Exp. Sta. for 1888, p. 19–56, 1889.

BANKS, N. Two new termites. Ent. News, v. 17, no. 9, p. 336–337, illus., November, 1906. *Cryptotermes cavifrons*, from Florida, and *Termopsis laticeps*, from Arizona, described.

BIDIE, G. White ants—termites eroding glass. Nature, v. 26, no. 675, p. 549, October 5, 1882.

BUCKLEY, S. B. Description of two new species of termites from Texas. Proc. Ent. Soc. Phila., v. 1, p. 212–215, May, 1862 (1863) *Hamitermes (Termes) tubiformans*, from Texas, p. 212, and *Eutermes cinereus*, from Texas, described.

CASEY, T. L. A new genus of termitophilous Staphylinidæ. Ann. N. Y. Acad. Sci., v. 4, p. 384–387, March, 1889.

CORYELL, J. R. The termite pest of the old world. Sci. Amer., v. 59, no. 10, p. 151, September 8, 1888, illus.

DERRY, D. E. Damage done to skulls and bones by termites. Nature, v. 86, no. 2164, p. 245–246, April 20, 1911.

DIXON, W. A. Protection from "white ants" and other pests. Nature, v. 85, no. 2148, p. 271, December 29, 1910.

---

a The references in the bibliography, as well as those in the footnotes, have been verified by the librarian of the Bureau of Entomology, Miss Mabel Colcord.

FITCH, A. Fourth Report on the Noxious and Other Insects of the State of New York. 1858, p. [8] or Trans. *Termes frontalis*. N. Y. State Agr. Soc., v. 17 (1857), p 694, 1858.

FORBES, S A. The white ant in Illinois. (*Termes flavipes* Kollar.) 19th Rpt. State Entomologist . . . of Illinois for 1893 and 1894, p. 190–204, 1895. Lists economic publications on termites.

GRANT, R. D. [Ravages of *Termes flavipes*.] Trans. Acad. Sci. St. Louis, v. 3, Jour. of Proc., p. cclxix, November 19, 1877.

HAGEN, H. A. Report on the Pseudo-Neuroptera and Neuroptera collected by Lieut. W. L. Carpenter in 1873 in Colorado. Ann. Rpt. U. S. Geol. Survey Ter. for 1873, by F. V. Hayden, p. 571–606, 1874.

White ants destroying living trees and changing foliage in Cambridge, Mass. Canad. Ent., v. 17. no. 7, p. 134–136, July, 1885.

The female of *Eutermes rippertii*. Psyche, v. 5, no. 157–159, May–July, 1889, p. 203–208

HAVILAND, G. D. Observations on termites; with descriptions of new species. Jour Linn. Soc. [London], Zool., v. 26, no. 169, p. 358–442, pls. 22–25, April 1, 1898.

Observations on termites or white ants. Ann. Rpt. Smithsn. Inst. for year ending June 30, 1901, p. 667–678, pl. I–IV, 1902.

HUBBARD, H. G. Notes on the tree nests of termites in Jamaica. Proc. Boston Soc Nat. Hist., v. 19, p. 267–274, December 26, 1877, printed March, 1878.

Insects affecting the orange, Washington, 1885. Chap. IX, p. 121–125, White ants or "wood-lice."

KELLOGG, V. L. Are the Mallophaga degenerate psocids? Psyche, v. 9, no. 313, p. 339–343, May, 1902.

KNOWER, H. McE. The development of a termite—*Eutermes* (*Rippertii?*). A preliminary abstract. Johns Hopkins Univ. Circ., v. 15, no. 126, p. 86–87, June, 1896.

The embryology of a termite, *Eutermes* (*Rippertii?*). Jour. Morphol. [Boston] v. 16, no. 3, p. 505–568, pl. 29–32, August, 1900.

A comparative study of the development of the generative tract in termites Johns Hopkins Hospital Bul., v. 12, no. 121–122–123, p. 135–136, 2 figs., Baltimore, April–May–June, 1901.

LINTNER, J. A. Entomology. Proc. Albany Inst., v. 2, p. 48–50, 1878.

OSBORN, H. On the occurrence of the white ant (*Termes flavipes*) in Iowa. Proc Iowa Acad. Sci., v. 5, for 1897, p. 231, 1898.

OSTEN-SACKEN, C. R. Extract from a letter by Baron R. Osten-Sacken, on the specimens of Termes found by him in California. Proc. Boston Soc. Nat. Hist., v. 19, p. 72–73, for January 3, 1877, June, 1877.

PACKARD, A. S. Third Report U. S. Entomological Commission, Washington. 1883, p. 326–329.

On the systematic position of the Mallophaga. Proc. Amer. Philos. Soc. v. 24, p. 264–272, 13 figs., September 2, 1887.

A Textbook of Entomology, New York, 1903.

PARKER, W. B. California redwood attacked by *Termes lucifugus* Rossi. Jour Econ. Ent., v. 4, no. 5, p. 422–423, October, 1911.

RILEY, C V , editor. Amer. Ent., v. 3 (n. s., v. 1), p. 15, January, 1880. Termitophilous insects collected by E A. Schwarz recorded.

RILEY, C. V., and HOWARD, L. O. Termites swarming in houses. U. S. Dept. Agr, Div. Ent., Insect Life, v. 6, no. 1, p. 35, November, 1893.

SCAMMELL, E. II. White ants. Knowledge, n. s., v. 4, p. 10–12, January, 1907. Protection of wood by a newly devised process.

SCHWARZ, E. A. Staphylinidæ inquilinous in the galleries of *Termes flavipes*. Amer. Ent., v. 3 (n. s., v 1), p. 15, January, 1880

Termitophilous Coleoptera found in North America. Proc. Ent. Soc. Wash., v. 1, no. 3, p. 160–161, 1889.

Scudder, S. H. Remarks upon the American white ant. Proc. Boston Soc. Nat. Hist., v. 7, p. 287–288, 1860.

Further injury to living plants by white ants. Canad. Ent., v. 19, no. 11, p. 217–218, November, 1887.

More damage by white ants in New England. Psyche, v. 6, no. 177, p. 15–16, January, 1891.

The fossil white ants of Colorado. Proc. Amer. Acad. Arts and Sci., v. 19 ` (n. s., v. 11), p. 133–145, 1884.

Smith, E. F. White ants as cultivators of fungi. Amer. Nat., v. 30, p. 319–321, April, 1896.

Snyder, T. E. Record of the finding of a true queen of *Termes flavipes* Kol. Proc. Ent. Soc. Wash., v. 14, no. 2, p. 107–108, pl. 3, June 19, 1912.

Changes during quiescent stages in the metamorphosis of termites. Proc. Ent. Soc. Wash., v. 15, no. 4, p. 161–165, December, 1913.

Stokes, A. C. The sense-organs on the legs of our white ants. *Termes flavipes* Koll. Science, v. 22, no. 563, p. 273–276, November 17, 1893.

Warren, E. Some statistical observations on termites, mainly based on the work of the late Mr. G. D. Haviland. Biometrika, v. 6, no. 4, p. 329–347, illus., March, 1909.

Wheeler, W. M. The embryology of *Blatta germanica* and *Doryphora decemlineata*. Jour. Morphol., Boston, v. 3, no. 2, p. 291–386, pl. 15–21, September, 1889.

The phylogeny of the termites. Biol. Bul., v. 8, no. 1, p. 29–37, December, 1904.

The fungus-growing ants of North America. Bul. Amer. Mus. Nat. Hist., v. 23, p. 669–807, pl. 49–53, September, 1907. Including Pt. IV, 1, The fungus-growing termites, p. 775–786.

RELATION OF TERMITES TO THE ORIGIN OF HOG-WALLOWS AND PRAIRIE MOUNDS.

Barnes, G. W. The hillocks or mound-formations of San Diego, California. Amer. Nat., v. 13, no. 9, p. 565–571, September, 1879.

Branner, J. C. Decomposition of rocks in Brazil—Burrowing animals—Ants. Bul. Geol. Soc. Amer., v. 7, p. 295–300, February 4, 1896.

Ants as geologic agents in the Tropics. Jour. Geol., v. 8, p. 151–153, February– March, 1900.

Natural mounds or hog-wallows. Science, n. s., v. 21, no. 535, p. 514–516, March 31, 1905.

Geologic work of ants in tropical America. Bul. Geol. Soc. Amer., v. 21, p. 449–496, fig 11, pl. 35, August 20, 1910.

Forshey and Copes. Mounds. Proc. New Orleans Acad. Sci., v. 1, no. 1, p. 18–20, March 1, 1854.

Hilgard, E. W. Physico-geographical and agricultural features of the State of California. U. S. Dept. Int., Census Off. 10th Census U. S., v. 6. Report on cotton production in the United States, Part II, p. 663–741. Hog-wallows, p. 676–677, 693.

Le Conte, J. Prairie mounds. Proc. Cal. Acad. Sci., v. 5, p. 219–220, January, 1874.

Hog-wallows or prairie mounds. Nature, v. 15, p. 530–531, April 19, 1877.

McCook, H. C. Note on mound-making ants. Proc. Acad. Nat. Sci. Phila., 1879, p. 154–156, 1880.

Turner, H. W. Hog-wallow mounds. U. S. Geol. Survey, Ann. Rept. 17, 1895–96, Pt. I, p. 681–683, pl. 33, 1896.

Veatch, A. C. Formation of natural mounds—Ant-hill theory. U. S. Geol. Survey, Prof. Paper no. 46, p. 55–56 and 58–59, 1906.

Wallace, A. R. Glacial drift in California. Nature, v. 15, p. 274–275, January 25, 1877.

INDEX.

○

U. S. DEPARTMENT OF AGRICULTURE,

BUREAU OF ENTOMOLOGY—BULLETIN No. 95.

L. O. HOWARD, Entomologist and Chief of Bureau.

PAPERS ON CEREAL AND FORAGE INSECTS

CONTENTS AND INDEX.

Issued February 7, 1913.

WASHINGTON:
GOVERNMENT PRINTING OFFICE.
1913.

U. S. DEPARTMENT OF AGRICULTURE,

BUREAU OF ENTOMOLOGY—BULLETIN No. 95.

L. O. HOWARD, Entomologist and Chief of Bureau.

PAPERS ON CEREAL AND FORAGE INSECTS.

I. THE TIMOTHY STEM-BORER, A NEW TIMOTHY INSECT.
By W. J. PHILLIPS, *Entomological Assistant.*

II. THE MAIZE BILLBUG
By E. O. G. KELLY, *Entomological Assistant.*

III. CHINCH-BUG INVESTIGATIONS WEST OF THE MISSISSIPPI RIVER.
By E. O. G. KELLY, *Entomological Assistant,*
AND
T. H PARKS, *Entomological Assistant*

IV. THE SO-CALLED "CURLEW BUG"
By F. M. WEBSTER, *In Charge of Cereal and Forage Insect Investigations.*

V. THE FALSE WIREWORMS OF THE PACIFIC NORTHWEST.
By JAMES A. HYSLOP, *Agent and Expert*

VI. THE LEGUME POD MOTH.
THE LEGUME POD MAGGOT
By JAMES A. HYSLOP, *Agent and Expert*

VII. THE ALFALFA LOOPER IN THE PACIFIC NORTHWEST.
By JAMES A. HYSLOP, *Agent and Expert*

WASHINGTON:
GOVERNMENT PRINTING OFFICE.
1913.

BUREAU OF ENTOMOLOGY.

L. O. Howard, *Entomologist and Chief of Bureau.*
C. L. Marlatt, *Entomologist and Acting Chief in Absence of Chief.*
R. S. Clifton, *Executive Assistant.*
W. F. Tastet, *Chief Clerk*

F. H. Chittenden, *in charge of truck crop and stored product insect investigations.*
A. D. Hopkins, *in charge of forest insect investigations.*
W. D. Hunter, *in charge of southern field crop insect investigations.*
F. M. Webster, *in charge of cereal and forage insect investigations.*
A. L. Quaintance, *in charge of deciduous fruit insect investigations.*
E. F. Phillips, *in charge of bee culture.*
D. M. Rogers, *in charge of preventing spread of moths, field work.*
Rolla P. Currie, *in charge of editorial work.*
Mabel Colcord, *in charge of library*

Cereal and Forage Insect Investigations.

F. M. Webster, *in charge*

Geo I. Reeves, W. J. Phillips, C. N. Ainslie, E. O. G. Kelly, T. D Urbahns, H. M. Russell, Geo. G. Ainslie, J. A. Hyslop, W. R. Walton, J. T. Monell, J. J. Davis, T. H. Parks, H. L. Viereck, R. A. Vickery, Henry Fox, J. R. Malloch, V. L. Wildermuth, W. R. McConnell, Herbert T. Osborn, Philip Luginbill, W. R. Thompson, Harrison R. Smith, E. G. Smyth, C. W. Creel, E. J. Vosler, R. N. Wilson, Vernon King, Philip B. Miles, E. H. Gibson, L. P. Rockwood, *entomological assistants*
Nettie S. Klopfer, Ellen Dashiell, *preparators.*
Chas. Petty, *collaborator.*

II

LETTER OF TRANSMITTAL.

U. S. DEPARTMENT OF AGRICULTURE,
BUREAU OF ENTOMOLOGY,
Washington, D. C., December 11, 1912.

SIR: I have the honor to transmit herewith, for publication as Bulletin No. 95, seven papers dealing with cereal and forage insects, and methods for their control. These papers, which were issued separately during the years 1911 and 1912, are as follows: The Timothy Stem-Borer, by W. J. Phillips; The Maize Billbug, by E. O. G. Kelly; Chinch-Bug Investigations West of the Mississippi River, by E. O. G. Kelly and T. H. Parks; The So-Called "Curlew Bug," by F. M. Webster; The False Wireworms of the Pacific Northwest, by James A. Hyslop; The Legume Pod Moth and The Legume Pod Maggot, by James A. Hyslop; The Alfalfa Looper, by James A. Hyslop.

Respectfully,

L. O. HOWARD,
Chief of Bureau.

Hon. JAMES WILSON,
Secretary of Agriculture.

PREFACE.

The articles included in this bulletin relate to insects more or less destructive to cereal and forage crops in the United States. They represent investigations largely completed during the fiscal year 1911–12.

The timothy stem-borer, the subject of Part I, has not during our observations been especially injurious, but is likely to become so should several favorable conditions result in increased abundance. The maize billbug (Part II) and the so-called "curlew bug" (Part IV) are two very closely related insects, the latter being especially destructive in Virginia and the Carolinas, while the former is occasionally quite injurious in the West. The paper relating to chinch-bug investigations west of the Mississippi River (Part III) is exceedingly timely and serves to make more clear the difference in conditions, as regards the chinch bug, between the country west of the Mississippi River and that lying east of it. Parts V, VI, and VII relate to species more or less destructive in the extreme northwestern portion of the United States, a section of the country somewhat peculiar in that it differs greatly in insect fauna from more southern and eastern sections of the country.

All of these papers relate to insects that the farmer must, to a greater or less extent, encounter and control in a successful carrying out of his business.

F. M. WEBSTER,
In Charge of Cereal and Forage Insect Investigations.

CONTENTS.

[1] The seven papers constituting this bulletin were issued in separate form on March 31, April 22, and December 14, 1911, and April 10, April 22, May 31, and October 16, 1912, respectively.

ILLUSTRATIONS.

ERRATA.

Page 2, line 18 from bottom, for *this year* read *in 1910*.

Plate I, facing page 8, after line at bottom, insert: *a, Work of larva in bulb of plan.: b, larva ascending stem to pupate; c, pupa in cell, the gallery plugged with frass below. Much enlarged.* (Original.)

Page 12, line 19 from bottom, before *corn borer* insert double quotation mark in place of single quotation mark.

Page 15, line 16, for *were* read *was*.

In Plate IV, facing page 24, the cut of figure 1 is upside down.

Page 46, line 16 from bottom, after *4* insert *per cent*.

Page 74, last line, insert superior figure 1 (¹) before *Bureau*.

Page 76, line 16 from bottom, for *littorale* read *aviculare*.

Page 77, line 12 from bottom, for *littorale* read *aviculare*.

Page 81, line 4, after *monograph* insert comma.

Page 81, line 5 from bottom, before *Dr.* insert *of*

Page 82, line 5 from bottom, for *littorale* read *aviculare*.

Page 83, line 8, for *littorale* read *aviculare*.

XII

U. S. DEPARTMENT OF AGRICULTURE,

BUREAU OF ENTOMOLOGY—BULLETIN No. 95, Part I.

L. O. HOWARD, Entomologist and Chief of Bureau.

PAPERS ON CEREAL AND FORAGE INSECTS.

THE TIMOTHY STEM-BORER,

A NEW TIMOTHY INSECT.

BY

W. J. PHILLIPS,

Entomological Assistant.

ISSUED MARCH 31, 1911.

WASHINGTON:

GOVERNMENT PRINTING OFFICE.

1911.

CONTENTS.

ILLUSTRATIONS.

PLATE.

TEXT FIGURES.

In 1905 the writer found numbers of larvæ inhabiting timothy at Richmond, Ind. Some time in the fall infested stems were collected and sent to the Department for rearing, but it was the same story—nothing issued.

Early in the spring of 1906 observations were begun with a view to rearing the adult. Infested stems were collected in May, and on June 8 the first adult appeared. Specimens were later submitted to the Department and were found to belong to the above species. Since that time they have been reared repeatedly.

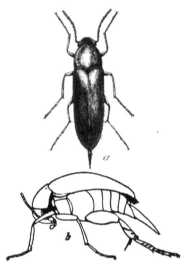

FIG. 1.—The timothy stem-borer (*Mordellistena ustulata*): *a*, Adult or beetle, dorsal view; *b*, same, lateral view. Greatly enlarged. (Original).

DISTRIBUTION.

The habitat of this insect has been given as the middle and southern United States. Adults have been captured as far east as Pennsylvania, and they have been reared from material collected in Indiana, Ohio, and Virginia. Timothy stems containing mordellid larvæ that were not identified, but which were probably *Mordellistena ustulata*, were found in Kentucky, Tennessee, Illinois, and, this year, Mr. T. H. Parks, of the Bureau of Entomology, found them at Chillicothe, Mo.

CHARACTER OF ATTACK.

As a rule the egg is deposited at or slightly below the center of the first or second joint in timothy, but much farther down the stem in other grasses. From here the larva bores into the center of the stem and then begins its downward journey to the bulb or root. It feeds upon the pith and the walls of the stem as it passes downward, and when it encounters a joint it tunnels completely through it, leaving a mass of detritus behind. Plate I is an illustration of its workmanship.

HOST PLANTS.

This species has been reared from timothy, orchard grass (*Dactylis glomerata*), quack grass (*Agropyron* sp.), and *Agrostis alba*, while larvæ that were supposedly this species have been found in bluegrass (*Poa* spp.) and cheat (*Bromus secalinus*).

CONTENTS.

ILLUSTRATIONS.

PLATE.

TEXT FIGURES.

U. S D A , B. F. Bui 95, Part I.   C. F I. I., March 31, 1911.

PAPERS ON CEREAL AND FORAGE INSECTS.

THE TIMOTHY STEM-BORER, A NEW TIMOTHY INSECT.

(*Mordellistena ustulata* Lec.)

By W. J. Phillips,
Entomological Assistant.

INTRODUCTION.

The writer's attention was first attracted, in 1904, to the interesting little insect which is the subject of this paper. On November 29 of that year, at Rives, Tenn., while examining timothy for joint-worms (Isosoma), a curious little larva, unknown at that time to the writer, was found tunneling the stems. In many cases it had traversed the entire length of the stem, from the top joint to the bulb. Although nothing was reared from this material it served to arouse interest. Since that time, however, it has been reared and some interesting facts learned concerning its habits and manner of living.

Thus far it has not proved a serious pest, having been found only in small numbers at any given point. In large numbers it would scarcely do any perceptible injury to the hay crop, although it could probably very materially lessen seed production. For this, as well as other reasons, it deserves more than passing notice.

HISTORY.

The adult (fig. 1) was described by Le Conte in 1862, but there is no reference in literature to its larval habits, although as early as 1877 it was known that larvæ of other species of this genus inhabited plant stems of different kinds.

During the early part of November, 1904, Mr. Geo. I. Reeves, of this Bureau, found larvæ tunneling timothy stems at Richmond and Evansville, Ind., and at Nicholsville, Ohio, but none was reared. In the latter part of the month the writer found a larva working in timothy stems at Rives, Tenn. Nothing could be reared, but in the light of recent observations it is very probable that they were *Mordellistena ustulata* in each instance.

In 1905 the writer found numbers of larvæ inhabiting timothy at Richmond, Ind. Some time in the fall infested stems were collected and sent to the Department for rearing, but it was the same story—nothing issued.

Early in the spring of 1906 observations were begun with a view to rearing the adult. Infested stems were collected in May, and on June 8 the first adult appeared. Specimens were later submitted to the Department and were found to belong to the above species. Since that time they have been reared repeatedly.

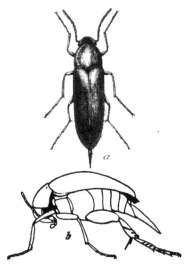

Fig. 1.—The timothy stem-borer (*Mordellistena ustulata*): *a*, Adult or beetle, dorsal view; *b*, same, lateral view. Greatly enlarged. (Original).

DISTRIBUTION.

The habitat of this insect has been given as the middle and southern United States. Adults have been captured as far east as Pennsylvania, and they have been reared from material collected in Indiana, Ohio, and Virginia. Timothy stems containing mordellid larvæ that were not identified, but which were probably *Mordellistena ustulata*, were found in Kentucky, Tennessee, Illinois, and, this year, Mr. T. H. Parks, of the Bureau of Entomology, found them at Chillicothe, Mo.

CHARACTER OF ATTACK.

As a rule the egg is deposited at or slightly below the center of the first or second joint in timothy, but much farther down the stem in other grasses. From here the larva bores into the center of the stem and then begins its downward journey to the bulb or root. It feeds upon the pith and the walls of the stem as it passes downward, and when it encounters a joint it tunnels completely through it, leaving a mass of detritus behind. Plate I is an illustration of its workmanship.

HOST PLANTS.

This species has been reared from timothy, orchard grass (*Dactylis glomerata*), quack grass (*Agropyron* sp.), and *Agrostis alba*, while larvæ that were supposedly this species have been found in bluegrass (*Poa* spp.) and cheat (*Bromus secalinus*).

DESCRIPTION OF THE DIFFERENT STAGES.

THE EGG.

(Fig. 2.)

Length 0.65 mm., diameter near center 0.16 mm. Color milky white. Acuminate-ovate, apparently smooth; one side convex and the other slightly concave; large end broadly rounded, small end acutely rounded.

Described from specimens dissected from gravid females.

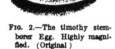

THE LARVA.

(Fig 3, a.)

FIG. 2.—The timothy stem-borer Egg. Highly magnified. (Original)

Length 6 mm., diameter 0.875 mm. Color varying from creamy to white, shading into a very faint tinge of salmon near center. None of the segments appears to be corneous, although each bears a number of setæ; last segment feebly bifid, ending in a two-pointed spear, and covered with stout bristles. Dorsal surface of abdomen with six pairs of fleshy tubercles which will be described later.

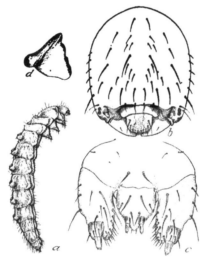

Head (fig. 3, *b*) narrower than thorax and arranged almost vertically with reference to it; ovoid, smooth, and bearing a number of large and small setæ, a very faint suture extending from the occipital area almost to the epistomal area.

Antennæ represented by two fleshy tubercles, situated laterad of and near insertion of mandibles; somewhat cone-shaped and bearing several small setæ on their summits.

Eyes represented by two tiny, slightly fused black dots situated slightly ventro-laterad of the antennæ.

Epistoma sublunate; anterior margin concave and posterior convex; about twice as broad as long and apparently inclosed by base of mandibles and a chitinous ridge extending from insertion of mandibles almost entirely across the face; ridge amber-colored.

FIG. 3.—The timothy stem-borer *a*, Larva, lateral view; *b*, head of same, dorsal view; *c*, maxillæ and labium of same; *d*, mandible of same. *a*, Greatly enlarged, *b*, *d*, more enlarged; *c*, still more enlarged (Original.)

Labrum tonguelike, lying directly over the mandibles, rectangular, longer than broad; distal margin rounded and densely fringed with setæ or bristles; upper surface pappose; two large setæ near center and two on each lateral margin, six smaller ones on distal margin.

Mandible (fig 3, *d*) short, very broad at base, almost as broad as long, tapering abruptly to a sharp point; outer face convex and smooth, with a small seta near center; ferruginous at base and black at extremity, strongly curved; inner face concave, with two small notches midway of superior margin and one small notch near extremity of inferior margin.

Maxillæ (fig. 3, *c*) inserted far to the rear, large, fleshy, curving considerably, thus inclosing the labium; extending considerably beyond tips of mandibles; distal extremity bearing the two-jointed palpus and the lacinia. Each maxilla bears a number of setæ, one large and two small ones occurring on outer face, a large one at outer and one at inner angle of base of palpus, and two small ones and one large one caudad of these. The lacinia is a brushlike organ bearing a fringe of stout bristles Maxillary palpi (fig. 3, *c*) two-jointed; first joint slightly obconical, about as thick as long, bearing several setæ on the outer face; the second joint is a slightly truncated cone, and much smaller than the first joint, and bears a number of minute setæ at the apex.

The *labium* (fig. 3, *c*) is a very simple organ inserted between bases of maxillæ, fleshy, rectangular; distal extremity sharply rounded and fringed with minute setæ, with two larger setæ at tip; four setæ forming a semicircle near center, the two in the center much the largest; a large seta at inner angle of base of each palpus. Labial palpi (fig. 3, *c*) very minute, two-jointed; first joint cylindrical; second joint almost cylindrical but much smaller than the first and slightly rounded at tip, bearing several minute setæ.

Prothorax as large as the two following segments combined; viewed from side triangular in form; not wrinkled or folded but finely striate; dorsally the posterior margin extends back for a considerable distance into the mesothorax. The mesothorax and metathorax lie at quite an angle with the abdominal segments; posterior margin of dorsum of mesothorax extending back to center of metathorax. Metathorax about same width as mesothorax, except on dorsum, where it is somewhat narrower.

Legs fleshy, cone-shaped, four-jointed; first joint very large and more like a projection of the thorax than a joint of the leg; second joint obconical, very short, and very much smaller than the first; third joint cylindrical, short, and very small; fourth joint the smallest of all, obconical, rounded at apex, and bearing three spines at tip; a whorl of spines at each joint on outer face; segments very imperfectly defined in most cases.

Abdomen composed of nine segments, all of which are broader than the mesothorax or metathorax. First six segments bearing on their dorsal surface two round, fleshy, somewhat retractile elevations or tubercles (M. Perris, in his Larves des Coléoptères calls them "ampoule ambulatoire.") These tubercles are almost circular in form and the apex is crumpled and folded and bears several small setæ. Dorsum of the seventh segment with a slight transverse ridge bearing a number of recurved bristles eighth segment bearing a number of bristles, which are more numerous near posterior margin, all directed backward First eight segments with a large fold extending their full length on each lateral face, most prominent near center of segments, at which points there are a number of bristles directed slightly to the rear. Ninth segment somewhat cone-shaped, densely covered with stout bristles, ending posteriorly in a two-pointed, chitinous projection; just below this, dorsally, are two chitinous spur or tubercles

Stigmata: Nine pairs of stigmata, one pair in mesothoracic region just above and slightly in front of insertion of legs and a pair to each of the first eight abdominal segments, very near the anterior margin and just above the lateral fold. They are circular in form, the thoracic being slightly the larger.

THE PUPA.

(Fig. 4.)

From the lateral aspect· Length 5 mm., diameter in thoracic region 1.125 mm. Pale cream color, somewhat acuminate-ovate, broadly rounded at head and thorax. Prothorax, from the lateral view, triangular, the dorsal surface being the base of the triangle, which is very broad and convex. Antenna passing upward at side of eye, between margin of prothorax and front femora, thence to dorsum, curving backward over base of wings. Wing-pads long and narrow, covering posterior legs, with the exception of the last three tarsal joints; front wings nearly covering hind ones.

From the ventral aspect (fig. 4): The front of head is in almost direct line with the body; mandibles small and not closed; palpi widely separated and extending beyond front tibiæ. Femora of first pair of legs directed dorsally, tibiæ resting on middle femora; first tarsal joints resting under tip of palpus; tarsal joints then extending caudad, almost parallel, except last two joints, which slightly diverge. Femora of second pair of legs parallel to femora of first pair, second tibiæ, however, forming a greater angle with their femora than tibiæ of first pair of legs; second femora and tibiæ resting upon wings for a part of their length; tarsal joints gradually converging until the last two, which nearly touch between wing pads. Third pair of legs covered by wing pads, with exception of last two joints and a part of third; last two parallel, touching, and extending to middle of sixth abdominal segment

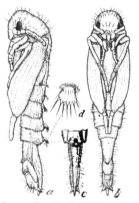

Dorsal surface of the first six abdominal segments almost flat, and in the third, fourth, fifth, and sixth segments the surface produced laterally into a fleshy fold projecting over side of abdomen, the anterior margin of which is rounded and the posterior square, giving the abdomen a notched appearance; in center of each projection laterally is a group of recurved bristles.

Seventh abdominal segment (fig. 4), from a lateral view, cylindrical anteriorly, posteriorly the dorsal surface is drawn out into a long cone-shaped projection, the tip of which extends beyond the last abdominal segment, resting between its two chitinous spurs; ventral surface

Fig. 4.—The timothy stem-borer: *a*, Pupa, lateral view; *b*, ventral view of same, *c*, ninth segment of same; *d*, setigerous tubercle of sixth segment. *a, b, c*, Greatly enlarged; *d*, more enlarged. (Original.)

extending backward into a fleshy fold or lip, beyond insertion of following segment.

Eighth segment (fig 4) somewhat cylindrical anteriorly, telescoping into seventh; dorsal surface extending backward into a large fleshy projection or lip, almost filling space between cone-shaped projection of preceding segment and Y-like chitinous projection of following segment; a deep lateral notch or incision, but ventral projection much shorter than dorsal.

Ninth segment (fig 4) smallest, telescoping into eighth, and extending posteriorly into a Y-like projection, the tips of which, inclining forward, are amber-colored, chitinous, and spinelike

Setæ· Numerous setæ on ventral surface of abdomen, on head, and on prothorax. First abdominal segments bearing a few setæ on dorsal surface; there is a small transverse ridge on dorsum of second segment, bearing a row of setæ; following four segments bearing each two recurved, fan-shaped, fleshy elevations or tubercles, increasing in size with each successive segment (fig. 4, *d*); each elevation simple and bearing a fringe of bristles directed to the rear.

Last three segments bearing a number of bristles; cone-shaped projection of seventh rather thickly studded and last segment densely covered with stout bristles, all directed to the rear.

THE ADULT.

(Fig. 1.)

The description by Le Conte is as follows:

Hind tibia with two oblique ridges on the outer face; ridges parallel, the anterior one extending almost across the outer face of the tibia; first joint of the hind tarsi with three, second with two oblique ridges; elytra ferruginous, with the suture and margin blackish; ferruginous, black limb of the elytra very narrow; abdomen, and sometimes the hind coxæ and pectus, blackish. 9–11.

LIFE HISTORY AND HABITS.

THE EGG.

Females have never been observed in the act of oviposition and the period of incubation of the eggs has never been determined. The latter would be rather difficult to obtain, as eggs that are deposited in living plant tissue rarely hatch after they have been exposed to the air.

As stated above, the egg is usually deposited at and slightly below the center of the first or second joint from the top, within the plant tissues.

The number of eggs that one individual is capable of depositing has not been ascertained. Upon dissection females have never been found to contain more than four fully developed eggs and several immature ones, but they probably deposit a much greater number than this.

THE LARVA.

Upon hatching, the young larva apparently destroys the tissue immediately surrounding it, thus forming a minute cell or cavity. It then eats its way into the center of the stem and starts downward, tunneling the joints as it reaches them, and at harvest time the earlier ones are below the fourth joint, where they will be out of danger of the mower. By fall they have reached a point just above the bulb.

THE MOVEMENTS OF THE LARVA IN THE STEM.

The manner in which the larva propels itself up and down the stem is very interesting. It can ascend or descend the stem, forward or backward, apparently with equal facility. The maxillæ, which extend beyond the mandibles, the true legs, the dorsal tubercles or feet, and the anal segment all play a part in its movements. In going forward the abdomen is advanced by means of the dorsal tubercles, which act as feet; the body is then braced by fixing the spines of the anal segment against the opposite wall of the stem; the maxillæ and true feet then advance the thorax and head. By executing these movements almost simultaneously the larvæ can move quite rapidly. In going backward the movements are reversed. The dorsal feet or tubercles and the anal spines enable the larva to support itself in the stem.

THE MOVEMENTS OF THE LARVA OUTSIDE THE STEM.

Naturally enough, as the larva seems peculiarly adapted for movement in a small hollow stem, when it is placed on a flat surface it appears wholly at a loss how to proceed. It arches its body and turns on its side, going through the same movements as though it were in a stem, but it moves very slowly. It then turns on its back and tries to walk on its dorsal feet. By bringing all of its knowledge of the different ways of walking to bear on the problem, it moves slowly, in a drunken way, to a protecting object, if any be near.

THE PUPA.

When ready to pupate, the larva (Plate I, *a*) reverses its position in the stem and ascends to a point anywhere between the first joint from the root and the first or second joint from the top, depending upon whether the timothy has been cut or not. Plate I, *b*, shows a larva ready to pupate, just below the second joint from the root. It probably locates most often just above the first or second joint from the root. It then seals up the stem above and below with detritus, making a cell of from 1 to 2 inches in length. It will reseal a stem if interfered with, but if its burrow be molested many times it will live for weeks and not pupate, finally dying.

After inclosing itself within this cell the larva becomes sluggish, contracts slightly, and thickens perceptibly in the thoracic region. It soon casts its larval skin and becomes a fully developed pupa. In Plate I, *c*, is seen a pupa in its cell just above the second joint from the root. After pupation it is a pale cream color, gradually changing to a brownish tint.

THE MOVEMENTS OF THE PUPA IN THE STEM.

The movements of the pupa in ascending and descending the stem are fully as complicated and interesting as in the case of the larva. In moving up the stem, the spurs of the last segment are planted firmly in the wall; the body is then bowed ventrally and the spines of the dorsal tubercles are brought forward and fixed in the wall; then by quickly releasing the anal spines, with the long pointed pygidium of the eighth segment, they and the dorsal tubercles act as levers and thus propel it up the stem. By executing these evolutions quickly they can move with considerable rapidity.

They can, apparently, descend with equal rapidity. By releasing the anal and dorsal spines they are lowered by gravity. If the stem be placed in a horizontal position, the pupa makes slow progress backward. The organs of locomotion are apparently not so well fitted for moving backward on a horizontal plane.

THE MOVEMENTS OF THE PUPA OUTSIDE THE STEM.

When removed from the stem and placed upon a flat surface, the pupa moves as uncertainly as the larva in the same position. It wriggles constantly, trying in vain to fix its "climbers" into something firm, whereby it can gain leverage and propel itself forward. It will fix the anal spines into the surface upon which it rests, but as there is no surface opposite and near it moves very slowly and uncertainly.

THE ADULT.

When ready to issue, the pupa is quite brown. The thin pupal envelope is ruptured along the dorsum of the thoracic region and the insect gradually forces its way out, after which it gnaws an irregular opening at some point in the stem and emerges.

The adult beetles are about 5 mm. in length, of a brownish color, and have pointed abdomens. From the lateral aspect they are somewhat crescent shaped. They are abroad from the latter part of May to the latter part of June, depending upon temperature conditions in early spring.

There is but one brood or generation during the year.

LIFE CYCLE.

LENGTH OF THE SEVERAL STAGES.

The larval stage covers a period of about 11 months. Nothing could be learned about the number of molts, as the larvæ will not develop if their galleries are disturbed.

The pupal stage varies from 11 to 16 days, depending, apparently, upon the temperature.

The adult beetles will live from 5 to 6 days in confinement, but they will probably survive a much longer period in the open.

HIBERNATION.

The insect hibernates in the larval stage. About the time freezing begins in the fall the larvæ are down to the bulb or crown of the root, where they are well protected from cold. They are nearly full grown by this time. Whenever a few warm days come, they apparently start feeding again. In the spring they burrow down into the juicy bulb, where they continue feeding until they become full grown.

THE TIMOTHY STEM-BORER (MORDELLISTENA USTULATA): LARVÆ AND PUPA IN STEMS.

PARASITIC ENEMIES.

This insect is apparently a very attractive host. Three species of parasitic Hymenoptera have been reared from it, all of them new, representing three genera—two braconids and one a chalcidid.

Messrs. H. L. Viereck [a] and J. C. Crawford [b] have kindly described these parasites, giving them the following names: *Heterospilus mordellistenæ* Vier., *Schizoprymnus phillipsi* Vier., and *Merisus mordellistenæ* Crawf. The descriptions appear elsewhere over the names of their respective authors.

In May of this year, *Mordellistena ustulata* was found to be very abundant at Wilmington, Ohio, in timothy; material was collected and sent in to the laboratory at La Fayette, Ind., for rearing. Two species of parasites were reared from it, *Heterospilus mordellistenæ* and *Merisus mordellistenæ*. *Schizoprymnus phillipsi* and *Heterospilus mordellistenæ* were reared from material collected at Richmond, Ind., in 1906 and 1908, respectively. The latter species and *Merisus mordellistenæ* were reared at La Fayette, Ind., in 1910.

It is very probable that the parasitic enemies keep the beetles pretty well in check and that this accounts for the appearance of the beetles in small numbers only in any given locality.

REMEDIAL MEASURES.

As this insect has never appeared in destructive abundance, so far as known, there has been no occasion to devise means of combating it. If a serious outbreak should occur, however, a short crop rotation should be adopted, allowing a field to remain in timothy sod not more than two or three years, thus preventing this stem-borer from becoming well established. The borders of the fields and waste places should be mowed frequently during the months of June and July. If this is done, the larvæ will not be able to reach maturity.

[a] Proceedings of the U. S. National Museum, vol. 39, pp. 401–408, 1911.

[b] Proceedings of the Entomological Society of Washington, vol. 12, no. 3, p. 145, 1910.

O

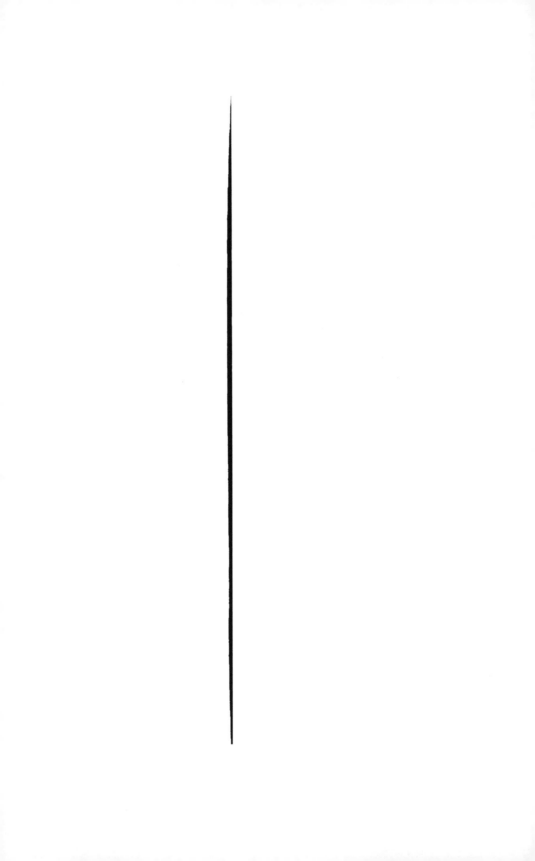

9770

U. S. DEPARTMENT OF AGRICULTURE,

BUREAU OF ENTOMOLOGY--BULLETIN No. 95, Part II.

L. O. HOWARD, Entomologist and Chief of Bureau.

PAPERS ON CEREAL AND FORAGE INSECTS.

THE MAIZE BILLBUG.

BY

E. O. G. KELLY,

Entomological Assistant.

ISSUED APRIL 22, 1911.

WASHINGTON:

GOVERNMENT PRINTING OFFICE.

1911.

CONTENTS.

ILLUSTRATIONS.

PAPERS ON CEREAL AND FORAGE INSECTS.

THE MAIZE BILLBUG.

(*Sphenophorus maidis* Chittn.)

By E. O. G. KELLY.

Entomological Assistant.

INTRODUCTION.

There are several species of the genus Sphenophorus that have been reported as being enemies to young corn in early spring. Heretofore these reports have always been made in connection with fields of grass, timothy sod, or lands recently reclaimed by drainage, and the depredations were on the first cultivated crop following these conditions.

Dr. S. A. Forbes [a] records eight species of Sphenophorus the adults of which are known to attack young corn. Dr. C. V. Riley, in the report of the Commissioner of Agriculture for 1881 and 1882, records Dr. L. O. Howard's observations on the habits and natural history of *S. robustus* and gives a description of the imago, larva, and pupa, Dr. Howard having found and reared these from specimens taken from young corn plants at Columbia, S. C. Dr. F. H. Chittenden, however, in a paper entitled " On the species of Sphenophorus related to *pertinax* Oliv., with descriptions of other forms," [b] with reference to *S. robustus*, says:

> This species ranges from Indiana and Michigan through Wisconsin, Minnesota, and western Kansas to California. It is a northern form, not occurring in the East. Nothing appears to be known of its habits, reference to *robustus* and its injuries to corn being due to a distinct species, my *S. maidis*.

Dr. Chittenden had before him, while describing *S. maidis*, among numerous other specimens, the one specimen from Columbia, S. C., reared by Dr. Howard. According to his statement *maidis* is a valid species and quite distinct from *S. robustus*.

---

[a] Twenty-third Report of the State Entomologist of Illinois, 1905.
[b] Proc. Ent. Soc. Wash., vol. 7, p. 57, 1905.

11

Dr. Chittenden informs the writer that since 1895 complaints have been made every few years in localities in Kansas, South Carolina, Georgia, and Alabama of injuries to corn by what he believes to be this species of billbug. The species has been quite generally confused with *Sphenophorus pertinax* Oliv. and *S. robustus* Horn, by both of which names it has been mentioned in economic literature, more especially by the latter. It is, however, quite distinct from either, in fact, different from any billbug known to inhabit the United States, and has only recently been described as new to science, although Dr. Chittenden states that it is by no means new as an agricultural foe.

The observations on the maize billbug (*Sphenophorus maidis* Chttn.) given herein were made by the writer between June and December, 1910.

HISTORY OF THE SPECIES.

The history of this species, the writer is informed by Dr. Chittenden, is, in brief, that it first attracted attention in Alabama as early as in 1854; again in the same State in 1880; in South Carolina in 1881; in Kansas in 1895; in 1901 it again did injury in Kansas, and in 1903 in Georgia. The fact that the insect is injurious to corn in both of the active stages, larva and beetle, indicates that it is a more or less permanent pest, whereas several of our equally common corn billbugs will eventually disappear with the reclamation by draining and the cultivation of the soil and the consequent destruction of their breeding places.

In the opinion of Dr. Chittenden, this is the species described and figured by Townend Glover in 1855 [a] as the "billbug" or 'corn borer (Sphenophorus ?)," since both description and figure do not apply to any other billbug known to breed in corn. Glover describes the beetle as from four-tenths to six-tenths of an inch in length, and of a reddish-brown or reddish-black color, and the rostrum or snout in the figure can not belong to any other Sphenophorus. None of the specimens which served as models of the drawing remains in the Government collections. The billbug was reported as very destructive to corn in many parts of the South and Southwest, more particularly along the Pedee River. Injuries were reported by Senator Evans, Gen. Fitzpatrick, and Col. Pitchlynn. Senator Evans's report is as follows:

The perfect insect eats into the stalk of the corn, either below or just at the surface of the ground, where it deposits its egg. After changing into a grub, the insect remains in the stalk, devouring the substance, until it transforms into the pupa state, which occurs in the same cavity in the stalk occupied by the grub. It makes its appearance the following spring in the perfect state, again to deposit its eggs at the foot of the young corn plants. These

[a] Agricultural Report of the Patent Office for 1854 (1855), p. 67, pl. 4.

Insects destroy the main stem, or shoots, thus causing suckers to spring up, which usually produce no grain, or, if any, of very inferior quality to that of the general yield. Swamp lands or low grounds are the places most generally attacked.

Senator Evans thus is, according to Chittenden, to be credited with the discovery that the larva develops in the stalk of corn below the ground, and not in decaying wood, as contended by Messrs. Walsh and Riley in later years. The insect was said to be very destructive in Alabama, from which State the specimens chosen for illustration doubtless came, and on the Red River in Arkansas. With little doubt it was the same insect operating in Arkansas, as it is now known that this species ranges between South Carolina and Missouri.

This insect was observed in the spring of 1881 by Dr. L. O. Howard, at that time assistant to Dr. C. V. Riley, Entomologist of the Department of Agriculture. Dr. Howard was at once sent to Columbia, S. C., to investigate the injury being done to corn by "billbugs," and the following account of these investigations is taken from the report of his observations:[a]

The species found near Columbia, S. C., is *S. robustus* [now *S. maidis*]. In the plantations along the bottom lands of the Congaree River much damage is done by the adult beetle every year, and the corn not infrequently has to be replanted several times, as the earlier plantings are destroyed. The beetles are first noticed in the spring after the corn is well up. Stationing themselves at the base of the stalk, and also burrowing under the surface of the earth slightly, they pierce the stalk and kill many plants ontright, others living to grow up stunted and dwarfed.

With *S. sculptilis* [*zeœ*], in spite of the damage it has done, the earlier stages remain unknown, Walsh surmising that the larva breeds on rotten wood, so situated that it is continually washed by water. With this statement in my mind I was prepared to doubt the statement of Mr. W. P. Spigener, of Columbia, who informed me that the "grub form of the billbug" was to be found in the corn, but a couple of hours in the field convinced me that he was right, my previous idea having been that he had mistaken the larva of *Chilo saccharalis* for the weevil grub. I searched a field on Mr. Spigener's plantation, which was said to be the worst point in the whole neighborhood for bugs, for some time before finding a trace of the beetle in any stage, but at last, in a deformed stalk, I found in a large barrow, about at the surface of the ground, a full-grown larva. After I had learned to recognize the peculiar appearance of the infested stalks I was enabled to collect the larvæ quite rapidly.

They were present at this date (Aug. 20) in all stages of larval development, but far more abundantly as full-grown larvæ. A few were preserved in alcohol and the remainder forwarded alive to the department, but all died on the way. Two pupæ were found at the same time; one was preserved in alcohol and the other forwarded to the department. The beetle issued on the way, and from this specimen we have been able to determine the species. From an examination of a large number of injured stalks it seems evident that the egg is laid in the

[a] Report of the Entomologist, Department of Agriculture, for 1881 and 1882, pp. 139–140.

stalk just at the surface of the ground, preferably and occasionally a little below. The young larvæ, hatching, work usually downward, and may be found at almost any age in that part of the stalk from which the roots are given out. A few specimens were found which had worked upward for a few inches into the first section of the stalk above ground, but these were all very large individuals, and I conclude that the larva only bores into the stalk proper after having consumed all available pith below ground.

The pupæ were both found in cavities opposite the first suckers, surrounded by excrement compactly pressed, so as to form a sort of cell.

Wherever the larva had reached its full size, the pith of the stalk was found completely eaten out for at least 5 inches. Below ground even the hard external portions of the stalk were eaten through, and in one instance everything except the rootlets had disappeared and the stalk had fallen to the ground.

In a great majority of instances but a single larva was found in a stalk, but a few cases were found where two larvæ were at work. In no case had an ear filled on a stalk bored by this larva. The stalk was often stunted and twisted, and the lower leaves were invariably brown and withered.

In one field, which had been completely under water for six days in January, the beetles were apparently as healthy as in fields which had remained above water.

INJURIES SINCE 1895.

The records of reports of injury which follow, received by Dr. Chittenden during the past decade, substantiate the observations of Dr. Howard made in 1881, and add as well to our knowledge of the life economy of the species.

In 1895 this billbug was destructive in three localities in Kansas, complaints all being made during the first week of May. At Cedar Vale immense damage was done, the insect "taking whole fields of corn, hill by hill." Similar injury was observed at Dexter and Leon, these reports having been made by Mr. Hugo Kahl in a letter dated July 27, 1898.

The following year Prof. F. S. Earle reported, June 6, injury by this species at Wetumpka, Ala., on the Coosa River, where there was great complaint of it as a destructive enemy of corn, especially on low-lying bottom lands. The insect was well known there as a billbug, and was not found on hilly land. It worked below ground, and when the stalks were not killed outright they put out an immense number of suckers. The beetles were most destructive to early plantings, corn planted after the middle of May being usually little injured.

In 1901 Mr. J. E. Williams, Augusta, Butler County, Kans., wrote, August 28, of injury to corn. Attack commenced as soon as the corn came through the ground, and the billbugs ate and dug down to the kernel and devoured that. In larger corn they bored into the stalk and wintered over in the old stalks, usually below ground. Whole fields were destroyed, the beetles remaining to continue their work on second plantings. The insect was known locally as the "elephant bug." September 6 Mr. Williams sent larvæ and adults and their work in the root-stalks of corn. He had observed that the eggs were deposited in the stalks, and that these serve for the winter quarters of the adults; that the beetles began work when the corn was about 4 or 5 inches high by inserting their beaks in the young stalks just above ground. By taking hold of the center of the corn and pulling it it came out, as it was nearly severed as from cutworm attack. Stalks that had been preyed upon by the billbug did not yield any amount of seed. No injury was observed to crops other than corn. Injury was only in lowlands, and the principal damage was accomplished before the woody outer shell of the stalk was formed. The beetles were active chiefly after dark, when they traveled, though slowly, from one place to another. They burrowed in the ground during the day. They were described as "cleaning up everything as they go, rendering the crop entirely worthless." September 17 another sending of larvæ, pupæ, and imagos were received from the same source. Out of 100 stalks examined by our correspondent only 10 were free from the ravages of this billbug. At this date of writing the beetles were deserting the corn.

In 1903 a report was received of injury by what was with little doubt this species at Griffin, Ga., although no specimens were received, as in all preceding instances cited.

DISTRIBUTION.

This insect has been reported, according to Chittenden,[a] from Augusta, Kans. (E. L. Williams); Riley County, Kans. (P. J. Parrott); Florence, Kans.; Dadeville, Ala. (S. M. Robertson); Wetumpka, Ala. (F. S. Earle); Columbia, S. C. (L. O. Howard); Ballentine, S. C. (J. Duncan); Texas (Ulke, 1 ex.); Michigan (Knaus). It has also been reported from Texas (T. D. Urbahns), and the writer found it at several points in Oklahoma and Kansas. Owing to the fact that representatives of the species have been taken in such widely separated localities, it is very probable that it occurs over the entire territory between South Carolina and Texas and northward to Kansas and Missouri.

FOOD PLANTS.

The adults attack young corn plants and probably some of the coarser grasses. Dr. Howard, and later the writer, found both adults and larvæ feeding on young corn. Mr. Urbahns found adults at base of swamp grass (*Tripsacum dactyloides*) in considerable numbers, and probably larvæ and pupæ of the species in this same grass (fig. 8). Mr. Urbahns found several Sphenophorus larvæ

---

[a] Proc. Ent. Soc. Wash., vol 7, No. 1, pp. 59–61, 1905.

in burrows in this swamp grass and two pupæ, but failed to rear them. Dr. Chittenden determined these pupæ as having adult characters of *S. maidis*.

DESCRIPTION AND LIFE HISTORY.

THE EGG.

(Fig. 5.)

Eggs were found by the writer in southern Kansas during June in punctures made especially for them (fig. 7, *b*) in young corn plants. These egg punctures, which the female makes with her beak, are scarcely visible on the outer surface of the stalk, being only a slit in the sheath of the plant, through which the beak, and later the ovipositor, are thrust, the sheath closing readily when the egg is deposited and the ovipositor withdrawn. The eggs are about 3 mm. long and 1 mm. thick, creamy white in color, elongate, and somewhat kidney-shaped, with obtusely rounded ends, being slightly more rounded at one end than at the other; the surface is smooth, without punctures.

Fig. 5.—The maize billbug (*Sphenophorus maidis*): Eggs. Enlarged three times. (Original.)

In the latitude of southern Kansas eggs were laid in the corn plants during the month of June, where they hatched in from 7 to 12 days, the young, footless grub thus finding itself surrounded with the choicest food.

THE LARVA.

(Fig. 6.)

The newly hatched larvæ are white, with a light-brown head, the head changing to darker brown within a few days. The color remains white in the full-grown larvæ, with the head chestnut brown. The length of full-grown living larvæ ranges from 15 to 20 mm. and the width from 4 to 5 mm.

The following description of the full-grown larva was made by Mr. E. A. Schwarz under the name of

Fig 6.—The maize billbug: Larva. Twice natural size. (Original.)

S. robustus, from the few alcoholic specimens collected by Dr. Howard at Columbia, S. C.:[a]

Length 12 mm.; color dingy white; head chestnut brown, with four vittæ of paler color, two upon the occiput, converging toward the base, and one along each lateral margin; trophi very dark, clypeus paler; body fusiform, strongly

---

[a] Report of the Entomologist, Department of Agriculture, for 1881 and 1882, p. 141.

curved, swelling ventrally from the third abdominal joint posteriorly, slightly recurved and rounded at anal extremity. Head large, oblong, obtusely angulate at base, sinuately narrowed anteriorly; frontal margin with a shallow emargination between the mandibles; upper surface with a median channel, the occipital portion deeply incised, with raised edges, continuing as a shallow impressed line to the middle of the front; on either side an engraved line, commencing upon the vertex, becoming deeper after crossing the branches of the Y suture, and terminating at the frontal margin in a bristle-bearing depression; sides and vertex with several long bristles arising in depressions; antennæ rudimentary, occupying minute pits on the frontal margin at the middle of the base of mandibles; ocelli a single pair, visible only as translucent spots upon the anterior face of the thickened frontal margin, outside of and closely contiguous to the antennæ from which they are separated by the branches of the Y suture, a few pigment cells obscurely visible beneath the surface; clypeus free, transverse, trapezoidal, with faint impressions along the base and at the sides; labrum small, elliptical, bearing spines and bristles, a furrow each side of the middle, forming three ridges, so that the organ, when deflected, appears three-lobed; mandibles stout, triangular, unarmed, with an obsolete longitudinal furrow on the outer face; maxillæ stout, cardinal piece transverse, basal piece elongate, bearing a palpus of two short joints, and a small rounded lobe, furnished at tip with a brush of spiny hairs, the lobe concealed by the labium; labium consisting of a large triangular mentum, excavate beneath, and a hastate palpiger, with a deep median channel; labial palpi divergent, separated by the ligula, of two joints subequal in length; ligula represented by a prominent rounded lobe, densely ciliate on the under surface. Thoracic joints separated above by transverse folds; the first wider, covered above by a transverse, thinly chitinous plate; the two following similar to the abdominal joints; abdominal joints forming on the dorsum narrow transverse folds, separated by two wider folds, the anterior fold attaining the ventral surface, the second fold confined to the dorsum, eighth and ninth abdominal joints longer, excavate above, without dorsal folds; beneath, the first three joints contracted, the succeeding joints enlarged, the terminal joint broadly rounded, with anal opening upon a fold at its base; sides of each joint presenting numerous longitudinal folds; stigmata very large, nine pairs; the first on the anterior margin of the prothorax, low down upon the sides; the remainder upon the sides of the first eight abdominal joints, above the lateral prominences, beginning upon the first joint at the middle of the side and gradually rising to a dorsal position upon the eighth joint; thoracic and last abdominal pairs large, oval; the intermediate pairs smaller, elliptical; all with chitinous margins of dark-brown color. The noticeable features of this larva are its cephalic vittæ, and conspicuous spiracles.

Upon issuing from the eggshell the young larvæ are about 5 mm. long and 2 mm. thick. They at once begin feeding on the tissues of the young corn at the bottom of the egg puncture (fig. 7, b), directing their burrowing inward and downward into the taproot. When they finish eating the tender parts of the taproot they direct their feeding upward, continuing until full grown, allowing the lower portion of the burrow to catch the frass and excrement (fig. 7, a). This burrowing of the taproot of the young growing corn plant is disastrous to the root system (Pl. I, figs. 1, 2); the roots, first dying at the tips, soon become of little use to the plant, allowing it to die or to become more or less dwarfed (Pl. II). The corn plants shown in Plate I were collected in Kansas and forwarded. in moist paper. to Washington, D. C., and photographed by the official photographer. Mr. L. S. Williams, and show the injuries more clearly, while Plate II, photographed in the field, illustrates the effect on the standing corn. Small plants, even those of less than one-half inch in diameter. are often recipients of eggs from which the larvæ, on hatching. burrow into the heart of the plant and cut off the growing bud. thus killing the top: they then direct the burrowing downward only to

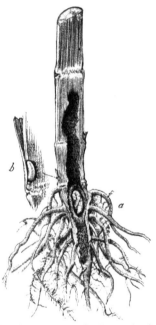

Fig. 7.—Corn plant showing result of attack by the maize billbug · *a*, Larval burrow containing pupa in natural position. *b*. egg puncture containing eggs *a*, Reduced two-thirds; *b*, enlarged (Original)

Fig. 8.—Swamp grass (*Tripsacum dactyloides*), attacked by the maize billbug. Reduced two-thirds. (Original.)

devour the stub, leaving themselves without food, and, being footless grubs, they of course perish. Plants of more than one-half in: diameter which become infested with larvæ make very poor growth. being very slender, rarely reaching a height of more than 2 or 3 feet before tasseling (Pl. II), and do not produce shoots or ears. Those that do not become infested until they are half grown may produce small ears. Each larva inhabits only the one burrow, and if, owing to any mishap. it becomes dislodged from it. it is powerless to reestablish itself. The larva does not become dislodged from the burrow

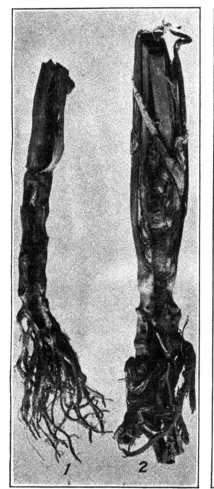

FIG. 1.—CORN PLANT INJURED BY THE ADULT OF THE MAIZE BILLBUG (SPHENOPHORUS MAIDIS), AFTERWARD ATTACKED BY THE LARVA. FIG. 2.—CORN PLANT SHOWING ON STALK THE EFFECTS OF FEEDING BY ADULT MAIZE BILLBUGS, AND EFFECTS ON ROOTS OF FEEDING BY THE LARVÆ. FIG. 3.—CORN PLANT, MUCH DISTORTED, SHOWING SUCKERS; FINAL EFFECTS OF FEEDING OF ADULT MAIZE BILLBUGS.

All figures about natural size. (Original)

CORN PLANTS SHOWING EFFECTS OF FEEDING OF MAIZE BILLBUG IN THE FIELD.

Plant at left not attacked, the two at right attacked by larva.   Reduced   (Original)

of its own accord. Sometimes there are three or four larvæ in the same plant, their burrows often running into each other, but this does not appear to discommode them in the least, as they can, and usually do, all mature. In badly infested fields two larvæ are quite often in the same plant, although one is the usual number and is sufficient to ruin the plant. The larvæ are easily managed in the laboratory; upon issuing from the eggshell they can be readily handled with a soft camel's-hair brush and placed inside a section of a cornstalk, where they will feed as readily as upon the growing plant. As soon as the section of plant is fairly eaten, and before decay sets in, the larvæ must be removed to fresh sections; keeping them thus supplied with fresh food they can be reared to maturity.

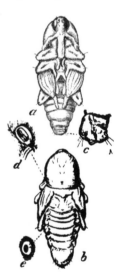

The length of the larval life ranges from 40 to 50 days, as indicated by laboratory observations and checked by collections in the field. They begin maturing and pupating by the 1st of August, pupation reaching the maximum by the 20th. and with the exception of a few stragglers all are mature and changed to pupæ by the 1st of September.

THE PUPA.

(Fig. 9.)

The larvæ, on finishing their growth, descend to the lower part of the burrow, to the crown of the taproot, cutting the pith of the cornstalk into fine shreds with which they construct a cell where they inclose themselves for pupation.

The newly issued pupæ are white, becoming darker after the fourth or fifth day, and continue to darken until just before the

Fig. 9—The maize billbug: Pupa. *a*, ventral view; *b*, dorsal, view; *c*, anal segment; *d*, thoracic spiracle; *e*, abdominal spiracle. *a, b*, Twice natural size; *c, d, e*, greatly enlarged. (Original)

adults issue. The adults are reddish black in color. The length of living pupæ ranges from 16 to 20 mm.

The following description of the pupa was made by Mr. E. A. Schwarz, of the Bureau of Entomology, under the name of *S. robustus*, from the single individual collected by Dr. Howard at Columbia, S. C.:[a]

Average length, 17 millimeters. Stout, rostrum reaching between first pair of tarsi. Antennæ but slightly elbowed and reaching not quite to bend of

---

[a] Loc. cit., p. 142.

anterior femora and tibiæ. Eyes scarcely discernible. Face with three pairs of shallow tubercles, the basal pair the largest, and each giving rise to a stiff, brown hair. Other minute piliferous tubercles, especially near the posterior dorsal border of the abdominal joints, being very stout on the preanal joint, or pygidium, where they form two series of quadridentate ridges.

To Mr. Schwarz's description the following may be added:

Lateral view: Body oblong, thickest at middle; thorax depressed anteriorly; abdomen cylindrical, tapering from seventh segment; thoracic pair of spiracles very prominent (fig. 9, d), first five pairs of abdominal spiracles prominent (fig. 9, c), the three on the rear segments not prominent. Elytra short, curving ventrally between middle and hind legs, reaching tarsi of hind leg, covering upper half of femur of hind leg; middle leg resting on elytra.   •

The pupæ occur mostly during the latter part of August and first part of September and are always to be found in their cells in the larval burrows near the crown of the taproot and nearly always below the surface. The pupal period is from 10 to 12 days.

THE ADULT.

(Fig. 10.)

The adults are very large, robust, and reddish black when newly issued, changing to black when older. The original description by Chittenden [a] is given here:

Body two-fifths as wide as long, of robust appearance because of the subquadrate thorax, which is nearly as wide as the elytra; general color black or piceous, moderately shining; alutaceous deposit on unelevated surfaces inconspicuous, appearing to be normally dark rufous or piceous velvety when the extraneous argillaceous covering does not persist.

Rostrum three-fifths the length of the thorax, considerably arcuate, strongly subequally compressed, apex prolonged at the posterior angle with an acute spine, producing the appearance of greater curvature of the inner surface, base feebly protuberant, moderately dilated; anterior face of apex broadly deeply concave; surface minutely punctate, more distinctly and densely at base, base moderately deeply channeled with distinct deep interocular puncture and short impressed line.

Thorax longer than wide, fully three-fourths as long as the elytra, sides usually widest just in front of middle, anterior third suddenly and very strongly arcuate and constricted at apex, posterior two-thirds or three-fourths subparallel, or gradually narrowing to the base which is feebly bisinuate. Vittæ feebly elevated, tending toward obsolescence, moderately finely but distinctly and sparsely punctate, more coarsely and densely at the ends; median vitta extending from a fine line and rapidly widening to a point just in front of the middle where it is broadly dilated, then more abruptly narrowed, extending in a narrower line to near the base; lateral vittæ sinuous with a tendency to become confluent with the median in the apical half, generally a little wider in basal half but narrower than the median, branch wide but ill-defined; interspaces and surface at sides coarsely foveate-punctate, punctures becoming confluent, especially posteriorly at sides. Scutellum deeply broadly concave.

[a] Proc. Ent. Soc. Wash., vol. 7, No. 1, p. 59, 1905.

Elytra little wider than the thorax; striæ usually deep and well defined, distinctly closely punctate; intervals with first, third and fifth elevated, with two or more series of rows of fine punctulation; first or sutural with basal third triseriately, posterior two-thirds biseriately punctulate; third widest and most elevated, with four or five rows of fine punctulations; fifth biseriately punctulate; seventh little or not at all more elevated than the remaining intervals; intervals 2, 4, 6, 8, as also 7, more coarsely and closely uniseriately punctulate. Pygidium deeply, coarsely and rather sparsely punctate, with sparse golden yellow hairs proceeding from the punctures and forming a short tuft each side, frequently abraded.

Lower surface coarsely and rather densely punctate, scarcely less strongly at the middle than at the sides, punctures largest at the middle of the metathorax. Punctures of the metepisterna (side pieces) more or less confluent. Second, third and fourth abdominal segments nearly uniformly punctured throughout, like the legs.

♂.—First abdominal segment very concave; pygidium truncate at apex.

♀.—First ventral scarcely different; pygidium narrowed and rounded at apex.

Aside from the differently shaped pygidium and the slightly shorter and less compressed rostrum there is little difference between the sexes.

Length. 10–15 mm., width, 4.5–6.0 mm.

The adults begin to issue about the middle of August and continue to do so until the middle of September. Some of them leave the pupal cell, but most of them remain there for hibernation. The adults that leave the pupal cell in the late summer disappear; continued search in every situation until December failed to reveal a single individual.

Fig. 10.—The maize billbug: Adult. Four times natural size. (Original.)

It is evident that they left the cornfield in which they developed, and it is very probable that they found their way to some dense, coarse grass (*T. dactyloides*), which is abundant in the locality. The adults hibernating in the pupal cells issue from them in late spring, about the time young corn is sprouting. The beetles are rarely observed on account of their quiet habits and because they are covered with mud—a condition which is more or less common among several species of this genus and is caused by a waxy exudation of the elytra, to which the soil adheres. The presence of the adults of this species in a cornfield is made evident by the withering of the top leaves of very young corn plants, the plants having been severely gouged. The adults kill the small plants outright and injure the larger ones beyond repair. After the plants grow 10 to 15 inches tall they do not kill them, but gouge out such large cavities in the stalks that they become twisted into all sorts of shapes (Pl. I, fig. 3). The attacked plants sucker profusely, affording young, tender growth for the beetles to feed upon, even for many days after the noninfested plants have become hard. The corn plants injured by *S. maidis* resemble somewhat corn plants injured by the lesser corn stalk-borer (*Diatræa saccharalis*), and are easily distinguished from plants injured by the smaller species of Sphenophorus owing to the fact that the punctures of the smaller species are not always fatal to the plants, which, however, in unfolding their leaves, show a row or series of rows of round or oblong holes in them.

The females issuing from hibernation feed on young corn for a few days before beginning to deposit their eggs. The egg punctures are made by the female in the side of the cornstalk (fig. 7, *b*) beneath the outer sheath. These egg punctures are not injurious to the plants, being only small grooves, about 5 mm. long and 3 mm. deep, in which the eggs snugly fit.

NUMBER OF GENERATIONS.

There is only one generation a year. The eggs occur throughout June, larvæ from early June until September, pupæ from the first part of August until the last part of September, and adults from the middle of August until the first part of August of the following year.

RECORDS OF DEPREDATIONS.

The depredations of this species have probably been confused with that of other species, the first and only known record of its attack on young corn being that made by Dr. Howard, at Columbia, S. C. During the season of 1910 both adults and larvæ were numerous in cornfields in lowlands in southern Kansas and northern Oklahoma, doing serious damage in some instances. They were frequently found in uplands, but not in injurious numbers.

REMEDIAL MEASURES.

The knowledge of the hibernating habits of the insect suggests an effective remedy in the pulling up and burning of the stubble, which is also the most practical means of destroying the lesser corn stalk-borer (*Diatræa saccharalis*). The beetles remain in the taproot of the corn plants until spring, allowing the farmer abundant time to destroy them. Care must be taken, however, in pulling up the infested stalks or else they will break off above the beetle, leaving the pest in the ground. The infested stalks, having a very poor root system, are easily pulled. Spraying the young corn plants with arsenical fluids at the time the beetles are making their attack is a very laborious procedure and not very effective.

○

U. S. DEPARTMENT OF AGRICULTURE,

BUREAU OF ENTOMOLOGY—BULLETIN No. 95, Part III.

L. O. HOWARD, Entomologist and Chief of Bureau.

PAPERS ON CEREAL AND FORAGE INSECTS.

.

CHINCH-BUG INVESTIGATIONS WEST OF THE MISSISSIPPI RIVER.

BY

E. O. G. KELLY,

Entomological Assistant,

AND

T. H. PARKS,

Entomological Assistant.

ISSUED DECEMBER 14, 1911.

WASHINGTON:

GOVERNMENT PRINTING OFFICE.

1911.

II

CONTENTS.

ILLUSTRATIONS.

IV

U. S. D. A., B. E. Bul. 95, Part III.   Issued December 14, 1911.

PAPERS ON CEREAL AND FORAGE INSECTS.

CHINCH-BUG INVESTIGATIONS WEST OF THE MISSISSIPPI RIVER.

By E. O. G. Kelly, *Entomological Assistant.*
and
T. H. Parks, *Entomological Assistant.*

INTRODUCTION.

Chinch bugs have long been a pest, and, especially so in the Middle Western States. During the last two years especially, Kansas, Oklahoma, and parts of Missouri and Illinois have suffered great losses from their ravages.

Owing largely to a lack of knowledge of the habits of the chinch bug, farmers are at a loss for remedies; the tried and successful tar lines and dust barriers, crudely used by them, to prevent the bugs from entering their corn, being often applied without success.

The "white fungus" or "fungus disease" of chinch bugs (*Sporotrichum globuliferum*), as it is commonly known among farmers, has been carefully observed and the conclusion reached that under ordinary farm conditions it can not be relied upon to afford immediate protection.

This paper has been written for the purpose of giving the farmers information regarding the habits of this insect, and the most effective methods of combating it. Field observations on this pest in Kansas, Oklahoma, and Missouri were begun during the spring of 1907 and continued till March, 1911, Mr. C. N. Ainslie and Mr. Paul Hayhurst making observations in 1907, and the senior author from the spring of 1908 to July, 1911, assisted by the junior author, who also did the photographic work, during the year ending with July, 1910.

DISTRIBUTION.

The chinch bug is widely distributed over the United States as well as in parts of Canada and in Mexico. The accompanying map (fig. 11) shows its distribution west of the Mississippi River. It is especially destructive over portions of Minnesota, Iowa, Missouri, Arkansas, Texas, Oklahoma, Kansas, Nebraska, and South Dakota,

23

and in parts of Illinois. Prof. T. D. A. Cockerell found a few specimens of both the long-winged and the short-winged forms at Mesilla
Park, N. Mex.; Messrs. E. A. Schwarz and H. S. Barber, of this
bureau, found a few short-winged forms at Hot Springs, Yavapai
County, Ariz.; Mr. George I. Reeves found some long-winged forms
in southwestern Washington; and Mr. Albert Koebele and Dr. P. R.
Uhler found a few at San Francisco and Alameda, Cal., and also in
Lower California. Prof. Herbert Osborn found the short-winged form
at Sault Ste. Marie, Mich., and Mr. Herbert T. Osborn found it at
Wellington, Kans.

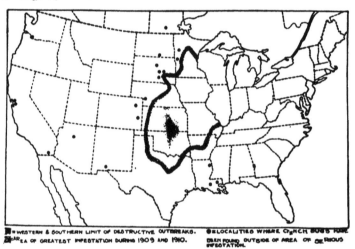

= WESTERN & SOUTHERN LIMIT OF DESTRUCTIVE OUTBREAKS.   = LOCALITIES WHERE CHINCH BUGS HAVE
AREA OF GREATEST INFESTATION DURING 1909 AND 1910.   BEEN FOUND OUTSIDE OF AREA OF SERIOUS INFESTATION.

FIG. 11.—Map showing distribution of the chinch bug west of the Mississippi River, 1911. (Original.)

DESCRIPTION AND NUMBER OF GENERATIONS.

Full descriptions of this insect are found in Bulletin No. 69 and in
Circular No. 113 of this bureau and will not be repeated here.

There are two principal generations annually in the Middle West:
the spring generation and the fall, or hibernating, generation, and a
partial third generation sometimes occurs in late fall to the southward.

MIGRATIONS.

The hibernating bugs (fig. 14) issue from their winter quarters as
soon as the sun warms up the grasses in the spring, and fly out to
green grasses and young wheat and barley, where they feed, mate,
and deposit their eggs. The eggs (fig. 13 a, b) begin to hatch in late
April and continue hatching until June, varying with the seasonal temperature and the latitude of the locality affected. The young bugs

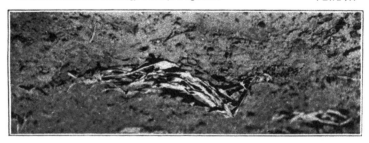

FIG. 1.—CORN PLANT KILLED BY CHINCH BUG. (ORIGINAL.)

FIG. 2.—CHINCH-BUG RAVAGES IN CORNFIELD IN SOUTHERN KANSAS, 1910　(ORIGINAL.)

FIG. 3.—CORNFIELD ADJOINING WHEAT FIELD FROM WHICH CHINCH BUGS MIGRATED IN
IMMENSE NUMBERS AT HARVEST TIME. (ORIGINAL.)

RAVAGES OF THE CHINCH BUG (BLISSUS LEUCOPTERUS).

hatching from these eggs constitute the spring generation (fig. 13 c, e, f, g) and are the ones that do such enormous damage to wheat and young corn. Some of this generation reach maturity as they are migrating from wheat to corn, but most of them reach the corn (see fig. 12) before maturing and do much damage thereto. It is because the immature bugs reach the young corn in such immense numbers and mass upon the plants that they do such widespread damage

FIG. 12.--Corn plant about 2 feet tall, infested by chinch bugs. (Original)

(see Pl. IV), their depredations ceasing as they reach maturity. Only on rare occasions is an entire cornfield devastated, and often the depredation is brought to an abrupt standstill within a few rods of the opposite margin of the field because the bugs have reached maturity and dispersed. During July and August the insects mate, and the eggs for the second generation are deposited about the corn plants, where the young, on hatching, find an abundance of easily accessible food. Some of this generation reach maturity (see fig. 14) before the corn becomes dry, and migrate to volunteer wheat, but most of them are forced to seek their food elsewhere. They usually find this in kafir cane fields, and among some of the grasses, where they reach maturity. From here they go to winter quarters before cold weather.

STATUS OF THE CHINCH-BUG PROBLEM IN KANSAS, MISSOURI AND OKLAHOMA.

That the seasons of 1907 to 1910 have been favorable for the development of chinch bugs in Kansas, Missouri, and Oklahoma is indicated by

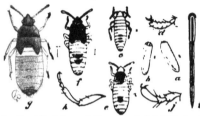

FIG. 13.—Chinch bug: *a, b*, Eggs; *c*, newly hatched larva, or nymph; *d*, its tarsus; *e*, larva after first molt; *f*, same after second molt; *g*, last-stage larva; the natural sizes indicated at sides; *h*, enlarged leg of perfect bug; *j*, tarsus of same, still more enlarged; *i*, proboscis or beak, enlarged. (From Riley)

reports of injury to crops received during these years.

The prevalence of the pest and its depredations depend upon meteorological conditions to a great extent, as has been discovered by entomologists in the past, and this fact is clearly brought out by the observations herein recorded and made during the four seasons from 1907 to 1910, inclusive.

The bugs were very numerous in wheat fields in the early spring of 1907 and deposited numbers of eggs on the young wheat plants. Much wheat had been destroyed by the "green bug" (*Toxoptera graminum* Rond.) during the spring, and many farmers had plowed up their fields and planted them to corn; sometimes they did not use the gang plow, but planted the corn with a combination lister and planter. There were large numbers of eggs and young of chinch bugs on this more or less dead wheat that was plowed under, and apparently very few were destroyed during the operation of preparing the ground for planting the corn. As soon as the corn plants pushed through the soil they were attacked by the young chinch bugs, with the result that a large amount of young corn was ruined by them.

In the fields of wheat that had not been plowed under in the spring there were also numbers of bugs, and at harvest time these found their way to other fields of young corn, where they inflicted considerable damage. The summer and fall were favorable for their

FIG. 14.—Chinch bug (*Blissus leucopterus*): Adult of long-winged form, much enlarged. (From Webster.)

development, and great numbers went into hibernation in grasses.

The spring of 1908 opened the first week of March with warm, dry weather and a deficiency of moisture which was alarming for this season of the year and which had a telling effect on young wheat. The chinch bugs flocked to the young wheat from their winter quarters, and very soon their ravages were in evidence. By the middle of April the plants were turning to a red color, and the stems were black and dry from their attacks, very few plants escaping, some plants having as many as 200 bugs massed on them. Very few eggs had been deposited up to the middle of April, but from this time on until May 1 eggs were abundant.

The eggs began to hatch about the 1st of May, and in one week's time the wheat fields were swarming with tiny red bugs, much of the wheat dying from their severe attack combined with the effects of lack of moisture. However, the situation was soon rapidly changed, for rains came, and by the 1st of June there was hardly a living bug to be found, either young or adult, but there were numbers of unhatched eggs left unharmed on the plants. The wheat partially outgrew this damage and made a fair crop, while the young corn escaped injury.

Many eggs hatched after the rains ceased, and the bugs from these eggs, together with the few that survived the drenching rains of May and June, succeeded in getting to the corn, where they bred in large numbers. By September, having left the Indian corn, which at this time was becoming dry, the young bugs were plentiful on cane and kafir. These bugs matured on the cane, kafir, and succulent grasses, and large numbers went into hibernation in the fall among the grasses.

The spring of 1909 was quite late in opening. A big snow on March 8 remained on the fields until March 20, and for several days afterwards the ground was alternately wet and frozen. During the last two days of March there was bright, warm sunshine, and the bugs began to move about. On April 3 the bunches of sedge grass, *Andropogon scoparius*, in which the insects hibernated, resembled living masses of crawling bugs, and before nightfall great swarms were flying; this continued during the next few days, and the wheat fields had now become badly infested. The chinch bugs commenced mating very soon after reaching the wheat fields, and in a few days egg laying began. Eggs were very numerous about wheat plants by April 22, some were hatching, and by May 1 young bugs were abundant. The massing in large numbers on single plants, which was so noticeable in the spring of 1908, was entirely lacking in 1909; seldom were there more than a dozen bugs on a plant, although nearly every plant was infested.

The entire month of April, together with the first ten days of May, 1909, were exceedingly dry and wheat suffered from lack of moisture. There was no noticeable injury to the young wheat from attacks of the old hibernating bugs and not much from the combined attack of both the old and young bugs, until the first week of May, when the wheat failed rapidly, owing to the great number of the insects and the lack of moisture. Here, again, heavy rains of four days (May

11–14) checked their ravages and killed great numbers of both the young and the old bugs. The rains were light from May 15 to 27, and during this interval the young chinch bugs became very abundant. On some wheat plants there were as many as 240 young bugs, while other wheat plants were free from them; infested wheat plants turned yellow and died, owing entirely to the effects of their attack. The heavy downpour of rain on the night of May 27 buried thousands of chinch bugs in the loose soil and again put a temporary check on their ravages. This rain was the last one of any importance until the night of June 22. By June 8 most of the bugs had changed to a brown color and few red ones remained among them; several eggs were observed on wheat roots. By June 14 the young chinch bugs were very abundant on the stems and leaves of wheat, and 300 to 500 bugs per tiller were counted on several plants, which were turning white and dying before the grain had matured. The ground was covered with moving insects but in no instance were the chinch bugs leaving the wheat fields. Five days later the wheat was prematurely "ripened" through insect attack and the hot sunshine. The "ripening" grain forced them to seek food elsewhere and on June 21 they began to migrate from wheat to corn. The grains in the heads of the severely attacked tillers were small and considerably shriveled, while grains in heads which had not been attacked were full size and plump. Great numbers which were migrating on June 22 were temporarily checked by a light shower, only to continue the movement as soon as the grass became dry. At night, however, a very heavy beating rain fell, burying millions of bugs of all ages and sizes in the mud. The mud-covered bugs nearly all died and those not dead by the next day were found stuck fast in the mud. The ground dried out by noon of June 24, and the bugs that were not killed by the rain continued migrating toward the cornfield. In an area of 4 square feet in a wheat field there were 2,411 dead chinch bugs, some of them buried one-half an inch in the soil. In an area of 10 square feet in the same field there were only 244 living chinch bugs, all of which were moving toward an adjacent cornfield where they collected in masses on the outer rows.

The weather was quite dry during the next few weeks and the bugs did considerable damage to young corn, some fields being almost devastated. The bugs began to transform to adults about July 5, and their ravages ceased within a week thereafter. There were hordes of adult bugs scattered throughout the cornfields, and these were mating and depositing eggs.

By the middle of August, when most of the corn had matured and dried, there were again great numbers of young bugs on the corn plants even under the outer husks below and about the bases of the ears; the greener plants were much infested. The corn now began to get too dry to afford them food and they began migrating to kafir, sorghum, and grass fields, killing all crab-grass and panic grass in the cornfields before they left them. Some bugs were observed in alfalfa fields but at no time were they feeding on the alfalfa.

During October and early November several chinch bugs were observed on young wheat and a few eggs and young larvæ of the third generation were observed. They did not damage the wheat during the late fall, and by the middle of November had left the wheat fields for winter quarters.

The kafir and cane were green until the middle of September, and great numbers of chinch bugs were feeding on these two crops at this time when the crops were harvested. They remained on these plants and were put into the shock, where they were observed on several occasions during the fall, but they all left before cold weather set in.

During early fall a few bugs were to be found in nearly every situation on the farm—some in corn husks, some in alfalfa fields under the leaves, and some among sparsely growing grasses, but most of them were in the bunches of clump-forming grasses along fences, roadsides, and railroad rights of way, in waste places, and in meadows. That the weather conditions of the summer and fall were the most favorable for the chinch bugs is indicated by the vast numbers that went into winter quarters. Some of the bunches of the red sedge grass (*Andropogon scoparius*) contained from 10,000 to 20,000 bugs, this grass affording them the most attractive and favorable hibernating quarters.

The severe cold weather of December, 1909, was disastrous to bugs that failed to reach some of the clump-forming grasses, and very few bugs in corn husks and among thin grasses survived the winter.

SOUTHERN KANSAS AND NORTHERN OKLAHOMA, 1910.

In southern Kansas and northern Oklahoma the spring of 1910 opened the last week of March with warm, dry weather. The chinch bugs began to migrate to the wheat fields; on March 24 the air was full of flying adults and by March 26 the wheat fields were badly infested. The weather turned much warmer about April 1, the chinch bugs began mating, and by April 10 were depositing eggs about the wheat roots; by April 20 the eggs were very abundant—hundreds on the exposed roots of every plant. Not many eggs hatched during April and the first 10 days of May on account of the cool weather, but they hatched profusely during the next few days, so that by May 20 the fields were overrun with myriads of very young chinch bugs.

Wheat throughout southern Kansas and northern Oklahoma had been severely winterkilled, which, together with the exceedingly dry spring, left the plants in a very weakened condition. The soil was so dry during April and May that the small amount of rainfall

was of little value. The ground was badly cracked, exposing the
wheat roots to the hot sun and drying winds and these, together
with the bugs, killed a great many plants. Wheat was so badly killed,
mostly by meteorological influences, that by May 1 thousands of
acres had been abandoned and corn planted in its stead. The young
bugs and eggs were little molested by the operation of corn planting
and as soon as the sprouting corn pushed through the soil hundreds
of bugs were ready for every plant. Some such fields were even planted
the third time only to furnish food for these hungry insects. The
bugs would crawl from beneath several inches of loose soil and be
ready to attack the young plants as soon as they appeared.

Such wheat as was not abandoned made a poor growth on account
of drought and was severely damaged by the chinch bugs, great
numbers of which were on every plant, so that by June 10 they had
sapped the life from them. As the wheat and grasses were killed
by the drought, the bugs were forced to abandon wheat fields and
hunt for food elsewhere. The corn, which had made very poor growth
up to this time on account of the lack of moisture, was very small
and weak when the bugs reached it, and this early attack by the young
chinch bugs resulted in the devastation of thousands of acres before
the bugs became mature. On reaching maturity they abruptly
dispersed and their depredations were brought to an end.

Young bugs of the second generation became numerous the first
week of August on the cornstalks under the sheaths and under the
outer husks of the ears. Corn leaves, stalks, and husks of ears were
dry by the middle of August and bugs were moving out of these into
fields of kafir and sorghum, both of which suffered under their
attack, many fields being laid waste by September 1. Owing to
the continued drought throughout the fall the Indian corn, kafir,
cane, and grasses died leaving the chinch bugs without sufficient food
supply, with the result that of the vast horde of bugs hatching in
August comparatively few survived to go into winter quarters.
During September and October a few young chinch bugs matured on
volunteer wheat, later depositing a few eggs for a partial third generation. No damage was done to wheat by these few insects, although
some of them remained on or about the wheat plants till cold weather
set in. There were not nearly so many chinch bugs in the bunches
of grass in the fall of 1910 as in the fall of 1909.

KANSAS, 1911.

As previously indicated, the fall of 1910 was very dry and wheat
failed to sprout in Kansas and Oklahoma. The winter of 1910–11
and the spring of 1911 continued dry, very little rain falling during
the entire period till the last of May, 1911. This extended drought

affected both crops and chinch bugs to such an extent that wheat was almost a failure and the numbers of the bugs were greatly reduced. The winter wheat which was seeded in September and October, 1910, was seeded in a very dry soil; very little of it sprouted. A few localities were favored with a shower of rain in September which sprouted the grain, and the bottom fields along the rivers and creeks contained enough moisture to sprout the grain and produce a fair crop. As a result of the drought, however, very little wheat matured on the uplands.

The chinch bugs went into hibernation in a very weakened condition because the grasses had dried up in the fall. Many of them did not reach their usual winter quarters but stopped in almost any place they could find. Many hibernated in trash in alfalfa fields where a small percentage succeeded in living through the winter. The death rate of hibernating bugs was greater than in any previous winter included in this study; in some of the bunches of Andropogon 60 per cent of the bugs were dead. When the first warm days of spring occurred the few bugs that were left were very inactive and the migration to green fields, which had been so noticeable in former years, was lacking. However, there was a migration during some very hot days in the latter part of April—over a month late. The dying of the bugs during the winter was probably due to two causes, one being the starved condition of the bugs when they went into hibernation and the other the hot days during the winter followed by the very cold days in the spring. During these hot days in the middle of winter the bunches of grass would be swarming with the bugs and the very next day the thermometer would register 8 to 10 degrees below zero. In spite of all these adverse conditions many of the bugs lived to infest the fields in the spring.

The failure of wheat caused the farmers to plant other crops in many of these fields, but some of them were left standing until May in the hope that a crop would be produced. The bugs reached the few fields that were left, damaged the plants considerably, and deposited numbers of eggs. Many wheat fields were seeded to oats after the wheat failed, thus leaving the few wheat plants growing in the fields for the young bugs to feed upon. At no time could young bugs or eggs be found on the oat plants. Notwithstanding the fact that numbers of farmers lost several plantings of corn in 1910 by planting the infested wheat field to corn, hundreds of acres of corn were planted in such fields in the spring of 1911, only to be destroyed later. There were areas of 8 to 10 square miles that did not contain an acre of wheat after the middle of May, all of the fields having been planted to other crops. In these localities where the wheat was missing the bugs were also missing, and where the wheat was plentiful the bugs were very plentiful and did considerable damage to adjacent corn.

It is very apparent that wheat has a decided effect on the presence of the chinch bugs, as indicated in the localities where the wheat was a failure this spring.

The dry weather had its effect on the fungus, *Sporotrichum globuli-ferum*, and it has not occurred in the fields this season (1911). Only once has it been possible to secure it in the laboratory. Continued search has been made for it in all kinds of places, especially after the rains in May and in July, beyond which latter month this record does not extend.

HIBERNATION.

At the beginning of the investigation the advisability of getting rid of the chinch bugs before they entered the young wheat in the early spring was very evident, for when once they have reached there they are not readily accessible. This led to a series of observations on their hibernating habits for the purpose of determining the places preferred by the bugs.

The current belief that most of the bugs pass the winter beneath corn husks, among cornstalks, in fence rows, under boards and rails, in heaps of rubbish, in straw stacks, along hedgerows, and in fodder shocks is not borne out by investigations in Kansas and Oklahoma following a severe winter. When the bugs are very abundant, as they were in Kansas, Missouri, and Oklahoma during the fall of 1909, a few may be found in any of these situations, especially in the early fall. The most of the bugs find their way to thick bunches of clump-forming grasses in waste places, in pastures and meadows, and along roadsides and railroad rights of way. During late fall and early winter great numbers of living bugs can be found in corn husks, fodder shocks, piles of kafir, cane, and in most any place covered with vegetation—even in alfalfa fields where they find no food. In the spring, however, very few living bugs but many dead ones can be found in such situations, indicating that most of them died there.

They find much better protection in the thicker and more dense, than in the thinner grasses and under trash in open fields. The bugs seem to prefer the thicker grasses, though they are quite often found in other situations, and after open winters, as the one of 1910–11, many living ones can be found under very thin protection. Many living chinch bugs were taken from trash collected in an alfalfa field and some were found in corn fodder and corn husks lying on the ground on February 24, 1911, at Wellington, Kans., though most of the chinch bugs are at this time in the thick bunches of sedge grass.

The situation in southern Illinois for the spring of 1911 was quite similar to that in southern Kansas; there were abundant clumps of Andropogon along roadsides, in fields, and in woodlands, and more chinch bugs were found in these clumps of grasses than in trash, under boards, or in any other place. There were not a great many bugs in any one place, however, and it would be rather difficult to make a conclusive comparison between the two widely separated localities, although the indications are that the majority of the bugs seek the thick grasses in which to hibernate.

The destruction of the hibernating chinch bugs is of much importance; it has been discussed frequently and is strongly recommended by entomologists throughout the United States. As the methods of farming, the prevailing crops, and the wild grasses vary in different localities where chinch bugs occur, the places of hibernation may also vary considerably.

Mr. C. L. Marlatt, of this bureau, in an article on the hibernation of chinch bugs,[a] has the following to say:

In nearly every account of the chinch bug which I have seen, stress has been placed on the hibernation of the adult in rubbish of any sort, such as piles of corn fodder, hay piles, straw piles, and dried leaves along hedgerows. In course of very careful investigation carried on in Kansas during a year of excessive abundance, I failed entirely to find any basis for the above supposition. Repeated careful search throughout the late fall and winter failed to discover a single living chinch bug in such situations.

Failing to find them in the situations noted, I carried the examination further and finally discovered what is probably the normal hibernating place of the chinch bug in the dense stools of certain of the wild grasses, such as the blue stem and other sorts. * * * So marked is this hibernating habit that it is reasonable to infer that it is the normal and ancient one of the species, the natural food plant of which, before the advent of settlement and growth of the cereals, must have been some of our native grasses.

Mr. H. W. Brittcher, in regard to hibernation,[b] says:

* * * they may frequently be found, more or less closely crowded, low down among the stems of clumps of wild rushes and grasses, often working their way down between the stems and the soil.

He recommends burning the sedges in which careful examination showed the bugs to be abundant. It must be remembered that in Maine it is the short-winged form that prevails and is there a grass as well as grain destroying insect.

In regard to hibernation,[c] Dr. S. A. Forbes, State entomologist of Illinois, says:

On the 7th of November a careful search was made in corn that had previously been badly infested by them, but none were to be seen upon the stalks or under the rubbish on the ground in the field; in the thickly matted grass adjacent only a single specimen was discovered by 15 minutes' search. On the 14th of this month the weather was cold and raw, and the ground was frozen about the hills of corn from an inch to an inch and a half in depth; a very few bugs were now found in the crevices of the ground, among the roots near the surface. At Champaign, on the 15th, I visited again the field of Bogardus and Johnson, making a careful search for hibernating individuals about the

---

[a] Insect Life, vol. 7, pp. 232–234, 1894.
[b] 19th Ann. Rept. Maine Agr. Exp. Sta., 1903, p. 42, 1904.
[c] 12th Ann. Rept. State Ent. of Ill., pp. 37–38, 1903.

stalks, under the weeds in the field, and beneath the rubbish collected about the hedgerows. Not a single specimen was found in these situations, although every temptation was afforded to hibernating insects, and many other species occurred abundantly. To what resorts the swarms which had developed in these situations had betaken themselves to pass the winter, I am not able to say.

Prof. Herbert Osborn,[a] in giving a summary of his observations on the chinch bug in Iowa in 1894, states:

In a great majority of cases, 90 per cent or more, the infested fields were directly adjacent to hedges or thickets or belts of timber, and in 75 per cent osage-orange hedges were the most available shelter. In about 13 per cent of cases the evidence showed hibernation in grass and weeds and in some of these cases there could scarcely be a doubt that the hibernating bugs were protected by a heavy growth of grass or weeds and that they moved from these directly into adjacent fields.

Prof. F. M. Webster, of this bureau, has probably given the chinch bug more attention than any other entomologist and has contributed more to our knowledge of the pest. His observations in Ohio and those made by Prof. Herbert Osborn in Iowa are at variance with those made in Kansas by Mr. Marlatt and by the writers. Prof. Webster offers the following explanation [a] of their hibernating habits in different localities:

In Kansas, where Mr. Marlatt made his observations, there was still too much prairie, and the species was doubtless still adhering to its ancient habits of hibernation. In southern Ohio the author has found it attacking the wheat in May, in small isolated spots over the fields, while there was nothing in the least to imply an invasion from outside, but the wheat had been sown in the fall among corn, and later the cornstalks cut off and shocked, remaining in this condition until the following spring. This occurred so frequently that there seemed no room to doubt that the attacks had been caused by adults wintering over in the corn fodder and that these left their winter quarters in spring to feed and breed on the grain growing nearest at hand.

The hibernating habits of the chinch bugs have been closely observed during two seasons in Kansas and Oklahoma, and the observations made indicate that the bugs hibernate there chiefly in dense clumps of sedge grass, principally those of different species of Andropogon.

The following data with reference to the hibernating habits of the chinch bugs were made by the writers in southern Kansas, and are given here to substantiate the above statement:

From Mr. Hayhurst's notes made in the fall of 1907: On October 26, at Winfield, Kans., he found active adult chinch bugs in stools of broom beard grass (*Andropogon scoparius*) in great numbers, always close to the ground. On November 1, 1907, at Newkirk, Okla., he again found many active bugs in the stools of forked beard grass (*Andropogon furcatus*) close to the ground. They were present in nearly every stool of this grass examined along roadsides and also on the open prairie where the grass had been cut.

---

[a] Bul. 69, Bur. Ent., U. S. Dept Agr., pp. 16–17, 1907.

At Wellington, Kans., the senior author, on October 3, 1908, observed adult chinch bugs flying and collecting in grass (Andropogon) down at the crown, and on October 12 he saw hundreds of bugs in some clumps of this grass along roadsides. During the early spring of 1909 hundreds of living adults were found in clumps of this same grass, where they had evidently hibernated. On November 30, 1909, in this same locality the authors observed some bunches of *Andropogon scoparius* containing 6,000 to 10,000 bugs each, and on November 29, 1909, they found a few living bugs in corn husks, but most all of the bugs were in bunches of grass. Again, on December 22 one bunch of Andropogon 2 inches in diameter was found to contain 1,508 bugs; many bunches were 8 to 12 inches in diameter, and contained as many as 15,000 to 20,000. Bugs were found only in grasses that grow in clumps; none was discovered in rubbish, old straw piles, fodder shocks, or in sorghum-cane piles. Several dead bugs were found in corn husks, but no living ones. A number of bugs collected at random from the base of some bunches of Andropogon was brought into a warm room to see how many would revive. There were 325 living bugs and 89 dead ones, or 21.5 per cent dead. None was found in bunches of reed grass (Calomagrostis), Sporobolus, blue grass (Poa), or crab grass (Digitaria).

The method for separating the living chinch bugs from the trash and dead bugs was a very simple one. The bugs and bunch of grass containing them were put on paper in an oblong box, and the box placed in front of a fire, where the living bugs would readily crawl out of the grass and trash to the corners of the box and beneath the paper. Care was taken not to heat the bugs to more than 100° F., as a higher temperature might kill them.

On December 24 another lot of chinch bugs, collected at random from clumps of Andropogon and brought into a warm room, gave 755 living and 234 dead, or 23.6 per cent dead. Again, on December 29 a third lot of bugs in clumps of Andropogon was brought into the laboratory, where the warmth of the room soon revived them, and they became active. Of this lot of bugs 20 to 25 per cent were found to be dead, probably from exposure to cold or perhaps from old age. On December 30, corn husks and the husks of fallen ears, which had not been harvested, were searched for living chinch bugs. Very few were found and none whatever about old straw piles. Very few bugs died during January, as indicated by observations during the last part of that month. Three lots of bugs were collected from bunches of Andropogon; the first contained 298 living and 72 dead bugs, or 19 per cent dead; the second contained 137 living and 35 dead bugs, or 20 per cent dead; the third, which was collected on February 12, 1910, contained a few more than 10,000 counted bugs, 20 per cent of which were dead.

The weather was somewhat drier, but agricultural conditions for the fall of 1910 were about as those of previous years, except that vegetation was practically all dry before frosts. Chinch bugs were not nearly so numerous as in the fall of 1909, clumps of Andropogon containing only 80 to 260 bugs each, whereas there were thousands the previous winter. During the early fall it was again observed that the corn husks were full of adult bugs, but by cold weather these were nearly all dead; the mild winter, however, permitted some of these bugs to live through, and some were alive on February 24, 1911.

From the foregoing data, covering four seasons, there can be little doubt that in Oklahoma, Kansas, Missouri, and probably southern Illinois these clump-forming grasses form the principal hibernating quarters of the pest. This definite knowledge of their habits puts a most practical and effectual weapon into the hands of the farmer, which he may apply months in advance, in defense of his crops.

The farmer can readily determine whether the grasses on or about his farm contain chinch bugs by pulling open the tufts of red sedge grass. (See Pl. V, fig. 3.) If the bugs are present in these clumps of grasses, it is of the utmost importance that they be burned. The habits of the hibernating generation and the migration of the spring generation offer the best opportunities for forestalling and preventing future ravages.

PREVENTIVE MEASURES RECOMMENDED.

DESTRUCTION OF CHINCH BUGS WHILE IN HIBERNATION.

The burning of grasses and rubbish about the farm to destroy chinch bugs has been often recommended and is doubtless the most effective measure to be taken against future ravages of the pest.

In the Southwest the chinch bugs are known to congregate in bunches of grass in late October and remain there till the warm days of early spring. It is only a matter of burning off these grasses at the proper time to effectually rid such places of the pest, and the grasses are generally sufficiently dry to burn readily by the 1st of November. The chinch bugs crawl deep down among the grass stems, a few of them even getting beneath the dust and débris, thus seeking protection from the freezes that are to come. It is very important that the grass be dry and yet burn slowly, so that the heat will thoroughly penetrate the dense grass and reach the bugs. It is not necessary for the fire to come into direct contact with the bugs in order to kill them, as they died very quickly in the laboratory when exposed to the heat of a flame from 12 to 20 inches distant, the fatal temperature being in these experiments about 111° F. Fall burning of the grasses among which the bugs are congregated has a

FIG. 1.—PILE OF SORGHUM CANES IN WHICH NO HIBERNATING CHINCH BUGS COULD BE
FOUND. (ORIGINAL.)

FIG. 2.—WASTE LAND ALONG STREAM IN FOREGROUND, SEDGE-GRASS MEADOW IN
BACKGROUND, CHINCH BUGS FOUND HIBERNATING IN BOTH. (ORIGINAL.)

FIG. 3.—CLUMPS OF RED SEDGE GRASS (ANDROPOGON SCOPARIUS) IN WHICH OVER 6,000
CHINCH BUGS WERE FOUND HIBERNATING DURING WINTER OF 1909-10. (ORIGINAL.)

HIBERNATION OF THE CHINCH BUG.

twofold value; first, it will kill large numbers of bugs directly; and, second, the bugs not killed by the fire will be left exposed to the winter freezes, which of themselves will in ordinary seasons kill many of them. On several occasions during fall and spring bugs were removed from the stubs of burned grass and the percentage of dead and live bugs obtained. On an average about 75 per cent were killed in the fall and about 63 per cent in the spring. In the spring about 20 per cent of the bugs, which hibernated in the clumps of grasses, were dead from exposure and other causes. From natural causes and burning in spring, there were about 83 per cent of the bugs dead. These percentages were obtained by actually counting the insects and are not from estimates. The fire can not reach all the bugs, even with the most careful burning, because of protection afforded by green or wet stems in early fall and late spring; therefore it is essential that the grass be burned during late fall or early winter. While this remedy is recommended above all others, its effectiveness is entirely dependent upon the farmers and their cooperation, and it is an easy matter for neighborhoods to combine in an effort to fight the pest in this manner.

DUST BARRIERS.

If the bugs are not killed out during hibernation, the main dependence of the corn grower must be in the destruction of the bugs as they migrate from the ripening small grain to enter the cornfields. As soon as the ripening grain compels the bugs to desert it and they start for the corn, a narrow strip of ground between the corn and wheat should be deeply plowed and thoroughly pulverized, making a dust bed. Then a short block or a triangular box should be made in form of a sled with the bottom fitted with a seat for a driver. This should be dragged back and forth in this dust bed until a deep groove or furrow has been made. If this furrow has been well prepared, and the weather continues dry, it will prove an impassable barrier to the progress of the bugs. In order that this kind of barrier may be successful, the block must be kept in constant use, from early until late and sometimes well into the night. Often, during the migration, the bugs travel all night. Slight showers render dust furrows useless and if a shower is pending, and under such weather conditions, it is best to employ the coal-tar barrier. Owing to the fact that dry weather often prevails on the plains during this season the furrow method under these conditions is especially recommended.

COAL-TAR BARRIERS.

The coal-tar barrier can be operated successfully in the Middle-Western States to prevent the bugs from migrating to young corn. The success of this measure depends on the farmer and his careful attention to the tar line. Preparations for the barrier should be begun in all cases where the bugs are found abundant in ripening grain. Frequently the soil is quite compact along the margins of wheat fields and if it is, a smooth path can be readily fixed at the edge of the field next to the corn. If the soil is not compact it is well to throw two furrows together, making a ridge, and with a heavy block make the top of it very smooth and compact. Along this smooth path, post holes 12 to 18 inches deep should be dug about every 20 or 30 feet. Get a supply of coal tar, or if it is more convenient, pine tar or crude oil, ready to use as soon as the bugs begin to travel. An old coffee pot with the spout pinched so as to allow a small stream to flow is convenient for putting the line of tar on the patch. By holding the vessel near the ground a narrow line no wider than a pencil can be made, and this is all that is necessary. The tar line should strike the post holes near the middle extending directly around the edge of the hole on the side next to the corn, leaving the edge of the hole next to the wheat free. When the chinch bugs reach the tar line they will not cross it but will turn aside and run or crowd each other into one of the post holes. As soon as the holes are partially filled with bugs a small amount of kerosene poured in them will kill them. Care must be taken in putting the line around the post hole so that the assembling mass of bugs does not crowd over it. When first applied the material will soak into the ground, but a hardened crust will readily form which will hold it until it slowly dries out. The line must be closely watched and renewed as often as the bugs begin to break over it. Wherever the soil is sandy and very loose a slight windstorm will cover the tar, making passageways for the bugs, and under such conditions the line must be renewed quite often. A man or boy can care for from 80 to 100 rods of the barrier, but he must stay with it from early till late. Ordinarily the bugs will have finished their migration from wheat in 10 days. This method is apparently costly and troublesome, but the actual expenditure of labor and money is insignificant as compared with the loss of the corn crop which may thus be prevented.

REMEDIAL MEASURES.

DESTROYING BUGS WHICH ENTER CORNFIELDS.

After the bugs enter the cornfields they will at first collect in masses on the plants of the first two or three rows and should be killed before they proceed farther. This can be done in two ways, by applying a gasoline torch and by spraying them with specially prepared solutions.

The flaming torch is not altogether satisfactory on account of the liability of damaging the plants. Great care must therefore be taken with the flame in order to use it safely and effectively. The flame must be in motion all of the time while on the plant, and generally one blast will cause all the bugs to fall to the ground, where they can be burned.

There are several spraying materials which can be used effectively against the bugs after they have congregated on the young corn, but, unfortunately, most of these are injurious to the plants. Kerosene emulsion of 5 per cent strength will generally kill the bugs and will not always injure the corn. The stock solution is made by boiling 1 pound of good lye soap in 1 gallon of water, adding this to 2 gallons of kerosene and stirring the mixture with a paddle for five to ten minutes. A better way to stir the mixture is to put the nozzle of the spray in the vessel and pump the liquid back into the vessel for five minutes. Dilute the mixture to a 4 or 5 per cent solution by adding soft water. Some of the proprietary spraying materials and cattle dips have been used to kill the bugs where they have become alarmingly abundant. One serious objection to these materials is that they are very injurious to the plants. However, it is sometimes better to sacrifice the first few rows and save the field than to let the bugs have their way.

UNSATISFACTORY REMEDIAL MEASURES.

GREEN-CORN BARRIERS.

Cutting the first half dozen or dozen rows of green corn and making a continuous pile along the last row cut is a method of creating a barrier very often employed. The green-corn barrier is made about the time the chinch bugs begin to enter the corn. The bugs are checked an hour or two by the corn pile, but readily pass on to the fresh, living plants.

The piles of corn plants afford shelter for the bugs, and often a quart of cast skins can be found in a heap under these piles. These cast skins have often been misleading to farmers, inducing them to believe that the bugs died from eating the sour juices of the cut corn or having died from disease. *A barrier of this kind is not to be recommended.*

PLOWING UNDER INFESTED CROPS.

As the chinch bugs in the Southwest do not hibernate in cornfields, plowing under of stalks and stubble in such localities will be of no advantage. It has been repeatedly shown that plowing under a crop of wheat, rye, or barley badly infested with young bugs is not effective unless the plowing is deep and very thoroughly done and the field is immediately afterwards harrowed and rolled.

In experiments conducted by Dr. Forbes in 1888 [a] bugs buried
with wheat at a depth of 6 inches were alive after five days and some
buried 5 inches came to the surface. The earth was packed over
these to imitate rolling.

E. M. Shelton,[b] from observations in Kansas, writes:

Chinch bugs plowed under with young wheat to a depth of 8 inches May 9–10—the
ground afterwards harrowed and repeatedly rolled—nevertheless emerged in enormous
numbers (some having apparently hatched in the earth), escaped from the plots and
attacked adjacent crops.

Because of the wheat being winter-killed in central and southern
Kansas and northern Oklahoma during the winter of 1909 and 1910
many of the wheat fields containing young and old bugs and eggs
were listed during April, at the time when the eggs were hatching.
In many cases corn was listed directly into the wheat ground, tearing up the young wheat, but not entirely destroying all of it between
the rows of corn. The eggs and young bugs were buried from 1 to
6 inches. The undestroyed wheat was soon covered with young
bugs, which afterwards attacked the corn as soon as it appeared
above the ground.

Even in fields where no wheat was visible after listing the corn
was entirely destroyed. There is every reason, the authors believe,
for presuming that the eggs hatched beneath the ground, and the
young, after feeding there, had found their way to the surface and
to the corn. Corn listed in fallow ground was free from bugs.
Planting of corn in wheat fields badly infested with chinch bugs is
not advisable, and is generally attended by the complete destruction
of the corn. When a badly infested crop is plowed under it should
be followed by a crop not affected by chinch bugs, such as cowpeas,
soy beans, alfalfa, or clover. Plowing the bugs under as a means of
destruction is not recommended unless in connection with a trap
crop, the work being thoroughly done and followed by harrowing
and rolling or otherwise packing the surface of the ground.

PARASITIC FUNGI.

EARLY OBSERVATIONS.

The susceptibility of the chinch bug to a contagious fungous disease
was first observed by Dr. Henry Shimer in Illinois in 1865. Since
that time two fungi have been found, which are credited with being
fatal to this insect. These have been determined as *Entomophthora
aphidis* Hoffman, and *Sporotrichum globuliferum* Speg. Of these two
fungi, Sporotrichum (known by farmers as "fungous disease" or
"white fungus") appears the most abundant in localities badly

---

[a] 16th Rept , State Ent. Ill., p. 45.
[b] Bul. No. 4, Kans. Agr. Exp. Sta.

infested with chinch bugs, and it is on this fungus that most observations have been made.

The genus Sporotrichum includes a large number of fungi, the most of which are purely saprophytic (i. e., living on dead animal or vegetable tissues). According to the best information obtainable, some of these are known to attack living tissues, causing their death, but afterwards developing rapidly on the body of the dead host. *Sporotrichum globuliferum* belongs to the latter class, and is known to occur on insects belonging to the orders Coleoptera, Lepidoptera, Hemiptera, and also on myriapods (centipedes and millipedes). It is credited with effectively attacking the elm leaf-beetle, the pupæ and adults of which are found covered with the fungus, especially in late summer of a moist season.[a]

Attempts have been made in most of the Central-western States to grow this fungus under artificial conditions, and then introduce it into fields badly infested with chinch bugs, where the fungus is not known to be present. With the possible exception of the Kansas experiments, made by Dr. Snow in the early nineties, these attempts met with little success. In most cases some unfortunate circumstance always arose to make the success of the experiment uncertain. (This fungus requires rather cool, moist weather for most rapid development, and is present in greatest profusion when the bugs are exceedingly plentiful and massed together.) In many of the early attempts at artificial introduction, Entomophthora was present as well as Sporotrichum, thus giving rise to some confusion concerning just which fungus was credited with actually killing the bugs.

The most important of these experiments are fully set forth in the reports of the State entomologist of Illinois and in Bulletins 15 and 69 and Circular 113 of this bureau, and therefore require no extended discussion here.

OBSERVATIONS BY THE WRITERS.

FIELD STUDIES IN KANSAS AND OKLAHOMA.

It is fully realized by the writers that the determination of the cause of a disease is most difficult, and that it requires extended laboratory research along many different lines. To state that this fungous disease is the cause of the mortality among the chinch bugs, without this extended laboratory investigation, would be entirely unscientific. The observations given here are published for what they are worth.

Observations on the habits and occurrence of the fungus were made in Kansas and Oklahoma during the spring of 1908 and 1909, and the spring and summer of 1910. Some additional data were obtained with respect to its behavior in the field among chinch bugs of all ages

---

[a] Conn. Agr. Exp. Sta., Bul. No. 155, 1907.

and under varying weather conditions. Most of the observations were made in Sumner County, Kans., where the fungus had probably been present among the bugs in the fields for a number of years. Considerable fungus was present during 1908 and 1909.

It again appeared among the chinch bugs in southern Kansas during April, 1910, and was first observed in the fields April 18 on the dead bodies of some adults lying on the ground at the base of young wheat plants. From this date the fungus gradually increased, dead adults covered with fungus being found almost every day. These were always on the soil, or slightly buried beneath the surface, about the roots of wheat.

During the first week of May the weather was cool and rainy, the mean temperature of the week being 59° F. and precipitation 0.75 inch for four days. The bugs, during this period, were sluggish and sought shelter under blades of wheat or any trash that would keep them off the ground. Succeeding this week of wet weather, followed two hot days, with a mean temperature each of 68.5° F. and 79.5° F. At the end of this period the following note was made:

A number of dead bugs were found lying on the ground, their bodies whitened with the fungus. One plant had 7 dead, fungus-covered bugs at its base. No young bugs found covered with the fungus.

From notes of May 18:

A great many old bugs are dead and covered with the Sporotrichum, but failed to find any young bugs covered with this fungus. The dead bugs are on the surface of the soil.

About this time the young bugs which had hatched from the eggs deposited in the wheat were massed together about the bases of the wheat plants. Where the wheat had winter-killed, and had been torn up and corn listed into the ground, the bugs which hatched on the wheat had gone to the nearest corn plants. Some stalks of these plants were red from the myriads of young bugs assembled upon them; in no case could any fungus be found on these young bugs, but there were usually a few overwintering adults to be seen about the base of these plants, and upon turning over a clod some of their dead bodies covered with Sporotrichum were usually to be found. Plenty of the fungus could be found at this time in wheat fields where the adult hibernating bugs were still present. Sporotrichum gradually became more abundant during the succeeding days of May and early part of June, the amount observed seeming to fluctuate with the weather, being most abundant while it was damp and cool and checked by a few days of dry weather. It developed most rapidly in wheat fields during the first 10 days of June, while the bugs, both young and old, were migrating from the wheat to the corn, the fungus being at this time abundant on dead adults in contact with the soil and in some cases on the bodies of young dead bugs.

At this time attention was transferred from the wheat to the corn-fields where very little fungus was noted, most of the old bugs being dead, and the young ones seemingly free from any fungus. Some of it could be found on the bodies of the adults about the base of the corn, just as observed about the corn which had been listed in the wheat infested with the bugs; none, however, could now be found above the surface of the soil, all fungus being either below or on the surface of the ground.

The bugs were now seriously damaging the corn, as will be seen by the following note made on June 15:

Bugs still migrating from wheat to corn. They are out for 40 and 50 rows on corn adjacent to wheat. No Sporotrichum is developing.

The farmers of Sumner County were informed of the presence of the fungus in this locality, through the press, and by interviews with them at the laboratory at Wellington. On June 24, two reports came from the farmers that the fungus was killing off the bugs. Visits were made to their farms, and the "dead bugs" proved to be only piles of cast pupal skins, which they had found beneath piles of green corn, and bundles of wheat and oats. This same mistake on the part of the farmers was made during the experiments of Dr. Snow in 1891. In reports of farmers to him we quote: "In some fields the bugs have been reported dead in bunches," but he continues "of the fields visited, no large bunches of white-fungus bugs have been found, * * * each bug had died by himself." [a] This confusion of the fungus-killed bugs with their cast-off pupal skins is one frequently made by farmers, and such reports in regard to the efficiency of the fungus are very apt to be erroneous.

On June 23 the bugs began to leave the badly infested corn, as the young had now developed wings. By July 1 they were so widely scattered over the corn as to give the appearance of having left the fields. In some instances this apparent disappearance of the bugs was credited to the fungus by those who had not been constantly watching them, several farmers reporting that the fungus had killed their bugs because there were so very few to be found.

Heavy rain fell on July 8 and 10, but little Sporotrichum could be found among the bugs in the corn. On July 19 a few dead adults located on corn leaves from 1 to 2 feet above ground were observed covered with Sporotrichum. This was the only case during the summer where the fungus was observed not in contact with the ground. At this time an examination of the wheat stubble was made and the following note made:

In pulling up wheat stubble large numbers of dead fungus-covered bugs were found in some places, the soil about the wheat roots being speckled with the fungus.

---

[a] 1st Report on Contagious Diseases of the Chinch Bug, F. H. Snow, 1897.

Particles of the fungus were found on dead adult bugs or a part of their bodies—some masses of white fungus covering only a broken abdomen, or a thorax with wing attached, by which the insect was identified. It appears that the fungus had developed rapidly on fragments of bugs and entire bodies of the bugs, especially when buried beneath the soil. The indications here were that much white fungus had made its entire growth on the dead bodies of the hibernating adults which had migrated to the wheat fields.

During the last week of July the weather was hot and dry and myriads of young bugs of the second generation, now feeding, sought shade under blades of corn, or beneath the clods at the base of the plant, but during this time no fungus developed. This condition prevailed until the middle of August, when wet weather again set in, and while this did not seem to increase the amount of Sporotrichum present, plenty of the fungus could be found in the soil around wheat stubble. However, this fungus seemed unable to infect the young bugs in the corn above ground, even where the cornfield joined the wheat stubble.

Sumner County, Kans., seemed to be about the center of the Kansas-Oklahoma chinch-bug infestation. With a view of finding out whether or not the fungus Sporotrichum was present in other places over the infested area, as well as to determine the extent of the infestation, a trip was made during June of 1910 through central and southern Kansas and central and northern Oklahoma, which represents pretty well the area infested during 1910 in this part of the Southwest. Twenty localities were visited, viz: Herrington, McPherson, Hutchinson, Pratt,* Dodge City,* Great Bend,* Sedgwick, Wichita, Winfield, Arkansas City, South Haven, Caldwell, and Caney, Kans.; Medford, Enid, Kingfisher, Elreno,* Chickasha,* Oklahoma City,* and Tulsa,* Okla. *Sporotrichum globuliferum* was found in every locality except those marked with an asterisk. The fungus covered bodies of the dead bugs were usually found lying upon the surface of the ground in wheat fields and in every case these were old migrant bugs, precisely as found in Sumner County, Kans. The seven localities where no Sporotrichum was found were all on the extreme outer edge of the infested area where there were very few bugs present. In every one of these seven localities the soil was very sandy, apparently not retaining moisture necessary for the development of the fungus.

ARTIFICIAL INTRODUCTION.—The fungus *Sporotrichum globuliferum* was already so abundant among the bugs in Sumner County during the summer of 1910 that no attempt was made upon the part of the authors to introduce it artificially. It seemed useless, since the amount already present so far exceeded any amount which could be introduced. However, a number of farmers anxious to try this secured some Sporotrichum from outside sources and introduced it into their fields. These attempts were made independently and, as far as could be learned, no satisfactory results followed.

In the neighboring counties of Harper and Cowley, Kans., and Kay County, Okla., farmers united in spreading the fungus over their fields, after having grown plenty of it in boxes on their farms. This experiment was tried very thoroughly and on a very large scale in Cowley County, Kans., where the farmers secured the fungus from boxes of chinch bugs at two central stations, and after having grown more of the fungus in boxes on their own farms spread the whole over their fields where the bugs were thickest. The distributing points were Arkansas City and Winfield, Kans., and about 700 farmers secured fungus from the two boxes at these points. This was done in early June, and was followed by rainy weather just after the fungus was placed in the fields. The precipitation record at Wellington showed an aggregate rainfall of 0.82 inch on June 6, 7, 8, 10, and 11. When Cowley County was visited on June 21 it had been dry for a week preceding. The central distributing points were visited, and it was learned that a large quantity of *Sporotrichum globuliferum* had been distributed from these boxes. No satisfactory results were reported to the central station. Many of the farmers were interviewed, and in almost every instance the lack of success was attributed to the dry weather, which prevailed between June 12 and 21. In some localities every farmer had used some of the fungus. Upon visiting the exact spots where the fungus had been placed in the fields many bugs were found covered with white fungus; however, the fungus was as easily located in places remote from any artificial importation. The fungus was always put out in places where the chinch bugs were massed together. In consequence of this massing more fungus would normally occur in these places than elsewhere, regardless of the source of infection. The damage caused by the chinch bugs was fully as great where the fungus was introduced as it was in places remote from these, and also fully as great in this locality as it was in Sumner County, where no fungus had been introduced.

The parties who carried on this experiment were so united in their work, and extended it over such a wide area, that it constitutes one of the most satisfactory field experiments ever carried out with this fungus. From the results obtained we arrive a step nearer the actual position this method should occupy. There can be hardly a doubt that Sporotrichum was present before the experiment was tried, and that, too, in sufficient quantity to inoculate the healthy bugs. Since this failed to hold the bugs in check, no artificial introduction could accomplish what nature failed in doing.

Sporotrichum has appeared during the past three years in this part of the Southwest, and has not been effective as a natural enemy, occurring chiefly among the old spent and therefore practically harmless bugs of the hibernating brood, they having already performed their mission of depositing their eggs. Since the fungus is so dependent on meteorological conditions, it can not be depended upon to exterminate the chinch bugs in this region. Farmers can accomplish a great deal more by employing methods more under their control.

OFFICE STUDIES IN WASHINGTON, D. C.

The following experiments were conducted at Washington, D. C., by the junior author who did the work on a table in an office room with no facilities for more elaborate experimentation. He entered into the investigation with a view of determining if the growth of this fungus is confined to the dead bodies of the chinch bugs and other insects, as the preceding field observations had, in most cases, borne out this supposition. These experiments led to a further study of the behavior of this fungus among living chinch bugs, and the results are here presented, in the hope that a better and more clear understanding may be had of what was observed to occur in the fields.

In these experiments the author is greatly indebted to Prof. F. M. Webster, of this bureau, under whose direction the work was done. Also to Dr. Flora W. Patterson, mycologist, Bureau of Plant Industry, for her valuable suggestions and kindness in determining the fungus.

Sporotrichum globuliferum was secured for these experiments by placing under a bell jar, on blotting paper kept constantly moist, dead bugs taken at Wellington, Kans., from beneath leaf sheaths of cornstalks, where they had died the previous autumn. The fungus was also secured by placing on moist blotting paper dead bugs collected in clumps of *Andropogon scoparius*, where the adults were hibernating but among which no fungus was observed. From 2 to 4 of all the dead bugs so treated became covered with the white fungus (fig. 15). The fungus was obtained also from live bugs collected from these same stools of *Andropogon*, confining the bugs in small cages of wheat kept well moistened. Some of these bugs died in the cages and on their bodies, lying upon the soil, the fungus appeared and was first observed eight days after the live bugs were placed in the cages.

During the spring of 1909, in the laboratory at Wellington, Kans., the senior author confined, on potted, growing wheat plants, numbers of living chinch bugs which were collected from tufts of *Andropogon scoparius*. The fungus appeared on several of the bugs which had died and were in the soil. He also obtained the fungus on adult dead chinch bugs collected from the tops of wheat blades and put on potted wheat plants. In these experiments the soil had been previously sprayed with a weak solution of corrosive sublimate for the purpose of killing any fungus spores that might have been present in the soil in these cages, thus eliminating that source of infection. This indicates that the spores of the fungus were present on the bodies of the living bugs and were only awaiting favorable conditions for germination and abundant growth. In all such cage experiments the fungus appeared only on the bodies of the chinch bugs that had died in the cages and were lying on the soil.

In the experiments at Washington, D. C., it was desired to obtain a better knowledge of this fungus and its relation to dead bugs, in order to ascertain to what extent, if any, it will grow saprophytically upon their dead bodies. Also to observe the rate of mortality among live chinch bugs, placed in cages and artificially inoculated, as compared with those under similar cage conditions not so inoculated. In the artificial inoculation, the fungus was applied by thoroughly dusting the bugs with the spores of the fungus by means of a small brush, or in other cases by shaking them about in a vial containing the fungus-covered chinch bugs.

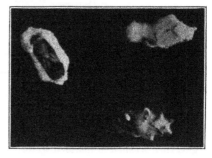

FIG. 15.—The chinch bug Adults covered with "white fungus" (*Sporotrichum globuliferum*) Enlarged. (Original.)

EXPERIMENTS WITH DEAD CHINCH BUGS.—To determine whether or not *Sporotrichum globuliferum* would make its entire growth upon the dead bodies of chinch bugs, a large number of live bugs, collected from tufts of *Andropogon scoparius,* and which died while confined in a vial, were inoculated artificially and their bodies placed on moist blotting paper. The results of these artificial inoculations compared with their checks are here given.

Inoculation experiments with dead bugs.

Artificially inoculated.			Checks		
Number bugs used.	Number developing fungus.	Per cent which developed fungus.	Number bugs used.	Number developing fungus.	Per cent which developed fungus.
25	5	20	100	2	2
25	1	4	200	2	1
25	5	20	25	0	....
25	4	16	25	2	8
40	5	12	25	0	....
			40	0	....
140	20	14 28	415	6	1.45

From this table it is seen that while the check cages gave but 1.4 per cent developing the fungus, this percentage was raised to 14.2 when the bugs were artificially inoculated.

Some live and apparently healthy bugs were then killed mecha ically by piercing them with a sharp needle, their bodies then bei inoculated with the fungus and placed on moist blotting paper.

Inoculation experiments with bugs killed mechanically.

Artificially inoculated.		
Number bugs used.	Number developing fungus.	Per cent which developed fungus.
25	4	16
35	3	8.57
25	3	12
85	10	11.76
Check.		
50	0	0

Others were killed with fumes of chloroform, allowed to dry ove night, then washed in distilled water, and treated as in the precedi experiment.

Inoculation experiments with bugs killed with fumes of chloroform.

Artificially inoculated.		
Number bugs used.	Number developing fungus.	Per cent which developed fungus.
20	6	30
25	5	20
20	6	30
65	17	26.15
Check.		
200	0	0

From these experiments with dead bugs it was found that a sma percentage of them could be made to produce *Sporotrichum globulifer* after inoculation, with every indication that the fungus made its e tire growth on their dead bodies.

EXPERIMENTS WITH LIVE CHINCH BUGS.—The cages used in th experiments consisted of a series of small flowerpots containing grow ing wheat, over each of which was placed a glass cylinder. In these cages the living chinch bugs were placed and cheesecloth stretched over the top to prevent their escape. All of the cages were kept on a table in the office, which was heated day and night, and were kept well moistened to promote the rapid development of the fungus. The live chinch bugs which had been dusted with the spores of the fungus were then placed in these cages and the results compared with check cages into which were placed live chinch bugs not previously inoculated. Most of the bugs were provided with plenty of food. Some, however, were placed in a cage containing no food to see if the rate of mortality and appearance of the fungus were to any extent dependent upon the food supply. These experiments are here given, with a record of each cage and its check.

Experiments with living chinch bugs.

CAGES CONTAINING PLENTY OF FOOD.

Cage No.	Number bugs used.	Treatment given.	Number of days experiment continued.	Number of bugs which died.	Number upon which visible fungus appeared.
a {63	25	Inoculated...............................	14	11	7
{62	40	Check....................................	14	0	0
b {63	15	After 14 days transferred from 62a to 63a........	24	15	11
{62	18	Taken from 62a and inoculated...............	24	17	7
80	9	Taken from 63a and placed in cage free from fungus.	24	0	0
82	25	Inoculated...............................	22	18	4
83	33	Check....................................	22	4	0
90	25	....do...................................	31	20	7
89	25	....do...................................	31	4	0
93	25	....do...................................	35	20	1
92	30	....do...................................	39	4	0
106	25	....do...................................	30	17	11
106a	25	....do...................................	30	5	0

CAGES CONTAINING NO FOOD.

120	29	Inoculated...............................	18	27	21
122	25	Check....................................	18	5	0

The fungus always appeared about the third day after the death of the chinch bug, and in no case did it appear sooner than six days after inoculation, the bugs gradually dying and some remaining alive 35 days. Living bugs removed from cage 63, where the bugs were dying and fungus developing, did not die when removed to another cage away from all infection, even though the moisture conditions were about the same. However, all of those remaining in cage 63 died and afterwards became covered with the fungus. This indicates that the rate of mortality can be checked if the bugs are removed from the presence of the fungus.

As between the inoculation and check cages there was a marked difference observed in the rate of mortality, for which no cause other than Sporotrichum seemed to be responsible, as the death rate was always greatest in the inoculated cages. The greatest mortality and most rapid development of the fungus appeared in cages where the bugs had little or no food. In the stock cages only a very small supply of the fungus could be secured until the food was removed, when it soon appeared abundantly upon their dead bodies lying on the soil.

CONCLUSIONS.—From these experiments it is apparent that *Sporotrichum globuliferum* will to a certain extent develop entirely upon the dead bodies of adult chinch bugs. However, the fungus was easily secured in the cages containing living bugs artificially inoculated, particularly when the bugs in these cages were given insufficient food supply.

The indications were also that this fungus is communicable to living chinch bugs, but is evidently not very effective in causing their death unless the bugs possess weakened vitality. This being the case, it would be most effective in nature against old spent adults which have laid their eggs and are comparatively harmless, and it is upon these bugs that the fungus always appeared in greatest abundance, as borne out by field observations during the past three years.

For this reason, as well as its dependence upon favorable weather conditions, its practical efficiency is very questionable; and since it is unable under favorable conditions of moisture and temperature to rapidly exterminate healthy bugs confined in a cage in the laboratory, little dependence can be put upon it to be used against the insects in the field when favorable moisture conditions are so apt to be lacking.

SUMMARY.

Injuries due to the chinch bug west of the Mississippi River are chiefly confined to the States east of the Rocky Mountains where wheat and corn are extensively grown, the most serious outbreaks during 1909 and 1910 occurring in southern Kansas and northern Oklahoma.

There are two generations each year, one during the spring, which attacks the wheat and corn, and one during the summer, which develops on the corn and hibernates. These last pass the winter as adults, and in the States west of the Mississippi River prefer for hibernation the dense clumps of red sedge grass in which they collect in the fall. Very few survive the winter in fallen ears or stalks of corn during severe cold winters, but may survive a mild winter. They fly from the sedge grass to fields of wheat during the first warm days of spring, where the eggs are deposited, and the young hatching therefrom feed with the adults upon the wheat until it has ripened, when they all march in a body to the nearest cornfield. The young become mature on the corn and lay eggs from which hatch the bugs that winter over as above stated. Weather conditions have much to do with the numbers that reach maturity, many young being killed during a period of wet weather attended by hard dashing rains. Dry weather is most favorable to their development in abundance. The injury to wheat and corn is often severe, sometimes resulting in almost complete destruction of the latter crop after serious injury to the former.

During the severe winter of 1909 and 1910 about 20 per cent of the bugs died normally in the clumps of red sedge grass, where they hibernate.

Experiments in Kansas made during the winter of 1909–10 showed that as high as 75 per cent of the hibernating chinch bugs could be killed by burning this grass. The best time to burn is in the fall, when the grass is as dry as possible. It is not necessary that the flame come in actual contact with the bugs.

The effectiveness of the burning is almost entirely dependent upon the cooperation of the farmers in infested localities. Neglect to destroy chinch bugs collected in these grasses will often result in serious injury, if not indeed a complete destruction of wheat, corn, cane, and kafir.

Next to burning, the dust and coal-tar barriers are the most effective remedies, and should be used while the bugs are migrating from wheat to corn. These barriers must be properly made, and demand constant attention to be of any value.

Many bugs can be killed while massed on the first rows of corn by applying a torch or spraying with kerosene emulsion or proprietary spraying materials.

Plowing under infested crops is not recommended unless the work is done very thoroughly and followed by a crop not susceptible to chinch-bug attack.

Barriers made of piles of green corn are of no value, and are not recommended.

The "white fungus," can not be depended upon to exterminate the chinch bugs. This fungus is very dependent upon moist weather conditions for its rapid development and diffusion. It can usually be secured by collecting live or dead chinch bugs from the sedge grass, and placing them under proper conditions of temperature and moisture. This fungus has many host insects, and is generally present where chinch bugs are found in destructive numbers.

Laboratory experiments show that this fungus is present in greatest abundance among old spent adults, or those bugs that are in a weakened condition. Also that it will grow upon dead bugs. This will partially account for the fact that the time of its greatest abundance in the fields occurs after the hibernating bugs have laid their eggs and are dying normally. These bugs have performed their mission of laying eggs, and are comparatively harmless.

Many bugs lived for weeks, and even months, in confinement in the presence of the fungus under conditions favorable for its development.

Attempts at artificial introductions of the fungus in the fields have so frequently resulted in complete failures that this method is not recommended. Vastly more may be accomplished by applying the same amount of time and labor to the application of practical, successful remedies.

○

U. S. DEPARTMENT OF AGRICULTURE,

BUREAU OF ENTOMOLOGY—BULLETIN No. 95, Part IV.

L. O. HOWARD, Entomologist and Chief of Bureau.

PAPERS ON CEREAL AND FORAGE INSECTS.

THE SO-CALLED "CURLEW BUG."

BY

F. M. WEBSTER,

In Charge of Cereal and Forage Insect Investigations.

ISSUED APRIL 10, 1912.

WASHINGTON:

GOVERNMENT PRINTING OFFICE.

1912.

CONTENTS.

ILLUSTRATIONS.

IV

PAPERS ON CEREAL AND FORAGE INSECTS.

THE SO-CALLED "CURLEW BUG."

(*Sphenophorus callosus* Oliv.)

By F. M. WEBSTER,

In Charge of Cereal and Forage Insect Investigations.

INTRODUCTION.

The so-called "curlew bug" (*Sphenophorus callosus* Oliv.) (fig. 16) is allied to the maize billbug, *Sphenophorus maidis* Chittn., the subject of Part II of this bulletin.   It is commonly known in the Carolinas as the curlew bug, sometimes as the "klew" or "clewbug," and the "kloobug," all probably contractions of "curlew bug."   The curlew is a shore bird, having a long, curved bill, while the insect, *Sphenophorus callosus*, which is provided with a long, curved snout, is found plentifully under rubbish along the shores of sounds and rivers, especially those of North Carolina.   Intelligent fishermen, who are familiar with it, claim that it is often found clinging to their fish nets spread in Albemarle Sound.

Fig. 16.—The "curlew bug" (*Sphenophorus callosus*): Adult. Four times natural size. (Original.)

The information herein given, in so far as it is original, has been accumulating for a number of years, some of the facts having been taken from the general correspondence of the bureau and others from the results of more or less fragmentary and more recent studies by several assistants engaged in the investigations of cereal and forage insects. While not complete in all of the scientific details, so much of this information is of economic value to the farmer, pointing out to him a practical method of prevention, and will also prove of assistance to station and State entomologists who may desire to study the pest in their own States, that it seems an injustice to withhold publication longer with the object of securing details of perhaps minor importance. The species was studied to some extent in Illinois several years ago by Dr. S. A. Forbes, State entomologist of Illinois, and his assistants, the results obtained by him being published later by Dr. Forbes under the name *Sphenophorus cariosus* Oliv. Information from that source is also embodied herein.

The author does not himself assume credit or responsibility, except where indicated in the text. Where so many individual investigators have contributed results of observations and studies, it becomes the duty of some one to act as spokesman and put the matter in shape for publication, and this duty the author has endeavored to fulfill.

In his paper relating to this and other species of the genus Dr. F. H. Chittenden,[1] of the Bureau of Entomology, came to the conclusion that *S. sculptilis* Horn is a synonym of *S. callosus* and that *S. sculptilis* Uhler is a synonym of *S. cariosus* Oliv.; also that in many cases specimens received by him showed that references to either of these three species related properly to *S. callosus*. It is upon these conclusions that many of the facts herein given are based.

HISTORY OF THE SPECIES.

The insect was first described by Olivier in 1807, from "Carolina," as *Calandra callosa*.[2] This locality, now somewhat vague, will, as shown in figure 17, apply almost equally well to both this species and *S. maidis*.

While it does not seem possible that this particular species of insect could have existed all of these years in that section of the country—one of the earliest settled, and therefore one of the first to be brought under cultivation—without doing injury to corn and rice, yet as a matter of fact over half a century did elapse before proof of its ravages, accompanied by specimens of the beetles, were in the possession of the Department of Agriculture.

BUREAU NOTES.

June 1, 1880, specimens of the beetle were received from Mr. E. T. Stackhouse, of Marion, S. C., who accused them of damaging young growing corn in his neighborhood.

May 27, 1884, a report was received from Prof. J. A. Holmes, of Chapel Hill, N. C., that this species was injurious to corn near Bayboro, N. C., where it is known as "kaloo bug"; also that the belief was prevalent there that the "insect winters in the rice stubble."

This species, specimens of which were kindly loaned Dr. Chittenden for identification by Messrs. Forbes and Hart, is recorded in the published notes of the State entomologist of Illinois as injuring corn at

---

[1] Proc. Ent. Soc. Wash., vol. 7, No. 4, pp. 166-182, 1906.
[2] Hist. Nat. des Ins., vol. 5, p. 92, pl. 27, fig. 416, 1807.

Pittsfield, S. C., June 4, 1888. The beetles when received were ovipositing in the box in which they were sent. The species was also found attacking young corn May 1, 1891, the beetle gnawing a large cavity in the stalk just below the ground. "When placed in a breeding cage the beetle made its way into the seed kernel of the stalk."

Brief mention has already been made of its occurrence in Illinois by Dr. Forbes in different papers.[1] It is obvious, as determined by Chittenden, that this entire account of *Sphenophorus cariosus* refers in reality to *callosus*. The illustration given in the publication first cited is of the latter species, as is also the description, and it should be further stated that the occurrence of *S. cariosus* in Illinois is doubtful.

June 10, 1889, Prof. A. J. Cook also reported injury to corn in North Carolina, though the source of his information is not clear.

July 20, 1893, Mr. S. L. Willard, Washington, N. C., wrote that this billbug, called "curlew bug" in that vicinity, was doing considerable damage to the corn crop. Rice also was injured, and it was stated that the insects had been noticed at least 8 years (or since 1886) in that neighborhood. The beetles were also operating in chufa (*Cyperus esculentus*), and some farmers had abandoned rice growing on account of the ravages of this pest.

July 8, 1895, we received word from Mr. B. A. Hallett, Mount Olive, N. C., that this species had completely destroyed the upland rice crop of that section and had greatly injured corn.

During July and August, 1895, Mr. A. N. Caudell reported injury to chufa or yellow nut-grass (*Cyperus esculentus*) at Stillwater, Okla., and August 24 sent pupæ.

May 11, 1896, we received report from Mr. R. J. Redding, Experiment Station, Ga., that this species was attacking fields of young corn in Jefferson County, Ga., by thousands.

May 22, 1897, Mr. Charles B. Guinn reported injury to fields of green corn at Georgia City, Mo.

August, 1898, Dr. Chittenden found, in low bottom-land on the banks of the canal at Glen Echo, Md., the pupa of this species at the roots of witch grass or tumbleweed (*Panicum capillare*). Transformation to imago took place on the 22d.

May 30, 1899, Mr. Edward Markham, jr., Kehukee, Pasquotank County, N. C., wrote that this species had made its appearance in that region, had spread rapidly, and that it was doing so much damage to corn that some crops were being abandoned. It was considered the worst pest of that vicinity—worse than all others combined. It was known locally as the "clue bug," a contraction of curlew bug. Beetles were received from this source as late as July 8.

---

[1] 16th Rept. St. Ent. Ill., f. 1887 and 1888, pp. 64, 68, 71, 1890. 22d Rept. St. Ent. Ill., f. 1903, pp. 19-21. Bul. 79, Univ. Ill. Agr. Exp. Sta., p. 453, 1902.

August 3, of the same year, Mr. James K. Metcalfe, Silver City, N. Mex., reported beetles as well as larvæ in stalks of corn growing near the Gila River in that vicinity. It was reported that the beetles deposited their eggs in the stalks near the ground; that the grubs ate all the lower part of the stalk and soon destroyed the plant. The beetles destroyed entire fields of corn, working in the usual manner.

August 6, 1900, Mr. G. L. Swindell, Swindell, Hyde County, N. C., reported that this beetle and its larva were injurious to rice—the larva by feeding on the roots and the beetles by attacking the stalks near the ground. The damage done by the beetle to corn in Hyde, Pamlico, Beaufert, and Tyrrell counties could hardly be estimated. In some years it amounted to almost total destruction and was noticeably worse near old rice patches. August 20 this bureau received crowns and roots of the infested rice, as well as larvæ and beetles. In nearly every case there was a cavity in the crown just above the roots, containing the larva and its castings.

During 1901 this species was reported by Mr. Franklin Sherman, jr., Raleigh, N. C., as the cause of general complaint in the eastern part of North Carolina of "billbugs," "klewbugs," "curlew bugs," etc. The species was identified from Elizabeth City and Goldsboro, N. C. Injury was not noticed in the western part of the State. April 23 a report was received from Mr. Robert T. Smith, Grant, Fla., of this species attacking corn, being most troublesome in early March, and infesting the young stalks just below the center, in such a manner that the central leaf often comes out entirely.

Prof. Franklin Sherman, jr.,[1] has given an account of this species with accompanying reports of its injuries to corn in 1902 and 1904 in low, swampy lands in the eastern sections of North Carolina. Injury was also reported by Mr. Sherman in portions of Bladen, Cumberland, Duplin, Moore, and Brunswick counties.

During June and July, 1904, Mr. Thomas J. Clark, Cliff, Grant County, N. Mex., reported billbug damage, and July 27 sent larvæ and imago of this species with the statement that the beetles began operations on young corn as soon as it came up; that they deposited their eggs in the roots; that the larvæ continued feeding in the stalk, and that after the corn began to tassel it would fall over and blight. The stalks, however, seldom developed to that extent before they were killed. He stated that cutworms were nothing to compare with them in their attacks on corn; also that the species was believed to be native to old Mexico and that it had been seen in that vicinity seven or eight years earlier.

June 26, 1906, information was received from Mr. F. B. Hopkins, in charge of the testing gardens at the Arlington Farms, Va., of injury

---

[1] "Insect Enemies of Corn," Bul. N. C. Dept. Agr., 1906, pp. 19–22.

' the beetle of this species to chufa. The larva was at this time)erating in the crown of the plant at the base of the leaves. It was)served that the insect was capable of doing considerable damage 'en at this early date, as crowns were then found completely honey->mbed. On June 28 Mr. I. J. Condit, detailed by Dr. F. H. Chittenden, . charge of Truck Crop and Stored Product Insect Investigations for ιe purpose of obtaining material, visited the infested locality and ob-.ined numerous specimens of hibernated beetles and larvæ. It was ιen estimated that the damage would amount to about 20 per cent of ιo crop. June 30 a larva was found nearly grown. Eggs were so obtained at that time and until the end of the first week of ugust. Search was made for the natural food of the insects, and ιis was found in Frank's sedge (*Carex frankii*).

May 22, 1907, Mr. R. I. Smith sent this species from Atlanta, Ga., ith the statement that these billbugs were reported as quite abun-ιnt and doing great damage to corn at Statesboro, Ga. They had ξen present at the date of that writing for about three weeks.

DISTRIBUTION.

As will be observed by consulting the accompanying map (fig. 17) ιis species covers a wider range than *Sphenophorus maidis*. It seems) center in point of abundance in eastern North Carolina, extend-

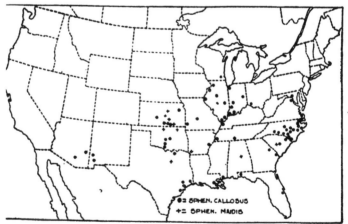

ιᴜ. 17.—Map showing distribution of the "curlew bug" (*Sphenophorus callosus*) and the maize billbug (*Sphenophorus maidis*). (Original.)

g southward to southern Florida, northward to Maryland, thence)rthwest to northwestern Illinois, southwest to extreme south-ιstern Arizona and northern Mexico, and eastward to the Gulf)ast.

It is essentially a lowland form, as its food plants clearly indicate, and will therefore especially interest the farmer whose. fields are of bottom, swamp, or other low-lying lands.

The following are the localities from which the species has been received:

Arizona: Tucson (Wickham), near Duncan (Cockerell). *Arkansas:* Helena (L. E Howard). *Florida:* McLellan (W. H. Gill), Grant (Robt. T. Smith). *Georgia:* Jefferson County (R. J. Redding), Statesboro (R. I. Smith). *Illinois:* From Pekin to Cairo, Savanna, Urbana, Metropolis (Forbes), Warsaw (Dr. Shafer), Rock Island (Det. Chittenden). *Indiana:* Lake, Vigo, Posey, Perry, Putnam, and Blackford counties (W. S. Blatchley). *Kansas:* Great Bend, Arkansas City (T. H. Parks), Wellington (Kelly and Parks), Douglas County (Det. Chittenden). *Kentucky:* Opposite Cairo, Ill. (Forbes). *Louisiana:* New Orleans (H. Soltau). *Maryland:* Glen Echo (Det. Chittenden). *Mississippi:* Gulf View (Det. Chittenden). *Missouri:* Georgia (Chas. B. Guinn), Atoka (Det. Chittenden). *New Mexico:* Cliff (T. J. Clark), Silver City (Jas. K. Metcalfe). *North Carolina:* Edenton (J. W. Mason), Elizabeth City (J. P. Overman), Hertford (W. T. Shannonhouse, Mrs. S. D. Jordan), Pineview (W. Barnett), Pyreway (Maj. Gore), Bayboro, Chapel Hill (J. A. Holmes), Washington (S. L. Willard), Mount Olive (B. A. Hallett), Kehukee (Edw. Markham), West Raleigh, Proctorville, Braswell (R. I. Smith), and Hyde, Pamlico, Beaufort, and Tyrrell counties; Swindell (G. L. Swindell), Bladen, Cumberland, Duplin, Moore, and Brunswick counties (Franklin Sherman). *Ohio:* Cincinnati and vicinity (Chas. Dury). *Oklahoma:* Stillwater (A. N. Caudell, C. E. Sanborn), Duncan, Anadarko, Pocasset, Hastings, Cement, Rush Springs, and Chickasha (A. L. Lovett), Duncan, Chickasha (T. D. Urbahns), Marlow (J. F. Davidson), Oklahoma City (T. H. Parks). *South Carolina:* Marion (E. T. Stackhouse), Pittsfield (Forbes and Hart), Rimini (C. R F. Baker). *Tennessee:* Appleton (P. Cox, Geo. G. Ainslie), Memphis (H. Soltau). *Texas:* Whitesboro (E. O. G. Kelly), Wallisville (W. L. McAtee), Alligator Head (J. D. Mitchell). *Virginia:* Norfolk (Popenoe), Arlington (F. B. Hopkins). *Mexico:* (Prof. Herrera).

FOOD PLANTS.

Dr. Forbes gives *Cyperus strigosus* as the natural food plant, in the roots of which it develops in Illinois. Mr. T. D. Urbahns found it developing in *Tripsacum dactyloides* at Plano, Tex., in July, 1909. At Appleton, Tenn., July 14, 1911, Mr. Geo. G. Ainslie found the infested fields in part grown up with weeds and a swamp Carex (*C. vulpinoidea*), but he was unable to find the beetle actually developing therein. (See Pl. IX, figs. 1, 2.) Mr. A. N. Caudell reported the larvæ injuring the roots of yellow nut grass (*Cyperus esculentus*) at Stillwater, Okla., in 1895. Dr. Chittenden reared the adult from a pupa found in the roots of *Panicum capillare* growing in low bottom lands along the canal near Glen Echo, Md., in August, 1897. Mr I. J. Condit found it breeding in Frank's sedge (*Carex frankii*) growing on the department farm at Arlington, Va. In Florida the insect develops from egg to adult in *Cyperus rotundatus*, while farther north, in the Carolinas, the common food plant is the "chufa" (*Cyperus esculentus*). To such a degree is this true in the latter locality that the insect is supposed by farmers to have been introduced with that plant. Quite in accord with the foregoing, Mr. J. G. Sanders reared adults March 30, and again April 25, 1908, from *Cyperus exaltatus*, introduced from Egypt and growing on the department farm at Arlington, Va.

The cultivated food plants are corn, rice, and peanuts, in importance according to the order given.

DESCRIPTION AND LIFE HISTORY.

THE EGG.

(Fig. 18.)

The egg appears to have been first observed by Mr. A. N. Caudell, who noted the female ovipositing at Stillwater, Okla., July 18, 1895. The egg was described as white, 1.5 mm. long and half as wide, oblong-oval in shape. Mr. E. O. G. Kelly, who studied the species carefully at Wellington, Kans., found eggs deposited June 17, 1911, to be "white," 1.5 mm. long, one-third as wide, and elliptical in form.

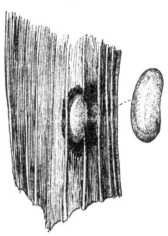

FIG. 18.—The "curlew bug": Egg as placed in stem of young corn plant. Greatly enlarged. (Original.)

Dr. Chittenden described the egg as found at Arlington, Va., as considerably larger, measuring 2.2 to 2.3 mm. in length and only 0.8 to 0.9 mm. in diameter. The outline is subreniform-elliptical, one side having a tendency to straightness along the greater portion of its length. The color is dull, slightly yellowish white. The surface is nearly smooth, with faint reticulation showing in very limited areas. The variation in size of the egg has also been observed by Mr. R. I. Smith, in North Carolina.

Mr. Kelly, in his studies, found eggs from June 16 to September 11, a period of nearly 3 months. The egg period varied from 4 to 6 days in June, in July 5, and from 6 to 8 days in September. In one case 58 eggs were secured from one female, and there was a possibility that she might exceed this number.

Mr. Vernon King and the author found ovipositing adults and half-grown larvæ on Harveys Neck, about 15 miles southeast of Hertford, N. C., on June 20, 1911. This would indicate that oviposition was in progress about June 1. Mr. Jas. A. Hyslop, of this bureau, and Mr. R. I. Smith of the North Carolina Agricultural

Experiment Station, found pupæ in the same locality November 4.
If we allow 8 days as the egg period and 37 to 41 days as the larval
period, as determined by Mr. Kelly in Kansas, the eggs, judging from
the records just mentioned, are deposited in North Carolina from
about June 1 to September 20, or during a period of approximately
4 months. Mr. I. J. Condit found a nearly full-grown larva at
Arlington, Va., June 30, 1906, which confirms in a general way the
preceding observations.

THE LARVA.

(Fig. 19.)

Quite naturally the larva of this species closely resembles that of
Sphenophorus maidis. The principal differences are brought out in
the illustration of *S. callosus*. The head (fig. 19,*b*), is more slender,
especially toward the vertex, the area between the Y sutures is nearly

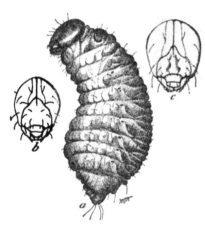

smooth and quite different in
outline from that of *S. maidis*
(fig. 19, *c*) drawn on the same
scale. In this latter species
the space is shallowly sculp-
tured, with the sutures more
sinuate. The larvæ of both
species vary greatly in size,
and it is doubtful if in this
respect they differ materially
from each other.

Mr. Caudell found the larvæ
among the matted roots,
where they form cells and
where they are frequently
seen completely embedded in
the chufa. These they hollow
out, leaving only the hull.
They are sometimes so nu-
merous that frequently as

FIG. 19 —*a*, Larva of the "curlew bug" (*Sphenophorus callosus*); *b*, head of same; *c*, head of larva of the maize billbug (*Sphenophorus maidis*). Enlarged. (Original.)

many as a dozen can be taken in a single bunch of roots.

Mr. I. J. Condit, on October 8, found the larvæ in chufa at Arlington
Farms, Va., usually from half an inch to an inch below the surface
of the ground. They seemed quite capable of subsisting upon the
dead, perfectly brown, and nearly rotten substance of the leaves,
stems, and crowns of the plant, but they also perforated the roots.

Mr. F. B. Hopkins observed, on June 26, also at the Arlington
Farms, Va., that the larvæ were operating in the crown of this same
plant, *Carex frankii*.

Mr. G. L. Swindell, in a communication dated August 6, 1900,
states that at Swindell, N. C., the larvæ injured the roots of rice by

THE "CURLEW BUG."

The portion of the Shannonhouse cornfield, Hertford, N. C., on which corn was grown in 1910; totally destroyed by the "curlew bug" *Sphenophorus callosus*, in 1911. (Original.)

THE "CURLEW BUG."

The portion of the Shannonhouse cornfield that was devoted to cotton in 1910, planted to corn in 1911 and uninjured by the "curlew bug."   (Original.)

THE "CURLEW BUG."

The dividing line in the Shannonhouse cornfield in 1911 between the portion (to the left of the man standing in the center) devoted to cotton in 1910 and the portion (to the right) where corn was grown in 1910. Note how exactly this corresponds, in point of damage by the "curlew bug," with the nature of the previous crop. (Original.)

feeding thereon. Later, roots sent to the bureau by Mr. Swindell contained larvæ in both the crowns and roots. In nearly every case there was a cavity containing a larva in the crown of the plant just above the roots.

Mr. S. L. Willard, Washington, N. C., under date of July 20, 1893, complained also of injury to rice, stating that the depredations had been observed in his neighborhood since 1886.

Under date of July, 1895, Mr. B. A. Hallet, of Mount Olive, N. C., complained that the insect had completely destroyed the upland rice crop of that section.

In August, 1910, Mr. J. W. Mason, of Edenton, N. C., through Representative J. H. Small, stated that the insect had attacked both corn and peanuts in his neighborhood, killing corn and seriously injuring peanuts.

Mr. R. I. Smith, of the North Carolina Agricultural Experiment Station, states that where rice is grown this grain appears to be its favorite food, as the insect is ten times more abundant in rice fields than in cornfields. The eggs are placed in the corn plant above the roots, as shown in figure 21, b. The larvæ work downward, eventually pupating at the lower end of the root, as shown in figure 21, c.

In nearly all of our records of injuries by this species, attention is called to the fact that its attacks are upon low or swampy land. The very nature of its food plants would indicate that the natural habitat of this species is in low or swampy lands. In a great many cases such lands are either subject to overflow or the plants are more or less submerged in water for considerable periods of time. While the insect is not aquatic, it most certainly is capable of living and developing on submerged plants without suffering material incon-venience therefrom. In the cornfields they are often found working several inches below the surface of soil thoroughly saturated with water. Farther on, it will be noted that the adult can also live sub-merged in water without apparent inconvenience.

INJURIES TO CORN BY LARVÆ.

While, as will be shown, corn is injured both by adults and larvæ, attacks by the latter are by far the most fatal to the plant. Good illustrations of a serious attack from a larva of this species are shown in Plates VI, VII, and VIII, from photographs of a field belonging to the Messrs. Shannonhouse, on Harveys Neck, N. C., along the shore of Albemarle Sound. While it is probably true that the insect is much more abundant in rice fields, it must be borne in mind that the area of rice culture is very small compared with that of corn. While local injuries to rice may be very severe, nevertheless the greatest losses from attacks of this insect most certainly fall upon corn grow-ers, especially those whose fields are on low or bottom lands. As illustrating this point, as well as the severity of attack, Mr. Swindell, who has been previously quoted, stated that in some years the loss amounted to almost total destruction. Mr. R. I. Smith, under date of May 22, 1907, stated that the species was doing great damage to corn at Statesboro, Ga. Under date of May 30, 1899, Mr. Edward Markham, Kehukee, N. C., complained that the insect was doing so much damage to corn in his neighborhood that in some instances the crop was being abandoned. In his estimation it was the worst insect pest observed in his community. Under date of August 3, of the same year, Mr. James K. Metcalf, Silver City, N. Mex., complained that the insects were destroying entire fields of corn, working in this same manner. At Arkansas City, Kans., June 22, 1910, Mr. T. H. Parks, of this bureau, found the species exceedingly abundant in fields of young corn growing along the Arkansas River. Mr. T. J. Clark, sr., of Cliff, N. Mex., under date of June 25, 1904, states that the larvæ are more destructive than cutworms. In his opinion the species is a native of old Mexico and had not been observed in his locality until about eight years previous. Mr. T. D. Urbahns, of this bureau, found that corn about Duncan, Okla., during June 18 and 19, 1909, had suffered very severely. In one case about 20 acres of lowland in the heart of a large field had been entirely destroyed. It had been replanted, but the second planting was also badly damaged. In another field an area, also of about 20 acres, in the heart of a still larger field had been completely destroyed. This land was wet and had been flooded during the previous summer. In a field near Comanche, Okla., consisting of about 100 acres of bottom land, Mr. Urbahns found that the crop had been very severely damaged, some of it having been twice replanted. This land had also been flooded the previous year. A field of about 60 acres in this same neighborhood, examined June 19, had been entirely destroyed. At Great Bend, Kans., July 7, 1910, Mr. Parks found that a field of 6 acres of corn had been damaged about 20 per cent by these insects. Under date of May 17, 1909, Mr. J. F. Davidson, of Marlow, Okla., complained that the insect had completely destroyed 100 acres of young corn on his farm. Some of this ground had been replanted a third time, with discouraging prospects of his being able to secure a crop of corn. He further states:

My land is all bottom and valley and adjacent to Little Beaver Creek, a stream 50 feet wide. The heavy floods of last spring caused this creek to overflow its banks and the water spread out over all the bottom land, inundating thousands of acres and destroying the growing crops; 40 acres of my land was thus under water seven different times. I assume the result is the billbug and its depredations this spring on all bottom lands, which are the best corn lands in this section of Oklahoma. Thousands of acres have already been completely destroyed by this pest, and farmers are now busy replanting, with slight hopes of securing a stand that will justify cultivation.

The larva on first hatching is, of course, very small, although, as the eggs increase in size after being deposited in the plants, the size of the larva will considerably exceed that of the newly deposited egg.

It burrows its way downward through the center of the lower stem into the main root or taproot and, unless this is entirely eaten away, probably finishes its development there. Although we have not observed it, it seems quite likely that under certain conditions it may transform to the pupa in the earth outside of the plant.

Studies carried out by Mr. Kelly at Wellington, Kans., have shown that the larval stage may occupy from 37 to 41 days—the latter period in most of his experiences.

THE PUPA.

(Fig. 20.)

Pupation in the crowns of chufa takes place normally in cocoons formed of dried castings, but in cases where the crowns have been much eaten away, the larva before transforming evidently falls out and pupates in the earth nearby, generally within an inch or less of the crown. In the earth a cell is made by the larva, which turns round and round, thus forming quite a distinct earthen cocoon. In one instance noted by Mr. Caudell, a pupal case or cocoon was found on chufa about 2 inches above the ground, indicating that the larva had floated to the surface of the water, the plants being submerged at the time.

In corn plants pupation takes place in the larval chambers as shown in figure 21, c.

A larva noted by Chittenden transformed to pupa August 22, and the beetles issued September 1, the period of the pupal stage having been about 9½ days. As the weather during this time was considerably over 80° F., and quite humid, 9 days is probably the minimum pupal period for this species.

Fig. 20.—The "curlew bug:" Pupa. Greatly enlarged. (Original.)

Of a number reared by Mr. Kelly at Wellington, Kans., during August, 1910, one pupated in 9 days, while three others pupated in 13 days.

THE ADULT.

(Fig. 16.)

This species was first described by Olivier in 1807.[1] In his paper entitled "New species of Sphenophorus with notes on described forms,"[2] Dr. Chittenden has fully discussed this species as follows:

This species was united by LeConte[3] and Horn with *cariosus* Ol., but wrongly so, as I shall attempt to prove. Olivier's description reads in substance as follows:

Body black with dark cinereous coating. Antennae brownish black, shining, cinereous at apex. Rostrum black, dark cinereous at base. Thorax uneven, "and

[1] Calandra callosa Olivier, Hist. Nat. des Ins , vol. 5, p. 92, pl. 28, fig. 416, 1807.
[2] Proc. Ent. Soc. Wash., vol. 7, No. 4, pp. 176-177, Mar. 9, 1906.
[3] Rhynch. N. A., p. 425.

one sees on the superior portion an elevation in the form of a cross, feebly marked "
Elytra uneven, feebly variolate, marked toward the apex with a callous point, nearly
spinose, blackish, shining.

Olivier's illustration is imperfect in that it is very crude, showing neither punctuation nor sculpture and the general impression is that of a shining species, which was
certainly not intended. The thorax is a little short, otherwise the form coincides
with the species which is figured herewith.

The cinereous base of the rostrum is an important character, as it signifies that a
considerable portion of the base is coated while in *cariosus* it is not. The cross-like
elevation of the thoracic disc is aptly described as feebly indicated, in fact it requires
a little imagination to discern it in many individuals; moreover, it is not shown in
Olivier's figure.

Among coleopterists in general the adults of this species are supposed to be covered by a coating, consequent upon the beetles coming in contact with the soil. This supposition is most certainly
erroneous, as adults secured by Dr. Chittenden from cocoons and
by others of the bureau from the chambers in the roots of corn before
they had come in contact with anything excepting the débris with
which the chambers are more or less filled, are found to possess this
coating.

Specimens secured by Dr. Chittenden are of a rich brown color
with velvety surface. It is only when the beetles become somewhat
abraded and this coating worn off of the elevations and the shoulders
and near the tip of the elytra that the callouses are formed, a character
upon which the specific named is based. Strictly speaking, the perfect insect has not been described, and it does not become "*callosus*"
until the insect has moved about and rubbed these points bare.

The adults evidently hibernate to some extent in corn in the chamber in which they have developed, but seemingly lower down than
in the case of *Sphenophorus maidis*. (See fig. 21, c.) They were found
very sparingly, by Mr. James A. Hyslop of this bureau and by Mr.
R. I. Smith, of the North Carolina Agricultural Experiment Station.
occupying this position in the cornfields of Harveys Neck, previously
mentioned, on November 1, 1911. The numbers found, however,
were far too limited to indicate that this can be true of even the
majority, the others probably wintering over either in or near the
surface of the ground. This was in the same field where Mr. Walton
had fruitlessly searched for them on September 5 and 6.

When we come to take into consideration the fact that the natural
coating with which the adults are covered is almost exactly the color
of the soil, with which it is, indeed, more or less begrimed, and that
the insects on being disturbed will draw up their legs and remain as
quiet as if dead, it will be seen that it is exceedingly difficult to detect
their presence in or on the surface of the ground, even by an expert
who knows exactly for what he is searching. Therefore that careful
search should not happen to reveal their presence is not especially to
the discredit of those who are engaged in trying to find them.

As will be observed, this method of hibernation is of the greatest importance from an economic standpoint, because hibernation in the lower stalk or roots would bring the inhabitants within reach by pulling up and burning these stalks during the winter or early spring. For data on the larger corn stalk-borer see Circular No. 116 of the Bureau of Entomology, entitled "The Larger Corn Stalk-Borer."

The beetles probably come forth from their hibernation quarters quite early in spring, as soon as the ground has become permanently warm from the spring temperature. Mr. Kelly found them under cornstalks of the previous year at Whitesboro, Tex., April 13, 1910. They evidently feed for a considerable time by puncturing the lower part of the stems of the plants. These punctures are quite different from the egg punctures, and the effect is often not so fatal as that occasioned by the downward burrowing of the larva. These punctures are usually made about or a little below the surface of the

Fig. 21.—The "curlew bug:" a, Corn plant attacked by adult insect; b, egg as placed in stem of young corn plant, enlarged at left; c, pupa and adult in root of corn in chamber eaten out by the larva, slightly reduced. (Original.)

ground, the beetle evidently searching for a point where the stem is tender and succulent.

If the punctures are made lower down on the plant, just above the root, the result is a throwing up of a number of tillers or suckers from the roots, the main stem itself having a stalky appearance, with the result that no ears are produced. In this respect the effect produced resembles to a degree that of an attack of the Hessian fly on a young wheat plant in the fall. This unusual development of tillers or suckers has sometimes been vulgarly termed "frenching," although it must not be understood that all of the difficulties known as "frenching" in corn have been due to the attack of these beetles.

If the puncture made by the beetle for the purpose of securing food has been made higher up the stem, food has been obtained from the unfolded leaves above the crown of the plant. When these leaves finally push forth, the puncture made by the beak of the beetle appears in the shape of transverse rows across the leaves, as illustrated in figure 21, at *a*. Frequently there will be a distorted growth on the stem, having much the appearance of galls or excrescences, as shown also in this figure.

While the damage done by the beetles in feeding is in many cases doubtless severe—if the corn plants are very young at the time of attack they are probably destroyed in this way—generally speaking the greatest damage is probably caused by the larvæ, especially in the East.

Attention has already been called to the fact that the larvæ can apparently live without difficulty for a considerable length of time in the stems of plants that are completely covered by water. This is surprisingly true in the case of the adult insect.

August 4, 1906, Dr. Chittenden collected adults of this species at Arlington, Va., and placed them in a jar of water with a few stalks of grass and chufa. The beetles attached themselves to these stalks under water and remained there. Two of the beetles were removed November 21 of the same year, and although they had been submerged during the entire period they were still "very much alive." Another instance has come to our notice which would indicate that not only can these beetles survive in fresh water, but also in salt water.

Mr. James Overton, a farmer and fisherman residing on the north shore of Albemarle Sound, informed the author that this species was frequently found by him clinging to his fish nets set in the water of the Sound, and that he found them abundantly under the débris along the shore. As Mr. Overton is perfectly familiar with the work of the insect in the cornfields, and was one of the first in his neighborhood to recognize it, there does not appear to be any reasonable doubt of the correctness of his statement. Indeed, farmers living on Harveys Neck are of the opinion that the pest first came to them from the South, having drifted across the Sound from the opposite shore. Mr. Overton, who resided on the southern shores of the Sound before taking up his residence on Harveys Neck, states that the insect was destructively abundant along the southern shore before it was observed in his present neighborhood. Whether this theory of the diffusion of the pest is correct or not, there does not appear to be any good reason why the insect might not drift about in the waters of the

Sound and be carried ashore by the tides. In at least two cases, each involving a different species of Sphenophorus, adults have been found along the sea beach in situations where they must have been submerged at each flow of the tide.

The length of life in the beetle stage is not definitely known, but Dr. Chittenden has observed overwintered adults as late as August 8, thus overlapping the appearance of the new generation of adults.

INJURY BY THE BEETLES.

It is rather difficult to separate out, in the correspondence of the bureau, injuries that have been caused by the adults, or beetles, of this species from those inflicted by the larvæ. That the larvæ are eminently capable of totally destroying young corn is very evident, but the following extract points more or less conclusively to the beetles themselves as being the authors of the injuries inflicted.

Mr. C. R. S. Baker, Rimini, S. C., June 26, 1909, in a communication to the Bureau of Statistics, stated that the beetles were killing young corn by puncturing the stalk to the heart, killing the plant precisely as with the "budworm." This particular field had been highly fertilized with guano and stable manure. Maj. Gore, of Pireway, N. C., May 12, 1910, stated that 30 acres of corn planted on new land had been literally eaten up by the beetles; presumably the new land was either very low or reclaimed swamp. Under date of May 21, 1910, Mr. W. Barnett, of Pireway, N. C., stated that a farmer in his neighborhood had lost half of his crop of corn from attacks of this beetle. Writing from Helena, Ark., June 2, 1911, Mr. L. E. Howard stated that these beetles were killing the corn, mostly young corn, but some as large as waist-high. Writing from McLellan, Fla., May 2, 1909, Mr. W. H. Gill stated that the pest had just made its appearance in Santa Rosa County and attacked young corn about a month old by boring in the stalk underground and killing the center. On June 1, 1910, a complaint was received from Mr. P. Cox, Appleton, Tenn., inclosing specimens of the beetle which he stated were destroying his corn crop and asked for an investigation of the trouble. July 14, Mr. G. G. Ainslie visited the locality and found that Mr. Cox's field consisted of about 40 acres lying in a creek bottom. Farther down the stream were two other fields of corn all of which had been damaged. It seems to be the plan in that particular locality to allow the land to go uncultivated every alternate year. During the season in which the land is idle there springs up a heavy growth of weeds and swamp grass. This particular field was plowed the latter part of March and replowed the last day of April, corn being planted soon after. The first planting came up quickly, but was utterly destroyed. The second planting, the date of which was not obtained, was also practically destroyed, and a little before the middle of June a third planting was made. On the lower depressions of the field (see Pl. IX, fig. 1), termed "swales" in that neighborhood, the corn from this planting was either small or missing, the size of the stalks being very irregular. In most cases the main stalk was aborted and suckers had been thrown up, sometimes a distance of several inches from the original plant. The main stalk was either missing altogether or had become so dwarfed and distorted as to be practically worthless. (See Pl. IX, fig. 2.) It lies prostrate on the ground, curled and twisted, being sometimes almost buried in the loose earth, and the beetles were still found attacking the plants.

RECENT INVESTIGATIONS OF THE BUREAU OF ENTOMOLOGY.

gations carried out in Kansas, Oklahoma, and Texas by Messrs. Kelly, Urbahns, and Parks, of the Bureau of Entomology, and in North Carolina by Mr. James A. Hyslop, also an assistant in the bureau.

Jordan, Hertford, N. C., accompanied by specimens of these beetles, stating that the insects take possession of and destroy whole fields of corn as soon as it comes up. Many farmers had been obliged that season to plow and plant their corn for the second time. The insects attack the plants by inserting their bill into the stalk near the ground, causing the plants to wilt in a few hours. The trouble had been noticed for several years and appeared to be rapidly on the increase. Apparently, unless some steps were taken for their protection, the farmers in that neighborhood would not be able to raise sufficient corn for their own use. Two days later a communication was received from Mr. William T. Shannonhouse, from the same post office, accompanied by specimens of the bettles. Mr. Shannonhouse complained that these insects attacked the corn from the time it was 3 or 4 inches high until it became 10 inches or a foot in height. Then they were found just below the surface of the ground puncturing the stalk, causing the death of the plant. Mr. Shannonhouse called attention to the fact that where corn had followed cotton crops no damage was apparent, but where the preceding crop had been corn the damage was in many cases very severe, often resulting in a total loss of the crop. In cases where the land had been planted to corn in alternate years, and during the intervening years to some other crop, no difficulty was experienced. The author, together with Mr. Vernon King, visited these fields in company with Mr. Shannonhouse on June 20 and made a careful examination of them. It was found that where cotton had been the previous crop, attack by Sphenophorus was hardly noticeable. Larvæ were abundant, ranging from newly hatched to half-grown, while eggs were being deposited on both small, tender plants, and on larger, more mature

FIG. 1.—VIEW OF CORNFIELD NEAR APPLETON, TENN., SHOWING DAMAGE BY THE "CURLEW BUG." (ORIGINAL.)

FIG. 2.—CORN PLANTS, SHOWING NORMAL PLANT AND THOSE DAMAGED BY THE "CURLEW BUG." (ORIGINAL.)

THE "CURLEW BUG."

ones. From one to several eggs were laid on each plant, either just beneath the surface of the soil, slightly above the roots, or from 2 to 3 inches above this point in the stem. Beetles usually rest on stems head downward, often partially hidden by soil around the plant, and frequently with the beak inserted into the tissues of the stem. At the bottom of slits made by the beak, and easily seen with the naked eye, there is often a white, elliptical egg, sometimes with one end transparent. Both males and females were common. On June 22, about 3 miles away across the Perquimans River, on the farm of Mr. R. L. Spivey, Mr. King observed the same work in a small patch of corn planted on spring-plowed land which bore corn the year before. Near Mr. Spivey's farm injury was also done to 3 or 4 acres of corn on the farm of Mr. J. T. Jackson, whose land also bore corn the year before. At this time only a few large plants of the first crop were standing. The land had been replanted, but only sickly plants were produced, as these had been attacked by Sphenophorus and by *Diatræa* sp., the latter of which is locally known as the "budworm." An adjoining patch of corn, planted on soil which bore corn last year, but cut early and the land fall-plowed, was seemingly growing.

On July 25 Mr. W. R. Walton visited the same locality and, in the same fields previously examined by Mr. King and the author, found larvæ, apparently nearly full grown, in the taproots and crowns of the plants. Although no longer feeding, they had not yet transformed to pupæ, nor were they yet in cells formed for pupation. Some of the less seriously injured stalks of corn were from 6 to 8 feet high, with one well-developed ear. No grasses or sedges in which the insect could develop could be found in the neighborhood, although Mr. Walton was told by Mr. Overton that masses of both of these plants, with heavy root-stalks, sometimes drift across Albemarle Sound from the South. Farmers in that neighborhood say the pest began its depredations along the north shore, and express the opinion that the pest came from the South. Later, September 5 and 6, Mr. Walton again visited the same locality to learn the condition of the pest, but although he examined and pulled up about 100 stalks of corn where the pest had been abundant earlier in the season, he could find no trace of it in any stage.

REMEDIAL AND PREVENTIVE MEASURES.

With these insects in full possession of a field, there does not appear to be any thoroughly practical and effective measure for preventing or overcoming their ravages. While throwing up the soil or hilling up the young plants with the cultivator might prevent the beetles themselves from puncturing the stems low enough down to cause the plants to sucker or become distorted, this is by no means assured.

We only know that the higher up the insect punctures the stem the more likely is the attack to result only in the transverse rows of holes across the leaves, as shown in figure 21 at *a*. In any case this ridging or hilling up would only form a possible slight protection against the injurious effects of the feeding of the beetles. Once the larvæ have started to burrow their way downward in the stem there is no way whereby they can be reached by any measure likely to seriously affect them. The beetles can not be trapped by inducing them to hibernate under piles of rubbish prepared for them especially for this purpose, because there is excellent proof that they pass the winter in fields entirely bare of vegetation. Late planting of the crop, as exemplified by repeated replantings, does not offer any encouragement in the way of preventing future injuries. As shown by the observations of Messrs. Walton and Hyslop, very few of the insects hibernate in the roots or old stalks, so that the pulling up and burning of these, as in destroying the larger corn stalk-borer, would not be of much value against this insect. They probably do not hibernate to any extent in their uncultivated food plants.

Fortunately, however, the farmer has within his reach two most practical and efficient measures of prevention. One of these is to entirely exterminate from his fields any of the natural food plants of this species. Indeed, he should by no means attempt to raise a crop of corn while any of this natural vegetation, upon which the insect can subsist, is still in existence. The other measure is to follow corn or rice with some crop upon which this insect can not feed and never to plant corn immediately after corn or rice. On the farms of the Messrs. Shannonhouse most convincing illustrations were afforded of almost complete protection by rotation of crops. While in no case was it possible to find a badly infested field of corn following cotton, there were plenty of illustrations of the disastrous effect of attempting to raise corn during successive years on the same ground.

One field offered such an excellent illustration of this phase of the problem that Mr. William T. Shannonhouse had it photographed, and these photographs are used for illustration in Plates VI to VIII. In 1910 the eastern portion of this field had been devoted to cotton, the western portion to corn. In the year 1911 the entire field was planted with corn; as a result that portion on which corn had been raised the previous year (see Pl. VI) was almost totally destroyed by this insect, while the other portion, where cotton had been grown (see Pl. VII), was almost entirely exempt from attack. In order to show the abruptness with which this injury terminated and the exactness with which this corresponded to the dividing line between the two previous crops, the farmer who had himself cultivated the field in 1910, and was therefore perfectly familiar with it, was induced to stand exactly upon the dividing line between the corn and the cotton. Plate VIII shows the area where this dividing line between the two crops of the previous year was located and the radical difference in attack by Sphenophorus between the two portions of the field. This field illustrates conclusively both the fact that the beetles winter in the fields where they develop and also that crop rotation is effective in preventing serious injury.

NATURAL ENEMIES.

Mr. W. L. McAtee, of the Biological Survey, has recorded the finding of *Sphenophorus callosus* in the stomach of the nighthawk (*Chordeiles acutipennis texensis*) at Wallaceville, Tex., August 4, 1907. This is the only exact record obtainable of the eating of this species by birds. In addition Dr. Chittenden has placed the following notes at the disposal of the author:

> Among the larvæ of this species in our rearing cages in late August and early September some years ago were some which had died, apparently of fungus attack, although here is a possibility that the fungus attacked the insect while dying or after death.
> In another instance, during the last week of August, larvæ of this same species were lying and specimens were referred to Dr. Haven Metcalf, a pathologist in the Bureau of Plant Industry, who stated that they were apparently free from fungi, and that while there was a possibility of the presence of a bacterial disease such presence could not be established at that stage. Examination, however, revealed the fact that the bodies of the larvæ were fairly reeking with nematodes, and it is not impossible that these are the cause of the insect's fatality.

On September 5 and 6 Mr. Walton found, in cornfields in North Carolina where the corn had been destroyed (see Pl. VIII), many exit holes of the predaceous maggots of a robber fly, *Erax lateralis*, between the rows of corn, and it is possible that these may have devoured some of the larvæ of Sphenophorus. Lamphyrid larvæ were noted, both by the author and by Messrs. King and Walton, about the infested hills of corn. Although these are known to be predaceous, none of us was able to catch them in the act of devouring the larvæ of the curlew bug.

9290

U. S. DEPARTMENT OF AGRICULTURE,
BUREAU OF ENTOMOLOGY—BULLETIN No. 95, Part V.
L. O. HOWARD, Entomologist and Chief of Bureau.

PAPERS ON CEREAL AND FORAGE INSECTS.

THE FALSE WIREWORMS OF THE PACIFIC NORTHWEST.

BY

JAMES A. HYSLOP,

Agent and Expert, Cereal and Forage Insect Investigations.

ISSUED APRIL 22, 1912.

WASHINGTON:
GOVERNMENT PRINTING OFFICE.
1912.

CONTENTS.

ILLUSTRATIONS.

U. S. D. A., B. E. Bul. 95, Part V.   C. F. I. I., April 22, 1912.

PAPERS ON CEREAL AND FORAGE INSECTS.

THE FALSE WIREWORMS OF THE PACIFIC NORTHWEST.

By James A. Hyslop,

Agent and Expert, Cereal and Forage Insect Investigations

INTRODUCTION.

Up to within the past five years, except for a few scattering notices, the species of Eleodes have been considered of only incidental, if of any, economic importance. The Tenebrionidæ, to which this genus belongs, are sometimes saprophagous, feeding on dead vegetable matter in the soil, and occasionally on dead animal tissue as well as on stored grain and other food products.

Superficially the larvæ resemble the true wireworms (elaterid larvæ), and on account of this resemblance and the similarity of their

Fig. 22.—The false wireworm, *Eleodes letcheri vandykei.* Adults in characteristic attitudes. Somewhat enlarged. (Original.)

depredations in the grain fields the two are often confused. On closer examination, however, Eleodes larvæ can be easily recognized; the antennæ are rather long and very conspicuously clavate, the body is not flattened, and the forelegs are long and stout. These larvæ can move with great rapidity as compared with true wireworms

The confusion of Eleodes with the true wireworms is unfortunate, as the preventive and remedial measures for the two pests are quite distinct, what is efficient treatment in one case being quite useless in the other.

HISTORICAL.

Among the earliest references to the economic importance of these beetles in this country is a note by Prof. Lawrence Bruner,[1] in which the species *Eleodes tricostata* Say is recorded as attacking cabbage

---

[1] Bul 20 (ol l series). Div. Ent , U S Dept Agr , pp 11-12, 1892.

plants at Lincoln, Nebr., doing even more damage than the cutworms. It was also said to have attacked other garden crops, but
these are not definitely recorded.

In 1895 Prof. C. V. Piper published an article in the Northwest
Horticulturalist in which he refers to Eleodes larvæ attacking garden
crops.

In 1898 Mr. Theo. Pergande [1] notes having received from McPherson, Kans., two larvæ of a tenebrionid with the statement that they
do serious damage to wheat in Salina County, Kans., by attacking
the grain when it becomes softened, destroying the germ. From one
of these larvæ an adult was reared which proved to be *Eleodes suturalis* Say. In the autumn of 1911 Mr. E. O. G. Kelly, of this office,
found the wheat in southern Kansas attacked by an Eleodes larva
which may prove to belong to this latter species.

In 1908 Mr. Myron Swenk,[2] assistant State entomologist of
Nebraska, reported *Eleodes opaca* Say as doing very serious damage
to wheat in Nebraska, in some instances 60 per cent of the seed
having been destroyed.

The larvæ were first found by the author in enormous numbers in
May, 1909, in a wheat field south of Pullman, Wash. The field was
entirely ruined and had to be reseeded, though these depredations
were not entirely due to the Eleodes, as a true wireworm, the larvæ
of *Corymbites inflatus* Say, was also very numerous.

On May 12 several adult *Eleodes pimelioides* Mann. were found at
a depth of about 4 inches below the surface in the field above mentioned, and more were found under boards and rubbish about the
fields. Many larvæ were placed in flowerpot rearing cages with
wheat as food, and on July 3 a pupa was found in one of these cages.
On July 20 an adult *Eleodes pimelioides* emerged. Later examination
of several collections very clearly indicates that this species is far the
more predominant in the Palouse country.

Other species known to occur in this region are *Eleodes obscura* Say
var. *sulcipennis* Mann., *Eleodes hispilabris* Say var. *lævis* Blaisd.,
Eleodes extricata Say, *Eleodes manni* Blaisd., *Eleodes humeralis* Lec.,
Eleodes schwarzii Blaisd., and *Eleodes nigrina* Lec.

In the spring of 1909, on examining an oat field at Govan, Wash.,
that had been almost completely destroyed, many tenebrionid larvæ,
Eleodes letcheri Blaisd. var. *vandykei* Blaisd., were found crawling
over the surface of the field. They had evidently been forced to
leave the ground by a heavy rain which fell the day before. On
digging in this field many more larvæ were found about ready to
pupate.

In the spring of 1910 the adults were found in enormous numbers
on the roadsides in the Big Bend region and in the middle of the

ummer they were found under the grain shocks in large numbers. n this region the 'species in enormous preponderance is *Eleodes etcheri vandykei*. *Eleodes pimelioides*, *Eleodes nigrina*, *Eleodes hisrilabris* var. *lævis*, and *Eleodes obscura* var. *sulcipennis* also occur; he first one rarely, the last three quite commonly.

The results of three seasons' work in the Pacific Northwest demontrate quite conclusively that the false wireworms are among the nost destructive insects to recently planted wheat and corn in this egion. They rank second only to the true wireworms (elaterid arvæ).

False wireworms are native and not introduced forms; the climatic onditions of the country are, therefore, ideal for their existence. The converting of enormous areas of the scantily verdured sagerush prairie into wheat ranches has afforded them a new and ncreased food supply and the destruction of the sage hen, badger, und horned toad has removed their normal foes.

DISTRIBUTION.

The genus Eleodes, to which the beetles treated in this paper belong, is very closely confined to the Upper and Lower Sonoran Zones. These beetles do not fly and are therefore more restricted n their distribution than insects which have a more active means of lissemination. The mass of the species occur in the Southwest, while several occur in the arid and semiarid regions of California, Oregon, Washington, and Idaho. A few species extend into the Carolinian Zone in Kansas, Nebraska, and Iowa, *Eleodes tricostata* having been collected as far east as Independence, Iowa.

Eleodes pimelioides, however, seems to be an exception to this general rule, and is very nearly confined to the northwestern portion of the Transition Zone, only occasionally being found in the Sonoran where this zone merges into the Transition. Specimens have been collected in the very humid coastal region of Washington, as well as in semiarid regions of this State, of Idaho, and of Oregon; in the Rocky Mountains at Helena, Mont., as well as at very nearly sea level on Vancouver Island. The species is predominant in the semiarid Transition of Washington and Idaho, the region commonly known as the Palouse country. The southernmost records of this species are Lake County, Cal.; Elko, Nev.; Wasatch, Utah; and Garland, Colo. It extends east to the middle of Colorado and north to Vancouver, British Columbia.

Eleodes letcheri vandykei has been collected at The Dalles, Oreg., by Messrs. Hubbard and Schwarz. Dr. E. C. Van Dyke has taken this species in Modoc County, Cal., and we have found it to be the predominant species in the Big Bend region of Washington. All of these localities are well within the Upper Sonoran Zone.

Eleodes opaca is apparently confined to the Plains region east of the Rocky Mountains, specimens having been collected in central and eastern Colorado, western Kansas and Oklahoma, northern Texas, all of Nebraska, and southern and eastern South Dakota.

Eleodes suturalis occurs over about the same region as *E. opaca*. with its variety *texana* Lec. extending southward into New Mexico and southern Texas.

THE WORK IN THE BIG BEND REGION OF WASHINGTON.

On May 28, 1909, an oat field at Govan, Wash., was examined. This field had been almost completely destroyed by true wireworms: besides these, many larvæ of *Eleodes letcheri vandykei* were found on the surface of the ground, evidently forced out by the unusually late heavy rains of the preceding day. These latter were by far too few in numbers to have destroyed the oats. Bluebirds (*Sialia mexicana occidentalis*) were noticed feeding in the fields in large numbers on the exposed Eleodes larvæ. Many of the larvæ were also found in the ground at a depth of from 3 to 5 inches, in small spherical cells. wherein they lay in a curled position. These were considerably softer and paler colored than those found in an active condition.

The work in 1910 started early in April when the false wireworms were to be found scatteringly throughout the grain fields, the grain having just sprouted.

Adults were first observed in 1911 on April 17, the day being quite hot, but the weather up to this time having been very cool. The beetles were to be seen at about 3 o'clock in the afternoon in great numbers along all the roadsides, where they were either awkwardly hurrying over the ground or nibbling at the foliage of the very young *Polygonum littorale*, which is very abundant in this region.

Adults of the larger species, *Eleodes obscura sulcipennis*, were usually to be found in or about the burrows of the ground squirrel (*Citellus townsendi*) and the badger.

When disturbed, the species of Eleodes have the ludicrous habit of standing still and elevating the abdomen so that the long axis of the body approaches the perpendicular instead of the nearly horizontal position it normally maintains while walking or at rest. The two beetles to the right in figure 22 are in this attitude. Thus they will remain motionless for several minutes and finally, if they are not further disturbed, they walk off. If one places the finger near the insect, an oily liquid is excreted from the anal aperture, which flows down over the elytra and abdomen. This liquid is pale yellow in color and makes a dark-brown stain; it has a very characteristic, strong, astringent, and offensive odor, and is evidently protective in function. Mr. Carl F. Gissler [1] describes this secretion and the glands from which it is secreted.

Many pairs were in coitu on the 17th of April, and on the 21st a female that was confined in a pill box laid four eggs. Between the 21st and 23d, when the female died, she laid 10 more eggs, which, however, failed to incubate. Females dissected in the laboratory were found to contain from 92 to 199 eggs. The eggs were found to lie on the ventral side of the abdomen and to extend upward and over part of the viscera, filling all the interstices about the alimentary canal. The eggs were so crowded in the abdomen that they were quite distorted. Anteriorly the eggs were smaller and were fastened to the anterior abdominal sclerite by fine filaments.

The mating season lasted about two weeks, but the adults were in evidence throughout May and June. Well-grown larvæ were also to be found at this time. Many of the adults in the rearing cages, as well as two individuals observed in the field, were seen to burrow into the ground. This is accomplished by digging with the front tibiæ, which are expanded and armed with spines for the purpose, the tarsi being folded back out of the way. The loose dirt is conveyed backward by the middle legs and piled up behind the beetle by the hind legs. When about one-fourth of an inch of dirt has accumulated the beetle backs out of the hole, pushing the earth out with the abdomen, the hind legs assisting in this process by keeping the earth piled behind the abdomen. On examining these burrows two or three eggs were found in each. The burrows are filled with earth after the beetles come out and are from 4 to 8 inches deep.

Rearing cages were established by sinking barrels to the surface level, filling them with earth, and fitting vertically onto the top a galvanized-iron cylinder 10 inches in height and the diameter of the inside of the barrel top, the open upper end being covered with wire mosquito bar, with an introduction hole made in the wire screen and corked.

On April 20 about 30 pairs of mated *Eleodes letcheri vandykei* were placed in this cage, which had previously been seeded with wheat and planted to *Polygonum littorale*.

By June 25 the beetles were all dead in the cage, probably due to abnormal conditions as well as age, though no living beetles could be found in the fields at that time. Small larvæ 4 to 5 mm. (about three-sixteenths inch) in length were then to be found in the cage.

On examining the cage on November 14 the larvæ were found to be about 14 mm. (nine-sixteenths inch) in length and at about a depth of 12 to 24 inches below the surface. The soil in the rearing cage was as dry as powder to a depth of nearly 2 feet, but the desiccation did not seem to affect the larvæ.

---

[1] Psyche, vol. 2, no. 58, 1879, p. 209.

July 4, 1911, the cage was again examined and the larvæ were found to be about 1¼ inches long.   From early in July to the middle of August it became necessary to be away from Govan, where this experiment was in progress.   On returning, August 16, the root cage was examined and two adults found about 6 inches below the surface. They were hard and had evidently emerged some days before this date.   When the boards which had been placed over the barrel to protect it during the winter were removed in the spring, a number of adults was found that had hibernated under this shelter.   On the above date and at a depth of about 20 inches a very young false wireworm (3.5 mm. long) was found; it was pure white and had evidently but very recently hatched.   This larva was undoubtedly the young of one of these accidentally introduced beetles.

The soil at this time was dry to a depth of 4 inches.   On June 25 a pupa was found in the field, placed in a pill box with dirt, and on the 30th an adult *Eleodes letcheri vandykei* emerged.   In the fields, where a farmer was plowing his summer-fallow—and it may be remarked that this is exceptionally late for working the summer-fallow in this country—pupæ were found turned out by the plow.   A little flock of Brewer's blackbirds (*Euphagus cyanocephalus*) were walking in the furrow a few yards behind the plow and picking up the upturned insects.

From the middle of July until the grain is harvested adults are to be found in large numbers under the grain shocks and bundles as they stand in the field, and also under grain sacks.   Most of the beetles are quite soft early in the season, but later become hard.

DESCRIPTIONS.

Eleodes letcheri vandykei Blaisd.

The egg (fig. 23) —The egg is bluntly oval in longitudinal section and circular in cross section, it measures 1 1 mm in length and 0.62 mm. in diameter, it is of a pure glistening white color and absolutely without sculpturing. Ovarian eggs measured 1.17 mm in length and 0.74 mm. in diameter.

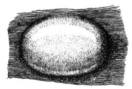

Fig. 23.—False wireworm, *Eleodes letcheri vandykei.* Egg. Greatly enlarged. (Original)

The larva (fig. 24).—Elongate, subcylindrical, convex dorsally, flattened ventrally. Yellowish, ventral surface paler, anterior and posterior margins of first thoracic and posterior margins of succeeding segments brown, head slightly brown, edge of mandibles black, base of mandibles brown; claws, spines on legs, and caudal segment dark brown; antennæ pale yellow. Anterior and posterior margins of first and posterior margins of succeeding segments with striate marginal bands; band on anterior margin of first segment broader and more coarsely striate. Caudal segment scutelliform, flattened dorsally and convex ventrally, bearing 18 stout spines on its margin—4 on each lateral margin and 10 on terminal margin; a slight space equal to that occupied by one spine separates the lateral from the terminal spines. Several long hairs and a basal row of short hairs on caudal segment. Head subquadrate, very convex, distance from base to labrum equal to one-half width of head, sides converging anteriorly, posterior angles rounded; two stout hairs on lateral dorsal surface; two oblong black eye-spots on lateral anterior part, a large one at base of antennæ, and a smaller one posterior and dorsad of this. Suture arising at base of each mandible flexed laterally and converging posteriorly to unite with the median suture at a distance from base of head equal to one-fourth distance from base to labrum; mouthparts usually directed ventrally. Labrum large, basal joint trapeziform, terminal joint rounded, bearing several hairs on margin, both joints margined anteriorly. Mandibles large, visible from above. Labium not covering mandibles, ligula broader than palpifer; labial palpi cylindrical, two-jointed, second joint papilliform; mentum larger than palpifer, submentum larger than mentum, all quadrangular and narrower at base than at anterior margin. Maxillæ larger than labium, stipes directed laterally, palpifer elongate and directed antero-medially; maxillary palpi at about middle of palpifer, three-jointed, first and second joints about equal in length, first stouter, third papilliform. Thoracic legs stout; first pair longer than width of thorax, second and third pairs one-third shorter; first pair very heavy, terminal hook as long as fourth joint, second joint bearing

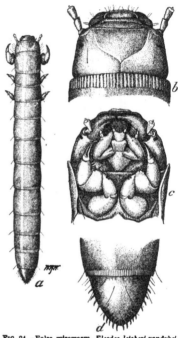

FIG. 24.—False wireworm, *Eleodes letcheri vandykei.* *a*, Larva, dorsal aspect; *b*, head, dorsal aspect; *c*, head, ventral aspect, *d*, caudal segment, dorsal aspect. *a*, Much enlarged, *b, c, d*, more enlarged. (Original.)

2 stout short spines on inner distal margin, third joint bearing 3 marginal spines and fourth joint bearing 4 such spines.

The pupa (fig. 25).—Length 11 mm., width 5.3 mm., arcuate dorsally, flattened to slightly concave ventrally. Entirely white when first pupated but eyes soon become black followed by tips of mandibles; just prior to emerging the elytra and dorsum of thorax become black. Head pressed to the prosternum. Pronotum very broad and protruding anterior to the head, making the latter invisible from above. Mesonotum very narrow and scutelliform, with indistinct transverse sulcus slightly anterior to middle. Metanotum sagittiform, about as broad at anterior margin as mesonotum. At the base of each elytron and of each secondary wing pad is a rounded swelling. Seven abdominal plates visible dorsally. Between dorsal and pleural abdominal

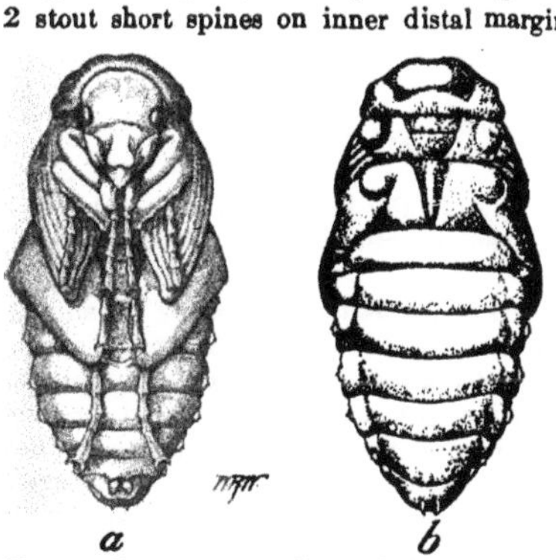

FIG. 25.—False wireworm, *Eleodes letcheri vandykei;* *a*, Pupa, ventral aspect; *b*, same, dorsal aspect. Much enlarged (Original.)

plates is a distinct depression forming a submarginal groove. The caudal segment bears a pair of posteriorly directed spines near posterior margin dorsally and a pair of median anal lobes ventrally. Head, legs, and antennæ free. Antennæ passing behind first and second pairs of legs and over wing pads. Elytra folded ventrally over the posterior legs. Eyes conspicuous. Pleural margin of abdominal segments bearing mammiform tubercles. Body without hairs or bristles.

The adult[1] (fig. 26).—More or less shining, elytra not pubescent. Antennæ with the third joint scarcely as long as the next two combined, fourth joint a little longer than the fifth, the latter slightly longer than the sixth, the latter and the seventh equal.

Pronotum usually widest at the middle, frequently widest just in front of the middle.

Elytra irregularly and quite densely muricately punctate, very minutely so on the dorsum, coarser on the sides and apex; from each puncture arises a rather short, stiff, curved, inconspicuous and semirecumbent seta. These are not evident on the inflexed sides.

Otherwise as in letcheri, but a little more robust.

Fig. 26.—False wireworm, *Eleodes letcheri vandykei*: Adult, dorsal aspect. Much enlarged. (Original.)

Measurements.—Males Length, 14.5–16 mm ; width, 5–6.5 mm. Females: Length, 15–16 mm.; width, 7.5 mm.

Genital characters as in *letcheri.*

Eleodes pimelioides Mann.

The egg.—Oval in longitudinal section and circular in cross-section; 1.34 mm. in length and 0 85 mm. in diameter; pure glistening white; without sculpturing of any kind.

The larva.—Elongate, cylindrical, convex dorsally, flattened ventrally. Yellowish, first thoracic and eighth abdominal segments brownish, ventral surface paler, anterior and posterior margins of first thoracic and posterior margins and anterior submarginal areas of succeeding segments brown, a distinct pale median vitta; head brownish posteriorly, edge of the mandibles black, base of the mandibles brownish; claws, spines on legs, and caudal segment brown; antennæ brownish. Anterior margin of first thoracic segment excavated, posterior margins on all segments except caudal

---

[1] The description of the adult given herewith is taken from "A Monographic Revision of the Coleoptera, belonging to the Tenebrionide Tribe Eleodiini, inhabiting the United States, Lower California and Adjacent Islands," by Frank E. Blaisdell, Sr. Bul. 63, U. S. Nat. Mus., p. 136, 1909.

depressed, faintly striate. Caudal segment scutelliform, slightly convex dorsally, margined laterally, tip curved slightly upward, bearing 18 acute spines on margin— 4 groups of 2 spines each on each lateral margin and 2 spines at tip. A number of hairs on dorsal surface and many on ventral surface. Head subquadrate, very convex, sides converging anteriorly, posterior angles rounded, 2 hairs on lateral dorsal and several hairs on ventral surface; 2 black eye spots on lateral anterior part of head, a large oblong one at the base of antennæ, and a smaller square one posterior and dorsal of this. Suture arising at the base of each mandible flexed laterally and converging posteriorly to unite with median suture near base of the head. Basal joint of labrum trapeziform, terminal joint rounded, trilobed, hairs on margin. Mandibles large, visible from above. Ligula broader than palpifer, labial palpi cylindrical, 2-jointed, second joint papilliform, mentum larger than palpifer, submentum larger than men-

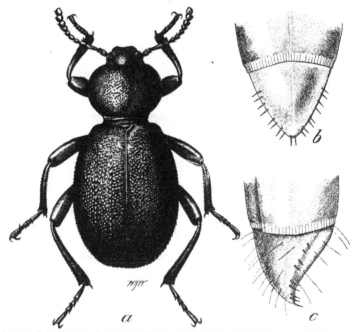

FIG. 27 —False wireworm, *Eleodes pimelioides:* a, Adult, dorsal aspect; b, caudal segment of larva, dorsal aspect, c, caudal segment of larva, lateral aspect. a, Much enlarged, b, c, more enlarged. (Original.)

tum. Legs 4-jointed. Second joint of anterior legs with 2 short stout spines on inner anterior margin; third joint of anterior legs with 3 longer spines on inner margin, and fourth joint with 3 still longer, stout spines; 2 spines on inner side and 1 spine on outer side of base of claw.

The pupa —The pupa of this species is very similar to that of *Eleodes letcheri vandykei*, from which it can be distinguished, however, by a pair of rather stout spines at the apex of the secondary wing pads.

The adult (Fig. 27).[1]—The following description is from page 384 Dr. Blaisdell's monograph previously mentioned in this paper

Moderately robust, ovate, feebly shining to opaque, about twice as long as wide; prothorax more or less strongly constricted at base, densely rugoso-punctate; elytra

---

[1] Originally described in Bul. Soc. Nat. Moscou, vol. 16, p. 274, 1843.

densely punctate, antennæ somewhat slende:
transversely oval, tenth more or less transverse
·ordate to transversely suboval, widest near the
der than long; sides evenly and quite strongly
late at middle, rounded in front and quite rap
ate at basal fourth, thence in each instance qu
angles; base equal to the length or in some mi
les obtuse, frequently not in the least rounded

·oadly oval to subquadrate, widest at or behind
than wide; disc more or less deplanate on the do·
·ally declivous posteriorly; surface densely
ged in rows on the dorsum or irregular througho
near the apex; when arranged in rows there are
·ed sparsely and irregularly between, always le
the tubercles are more or less rounded and shi
less opaque.   Otherwise as in *cordata*.
·o joints of the protarsi with tuft of yellowish
he second joint is rather small; tuft on the first j
ts somewhat long and truncate at tips.   Otherw
joint of the anterior tarsi distinctly thickened a
ata.

Males: Length, 12–14.5 mm.; width, 6–7 mm
th, 6.5–8 mm.

·ers, male.—As in *cordata*.

tal segment triangular, surface quite plain and s
al plate oblong, feeble or scarcely narrowed apic
pical margin nearly transverse to feebly obliq
face of the apex, angle obtuse and more or less
rfaces.—Submarginal groove distinct and well
al border of the dorsal plate.   Otherwise as in

SEASONAL HISTORY.

The following life history was worked out for *Eleodes letcheri vandykei* at Govan, in the Big Bend region of Washington, and unless otherwise indicated the data refer to this species.

The adults emerge from hibernation in the early spring, about the middle of April, and after feeding for a short time on the leaves of various weeds, principally *Polygonum littorale*, mate and start oviposition. The eggs are deposited a few at a time in the ground, the adult female burrowing down through the soft dust to the moist soil below, usually to a depth of from 2 to 4 inches. The average number of eggs laid by one female is probably about 150. Five specimens of female *Eleodes letcheri vandykei* that were collected on April 30, 1911, were dissected and found to contain 199, 138, 161, 157, and 92 eggs, respectively. Most of these eggs were full-sized and probably mature, though one female contained 45 eggs and another 91 eggs which were about one-third full size. A female *Eleodes pimelioides*, collected May 1, 1911, at Pullman, Wash., was found to contain 167 eggs and 2 females of *Eleodes nigrina* collected late in April contained 96 and 58 eggs, respectively.

The eggs hatch in about 18 days, the recently emerged larvæ being cream-white, but rapidly assuming the normal amber-yellow color.

The larvæ feed throughout the ensuing summer, usually on decaying vegetable matter, hibernate, and resume feeding as soon as the soil becomes warm enough the following spring, but this time disastrously to the spring-sown grain. In June the larvæ transform to pupæ, and early in July the newly emerged adults commence to appear. They are quite soft on first emerging and take two or three days to become thoroughly hardened. These adults feed during the remainder of the summer, congregating in large numbers under the grain sacks, shocks, and any convenient shelter. They eat a small amount of grain and other vegetable matter and go into hibernation without mating. In the spring they resume activity and mate, thus completing the life cycle. They hibernate under boards, in squirrel holes, and in the ground. Prof. W. T. Shaw, of the Washington State College, in digging out burrows of a ground squirrel (*Citellus columbianus*), found specimens at a depth of 6 feet below the surface in the burrows. I have dug out the hibernating beetles at a depth of about 6 inches in the soil in wheat fields and also in barrel root-cages.

Larvæ of *Eleodes suturalis* were received by Mr. Theo. Pergande [1] on October 26, 1898, from McPherson, Kans. These pupated before May 19 and adults emerged May 30. From this note it would seem that *Eleodes suturalis* varies from *Eleodes letcheri vandykei* in its life

---

[1] Bureau of Entomology Notes, No. 8186.

history, hibernating as mature larvæ or pupæ and transforming to adults much earlier in the season than the latter.

The adults of the species herein treated seem normally to live but one season, but Dr. F. E. Blaisdell records keeping adults of *Eleodes dentipes* in confinement for over four years.

NATURAL ENEMIES AND PARASITES.

The hard chitinous integument, together with the offensive secretions, of these beetles render them almost immune to attack by birds. Several western vesper sparrows (*Poœcetes gramineus confinis*), two horned larks (*Otocoris alpestris* var.), a killdeer plover (*Oxyechus vociferus*), a "billy" owl (*Speotyto cunicularia hypogæa*), and a Brewer's blackbird (*Euphagus cyanocephalus*) were shot while feeding in the grain fields and the stomach contents examined. these failed to show any evidence that the birds had fed on adult Eleodes.

Of the domesticated birds, chickens and ducks eat adult Eleodes in large numbers. Twenty-five beetles were fed to one hen and were eaten very greedily. Young turkeys would not eat these insects. They would seize the beetle and immediately drop it and shake the head violently as though they disliked the taste, and after two or three similar experiences would learn to recognize these insects and would not touch them.

A large number of these beetles were fed to confined pheasants. and though the conditions were very abnormal, the results may be suggestive. Reeves pheasant (*Phasianus reevesi*) and the silver pheasant (*Gennæus nychthemerus*) ate the beetles freely, while the golden pheasant (*Chrysolophus pictus*) and the Lady Amherst pheasant (*Chrysolophus amherstæ*) refused even to notice the beetles. However, these birds seemed quite annoyed by our presence, and might have eaten the beetles had they not been frightened. No Chinese pheasants (*Phasianus torquatus*) were available, so we can not say whether or not these birds would be of any value as enemies of Eleodes.

From several sources we were informed that the sage hen (*Centrocercus urophasianus*) feeds largely on these beetles, the crop at times being gorged with the black chitinous fragments. In the records of the Bureau of Biological Survey of this department the following birds are listed as feeding more or less extensively on adult Eleode.

California shrike (*Lanius ludovicianus gambeli*), road-runner (*Geococcyx californicus*), Lewis's woodpecker (*Asyndesmus lewisi*), western crow (*Corvus brachyrhynchos*), bronzed grackle (*Quiscalus quiscula æneus*), red-headed woodpecker (*Melanerpes erythrocephalus*), curve-billed thrasher (*Toxostoma curvirostre*), hairy woodpecker (*Dryobates villosus* var), western mocking bird (*Mimus polyglottos leucterus*), western robin (*Planesticus migratorius propinquus*), the field plover (*Bartramia longicauda*), the mallard (*Anas platyrhynchos*), and the baldpate (*Mareca americana*

Dr. Blaisdell [1] refers to the ground owl (*Speotyto cunicularia hypogæa*) as one of their enemies, and further states that the butcher bird impales them on thorns.

It is very generally known among the farmers of the wheat regions of the Pacific Northwest that the Brewer's blackbirds (*Euphagus cyanocephalus*) follow the plow and eat the "white worms" (*Eleodes* pupæ) when the summer-fallow is being worked. The birds are to be seen walking in the furrows and flying away with their beaks filled with the soft white pupæ.

The western bluebirds (*Sialia mexicana occidentalis*) were seen at Govan, Wash., in large flocks feeding on the larvæ which had been driven to the surface by an unusually heavy rain.

The stomachs of several horned toads (*Phrynosoma douglasii douglasii*) were examined and found to contain fragments of Eleodes larvæ, but several of these toads kept in captivity refused to eat the adult beetles, though they would feed voraciously on other beetles. These little horned toads, or, as they are locally known, sand toads, are without doubt one of the most valuable animals in the western dry-farming regions. In the Southwest a larger species (*Phrynosoma cornutum*), with long stout spines on the head, supplants the former species. These toads move very rapidly and eat enormous numbers of insects. The garden toad (*Bufo* sp.) is recorded in the files of the Bureau of Biological Survey as feeding on Eleodes. Dr. Blaisdell [2] gives the skunk as a natural enemy of these beetles.

In the files of the Bureau of Entomology there is a note (No. 8186) by Mr. Theo. Pergande, wherein he records having received two larvæ from McPherson, Kans. These pupated, and later one of these pupæ was killed by an ant (*Tetramorium cæspitum*).

Another of Mr. Pergande's notes [3] records receiving an adult of *Eleodes suturalis* from Mr. C. E. Ward, of Belvidere, Nebr. This beetle was placed in a cigar box, and on examining the box on the following morning a large number of larvæ were noticed crawling about. These larvæ later spun cocoons around the edge of the box and were believed to be microgasterid parasites that had issued from the beetle. The adult parasites were later determined as *Perilitus* n. sp., and these are preserved in the National Museum collections.

The author found an adult beetle with the abdomen nearly filled by a nematode worm, but lost the specimen, making further determination impossible. Mr. Myron Swenk [4] records a disease, probably caused by bacteria or a fungus, that attacks the larvæ. The

[1] Bul. 63, U. S. Nat. Mus., p. 29, 1909.
[2] Loc. cit., p. 29.
[3] Proc. Ent. Soc Wash , vol. 2, pp. 211, 219, 1892
[4] Jour. Econ. Ent., vol. 2, p. 335, 1909.

first symptom of this disease is a small red spot on one of the body segments, usually on the first thoracic or the terminal abdominal segment. This spot enlarges, finally encircling the body, and within a very short time the insect succumbs. This disease was so prevalent as to interfere with much of Mr. Swenk's experimental work.

REMEDIAL AND PREVENTIVE MEASURES.

If a field is well stocked with false wireworms at the time wheat is sown, remedial measures are of little avail, as was demonstrated by our experiments carried out in the Big Bend country of Washington. The insects are well adapted to the present agricultural practices of the spring-wheat growers in the Pacific Northwest. Here the plowing of summer-fallow land is commenced as early as possible in the spring, which in the average season is in April. Those who can spare teams and men often commence while the seeding of the crop in other fields is in progress. The most progressive farmers then disk their fallow land in June so that this will be well finished when haying commences.

By slightly modifying this procedure an enormous number of these beetles would be destroyed. Instead of plowing early in the spring and disking in summer, reverse the process. Disk as early as the land can be worked and the apparatus is available, which will usually be in April. This will conserve the moisture fully as well as plowing. Then plow as late as possible; if the land has been well disked and the men and horses can be spared, it is well to defer this plowing to late July and early August. At this time the beetles are in the pupal, or, as they are commonly called, "white-worm," stage. They can not move through the ground as can the active larvæ, but can merely squirm when irritated. The plowing, which should be deep to be effective, turns out great numbers of these pupæ, and they are either eaten by birds or killed by the burning sun. Many more are destroyed by being crushed or suffocated in the broken pupal cells. Aside from killing many Eleodes pupæ, this practice of late plowing the summer-fallow would greatly aid in weed eradication. The early disking would not bury the weed seed to retard germination; all the seed would develop; then the late plowing would destroy the entire crop of weeds. If the weeds start very early, a second disking may be necessary, as weeds very rapidly deprive the soil of its moisture.

Concerted effort and very thorough work are absolutely essential to render this treatment appreciably effectual. The cooperation of all the farmers over a considerable area is advised, as the adult beetles walk rapidly and will readily reinfest a well-treated ranch, coming in from an adjoining, poorly worked field or pasture.

This treatment is by no means advocated for those farmers who find it impossible to disk their summer-fallow, as leaving the land untouched until July would cause all the accumulated winter's moisture to evaporate, and the plowing would simply be stirring the dust and be of no value whatever.

In the spring of 1910 a series of experiments was carried out at Wilbur, Wash., to determine the value of certain substances alleged to be useful as poisons or repellents against elaterid larvæ. Eleodes larvæ were also quite numerous in the fields where these experiments were carried out; hence mention of the results, though relating principally to another insect, may not be out of place here.

Seed in bulk was treated with the following substances: Lead arsenate, at the rate of two-thirds of a pound per bushel of seed, dissolved in water; strychnine sulphate, at the rate of two-thirds of an ounce per bushel of seed, dissolved in water; coal tar applied until seed was all coated, then sanded until dry. The substances were stirred into the grain thoroughly with a wooden paddle and then allowed to dry several days.

The experiments were sown in strips with a wheat seeder 11 feet wide and one-half of a mile long. Untreated check strips were planted between each treated plat.

Just after sprouting the percentage of damage done by insects was estimated by counting the damaged and undamaged seed in several areas of 1 square yard each in each plat. The results were entirely negative as all the plats, including the checks, were about equally attacked.

These treatments, even had they been found efficient, would have been impracticable from an economic standpoint. The poisons were too expensive and the application too expensive and laborious, and, in addition, the coal-tar treatment, even after drying several days, so clogged the seed cups on the seeder as to cause very uneven distribution of seed.

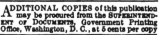
ADDITIONAL COPIES of this publication may be procured from the SUPERINTENDENT OF DOCUMENTS, Government Printing Office, Washington, D. C., at 5 cents per copy

9790

U. S. DEPARTMENT OF AGRICULTURE,

BUREAU OF ENTOMOLOGY—BULLETIN No. 95, Part VI.

L. O. HOWARD, Entomologist and Chief of Bureau.

PAPERS ON CEREAL AND FORAGE INSECTS.

THE LEGUME POD MOTH.
THE LEGUME POD MAGGOT.

BY

JAMES A. HYSLOP,

Agent and Expert, Cereal and Forage Insect Investigations.

ISSUED MAY 31, 1912.

WASHINGTON:
GOVERNMENT PRINTING OFFICE.
1912.

CONTENTS.

ILLUSTRATIONS.

IV

PAPERS ON CEREAL AND FORAGE INSECTS.

THE LEGUME POD MOTH.

(Etiella zinckenella schisticolor Zell.)

By JAMES A. HYSLOP,

Agent and Expert, Cereal and Forage Insect Investigations.

HISTORICAL.

The first mention of the legume pod moth (*Etiella zinckenella schisticolor* Zell.) as of economic importance in the United States is found in the unpublished notes of Mr. Theodore Pergande [1] in the files of the Bureau of Entomology, and refers to a number of lima-bean pods sent in by Mr. Albert Koebele from Eldorado County, Cal., on July 21, 1885. The larvæ were reported as doing considerable damage to the bean crop in that region. They left the pods and, if not full grown, entered other pods to continue feeding or, if full grown, constructed slight cocoons in the bottom of the rearing jar and pupated. Adults emerged September 2 and September 15.

A similar reference,[2] from the same source, refers to a number of pods of *Crotolaria incana* collected by Mr. E. A. Schwarz at Cocoanut Grove, Fla., on May 9, 1887, and sent to the Bureau of Entomology. The larvæ were eating the seed, and from these on May 16 an adult emerged. From a second lot of seed pods from the same source received on June 1, three more moths emerged on the 24th.

On May 2, 1896, a number of pods of *Astragalus* sp. were received by the bureau from Mr. C. L. Marlatt,[3] the material having been collected at Neucestown, Tex., infested with larvæ of *Etiella zinckenella*. On May 13 a braconid parasite issued, and on June 5 an adult moth emerged.

A single specimen was received from Mr. E. E. Bogue,[4] of Stillwater, Okla., on August 17, 1896, which he had previously reared from the seed pod of *Crotolaria sagittalis*.

Dr. F. H. Chittenden,[5] of this bureau, has published a paper on this insect in which the records made by Mr. H. O. Marsh are incor-

---

[1] Bureau of Entomology Notes, No. 3819.
[2] Bureau of Entomology Notes, No. 4129.
[3] Bureau of Entomology Notes, No. 7044.
[4] Bureau of Entomology Notes, No. 7173.
[5] Bul. 82, Pt. III, Bur. Ent., U. S. Dept. Agr., p. 25, 1909.

89

porated. Mr. Marsh found the larvæ attacking lima beans at Santa Ana, Garden Grove, Anaheim, and Watts, in California. At Garden Grove they had destroyed 40 per cent of the crop.

SYNONYMY AND DISTRIBUTION.

The species *Etiella zinckenella* was described by Treitschke [1] in 1832, and the variety *E. zinckenella schisticolor* was described as *E. schisticolor* in 1881 by P. C. Zeller [2] from two specimens, a male and a female, collected from "very different parts of North America." The male was from California and was collected October 8, but of the female he has nothing to say. He also refers to specimens of *E. zinckenella* examined by him from Sierra Leone, West Africa; Madagascar; Honda, Colombia, South America; and "Carolina" in this country. Later Rev. G. D. Hulst [3] redescribed this species under the name *Etiella villosa*, and gave Colorado and Califorina as the habitat. Dr. H. G. Dyar in his catalogue gives Arizona as an addition to the habitat.

The typical *E. zinckenella* is represented in the National Museum collection by specimens from Hampton, N. H.; Weekapaug, R. I.; Key West and Archer, Fla.; Oxbow, Saskatchewan; Texas; Stillwater, Okla.; and Denver, Colo. The variety *E. zinckenella schisticolor* is represented by specimens from Stockton, Utah; Springfield, Idaho; Eldorado, Clairmont, Alameda, and San Diego, Cal.; Nogales, Ariz.; and Pullman, Wash. It will be noted that all the specimens of the variety were collected west of or in the Rocky Mountains.

Etiella zinckenella schisticolor differs very slightly from the typical form. It has a suffusion of gray scales on the primaries as its chief distinctive character. A number of specimens from Florida, one specimen from Rhode Island, and one from New Hampshire very closely resemble the European specimens of *E. zinckenella*.

A possible explanation of the above facts may be that the variety *schisticolor* is a native of the Pacific slope of this continent, while the forms found in the eastern United States are the typical *E. zinckenella* recently introduced into this country from the Old World or South America.

FOOD PLANTS.

Larvæ of *Etiella zinckenella* have been recorded as feeding on the seed of several species of leguminous plants. In California (Eldorado County) Mr. A. Koebele [4] found them doing considerable damage to lima beans and they were recently found by Mr. H. O. Marsh,[5] of this bureau, working on the same crop in that State. Mr. E. E.

[1] Die Schmetterlinge von Europa, von Friedrich Treitschke, 9 Band, p. 201, 1832.
[2] Horæ Societatis Entomologicæ Rossicæ, vol. 16, p. 177, 1881.
[3] Ent. Amer., vol. 3, p. 133, 1887.
[4] Bureau of Entomology Notes, No. 48 K.
[5] Bul. 82, Pt. III, Bur. Ent., U. S. Dept. Agr., p. 25, 1909.

Bogue [1] found the larvæ in the seed pods of the common rattlebox (*Crotolaria sagittalis*) at Stillwater, Okla., and Mr. E. A. Schwarz [2] found them in the pods of a tropical species of this genus (*Crotolaria incana*) at Cocoanut Grove, Fla. Mr. C. L. Marlatt [3] records finding the larvæ in the seed pods of milk vetch (*Astragalus* sp.). They are also recorded by Herrich-Schäffer [4] as feeding in the seed pods of *Spartium junceum* near Vienna, Austria.

FIG. 28.—The legume pod moth (*Etiella zinckenella schisticolor*): Egg. Greatly enlarged. (Original.)

During 1910 and 1911 the author reared the species from the pods of common lupines (*Lupinus* spp.) and Canada field peas at Pullman, Wash.

DESCRIPTION.

THE EGG.

(Fig. 28.)

Egg glistening white, bluntly elliptical in outline and circular in cross section, measuring 0.58 mm. in length and 0.31 mm. in diameter. Chorion very delicate, colorless, and with fine irregular corrugations on the surface.

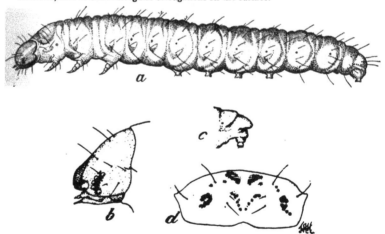

FIG. 29.—The legume pod moth. Larva. *a*, Side view; *b*, lateral aspect of head; *c*, caudal segments; *d*, pronotal shield. *a*, Enlarged, *b*, *c*, *d*, greatly enlarged. (Original.)

THE LARVA.

(Fig. 29.)

Full-grown larva from 12 to 17 mm. in length and from 2 5 to 3.5 mm. in diameter. Head yellow, black patch over ocellar area, mandibles and tip of labrum black; five ocelli arranged in an anteriorly directed semicircle at base of antennæ. Dor-

[1] Bureau of Entomology Notes, No. 7173.
[2] Bureau of Entomology Notes, No 4129.
[3] Bureau of Entomology Notes, No 7044.
[4] Syst. Bearb. der Schmett. von Europa, vol. 4, p. 72, 1849.

sum ruddy pink, except the pronotum (immature larvæ are evenly pale green or cream colored, with head and pronotal shield black or brown), the pleural and ventral surfaces pale green or cream white. Head pale yellow. Tips of thoracic legs and head brownish yellow. Five pairs of prolegs situated on segments 3, 4. 5, 6, and 9; the last pair can be retracted, so as to be almost invisible. Pronotum yellowish, with brown or black markings as follows: Two medial pairs of dot patches,

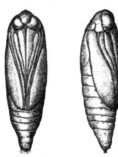

the posterior pair nearly touching the posterior margin and somewhat oblong in outline, the anterior pair approaching the anterior margin and linear in outline, being for the most part made up of a single row of dots; both pairs converging anteriorly, the anterior rows forming a V. Posterior margin bearing four bristles. Two pairs of bristles flanking the anterior median patches of dots, the posterior bristle in each pair short.

THE PUPA.

(Fig. 30.)

Pupa from 6 to 10 mm. in length, amber-yellow, with the tip of the abdomen, the edges of the abdominal segments, and margins of the wings outlined in brown. Fourth, fifth, and sixth abdominal segments each bearing a pair of short spines on the ventral surface. Abdominal segments 2 to 9 bearing well-defined brown spiracles. Terminal abdominal segment provided with a transverse row of 6 hooked bristles and a pair of lip-shaped tubercles on its dorsal surface.

FIG 30.—The legume pod moth: Pupa. Greatly enlarged. (Original.)

THE ADULT.

(Fig. 31.)

"The adult expands 24–27 mm. Labial palpi russet-gray above. gray below. Maxillary palpi yellowish, brown on end. Head, collar and fore-thorax orange fuscous. Thorax behind fuscous gray. Abdomen fuscous; fore wings mouse color, consisting of bluish gray, overlaid partly with fuscous A broad white stripe extending from base along costa to apex. Extreme edge of costa of ground color broadening outwardly just beyond middle and fading away toward apex A dull yellowish basal stripe reaching from white costal stripe to inner margin, edged inwardly with a row of maroon-brown scales, the scales being longer than usual. Hind wings fuscous, deepening outwardly, with dark marginal line. Beneath even glistening, very light fuscous." (Hulst, Trans. Amer. Ent. Soc., vol. 17, p. 170, 1890.)

FIG 31.—The legume pod moth: Adult. Enlarged about 2½ diameters. (Original.)

SEASONAL HISTORY.

On July 26, 1911, an eggshell was found on the outer surface of the calyx of a well-filled though still green lupine seed pod. The larva, which had evidently emerged only a few minutes before it was discovered, measured 1.2 mm. in length and was found just inside the pod at the terminus of a burrow which led from just in front of the egg through the calyx and pod wall. The hole through which the larva emerged from the eggshell was terminal, round, and very neatly cut. A few days later two eggs were found on the calcyces of field peas; these were brought into the insectary but failed to hatch.

During late July and early August, 1911, larvæ in all stages of development, from very small specimens, evidently just hatched, to those which were mature and spinning cocoons, were found in both the pods of field peas and lupines. The larvæ on first emerging are pale green or cream colored, the pronotal plate and head being entirely black or brown; with the first molt the pronotal plate assumes the characteristic pattern described elsewhere in this paper, but the body does not assume the rosy tint as described until nearly mature.

The larvæ feed for about three weeks, only partly consuming the peas, as is seen in figure 32, destroying them as seed, besides greatly reducing their weight as stock feed. The pod always contains a mass of frass held together by a loosely constructed web. The larva will leave one pod and enter another if the food supply is exhausted, or if

Fig. 32.—The legume pod moth: Larva feeding in a pod of field pea. Enlarged. (Original.)

for any other reason the pod becomes uninhabitable. When mature, if the peas are still unharvested in the field, it emerges from the pods and enters the ground to pupate, or if the pods have been harvested it spins a tough silken cocoon in the nearest available sheltered place.

Larvæ that become mature during the warm weather of early August, out of doors, or later under laboratory conditions, pupate immediately and emerge as adults in about six weeks. Adults have been obtained in our laboratory on August 5 and as late as August 28. Whether these lay eggs which pass the winter successfully, or whether they hibernate as adults, is still undetermined. Larvæ that reach maturity in late September, when the nights are cold, spin their cocoons and hibernate therein as larvæ, pupating in the spring and emerging at the time the earliest lupines are setting seed.

On the lupines there are very probably two generations a year. The moths of the first generation, coming from hibernating larvæ, lay all their eggs on the lupines, as the field peas are just commencing to grow. The offspring of this generation mature late in July and, finding the field peas ripening, very naturally turn their attention to these large areas of suitable food as well as to their natural food, the later lupines.

Mr. C. L. Marlatt [1] reared adults on June 5, 1896, from larvæ that were collected May 2 of that year at Nuecestown, Tex. Mr. Albert Koebele [2] reared adults on September 2 and 15, 1885, from larvæ collected June 21 at Rattlesnake Bridge, Eldorado County, Cal. Mr. Koebele also noticed the entrance holes of young larvæ and the exit holes of older larvæ in the pod husks. He says, "The larvæ * * * spun a web on the bottom of the jar in which they pupated."

Mr. Theodore Pergande [3] received a number of larvæ from Mr. E. A. Schwarz collected at Cocoanut Grove, Fla., on May 9, 1887, and on May 26 reared an adult from this material. More material from the same source was received on June 1, and on the 24th three more adults emerged.

Mr. H. O. Marsh obtained adults from January 9 to February 25, 1909, from larvæ collected October 22, 1908, at Anaheim, Cal. [4]

These observations indicate that two generations a year is characteristic of this species, the adults of the first appearing in early June and those of the second in September. The adults obtained in January and February were reared under laboratory conditions, which very probably accelerated their development. In the more southern parts of its range this species may have more than two generations.

FIELD WORK.

July 21, 1909, while examining the seed pods of the common lupine (*Lupinus* sp.) many were found to contain lepidopterous larvæ. In such pods the seeds were always more or less destroyed. The pods also contained a mass of frass which was held together by a loosely constructed web. A few days later, on examining the collected material, several of the pods were found with newly eaten holes in the sides and two larvæ were found with half their bodies within fresh pods.

On August 7 one of the larvæ, very plump, was found still in the pod, it having in the meantime become suffused with a rosy color. This larva had constructed a loose silken cocoon, through which its body could be easily seen. Ten days later the larva pupated and emerged as an adult (*Etiella zinckenella schisticolor*) September 28.

In the rearing cages with solid bases most of the larvæ left the seed pods and spun their cocoons among the litter and dirt in the bottom of the cages. In cages with bottoms of earth the larvæ always burrowed 2 or 3 inches below the surface to pupate.

On August 1, 1910, Mr. M. W. Evans, of the Bureau of Plant Industry, told the author of a larva that he was finding in the field-pea

---

[1] Bureau of Entomology Notes, No. 7044.
[2] Bureau of Entomology Notes, No. 48 K.
[3] Bureau of Entomology Notes, No. 4129.
[4] Bul. 82, Pt. III, Bur. Ent., U. S. Dept. Agr., p. 25, 1909.

pods in his experimental plats at Pullman, Wash. On examining some of these larvæ it was found that they resembled those of *Etiella zinckenella schisticolor* that had been found in lupine-seed pods the previous year, differing only in being larger, measuring about 17 mm. in length, while those from lupine measured only 13 mm., due, without doubt, to the difference in food plant.

The following day a number of the larvæ were collected from the field-pea pods and placed on earth in a flowerpot, into which they immediately burrowed.

From a larva placed in a pill box an adult emerged on August 27 and on the same day two moths emerged from the earth in the flowerpot.

EXPERIMENTAL WORK CARRIED ON DURING THE SEASON OF 1910.

In the spring of 1910 the Bureau of Plant Industry planted over 100 varieties of field peas at Pullman, Wash These were planted

Fig. 33.—The work of the legume pod moth (upper row) compared with that of the pea weevils (Bruchidæ) (lower row). Enlarged. (Original.)

in plats 1 rod square, in order to study development and adaptability of the various varieties under semiarid conditions. Mr. M. W. Evans has very kindly permitted the use of his field notes, and of the crop when harvested from these plats, which has greatly facilitated this investigation. These notes indicated the time of planting, the time the first flowers of each variety came into bloom, the time of maximum blooming, and the time of last blossoms. The seed harvested from each plat was kept in individual packages.

By actual count the writer determined the percentage of damage to each variety of field peas by these moths. The "worms" do not usually consume the entire seed, but so far destroy it as to render germination impossible. Seed thus damaged is easily distinguished from that attacked by Bruchus. The latter makes a very smooth round exit hole, while the former gnaws into the seed very irregularly. (See fig. 33.)

Figure 34 is a plan of the experimental plats and shows very clearly that the attacks of the pod moth were not restricted to any one part of the field, but were more or less promiscuously distributed. The plantings were made on a gentle slope, the upper side being to the south, thus giving a northerly exposure. The vacant rectangle in the center of the plan indicates a wheat-straw stack. All sides of the field were bounded by grain fields, and the roadside was practically

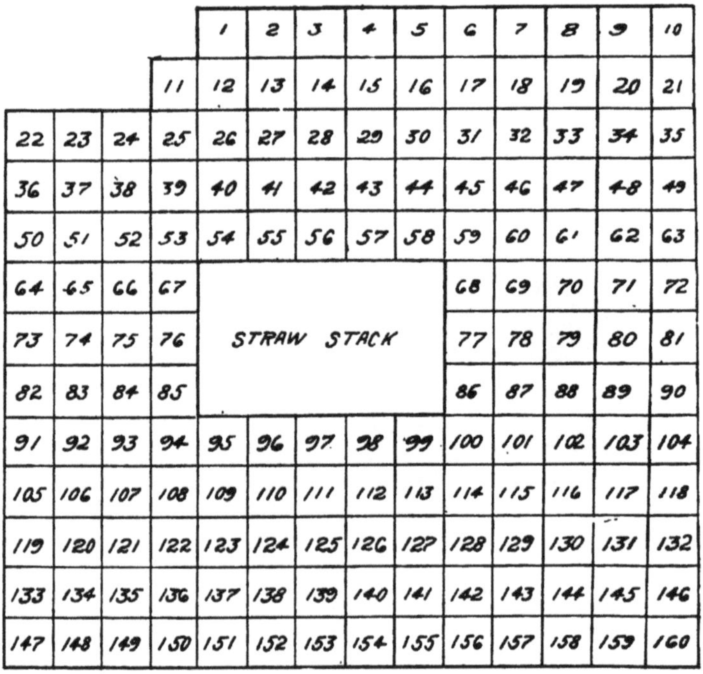

Fig. 34 —Planting plan of plats used in investigations of the legume pod moth during the season of 19:·
(Original.)

without weeds of any kind. The plats were separated from each other by strips of oats 1 yard in width.

Field peas vary greatly in texture of the seed, time of blooming. time of maturing seed, and adaptability to semiarid conditions. The variability, however, is confined to the varieties to a large extent. the individuals of a variety being quite uniform in response to given conditions. This fact at once opened the question of a worm-resistant variety. The results arrived at by this investigation are very suggestive. The actual records made in the investigation are to be found in Table I; but to make the results more readily available the variations in percentage of damage under several variables are graphically illustrated by diagrams.

TABLE I.—*Record of experimental work on the legume pod moth for the season of 1910.*

Plat No.	B. P. I. No.	Began blooming.	Full bloom.	Blooming ended.	Total number of seed.	Number of seed damaged.	Per cent damaged.
51....	21289	June 14	June 17	July 6	21,802	0	0.0
97....	24177	June 16	June 21	...do....	14,098	11	.0
102....	21288	June 17	June 23	July 9	19,920	0	.0
26....	11097	...do....	June 21	July 12	7,506	0	.0
20....	11112	...do....	June 23	...do....	24,722	15	.0
43....	22639	June 19	June 24	...do....	9,010	0	.0
46....	24895	...do....	June 25	...do....	13,828	10	.0
31....	24179	...do....	June 26	...do....	19,508	8	.0
59....	22077	June 14	June 21	July 14	14,517	0	.0
2....	23331	June 21	June 26	July 16	6,662	2	.0
100....	22007	June 16	....	July 17	15,484	4	.0
1/2 acre	20466	June 28	July 1	...do....	22,117	4	.0
8....	22036	June 23	..do....	July 24	15,539	14	.0
16....	21709	June 26	...do....	...do....	24,581	11	.0
101....	21605	June 10	June 16	July 3	26,431	36	.1
4....	23290A	June 26	July 1	July 7	11,781	21	.1
63....	22638	June 10	June 17	July 12	30,276	36	.1
149....	23525	June 19	June 26	...do....	10,500	20	.1
5....	22640	June 26	July 1	July 26	13,625	25	.1
150....	23290H	July 1	July 6	....	19,413	33	.1
72....	17006	....	...do....	Aug. 3	23,707	37	.1
19....	18455	June 24	July 1	July 14	11,310	30	.2
32....	24178	June 21	....	July 22	12,694	31	.2
73....	24940	..do....	June 26	...do....	13,063	28	.2
70....	20467	June 26	June 30	..do....	18,495	45	.2
42....	24262	June 16	June 23	July 17	9,792	33	.3
48....	19389	June 23	June 28	July 24	22,960	80	.3
13....	22040	July 1	July 9	Aug. 3	19,039	65	.3
41....	17486	June 26	July 3	July 24	13,252	66	.4
7....	22045	..do....	July 6	July 26	14,522	72	.4
55....	12888A	June 30	..do....		10,477	43	.4
10....	19786	.do....	July 3	July 26	11,939	59	.4
103....	20382A	.do....	July 6	July 29	20,335	85	.4
44....	19290	June 19	June 24	July 6	3,346	21	.6
17....	21290	June 21	June 30	July 17	22,372	136	.6
54....	16439	June 23	....	...do....	24,937	169	.6
148....	23547	June 26	June 30		14,585	95	.6
49....	17483	June 30	July 6	July 26	33,529	229	.6
104....	19788A	...do....	...do....		16,509	107	.6
61....	16437	June 26	...do....	July 29	9,350	63	.6
35....	22043	.do....	...do....	July 26	20,588	145	.7
14....	22038	July 3	...do....	.do....	20,255	145	.7
98....	23850	June 19	June 26	July 17	13,971	138	.9
76....	18456	June 23	...do....		13,639	131	.9
25....	16130	...do....	July 1	Aug. 3	2,279	21	.9
80....	10274A	July 9	July 17	Aug. 14	10,290	122	1.1
56....	12887	June 28	June 30	July	9,848	119	1.2
87....	23851	..do....	...do....	July	21,895	292	1.3
60....	17483A	June 28	July 6	Aug.	9,400	127	1.3
90....	20381	July 17	July 24	Aug.	8,887	118	1.3
69....	23847	June 28	July 6	Aug.	19,007	302	1.5
58....	22078	June 23		July	8,444	164	1.9
160....	16436	June 24	June 30	July	18,724	382	2
37....	22044	July 1	July 9	Aug. 22	18,555	411	2.2
12....	22041	July 3	July 12	July 9	12,053	278	2.3
77....	19787	July 2	July 9		14,699	389	2.6
82....	19709	July 3	July 14	Aug. 9	14,713	387	2.6
15....	22037	...do....	July 9	July 26	22,148	683	3
40....	19785	June 26	July 6	...do....	10,447	346	3.3
159....	16437A	June 28	...do....	...do....	7,105	244	3.4
36....	22046	July 1	...do....	...do....	11,191	391	3.4
38....	22042	July 3	July 9	...do....	13,822	502	3.6
88....	22049	...do....		July 29	23,573	893	3.7
99....	23848	June 26	July 3	Aug. 3	15,066	563	3.7
1/2 acre	16819	July 3	July 9	..do....	6,581	281	4.2
156....	17483E	July 9	July 14	July 29	10,732	821	7.6

The diagram, figure 35, shows the maximum (solid line) and minimum (dotted line) percentage of seed damaged in all varieties which commenced blooming at any given time. It shows that varieties

which began blooming during and after the last week in June were
decidedly the most severely attacked.

Figure 36 shows similar data on all varieties in full bloom at any
given time, besides very clearly showing that such varieties as were
in full bloom between the first and last weeks of July were the most
severely attacked.

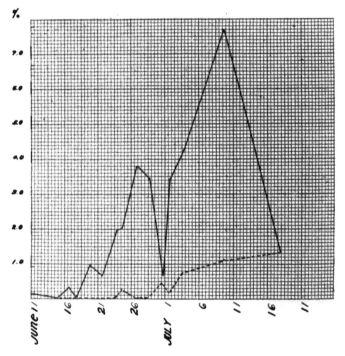

Fig. 35 —Diagram showing maximum and minimum damage done by the legume pod moth to varieties
of peas commencing to bloom on a given date in 1910. (Original.)

Figure 37 shows similar data on all varieties which ceased blooming on any given date, and indicates that such varieties as had
ceased blooming before the middle of July were only slightly damaged.

As the time the plants are in full bloom is very evidently of the
greatest significance, Table II and figure 38 were arranged to show
the mean percentage of damage done to all varieties in full bloom
at any given time, which indicates very conclusively that varieties
which were in full bloom from the 1st to the middle of July are by
far the worst damaged by the legume pod moth.

TABLE II.—*Mean percentage of damage done by the legume pod moth to all plats in full bloom on a given date in the season of 1910.*

Date of full bloom.	Mean per cent of seed damaged.	Number of plats.	Date of full bloom.	Mean per cent of seed damaged.	Number of plats.
June 16	0.1	1	July 1	0.2	7
June 17	.1	2	July 3	2.1	2
June 21	.0	3	July 6	1.1	15
June 23	.1	3	July 9	2.5	8
June 24	.3	2	July 12	2.3	1
June 25	.0	1	July 14	5.1	2
June 26	.4	6	July 17	1.1	1
June 28	.3	1	July 24	1.3	1
June 30	1.0	6			

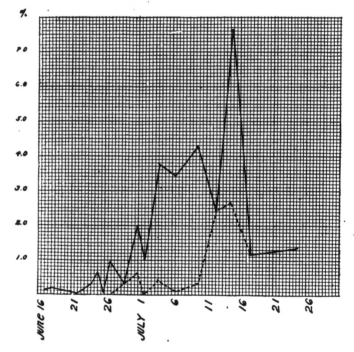

FIG. 36.—Diagram showing maximum and minimum damage done by the legume pod moth to varieties of peas in full bloom on a given date in 1910. (Original)

EXPERIMENTAL WORK CARRIED ON DURING THE SEASON OF 1911.

Sixty-seven varieties of field peas were planted in 1911 on the farm of the State College at Pullman, Wash. The field selected was in a draw, or ravine, bordered on three sides by grain fields and on the fourth side by a clean cultivated orchard. The plats were 1 square rod in area, as they were the year before, and separated from each other by strips of oats 1 yard wide. Two plats of each variety were planted side by side. (See fig. 39.) Plate X shows a part of the experimental plats used this season. The moths were not quite so destructive this season as last, 2.5 per cent being the greatest damage done to any variety this year, while in 1910 as high as 7.6 per cent

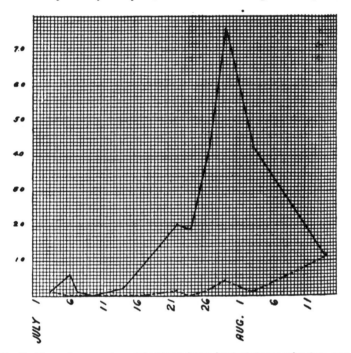

Fig. 37.—Diagram showing maximum and minimum damage done by the legume pod moth to varieties of peas which ceased to bloom on a given date in 1910. (Original.)

of one variety was destroyed. However, the results obtained are as conclusive as those recorded last year.

Table III gives the data of this season's work and is self-explanatory. Table IV is arranged to show the mean percentage of damage done to all plats which came into full bloom on any one date. Figure 40 graphically illustrates these results and very clearly shows that varieties which were in full bloom prior to June 28 were practically unmolested.

EXPERIMENTAL PLATS USED IN INVESTIGATIONS OF THE LEGUME POD MOTH DURING THE SEASON OF 1911. (ORIGINAL.)

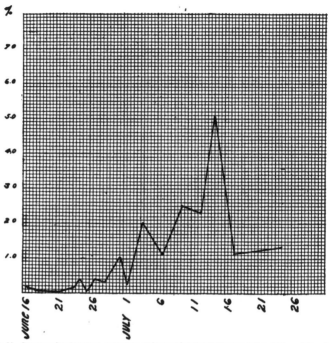

FIG. 38.—Diagram showing mean percentage of damage done by the legume pod moth to varieties of peas in full bloom on a given date in 1910. (Original.)

10	10	11	11	30	30	31	31	50	50	.51	51		
9	9	12	12	29	29	32	32	49	49	52	52		
8	8	13	13	28	28	33	33	48	48	53	53		
7	7	14	14	27	27	34	34	47	47	54	54	67	67
6	6	15	15	26	26	35	35	46	46	55	55	66	66
5	5	16	16	25	25	36	36	45	45	56	56	65	65
4	4	17	17	24	24	37	37	44	44	57	57	64	64
3	3	18	18	23	23	38	38	43	43	58	58	63	63
2	2	19	19	22	22	39	39	42	42	59	59	62	62
1	1	20	20	21	21	40	40	41	41	60	60	61	61

FIG. 39.—Planting plan of plats used in investigations of the legume pod moth during the season of 1911. (Original.)

TABLE III —*Record of experimental work on the legume pod moth for the season of 1911.*

Plat No.	B. P. I. No.	Date of planting.	Began blooming.	Full bloom.	Blooming ended.	Total number of seed.	Number of seed damaged.	Per cent damaged.
1	24995	Apr. 15	June 17	June 28	July 15	9,490	0	0 0
2	39946	..do....	June 7	June 12	July 7	33,529	0	.0
3	22639	..do....	June 16	June 19	July 15	23,797	5	0
5	19906	..do....	..do....	..do....	July 7	2,279	2	0
12	29948	Apr. 17	June 22	June 28	July 15	28,417	7	0
38	22628	Apr. 18	June 12	June 16	..do....	19,899	4	0
57	21288	..do....	June 19	June 28	..do....	22,462	8	0
65	3182	..do....	June 22	..do....	..do....	15,745	9	0
4	17483B	Apr. 15	June 25	July 8	July 30	18,725	20	1
8	23850	..do....	June 16	June 28	July 21	16,360	18	1
14	22547	Apr. 17	June 22	..do....	July 15	19,725	21	.1
6	22540	Apr. 18	June 19	July 8	July 23	10,290	24	.2
21	22290D	Apr. 17	..do....	July 3	July 15	22,822	46	.2
9	22044	Apr. 15	June 28	July 15	Aug. 1	14,463	104	.3
17	22042	Apr. 17	..do....	July 13	July 29	8,496	68	.8
20	23414	..do....	June 26	July 7	July 15	16,279	131	.8
30	29372	..do....	June 19	June 28	..do....	8,726	88	1 0
52	22290H	Apr. 18	June 28	July 10	..do....	8,909	90	1 0
58	19389	..do....	June 22	July 7	..do....	10,080	101	1.0
10	17486	Apr. 15	June 28	..do....	July 21	13,772	161	1 1
13	29365	Apr. 17	July 12	July 15	Aug. 1	21,795	242	1 1
39	25439	Apr. 18	June 28	July 7	July 21	20,224	223	1 1
56	34179	..do....	June 19	July 1	July 10	13,406	148	1 1
62	11097	..do....	..do....	June 28	..do....	31,360	377	1 2
59	24940	..do....	June 22	July 5	July 15	22,259	268	1 2
54	22290E	..do....	June 28	July 7	July 21	24,165	290	1 2
51	22049	..do....		July 15	..do....	12,514	151	1 2
47	21709	..do....	June 22	June 28	July 15	3,285	42	1 3
48	22077	..do....	June 12	..do....	..do....	14,302	201	1 4
34	16130	Apr. 17	June 28	July 8	July 26	23,657	339	1 4
44	16486A	Apr. 18		July 7	July 21	10,296	154	1 5
22	22037	Apr. 17	July 12	July 15	Aug. 1	19,761	299	1 5
23	17483	..do....	July 7	July 13	July 30	12,814	193	1 5
63	11112A	Apr. 18	June 16	July 7	July 23	13,164	211	1 6
19	17006	Apr. 17	July 7	July 12	July 21	14,352	243	1 7
43	22042	Apr. 18	..do....	July 15	Aug. 1	11,828	201	1 7
26	22046	Apr. 17		..do....	July 30	13,141	224	1 7
16	29309	..do....	June 26	July 7	July 15	10,112	184	1 8
31	2203	..do....	July 7	July 13	July 21	15,373	277	1 8
35	22036	..do....	June 26	July 7	..do....	20,868	376	1 8
67	11112	Apr. 18	June 19	..do....	July 15	16,234	294	1 8
28	29367	Apr. 17	June 24	..do....	..do....	9,399	188	2
42	21290	Apr. 18	..do....	July 5	..do....	14,588	292	2.0
33	29371	Apr. 17	June 28	July 7	July 21	26,239	527	2 0
46	25917	Apr. 18	July 7	July 15	..do....	22,859	457	2
66	3179	..do....		July 10	July 26	13,509	272	2 0
61	22048	..do....	Ju.y 7	July 15	Aug. 1	19,140	383	2
11	22639A	Apr. 17	June 28	July 7	July 22	9,277	190	2
18	17483F	..do....	July 12	July 15	..do....	15,379	323	2 1
24	29328	..do....	June 22	July 7	July 15	13,528	285	2
27	19786	..do....		..do....	July 25	10,621	223	2
30	27003	..do....	July 7	July 13	July 26	22,266	408	2
40	16437A	Apr. 18	June 28	July 7	July 21	8,776	185	2
53	12887	..do....	..do....	July 7	July 15	29,165	613	2
60	22041	..do....	July 7	July 18	July 25	20,366	428	2
55	29370	..do....	..do....	July 15	..do....	7,383	162	2
49	23847	..do....	..do....	..do....	July 21	18,206	401	2
45	20467	..do....	June 28	July 7	..do....	14,602	322	2
25	19788	Apr. 17	July 7	July 15	Aug. 1	11,209	259	2
50	27004	Apr. 18	..do....	..do....	July 27	10,370	239	2
64	3184	..do....	June 28	July 5	July 15	14,933	359	2
41	22040	..do....	July 7	July 13	July 21	16,498	397	2
15	17483C	Apr. 17	..do....	July 15	Aug. 1	1,871	45	2 4
32	26819	..do....		July 18	..do....	24,470	612	2
29	19787	do....	July 7	July 15	..do....	15,473	387	2
7	19709	July 15	...	do....	do....	7,506	192	2
37	16436C	July 18	June 28	July 7	July 21	21,791	557	2 8

TABLE IV.—*Mean percentage of damage done by the legume pod moth to all plats in full bloom on a given date in the season of 1911.*

Full bloom.	Mean per cent of seed damaged.	Number of plats.	Full bloom.	Mean per cent of seed damaged.	Number of plats.
June 12	0.0	1	July 7	1.7	19
June 16	.0	1	July 8	.8	3
June 19	.0	2	July 10	1.5	2
June 28	.5	10	July 12	1.7	1
July 1	1.1	1	July 13	1.8	6
July 3	.2	1	July 15	1.9	15
July 5	1.9	3	July 18	2.5	1

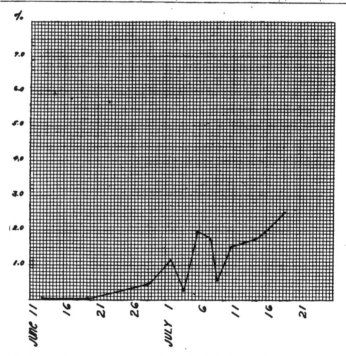

FIG. 40.—Diagram showing mean percentage of damage done by the legume pod moth to varieties of peas in full bloom on a given date in 1911. (Original.)

ARTIFICIAL DISSEMINATION.

On examining sacks of seed peas, September 14, 1911, a larva was found enclosed in a very tough silken cocoon. On October 24 another larva was found in a second sack. Mr. Evans also found a hibernating larva in a seed sack and kept it alive on his desk during the greater part of the winter. It seems as if this insect could very easily be introduced into regions where it does not at present occur, by being shipped with the seed field peas.

PARASITES.

Two hymenopterous parasites were reared from the larvæ of *Etiella zinckenella schisticolor* during the investigations at Pullman, Wash., viz, *Pseudapanteles etiellæ* Vier. and *Microbracon hyslopi* Vier.

Dr. F. H. Chittenden [1] records having reared *Bracon* sp. (determined by Viereck) from *Etiella schisticolor* on October 19, 1908, at Santa Ana, Cal., and Mr. C. L. Marlatt [2] records rearing a braconid from the larva of this moth at Nuecestown, Tex., on May 13, 1896.

REMEDIAL AND PREVENTIVE MEASURES.

The legume pod moth is readily controlled by preventive measures, and for this reason there have been no experiments with remedies. The transportation of the hibernating forms in sacks of seed, and the consequent dissemination of the pest, may be prevented by fumigation of the seed with carbon bisulphid.

Owing to the presence of the native lupines, extermination of the pest is impossible, but by planting such early varieties of field peas as come into full bloom before the last week in June it may be practically eliminated as a factor to be dealt with in seed growing in the Pacific Northwest. The date of planting, however, will vary in different localities and under different conditions.

---

[1] Bul. 82, Pt. III, Bur. Ent., U. S. Dept. Agr., p. 28, 1909.
[2] Bureau of Entomology Notes, No. 7044.

THE LEGUME POD MAGGOT.

(*Pegomya planipalpis* Stein.)

By JAMES A. HYSLOP,

Agent and Expert.

GENERAL ACCOUNT.

About the middle of July, 1909, a large number of larvæ of *Pegomya planipalpis* Stein were found leaving the pods of lupines that had been placed in rearing cages. On the 28th two pupæ were found in one of the cages. Within the next few days many more larvæ left the pods and pupated. A number of these puparia placed in a glass vial during the autumn of 1910 were kept in the field laboratory all winter. May 11 of the following year the first adult emerged and from that date others emerged daily throughout the remainder of the month. By a number of experiments it was found that humidity greatly facilitated the emergence of these flies.

These flies were first believed to be scavengers, feeding on the frass and decaying seed of the lupine and field peas in the wake of the legume pod moth. However, investigations in 1910 proved that the insect, though often found with Etiella, was quite capable of independently infesting seed pods and was itself an actual seed destroyer. Many pods were found to contain from one to three of these larvæ.

Dr. F. H. Chittenden,[1] of this bureau, notes this species as attacking radishes at San Francisco, Cal.

The larvæ molt at least twice, as two pairs of pharyngeal hooks were found in a pod with one larva. Though several of these dipterous larvæ were found in field-pea pods with the head capsules of larvæ of the legume pod moth, we hardly believe this species to be parasitic, as larvæ confined in small vials with pod-moth larvæ would not attack the latter.

In cages with earth in the bottom the pupæ were always to be found below the surface at distances ranging from 1 to 3 inches. The larvæ contract just before forming a puparium. The puparium is at first creamy yellow, turning brown at the ends first and finally becoming entirely ferruginous. A larva that contracted on the morning of July 31, 1911, assumed the usual puparium form by 9.30 a. m. of the same day. It was still pale yellow, but by 2.30 p, m. it had become brownish at the ends and deep orange-yellow at the middle, while next morning the puparium was uniformly ferruginous brown.

---

[1] Bul. 66, Pt. VII, Bur. Ent., U. S. Dept. Agr , p. 95, 1909.

DESCRIPTION.

THE LARVA.

(Fig. 41.)

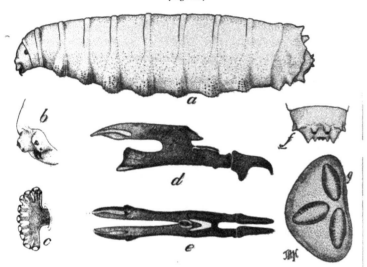

FIG. 41.—The legume pod maggot (*Pegomya planipalpis*). Larva: *a*, Side view; *b*, oblique aspect of head; *c.* thoracic spiracle, *d*, *e*, pharyngeal hooks, lateral and dorsal aspect; *f*, dorsal aspect of caudal segment; *g* anal spiracle *a*, Much enlarged, *b–g*, highly magnified. (Original.)

Larva cream-white, 7 9 mm. in length and 1.9 mm. in diameter. Broadly blunt posteriorly, conically tapering to anterior end. Hook jaws black. At base of second thoracic segment a pair of fan-shaped thoracic spiracles which are pale yellow and ten-lobed Anal spiracles on rather long papilliform tubercles, spiracular orifices of each spiracle three in number and arranged to form a letter T, the stem directed laterad and slightly ventrad Four small tubercles below the spiracles are arranged in a row across the end of the caudal segment, the outer pair the larger In front and to the side of this row is a pair of larger tubercles and in front of this pair is a ring of twelve tubercles around the segment, the two dorso-lateral and the two ventro-lateral tubercles large and conspicuous, the others smaller. Ventral swellings on segments 3 to 9, inclusive, armed with many small spinous papillæ

THE PUPARIUM.

(Fig. 42.)

FIG. 42.—The legume pod maggot Puparium. Much enlarged. (Original.)

Puparium ferruginous, dark brown at ends. Cylindrically oval, finely wrinkled. Hook jaws of larva visible. Length 6.4 mm.; width 2.3 mm

THE PUPA.

(Fig. 43.)

White, head large, front protruding. Legs and wing pads free. Third pair of legs under wing pads except tarsi.

THE ADULT.

(Fig. 44.)

The following is a translation of the original description by P. Stein, published in the Berliner Entomologische Zeitschrift, volume 42, page 234, 1897.

Pegomya (Chortophila) planipalpis. ♂ ♀. Size, shape, and color similar to *Ch cilicrura* Rd Cinereous; eyes cohering closely on inside, frontal triangular stripe practically straight, reddish; front and epistoma laterally moderately prominent, cinereo-rufous, peristome moderately broad, grayish; antennæ black, third joint twice as long as second, base of arista thickened, very slightly pubescent,

Fig 43.—The legume pod maggot: Pupa. Much enlarged. (Original)

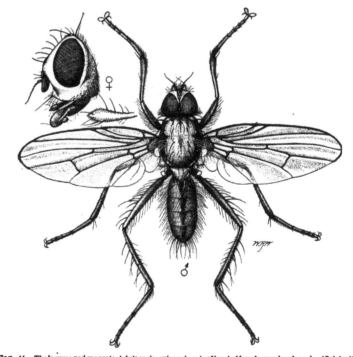

FIG. 44.—The legume pod maggot: Adult male, enlarged, side of head of female, much enlarged. (Original).

palpi black, apex a little dilated; thorax cinereous, less so on the sides, median line narrow and lateral stripes hardly perceptibly wider; abdomen elongate, depressed, median stripe and incisures narrowly black; tarsi black; pulvilli and ungues, the anterior ones moderately and the posterior one slightly, elongate; wings nearly hyaline, longitudinal veins 3 and 4 parallel, anterior cross-vein perpendicular and nearly straight, costal spines small; squamæ equal and white, halteres yellowish Female yellowish-gray; eyes with frontal stripe broad, dirty reddish, and quite broadly separated with yellowish gray; palpi distinctly dilated at apex; thorax nearly immaculate; abdomen oblong, median stripe and small areas shining indistinct yellowish brown; base of wings yellowish   Size ♂ 5, ♀ 6 mm.

PARASITES.

Pegomya planipalpis is attacked by two chalcidid parasites. One (*Holaspis* n. sp.) belongs to a genus of which there are two known species, *Holaspis parellina* Boh. and *H. papaveris* Thoms.. recorded [1] as parasitic on *Cecidomyia* spp.   The other parasite (not yet determined) reared from the legume pod maggot is probably also new to science.

REMEDIAL AND PREVENTIVE MEASURES.

This maggot has not as yet become a serious factor in field-pea seed growing in the Pacific Northwest. The dissemination as hibernating puparia can be readily prevented by fumigation. As is the case with the legume pod moth, it can not be exterminated because of the native lupines.

---

[1] Dalla Torre, Catalogus Hymenopterorum, vol. 5, p. 291, 1898.

U. S. DEPARTMENT OF AGRICULTURE,

BUREAU OF ENTOMOLOGY.—BULLETIN No. 95, Part VII.

L. O. HOWARD, Entomologist and Chief of Bureau.

PAPERS ON CEREAL AND FORAGE INSECTS.

THE ALFALFA LOOPER.

BY

JAMES A. HYSLOP,

Agent and Expert.

ISSUED OCTOBER 16, 1912.

WASHINGTON:
GOVERNMENT PRINTING OFFICE.
7 1912.

CONTENTS

ILLUSTRATIONS.

U. S. D. A., B. E. Bul. 95, Part VII.   Issued October 16, 1912.

PAPERS ON CEREAL AND FORAGE INSECTS.

THE ALFALFA LOOPER IN THE PACIFIC NORTHWEST.

(*Autographa gamma californica* Speyer.)

By J. A. HYSLOP,
Agent and Expert.

INTRODUCTION.

The first record of this moth of economic importance is an unpublished note by Mr. Theodore Pergande[1] made June 29, 1895, wherein he records having received from Mr. E. W. Baker, of Grand Junction, Colo., a few specimens of the larvæ of a Plusia, determined on the note as "*Plusia gamma* (?)," with the statement that the larvæ do much damage to the leaves and blossoms of alfalfa. Material was not preserved, so actual specific determination is impossible. As *Plusia gamma* L. is a European species, presumably it was *P. gamma californica* Speyer, now known as *Autographa gamma californica.*

The depredations of this species have not as yet been sufficiently serious to cause damage in the Palouse region of Washington and Idaho, although its attacks on alfalfa and clover have attracted the attention of many ranchers. The larvæ are usually very numerous in the early spring and gradually increase in numbers until the first hay cutting, when they appear to reach the maximum. They do not seem to be inconvenienced by the removal of the hay crop, but immediately turn their attention to the young second growth, on which larvæ are to be found throughout the summer until the early frosts.

The alfalfa looper in this locality has been held in check by a number of parasites and a disease, but any change in environmental conditions which might tend to reduce the efficiency of these natural checks or accelerate the reproduction of these moths would undoubtedly cause a serious outbreak, such as occurs periodically with the highly parasitized white-marked tussock moth, *Hemerocampa leucostigma* S. & A. It does, however, offer an excellent illustration of the statement, so often made, that many injurious insects are held in check by their parasites. Such cases as the above justify the arti-

---

[1] Bureau of Entomology Notes, No. 6692.

ficial introduction of parasites as one of the efficient measures to be taken in the control of a serious pest.

That *Autographa gamma californica* may appear in enormous numbers is evidenced by one of the earliest biological records on this species. A note in the Bureau of Entomology files, made by Mr. Koebele[1] in 1886, states that on the morning of April 30 he examined a mass of material collected from within an electric light globe at Los Angeles, Cal. Of 4,161 moths examined, 2,005 were *Autographa gamma californica*. He further states that the larvæ were numerous on a variety of plants. In anticipation of such an outbreak the biological notes and other data at hand in this office are herewith published.

DISTRIBUTION.

Specimens of this moth (*Autographa gamma californica*) in the United States National Museum were collected in Los Angeles County, Kern County, Placer, Alameda, and Fresno, Cal.; Colorado; Nevada; Seattle, Pullman, and Easton, Wash.; and along Kaslo Creek, in British Columbia. Mr. T. H. Parks, of this office, has collected larvæ of this species at Cokeville, Wyo., and Idaho Falls and Blackfoot, Idaho, in all cases feeding on alfalfa.

SEASONAL HISTORY.

This insect, in the Palouse region of Washington, passes the winter as hibernating pupa and probably also as the adult moth, since much-battered adults are to be seen early in the spring. Late in May and throughout June the adults are to be seen in the alfalfa and clover, darting rapidly away when disturbed. They are active in bright sunlight, feeding on the nectar from the clover and alfalfa blossoms. The flight, though short, is very direct and so rapid as to render the insect almost invisible.

May 2, 1887, Mr. Koebele[1] records observing one of these moths, at Alameda, Cal., ovipositing on *Malva rotundifolia* at 3 o'clock in the afternoon. Definite data relative to the length of the egg stage of this species have not been obtained, though several female moths were confined for that purpose. They fed greedily on sugar sirup but refused to oviposit. However, Mr. E. O. G. Kelly, of this bureau, captured a female of *Autographa brassicæ* Riley in an alfalfa field at Wellington, Kans., on October 27, 1909. This moth died the following day, after laying eight eggs. These began hatching on November 2 and were all hatched the next day. This limits the egg stage of *Autographa brassicæ* to seven days, and this is very likely the time of incubation of the other species of this genus.

---

[1] Bureau of Entomology Notes, No. 95 K.

Early in June the young larvæ become numerous in the fields, walking very much as do geometrid larvæ or "measuring worms." This is due to the fact that the larvæ have prolegs on only the fifth, sixth, and ninth abdominal segments. If disturbed they curl up and drop to the ground, the older larvæ lying there tightly curled up and refusing to move when irritated, but larvæ of the second and third instars when touched alternately straighten out and curl up very suddenly, thus jumping about spasmodically. The larvæ while young feed upon the epidermis of the leaves, skeletonizing them and giving to the attacked plant a brownish appearance. The older larvæ—that is, after the third molt—eat from the edge of the leaf toward the midrib, entirely consuming the leaves. The larval period lasts about two weeks, there being five molts with periods of about three days elapsing between each. When ready to pupate the larva spins a loose white silken cocoon (Pl. XI, fig. 2) among a number of leaves, usually well up in the plants, incorporating two or three leaves in its structure. The larva completes the cocoon in about half a day and, at least in the case of specimens reared in our insectary, pupates the day following that on which the cocoon is completed. The length of the pupal stage of specimens reared in our laboratory was very uniformly 10 or 11 days. Dr. H. G. Dyar[1] gives 12 days as the length of this period, and Mr. Koebele[2] records the length of the pupal stage as from 10 to 15 days. He gives an exact rearing record[3] wherein he mentions a larva collected at Piedmont, Cal., February 24, 1888, which pupated on March 5, the moth emerging March 22, making a pupal period of 17 days. Dr. F. H. Chittenden, of this bureau, gives from 6 to 22 days as the pupal period of a closely related species, *Autographa brassicæ*.

Thus the time elapsing from egg laying until the adult emerges covers a period of from 26 to 48 days, probably being about 30 days in the Palouse country of Washington.

The first adults of the second generation appear in early July, and adults continue quite numerous throughout this month, belated individuals having been collected as late as August 3. There are two generations, and probably three in the case of the earlier appearing individuals, and larvæ in all instars are to be found in the field as late as the end of August, but these very late larvæ probably succumb during the winter.

Mr. T. H. Parks, of this office, records finding the larvæ of this species about half grown in the alfalfa fields about Salt Lake City, Utah, as early as May 22, in 1911. Larvæ were found throughout June. The first pupa found in this locality was obtained on June 5

---

[1] Entomologica Americana, vol. 6, p. 14, 1890.
[2] Bureau of Entomology Notes, No. 95 K.
[3] Bureau of Entomology Notes, No. 389 K.

from a larva in one of the rearing cages. Mr. Parks records obtaining an adult in his cages on July 7, and observed another adult in the alfalfa fields near the Salt Lake City Field Laboratory on August 2. On August 23 one of these moths was found under a board lying along a fencerow adjoining an alfalfa field near Salt Lake City.

FOOD PLANTS.

Mr. Koebele[1] records the larva of this species as feeding on cabbage, barley, and elder (*Sambucus* sp.) at Los Angeles, Cal.; on dock (*Rumex* sp.) at Piedmont,[2] Cal., and records collecting a female while ovipositing and also larvæ while feeding on *Malva rotundifolia* at Alameda, Cal. We have collected the larvæ and reared adults from red clover, alfalfa, and garden peas at Pullman, Wash.

DESCRIPTIONS.

THE EGG.[3]

" Egg hemispherical, rounded at the base, the apex with a rounded depression. Finely creased vertically; color pale yellow."

THE LARVA.

First instar (?).—Body slender, pale creamy white, with long black hairs; head conspicuously large, shining black; thoracic legs blackish; only three pair of prolegs, situated on abdominal segments 5, 6, and 9, prolegs concolorous with body. Length, 1.8 mm.

Second instar.—Body segments 7 and 8 enlarged, 9 small; color pale green, marked with cream-colored longitudinal lines as follows: A subdorsal line, very fine and wavy; a stigmatal line, broader, straight, sharply defined dorsally, and fading out ventrally; segments ornamented with transverse row of black papillæ bearing black hairs; head cream-color; thoracic legs cream-color, with tips of claws ferruginous. Length, 3 mm. to 5 mm.

Third instar.—Body papillæ white, with black dots at base of hairs; three longitudinal lines in subdorsal space, the more dorsal one fine, clearly defined and wavy, the middle one broad and indistinct, and the third one about as fine as the first but less wavy; stigmata on first thoracic and first to eighth abdominal segments, pale, with oval black margins, that on eighth abdominal segment much larger than others; mandibles pale reddish brown; eyes with series of six black dots arranged in a ventrally directed semicircle near the base of the antennæ. Length, 6 mm. to 9 mm.

Fourth instar.—Body darker green, papillæ in two transverse rows, the papillæ of one row alternating with those of the other; head green, paler than body, mandibles and palpi brownish; thoracic legs infuscate; marginal hooks of prolegs ferruginous-brown. Length, 10 mm. to 14 mm.

---

[1] Bureau of Entomology Notes, No 95 K.
[2] Bureau of Entomology Notes, No. 389 K.
[3] Entomologica Americana, vol. 6, p. 14, 1890.

THE ALFALFA LOOPER AND ITS PARASITES.

Fig. 1.—Cocoon cluster of *Apanteles hyslopi*. Fig. 2.—Cocoon of alfalfa looper (*Autographa gamma californica*). Fig. 3.—Larva of alfalfa looper with cocoon of *Microplitis alaskensis*. Fig. 4.—Pupa of alfalfa looper. Fig. 5.—Adult alfalfa looper. Fig. 6.—Adult alfalfa looper at rest. All enlarged. (Original.)

Fifth instar (fig. 45).—Body dark olive-green, dorsal edge of stigmatal line opposed by an almost black line which fades into the general body color dorsally; head brilliant green, mandibles and palpi brown, a black oblong patch on posterior margin of eye extending from near median line almost to base of mandibles; thoracic legs almost black. Length, 21 mm. to 28 mm.

THE PUPA.

(Pl. XI, fig. 4.)

"Pupa depressed somewhat above the wing cases at back of the thorax, the eyes prominent, the tongue case projecting below the wing cases, forming a round prominence over the first abdominal segment. The cremaster is short and blunt, and the hooks with which it is furnished are fastened in the silk of the cocoon. Wing cases slightly creased. Color brownish-black, but paler at the joinings of the parts and between the abdominal joints. In occasional instances the whole pupa is pale."

The above description of the larval stages of this moth agrees in substance with that published by Dr. H. G. Dyar[1] for this species in California. The larvæ, however, are not constant in the matter of coloration. Adult larvæ were found without the characteristic black markings, while others had the entire head black. Some larvæ were of the same pale-green color in the fifth instar as in the fourth, while others were almost white.

That we had the first instar is only an assumption

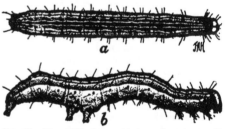

Fig. 45.—The alfalfa looper (*Autographa gamma californica*): *a*, Larva, dorsal aspect; *b*, same, lateral aspect. Enlarged 2½ diameters. (Original.)

from the number of subsequent molts obtained. Dr. Dyar's description of the pupa is used, as it agrees identically with those found in the Northwest. The pupa is dark olive-green, with brown shadings and pale intersegmental bands when first formed, but soon becomes uniformly brown.

THE ADULT.[2]

(Pl. XI, figs. 5, 6.)

The following is a translation of Speyer's original description:

Two California males from Moschler's collections differ from the European *gamma* (which I can, however, only compare with native specimens) in the following points: The color of the dorsal surface of the forewings is a clear, light-blue gray, except the punctuation which is rose colored or rust colored; this is exhibited in all the *gamma* found here by me, and also in several distinct variations.

---

[1] Entomologica Americana, vol. 6, p. 14, 1890.
[2] Stett. Ent. Zeit., vol. 36, p. 164, 1875; syn: *russea* Hy Edwards.

The *gamma* mark is somewhat differently shaped: Both its upper arms diverge, thus cutting off a broader equilateral triangle, through the upper median border, between them. The outer arm of this figure forms at its inner side an obtuse angle and is directed parallel with the hind edge (of the wing) while in the European species it is more basally directed. The lower arm of the *gamma* sign runs nearly horizontal. In this direction the black enclosed part, supported by the two outer arms, is more snout-shaped while in the German forms this form is approximately a rectangular space, rounded at its apex. The ring spot is surrounded by a whitish margin, elongate, and inclined very obliquely basally. The posterior diagonal band runs from the subcostal to the inner branch of the median (branch 2), not in a smooth curve as in *gamma* but in an unbroken straight line and is not so strongly curved basally opposite the lower arm of the *gamma* sign as in the latter. On the under surface, as in the French forms, the wings are clear white with, especially on the forewings, sharp black mottlings. The remainder of the border of the underside, as is also the color of the hind wings, is lighter colored than the European forms. This form may be easily separated from moderately light varieties in color and design by these important distinctions—the *gamma* sign and the posterior diagonal band.

Whether we have to do with a distinct species, a local variety, or merely an accidental variety, must be proved by more extensive comparisons.

Specimens of *gamma* from the Atlantic States, where they are said to be indigenous by Ruhordem, Koch, and Grote, I have not yet seen. They are, when the *californica* form is considered with them, spread over the whole northern hemisphere from Greenland to Abyssinia. They should be found even in New Holland [Australia], cf., my Geographical Distribution of the Butterflies of Germany, etc., Volume II, page 219.

PARASITES.

This moth is severely parasitized. At the field laboratory in Pullman, Wash., we obtained five hymenopterous and two dipterous parasites and observed a disease during the seasons of 1909 and 1910.

On July 12, 1909, two small larvæ of the alfalfa looper, measuring about 14 mm. in length, had contracted to 8 mm., became turgid, quite hard, and changed to a rich ferruginous brown. On July 22 of the same year two specimens of *Rhogas autographæ* Vier. (fig. 46) emerged. On close examination of this improvised puparium it is found to be almost entirely made up of the abdominal segments 7 to 12 of the lepidopterous larva (see fig. 47). The thoracic and anterior abdominal segments 1 to 6 contract to form an almost black annulated cap over one end of the puparium, with the transparent head shield terminating this cap. Segment 13, bearing the anal prolegs, terminates the other end of the puparium and is also transparent. Segments 9 and 10 each bear the transparent skin of their respective prolegs. The hair papillæ on segments 7 to 10 are pale and collected in a darker elevated band around the middle of their respective segments. The adult in emerging gnaws a smooth, circular hole through the dorsum of the eleventh and twelfth segments.

On July 29, 1909, a larva was observed with a cocoon fastened between the middle and anal prolegs (Pl. XI, fig. 3). The moth larva was still alive, though unable to move from the cocoon, and died the following day. On August 1 an adult hymenopteron, *Micro-*

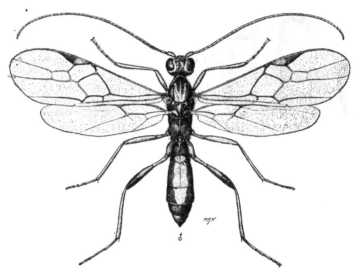

FIG. 46.—*Rhogas autographæ*, a hymenopterous parasite of the pupal stage of the alfalfa looper. Greatly enlarged. (Original.)

plitis alaskensis Ashm., emerged from this cocoon. The cocoon of this parasite (fig. 48, *c*) is pale green, 5 mm. long, cylindrically ovoid, and slightly pointed at the anterior end. In emerging the adult very neatly cuts a cap from the anterior end, this cap often remaining fastened to the cocoon by a few threads. On August 12 another larva was found bearing one of these cocoons, and on August 14 an adult parasite emerged.

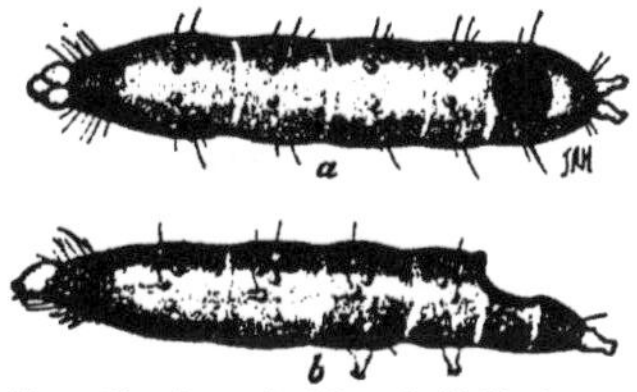

Microplitis n. sp., determined by Mr. H. L. Viereck, was reared from the larvæ of these moths on June 28, 1910. This parasite spins a tan-colored cocoon which measures 3.53 mm. in length (fig. 48, *a*).

FIG. 47.—Larval skin of alfalfa looper from which *Rhogas autographæ* has issued. *a*, Dorsal aspect; *b*, lateral aspect. Enlarged 6 diameters. (Original.)

Sargaritis websteri Vier. is one of the most numerous parasites of the alfalfa looper at Pullman, Wash. The first specimen obtained emerged on August 14, 1909. The following year specimens emerged June 21, July 5, and July 25. This species spins a cocoon (fig. 48, *b*) which is bluntly oval, mottled with brown, and measures 6.53 mm. in length.

On August 27 a dead larva of the alfalfa looper with a mass of hymenopterous cocoons fastened to it (Pl. XI, fig. 1) was found in an alfalfa field in Pullman, Wash. The cocoons were enveloped in a loose, white, silken ball 18 mm. in diameter. On August 29, 34 specimens of *Apanteles hyslopi* Vier. (fig. 49) emerged from this mass.

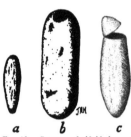

Fig. 48.—Cocoons of alfalfa looper parasites: *a*, *Microplitis* sp.; *b*, *Sargaritis websteri*; *c*, *Microplitis alaskensis*. Enlarged 5 diameters (Original.)

Ameloctonus n. sp., determined by Mr. H. L. Viereck, was reared from the larva of this moth at Salt Lake City, Utah, in 1911, by Mr. T. H. Parks, of this office. The cocoon of the parasite was spun on August 17, and the adult parasite emerged on August 23.

One of the alfalfa looper larvæ in the insectary rearing cages started to spin a cocoon on January 11, 1909. On examining the cocoon two days later it was found to contain puparia of *Plagia americana* Van der Wulp, one entirely and one partly within the dried larval skin. On July 23 one adult emerged,

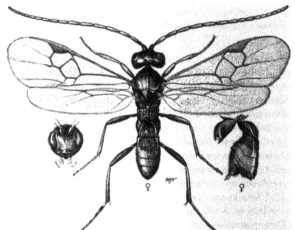

Fig. 49.—*Apanteles hyslopi*, a hymenopterous parasite of the alfalfa looper. Greatly enlarged. (Original.)

and on either July 24 or 25 another emerged (fig. 50). Several more of these flies were reared, and the number that would emerge from a larva was always directly associated with the size of the larva. A very small larva in our cages produced one fly, a medium-sized larva two, and a full-grown larva produced five of these parasites.

Two specimens of a dipterous fly, *Phorocera saundersii* Will. (fig. 51), were reared from a larva of the alfalfa looper on May 18, 1910.

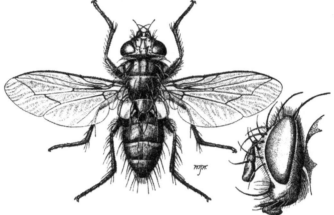

Fig. 50.—*Plagia americana*, a dipterous parasite of the alfalfa looper. Greatly enlarged. (Original.)

Mr. Koebele [1] records having reared 14 parasitic flies from a single larva of this moth at Los Angeles; but as the material is not now available, determination is impossible. These were probably not flies but Hymenoptera and very likely *Apanteles hyslopi* Vier. On

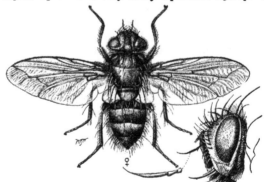

Fig. 51.—*Phorocera saundersii*, a dipterous parasite of the alfalfa looper. Greatly enlarged. (Original.)

the same note he also records rearing two flies from a larva of this moth at Alameda, Cal.

On July 8, 1909, two ants (*Formica rufa obscuripes* Forel) were found at Pullman, Wash., dragging a young larva of this insect that was alive and struggling.

---

[1] Bureau of Entomology Notes, No. 95 K.

Mr. T. H. Parks, of this office, also made a similar record near Salt Lake City, Utah, on August 16, 1911, wherein he observed a "number of large ants (*Formica subpolita* Mayr) dragging one of these half-grown larvæ to their nest in a field of alfalfa. The larva fought to free itself but was finally overcome and killed by the ants."

DISEASES.

Early in July, 1909, many of the larvæ in our rearing cages at Pullman, Wash., were killed by a disease. First, they became sluggish and contracted, and then turned dark brown, often being reduced to a black purulent mass. In other cases they became mummified. Specimens of these diseased larvæ were sent to Dr. Flora W. Patterson, Mycologist of the Bureau of Plant Industry, who reported that while she could find no fungi, the specimens were swarming with bacteria.

A note made by Mr. Theodore Pergande[1] February 21, 1883, records a fungous disease (*Botrytis rileyi*), having been found parasitic on *Plusia brassicæ*, by Mr. W. G. Farlow, of Cambridge, Mass.

REMEDIES AND PREVENTIVES.

Attention has already been called to the fact that in this species we have an illustration of the influence of natural enemies of a pest in protecting the interests of farmers by keeping the insect so reduced in numbers as to prevent injury to his crops. So effectually was this being done in the case of the present species as actually to prevent the conducting of experiments for warding off such injuries; hence no recommendations can be given. The time may come, however, when these natural enemies may themselves suffer reverses and temporarily fail to hold the pest in check; then the information here given will become of the utmost importance as a basis for experiments with restraining measures.

---

[1] Bureau of Entomology Notes, No. 294ᵃ.

INDEX.

124   PAPERS ON CEREAL AND FORAGE INSECTS.

Lightning Source UK Ltd.
Milton Keynes UK
UKHW010333120219
337137UK00004B/269/P